The AMTE Handbook of Mathematics Teacher Education: Reflection on Past, Present and Future—Paving the Way for the Future of Mathematics Teacher Education
Volume 5

A Volume in:
The Association of Mathematics Teacher
Educators (AMTE) Professional Book Series

Series Editor:
Babette M. Benken

The Association of Mathematics Teacher Educators (AMTE) Professional Book Series

Series Editor:
Babette M. Benken
California State University, Long Beach

Series Books

The Mathematics Teacher Education Partnership: The Power of a Networked Improvement Community to Transform Secondary Mathematics Teacher Preparation (2020)
W. Gary Martin, Brian R. Lawler, Alyson E. Lischka, & Wendy M. Smith

Building Support for Scholarly Practices in Mathematics Methods (2017)
Signe E. Kastberg, Andrew M. Tyminski, Alyson E. Lischka, & Wendy B. Sanchez

Elementary Mathematics Specialists: Developing, Refining, and Examining Programs That Support Mathematics Teaching and Learning (2017)
Maggie B. McGatha & Nicole R. Rigelman

Cases for Mathematics Teacher Educators: Facilitating Conversations About Inequities in Mathematics Classrooms (2016)
Dorothy Y. White, Sandra Crespo, & Marta Civil

The AMTE Handbook of Mathematics Teacher Education: Reflection on Past, Present and Future—Paving the Way for the Future of Mathematics Teacher Education
Volume 5

Babette M. Benken

INFORMATION AGE PUBLISHING, INC.
Charlotte, NC • www.infoagepub.com

Library of Congress Cataloging-In-Publication Data

The CIP data for this book can be found on the Library of Congress website (loc.gov).

Paperback: 979-8-88730-541-7
Hardcover: 979-8-88730-542-4
E-Book: 979-8-88730-543-1

CONTENTS

SECTION I
PROFESSIONAL LEARNING FOR
MATHEMATICS TEACHER EDUCATORS

SECTION II
PREPARING TEACHERS TO TEACH SPECIFIC AREAS OF MATHEMATICS AND APPLIED MATHEMATICS

SECTION III
UTILIZING TECHNOLOGY TO SUPPORT THE LEARNING OF MATHEMATICS

SECTION IV
RECONCEPTUALIZING TEACHER
PREPARATION PROGRAMS

FOREWORD

Megan Burton and Enrique Galindo

On behalf of the Association of Mathematics Teacher Educators (AMTE) Board of Directors, we are pleased to present Volume 5 of the AMTE Professional Book Series, *Reflection on past, present and future: Paving the way for the future of mathematics teacher education*. With over 1,000 members, AMTE focuses on preservice teacher education and the professional development of K–12 teachers in mathematics. Therefore, examining what is known about how to best prepare mathematics teachers while also exploring ways to push the field forward is key to our mission. As the leaders in the largest professional organization focused on the improvements of mathematics teacher education, the AMTE Board advocated for this volume as an opportunity to reflect on the past, current and future of mathematics teacher education in relation to research and practice.

This AMTE Handbook for Mathematics Teacher Education shares current understanding of effective practice, but also sheds light on and challenges us to consider new directions for mathematics teacher education in the future. The 95 contributing authors range from graduate students to those who have served as leaders in the field in multiple ways for many years. Authors include K–12 teachers, school administrators, district leaders, graduate students, higher education faculty, and professional development facilitators. The chapters share efforts,

The AMTE Handbook of Mathematics Teacher Education: Reflection on Past, Present and Future—Paving the Way for the Future of Mathematics Teacher Education, Volume 5
pages ix–x.

research-based best practices, opportunities, and challenges across a variety of settings and contexts.

This collection of 22 chapters examines a variety of topics, such as teacher knowledge, balancing content and pedagogy, teacher education standards, rural contexts, field placements, tools for instructing and learning, partnerships, issues of equity, justice-oriented mathematics teacher education, trajectories, and preparing teachers for essential topics such as geometry, data science and statistics. There are single case studies shared along with reviews of programs, self-studies, and other ways of communicating about mathematics teacher education.

The AMTE Board appreciates the chapter authors who shared their work with the mathematics education and teacher education communities. We also want to thank the Book and Series Editor and Vice President for Publications, Babette Benken. In addition, the reviewers, which include several board members and members of the Publications Division, who volunteered time to ensure this handbook would be an asset to our field. Finally, we want to thank the members of the Publications Review Committee, who also provided essential contributions to the review of these chapters.

We are excited that AMTE is able to share the ideas contained in this volume with the mathematics teacher education community. We hope you find it useful in your own practice and research and hope that this volume will prove to be an essential handbook for mathematics teacher education, as we all work to listen, learn, and grow as individuals and as a field.

Listening and Learning,

—Megan Burton
Past President, Association of Mathematics Teacher Educators

—Enrique Galindo
President, Association of Mathematics Teacher Educators

PREFACE

Babette M. Benken

This new volume of the Association of Mathematics Teacher Educators (AMTE) Professional Book Series is a critical and timely resource that paves the way and guides the future of mathematics teacher education. With mathematics achievement scores continuing to decline across the U.S. (NAEP, 2023) and an epic and ongoing teacher shortage continuing to grow (Franco & Patrick, 2023), a researched-based vision for preparing and supporting teachers is needed. The collection of work in this AMTE Handbook for Mathematics Teacher Education reflects on research and what we know about how best to prepare and support both mathematics teachers and mathematics teacher educators and presents what is happening in the field. Examples included in the 22 chapters highlight how we are preparing teachers across multiple contexts (e.g., within district, in content courses for the major) and grade ranges (K–20+) with many explicitly focusing on access, equity, and inclusion. Further, all chapters provide relevant connections to the AMTE *Standards for Preparing Teacher of Mathematics* (2017). Most importantly, all chapters explore what we do not yet fully understand and where we are going–in essence, how we can move the field forward.

This handbook is divided into four sections. The first section of the handbook describes models for the professional learning of mathematics teacher educators (MTEs). Some chapters provide specific course-based recommendations and oth-

The AMTE Handbook of Mathematics Teacher Education: Reflection on Past, Present and Future—Paving the Way for the Future of Mathematics Teacher Education, Volume 5
pages xi–xii.

ers present cases that illustrate how MTEs can use practice-based reflection to grow professionally. Further, some chapters focus on the professional learning for doctoral students and graduate student instructors.

The second section of the handbook explores preparing teachers to teach specific areas of mathematical content. Two chapters focus on applied mathematics (i.e., statistics, data science) and the other two chapters focus on what should comprise advanced courses that prospective secondary mathematics teachers typically take (e.g., geometry).

The chapters in the third section of the handbook communicate how we can utilize technology to support the learning of mathematics. They outline suggestions for how we can support the practical and theoretical challenges MTEs face, including how to integrate technological tools (e.g., a video-tagging tool, virtual manipulatives) into professional development and methods courses.

The fourth section of the handbook presents models for how we can reconceptualize elementary and secondary teacher preparation programs in a variety of contexts (e.g., rural areas). Some chapters outline structure and content for specific program courses (e.g., methods courses, math content courses), while others offer recommendations for innovative program designs and routes to licensure. Further, some chapters explore the role of the field experience in mathematics teacher preparation.

This volume provides valuable examples and vision for how to implement and achieve the goals of AMTE's Standards (2017), as well as presents research-based practices that can broaden our vision and pave the way for the future of mathematics teacher education. The Association of Mathematics Teacher Educators hopes that you find this forward-thinking *Handbook of Mathematics Teacher Education* to be a useful resource as you look to improve and/or expand your efforts to prepare and support teachers of mathematics.

REFERENCES

Association of Mathematics Teacher Educators. (2017). *The standards for preparing teacher of mathematics*. https://amte.net/standards

Franco, M., & Patrick, S. K. (2023). *State teacher shortages: Teaching positions left vacant or filled by teachers without full certification*. Learning Policy Institute. https://learningpolicyinstitute.org/product/state-teacher-shortages-vacancy-resource-tool

National Assessment of Educational Progress (NAEP). (2023, June 21). *New data on pandemic-era learning from the nation's report card shows steep declines in math, falling scores in reading* [Press release]. https://www.nagb.gov/news-and-events/news-releases/2023/new-data-on-pandemic-era-learning.html

SECTION I

PROFESSIONAL LEARNING FOR MATHEMATICS TEACHER EDUCATORS

CHAPTER 1

JUSTICE AND THE MATHEMATICS CLASSROOM

Realizing the Goals of the AMTE Standards for Preparing Teachers of Mathematics

Shandy Hauk
San Francisco State University

Marilyn E. Strutchens
Auburn University

Dorothy Y. White
University of Georgia

Jennifer Bay-Williams
University of Louisville

Jenq Jong Tsay
University of Texas Rio Grande Valley

Billy Jackson
University of Illinois, Chicago

This chapter is an introduction to justice in the post-secondary context of mathematics courses for prospective teachers. The chapter is a research-to-practice report (i.e., it describes an aspect of instruction and discusses how it is informed by, connects to, or is illustrative of findings from research). While the reader might be any type of mathematics teacher educator, the focus here is supporting those who teach mathematics content courses for elementary school teacher candidates. In addition to having an effect on discipline-specific knowledge, college mathematics classes contribute to the ways candidates communicate in/with/through mathematics in

The AMTE Handbook of Mathematics Teacher Education: Reflection on Past, Present and Future—Paving the Way for the Future of Mathematics Teacher Education, Volume 5
pages 3–27.

working with children. The chapter includes discussion of the keys of mathematical literacy: mathematics *with*, *about*, and *for* justice and examples of what the ideas look like in practice. The examples include information from research and a reference case presented as the accumulation of experiences for Kara Thomas and Dr. Rhodes. The case is a means for exemplifying issues such as equity, agency, and identity in the mathematics classroom.

Mathematical literacy is a civil right...the idea of citizenship now requires not only literacy in reading and writing but literacy in math and science.
(Moses, 2001, p. 11)

It is impossible to struggle for civil rights, equal rights for blacks, without including whites. Because equal rights, fair play, justice, are all like the air: we all have it, or none of us has it. (Angelou, 1990)

The function of education, therefore, is to teach one to think intensively and to think critically. But education which stops with efficiency may prove the greatest menace to society. (King, 1947)

INTRODUCTION

Put simply, teaching with justice means instruction that is "lively, accessible, and personally meaningful for students" (Gutstein & Peterson, 2013, p. 1). This is true across the grades, up to and including university. For undergraduates who seek to become school teachers, the knowledge and skill for supporting children's learning of mathematics is personally meaningful. Thus, a central question in mathematics teacher education is: How do people with advanced degrees in mathematics contribute to the preparation of prospective school teachers so that children will learn to be mathematically literate citizens?

There is nothing in the training of mathematicians that prepares them to lead such a literacy effort. Yet the literacy effort really cannot succeed unless it enlists the active participation of some critical mass of the mathematical community. The question of how we all learn to work across several arenas is unsolved. Those arenas are large and complicated. (Moses, 2001, p. 16)

Starting from the foundation that systems of meaning (discourses) infuse the work in every arena of teaching, this chapter:

1. explores teaching mathematics with justice through the educational journeys of an elementary teacher candidate and a university mathematics faculty member,
2. unpacks the meaning of mathematics *with*, *about*, and *for* justice,
3. provides guidance for teaching mathematics with justice,
4. offers readers some suggestions for next steps in professional learning.

When mathematics is lively, accessible, and connected to meaningful contexts, learners develop positive mathematics identities and a sense of mathematical agency (see Table 1, Aguirre et al., 2013a). This chapter highlights specific instructor/teacher actions that impact teacher candidates' conceptions of mathematics and of themselves as learners. We offer the story of Kara's educational journey from middle school through teacher preparation to illustrate these ideas. The case story for Kara is an intentional combination of multiple research studies and experiences of the authors to illuminate mathematics *with* justice (Hauk, 2019; Jackson et al., 2022; Strutchens, 1993; White et al., 2016).

THE CASE OF KARA: PART 1

To begin, read the first excerpt from the case of Kara and reflect on how the story demonstrates just (and unjust) mathematics teaching.

Kara Thomas in Middle School—Grade 6

Kara Thomas talked with a researcher, Dr. Brown, during a study among students in the middle grades (grade 6, 7, and 8). Kara was in the gifted program at school. Kara said she liked reading and "really liked mathematics." When asked how she would rate herself as a mathematics student, Kara said "9, because like mathematics comes pretty easy to me." Kara attributed her success to a natural ability to do mathematics and to her mother and grandmother—both were very supportive and encouraging about Kara's mathematical ability. Kara did well in elementary school. She swiftly memorized and recalled many mathematics facts and skills. However, Kara did not have opportunities to apply these facts and skills in new or unique situations. Though Kara said she believed mathematics was "useful in the real world," when asked for examples, the only illustrations that came to mind for her were of using mathematics at a grocery store.

When Kara entered middle school, she was placed in an accelerated pre-algebra class for sixth graders. Kara struggled to make her usual As and Bs in the class. By the end of the first quarter, her grade was a high C. Kara's view of mathematics and herself as a doer of mathematics was challenged. She had never before had to articulate her thinking or "make an argument to explain an answer" in mathematics. She saw her high C as a bad grade. Kara wanted to withdraw from the class and convinced her mother to have her moved to a different sixth grade mathematics class. Kara's mother agreed to withdraw Kara from the accelerated class because she wanted Kara to continue liking mathematics. The new class was using a grade 7 textbook and her mother felt Kara was still ahead of grade level.

In her new class, Kara was quiet and completed tasks. She did not have a particular group of students that she liked to sit with in class and negative peer pressure did not seem to be a problem for Kara. Kara was doing well in the class without studying. She usually did her homework during class time. Kara said that she hardly ever studied at home and that she did not really have to pay attention to understand what was going on in the class. In fact, an avid reader, she often read or worked on writing assignments in her mathematics class. During one interview, Dr. Brown asked Kara about how often she asked questions in class:

Kara: Never.
Dr. Brown: Never, okay. Why do you not ask questions?
Kara: Most of the time I understand what we are doing.

Kara's comments were confirmed by her teacher. He said, "She doesn't interrupt and ask questions. That's because she usually understands already." This teacher was the same teacher Kara had for the pre-algebra class from which she had withdrawn. In fact, the teacher said, "There may have been a gap that Kara needed filling to survive the pre-algebra class, it's better that she moved down to a more appropriate class." When Dr. Brown asked Kara if her teacher encouraged her to do higher level mathematics, she replied:

Kara: Yes, he had me studying pre-algebra in the first quarter.
Dr. Brown: Okay, and then what happened?
Kara: I quit.
Dr. Brown: Why?
Kara: Because it was too hard. I was making bad grades.

There are several things to notice in this glimpse into Kara's middle school mathematics learning. First, Kara was identified early as a gifted student yet her experiences with mathematics had promoted the views that (a) mathematics is algorithmic where memorizing procedures is the key to success, (b) high grades are the measure of success (e.g., not sense-making or a feeling of understanding), and (c) struggling when mathematics is "hard" is a reason to quit. Second, consider Kara's teacher, for whom silence was evidence of competence. Kara *not* asking questions (which would "interrupt" class) was evidence for the teacher that Kara understood mathematics and was an indication of Kara being in an "appropriate" class. Kara enjoyed reading and writing, but had few opportunities to learn how to communicate with and about mathematics beyond drill and recall.

There was a miscarriage of justice for Kara: she was not challenged to meet high expectations nor supported to reach them. Teachers were complicit in it. Her caregivers were complicit in it. Kara did not have the opportunity to learn mathematics in a way that would allow her to see it as a tool to use to read the world, to see mathematics as more than something that one memorizes in order to do well in school (Gutstein, 2006a). The injustice of low expectations was demanded by, and built upon, conflicting status quo mandates such as "mathematics is supposed to be hard" and "good teaching should make students like and be comfortable with mathematics." Whether or not a student is gifted, worthwhile learning— knowledge growth that is flexible, extensible, persistent—involves struggle. It includes missteps and stumbles to generate complex cognitive and socio-emotional connections among existing mental structures.

The researcher Dr. Brown's particular arena was factors impacting sixth grade African American students' mathematics performance. As a reader, this statement may influence the interpretation of the previous pages. Pause for a moment and ask yourself why this might be the case. While the details of Kara's story are unique to Kara, the school environment she experienced as well as the caregiver decisions and the teacher perspectives that shaped her early opportunities to learn occur in various combinations for many students across the United States (Abedi & Herman, 2010; Carter & Welner, 2013).

As noted in the *Standards for Preparing Teachers of Mathematics* (Association of Mathematics Teacher Educators [AMTE], 2017), it is essential to prepare school teachers to support children in building skills for learning, for handling frustration, and for engaging in productive interactions with other people while making sense of mathematical ideas. We began with Kara's middle school experience as a way to bring up the question: What does justice look like in mathematics classrooms? Addressing that question is the foundation for understanding the need for attention to justice in mathematics content courses for future school teachers.

WHAT IS JUSTICE IN THE CONTEXT
OF A MATHEMATICS CLASSROOM?

Equity is a significant component of justice (although certainly not the only one). For the authors, equity is a verb and not a noun. Justice and equity work together to form a kind of calculus of accumulating moments with and among people, curricula, and policies, in the pursuit of liberation. A joint position statement by TODOS: Mathematics for All and the National Council of Supervisors of Mathematics (2016), describes justice in mathematics learning in K–12 and the importance of acknowledging, acting, and being accountable for next steps in seeking a just mathematics education for learners. Based on this position statement, and related research (e.g., Aguirre, Turner et al., 2013; Celedón-Pattichis, et al., 2018; Gutstein & Peterson, 2013), Table 1.1 offers examples of components in a just mathematics classroom, along with a brief description of each. As you review the list, reflect on Kara's pre-algebra experience and the extent to which these components were a part of her mathematics learning.

In the pre-algebra class, Kara had encountered for the first time an expectation that she explain her reasoning out loud and in writing. She had not had an opportunity to learn to do so in her previous schooling. Nonetheless, the school, program, and teacher all expected the gifted Kara to be skillful at it. She experienced the demands of status quo authority that include high status accorded to individual work, that student work can (or should be) perfect (and fast), that there is one right way (and a right to comfort in getting there), and that to make a mistake is to be a mistake (Okun, 2021). There was no scaffold planned into the pre-algebra class that included initiating and rapidly developing proficiency in mathematical discourse. Kara's only apparent option was to withdraw.

TABLE 1.1. Examples of Components of Justice in Curriculum and Instruction

Component	Description
Variety in Cognitive Challenges	A just classroom includes challenging tasks with multiple possible points of initial engagement, variety in forms of interaction with ideas (e.g., asking students to make and test conjectures, compare, generalize, explain, or critique the reasoning of others). Variety also means students have regular opportunities to tackle novel problems as well as some practice with basics (e.g., algorithms, procedures).
Relevant Contexts	By definition, story problems have contexts. But many story problems are set in contexts that are shallow or completely removed from students' experiences. For example, authentic societal contexts can leverage mathematics to understand a social justice issue (e.g., exploring living wages algebraically, Dean, 2013). In addition to social justice contexts, mathematical applications or models for what is familiar and interesting to students provide opportunities to read the world with mathematics.
Mathematical Discourse	Mathematical discourse involves students discussing important mathematical ideas. This may include justifying a method or whether or not something is always or never true, or contrasting two problems/methods. When students dialogue about mathematics, they make sense of ideas and their own agency advances in positive ways. Note that for this to occur, tasks must be adequately challenging and relevant.
Mathematics Identity	The way a person engages in a mathematics classroom ends up shaping their identity across a continuum from a strong negative identity (I have anxiety) to a strong positive identity (I am really good at this subject).
Mathematical Agency	This is a person's sense of being able to figure out mathematics. Aguirre, et al. (2013b) refer to it as "Identity in Action." When people solve challenging problems, they feel good about it and this has a positive impact on perceptions of their voice and choice in what is being learned.
Responsive Instruction	This means that the teacher adapts instruction to support mathematical thinking, connecting to what students know or following up on a question posed in a classroom setting (sometimes called a teachable moment). Sometimes, the phrase culturally responsive is used, meaning that instruction attends to students' cultural, linguistic, and community-based knowledge.

HOW ARE TEACHING MATHEMATICS *WITH* JUSTICE AND *ABOUT* JUSTICE DIFFERENT?

As noted in a Benjamin Banneker Association (2017) position statement, there are at least three ways mathematics and justice are intertwined in a classroom. In particular, teaching mathematics *with* justice concerns the nature of classroom interactions, where the discourse, norms, and habits of classroom conversations endorsed by the instructor encourage equitable participation and status. Teaching mathematics *about* (social) justice means a lesson is planned and purposeful in looking at serious or provocative (social) issues using mathematics. Teaching and learning mathematics *for* justice are anchored by the idea that mathematics is a means to challenge the status quo, that mathematical activity can be part of actions that transform social, political, and economic conditions to reduce injustice (Hauk et al., 2022).

Larnell et al. (2016) discussed two paradigms in the teaching and learning of mathematics using a social justice lens. One paradigm is based in Freire (1970/1993): mathematics is a tool to critically investigate, critique, and address social/societal issues. The other is based in Moses (1994): mathematical teaching and learning are the foundation for participation in, and transformation of, the majority society status quo. The Freirean perspective looks at mathematics *about social justice* while Moses' is concerned with mathematics *with* and *for* social justice, and more broadly, *with justice* and *for justice* in many arenas (political, economic, environmental; Moses, 2001). Thus, the distinction between "about social justice" and "with justice" is an important nuance in communicating meaning. Natural questions arise: What are the mathematical content and pedagogical foci in a just and effective learning experience? What does it take to teach mathematics with, about, and for justice?

The answers have significant implications for instructional practice. As noted previously, the focus of this chapter is the first question and what is already known about good teaching that is also just (e.g., Table 1.1). For the second question, college instructors (like many teacher candidates) may confound teaching mathematics with justice with teaching mathematics about social justice. Research- and practice-based evidence points to the importance of scaffolds for teachers at all levels who seek to include highly charged events of the day in teaching the application of mathematics to questions about social justice (Downing & Black, 2020; Gewertz, 2020).

Mathematics teaching and learning with, about, and for social justice is not only possible, it is realized daily in some classrooms. The nature of that realization has been described in several ways, including through a contrast between social justice goals and mathematics goals (Gutstein, 2006a; see Table 1.2).

For instance, in reading the world with mathematics, context matters. How would a problem given to first graders asking them to count the number of feet in a collection of animals interact with the funds of knowledge brought to the classroom by children in urban versus rural contexts? What is personally meaningful to the children will differ. To cast the problem in terms of cows and chickens for rural students could be appropriate, but many children who have grown up in a city may have everyday experiences of other animals. If the problem is changed to count the number of legs of the dogs and pigeons on a particular street, then the urban children can relate but the rural children may not. Consideration of reading the world with mathematics such as this is available in the book *What Is It About Me You Can't Teach? Culturally Responsive Instruction in Deeper Learning Classrooms* (Rodriguez et al., 2016).

TABLE 1.2. Mathematics and Social Justice Goals (Gutstein, 2006a)

Mathematics Pedagogical Goals	Social Justice Pedagogical Goals
Reading the mathematical world	Reading the world with mathematics
Succeeding academically	Writing the world with mathematics
Developing agentic mathematical identities	Developing agentic cultural and social identities

Indeed, mathematics about social justice can be taught in unjust ways. Contrary to the belief among some in the mathematics education community, attempting to teach mathematics about social justice does not guarantee, de facto, the humanizing of students or an instructional responsiveness to the people in the room. Teaching mathematics with and for justice demands responsiveness by definition, as the transformation of the status quo requires decentering the instructor as authority (i.e., defies the status quo). In increasing complexity, teaching mathematics with justice, about social justice, includes goals that (Gutstein, 2006b, 2012):

1. Engage students in critical mathematics through a pedagogy of questioning.
2. Facilitate student's development of mathematical power (as defined by the National Council of Teachers of Mathematics [NCTM], 2020).
3. Use problems that motivate students to study and use mathematics.
4. Cultivate students' development of a sense of agency.
5. Incorporate students' life experiences directly into the curriculum.
6. See and encourage students to see mathematics in life daily.
7. Help students to develop sociopolitical consciousness.

One example of the outcomes of teaching with justice and about social justice is the following report from a 9th grade student (Gutstein & Peterson, 2013):

> I thought math was just a subject they implanted on us just because they felt like it, but now I realize that you could use math to defend your rights and realize the injustices around you. Now I think math is truly necessary and, I have to admit, kinda cool. It's sort of like a pass you could use to try to make the world a better place. (p. 1)

THE CASE OF KARA: PART 2

Kara reported on her continued learning of mathematics in high school. Her experience was not aligned with teaching mathematics with justice. All the same, it influenced her understanding of the content and what it meant to do mathematics.

Kara Thomas in High School

Through the rest of middle school and her first three years of high school, Kara continued to learn mathematics by memorizing algorithms and formulas, focusing on which procedure to use to solve particular types of word problems (e.g., "I know that when I see 'times more' you have to multiply and for 'less than' you have to take away."). She used mnemonic devices without understanding the underlying concepts or where the device might break down. For example, she quickly used "the FOIL trick" (first-outer-inner-last) to find the product of two binomials:

$$(x + 1)(x + 2) = x^2 + 2x + x + 2 = x^2 + 3x + 2$$

but did not connect the "trick" with the distributive property. In her senior year of high school, Kara took a course for Advanced Placement Statistics (AP Stats). Here is what she had to say about this experience:

Our "textbook" for the class was a packet of problems for each section that we worked on in class. This teacher's teaching style was for us to come to class with the packet and work on the problems with other students in the classroom. He asked us questions about the problems and asked us what we thought about how to solve the problems. When we asked a question, he asked us things like "what do you think" and "what do you know about this topic." There were no other textbooks and he did not lecture or show us things to help us with the concepts. He just said to ask our peers for help or told us to think about our thinking. This resulted in learning a lot of the math completely on my own with the help of some YouTube videos and my peers who were just as lost as I was. I think learning from peers and listening to their strategies are important when learning math, but because no one was taught the information, this interaction with my peers was not very useful.

Kara's reflection on her AP Stats experience is in alignment with how she had been acculturated into learning mathematics throughout her K–12 education. Kara was successful in high school most of the time because she could memorize what she needed for tests. As in sixth grade, Kara's high school AP Stats experience was a rare disruption to the way she had always succeeded in doing mathematics. Unlike sixth grade, she stayed in the class. For Kara, the experience of AP Stats was not positive because the teacher relied heavily on students' reasoning and making sense of mathematics and an expectation that students were *already* effective at communicating about it with each other. Like her sixth grade experience, Kara reported a high school experience in AP Stats that did not include support from the teacher for how to succeed in meeting the mathematics discourse expectations.

In high school, Kara loved her English classes and felt successful in most of the mathematics and science classes she took. She also liked working with the young children at her church. For college, Kara decided to pursue becoming an elementary school teacher.

WHY JUSTICE IN CONTENT
COURSES FOR PRESERVICE TEACHERS?

Teaching mathematics with justice is actually an expectation in standards and related accreditation requirements in teacher education. National policy docu-

STANDARD C.4. SOCIAL CONTEXTS OF MATHEMATICS TEACHING AND LEARNING

Well-prepared beginning teachers of mathematics realiz[e] social, historical, and institutional contexts of mathemati[cs] teaching and learning and know about and are committe[d to] critical roles as advocates for each and every student.

Indicators include

C.4.1. Provide Access and Advancement
C.4.2. Cultivate Positive Mathematical Identities
C.4.3. Draw on Students' Mathematical Strengths
C.4.4. Understand Power and Privilege in the History of Mathematics Education
C.4.5. Enact Ethical Practice for Advocacy

FIGURE 1.1. AMTE (2017) Standards Example: Standard C.4 (Social Contexts of Mathematics Teaching and Learning)

ments include touchstone standards about the contexts of mathematics teaching and learning (see Figures 1.1 and 1.2).

Recent major standards documents, like those in Figures 1.1 and 1.2, have attended to justice more so than previous standards. In U.S. education, and more broadly, the last decade has seen increasing awareness that mathematics curriculum and instruction, from kindergarten through graduate school, have been shaped by and perpetuate economic-, linguistic-, gender-, and race-based systemic inequities (NCTM, 2020; Sensoy & DiAngelo, 2017). Concurrently, education at all levels is moving towards an orientation that is both responsive and sustaining to the variety of learner cultures and funds of knowledge (Gay, 2018; González et al., 2005;

Standard 3: Knowing Students and Planning for Mathematical Learning		
Candidates use knowledge of students and mathematics to plan rigorous and engaging mathematics instruction supporting students' access and learning. The mathematics instruction that is developed provides equitable, culturally responsive opportunities for all students to learn and apply mathematics concepts, skills, and practices.		
3a) Student Diversity Candidates identify and use students' individual and group differences to plan rigorous and engaging mathematics instruction that supports students' meaningful participation and learning	**3b) Students' Mathematical Strengths** Candidates identify and use students' mathematical strengths to plan rigorous and engaging mathematics instruction that supports students' meaningful participation and learning.	**3c) Students' Mathematical Identities** Candidates understand that teachers' interactions impact individual students by influencing and reinforcing students' mathematical identities, positive or negative, and plan experiences and instruction to develop and foster positive mathematical identities.

FIGURE 1.2. NCTM (2020) Standards Example—Middle Level Mathematics Teacher

Ladson-Billings, 1995; Paris, 2012; Turner et al., 2019). For the expectations in Table 1.2 to become reality in teacher preparation, justice must be addressed in both methods and mathematics courses. It is impractical to expect to achieve the standards if teacher candidates have never seen such practices implemented in their own mathematical learning experiences. The content courses for prospective teachers provide a natural setting in which to model teaching mathematics with justice.

An additional reason to attend to justice is that the ways in which students learn content can reflect the norms of the discipline itself without being constrained by them. Mathematicians do not replicate memorized procedures. Mathematicians do not stay silent when they understand an idea. And, as Danny Martin noted (2012):

> Despite the tensions, I am convinced that a focus on mathematics content knowledge alone is not in the best interest of the students or of the children they will teach. "We'll focus on the math, you'll get that other stuff in education" is insufficient. Such a compartmentalized approach to educating and developing elementary school teachers whose responsibility it will be to educate the whole child seems contradictory. Moreover, there exist very few examples of highly skilled, human services, professional work where knowledge of those who are served and the knowledge needed to serve them are artificially separated. To the degree that math departments perpetuate such separation, they reinforce that teaching mathematics to children is mostly about teaching mathematics and less about teaching children. (p. 19)

Recall Gutstein and Peterson's (2013) description involves rethinking teaching to "make mathematics more lively, accessible, and personally meaningful for students" (p. 1). Transitioning from having quiet students do what is shown to them to a lively classroom engaging students in accessible and challenging content is a process. It begins with a commitment and grows through cycles of change as an instructor learns from how lessons play out. To illustrate the professional learning process for a mathematics teacher educator, we turn to the experiences of Kara when she was an undergraduate in Dr. Rhodes' *Geometry for Teachers* course.

THE CASE OF DR. RHODES: PART 1

A PhD mathematician, Rhodes had taught mathematics content courses for prospective teachers for about 10 years before Kara enrolled in his course.

Dr. Rhodes: Start by Decentering
When asked about his early days in teaching future elementary school teachers, Dr. Rhodes reflected on developing his "decentering" of himself. As he stated:

> I have, over the years, changed how I see student questions like: "When am I ever going to use this?" and "Is this mandatory?" or "Is this going to be on the test?" Years ago, I was annoyed and then insulted by such questions. They seemed confrontational to me. I saw myself and my discipline as US and students as THEM. Then, I shifted to a perspective that meant my job was to enculturate students—get them to see and value the academic way, to find commonality in what they valued and what the academy or the department or,

well mathematicians (US) valued and leverage that. I listened for universals in the questions about "when am I ever going to use this" and "is it on the test" and then I said yes (or no) it was on the test or that I wasn't sure when they were going to use it, but that the point of mathematics learning was to give them tools they could reach for when needed and I could not predict precisely which tool they would need. After all, they were all attending college, there was a common goal there, in the "success in college" realm, to acquire tools for later success.

Then, some time and experiences happened and I started paying attention to more subtle similarities and differences in the questions, restating the questions a bit and asking for confirmation or refinement from students. I accepted the fact that there were some commonalities, but that my tool box analogy might not be enough. At some point, I decided that the next time I got the "Is this on the test?" question, I would say "Thank you for asking! I am wondering about what the answer means *for you*. I really am wondering. Everyone, take out a piece of paper, don't put your name on it. Please think for a moment and write a bit about what it means if I say 'yes, this is on the test,' what does that mean for you? What are the consequences for your next actions, decisions, thoughts, feelings? That's number one. Number two, if I say 'no,' what does that mean for you?"

Well, of course, someone asked. I did what I planned and I collected the answers. I don't know if other people's students will say the kinds of things mine did, but they were certainly interesting. Some said "yes" would help them decide what to study, because they had limited time for studying, prioritizing what was useful, and that I (the instructor) was the expert and they wanted expert advice on how to prioritize. Some said a "no" would help them know how stressed to feel about what was about to happen regarding the topic. This is all to say, when I stopped minimizing differences and started unpacking the details, I learned about student experiences in ways that felt useful to me. Notice, I had to PLAN for that moment, ahead of time. Many times, I have left a classroom and three hours later, go: Oh, now I know what I should have said!

Dr. Rhodes' story offers insights for those who might be new to thinking about justice in their own classrooms. His journey demonstrates how learning to teach with justice is a process, with pitfalls and joys along the way. It also shows how instructors can use regular reflection about their own practice to grow as a just teacher educator.

WHAT DOES MATHEMATICS WITH JUSTICE LOOK LIKE IN A COURSE FOR FUTURE TEACHERS?

Whoever the learner is, learning includes work to create, connect, and revise old and new knowing, thinking, and interacting with mathematics. To do so requires opportunities to learn. Offering opportunities to learn that are *seen by the learner* as opportunities is part of teaching with justice. Thus, the first part of Kara's story, up through high school, has immediate parallels in post-secondary settings. When prospective teachers reach college and begin their mathematics coursework, what are the expectations regarding what they already know and will gain in knowledge? What are the scaffolds and associated opportunities to learn in the courses for prospective teachers to support meeting those expectations?

As many have noted, the social, cultural, and anthropological aspects of teaching are frequently ignored in content courses (e.g., Jett, 2013; Ramirez & McCollough, 2012). Mathematics is often treated as devoid of the human experience and condition, which does more than simply ignore the individual mathematical identities of students, it steals them (Jett, 2013).

For Dr. Rhodes it was important to encourage and make explicit the self-perceptions of the prospective teachers he taught. Course activities were created to support future teachers to develop a deep understanding of mathematics through problem solving, making and testing conjectures, and pausing to unpack and discuss how to identify and talk about the exploration of mathematical ideas they and others had. Anticipating that students might enter with a belief that mathematics was about getting right answers and getting them fast, Rhodes' course was designed to support students in challenging that belief. During the first few weeks of the course, Kara hit some roadblocks like she had experienced in AP Stats. But, as the semester progressed, Kara began to assert herself and her abilities through in-class talk and written assignments.

THE CASE OF KARA AND DR. RHODES

Kara's experience in the *Geometry for Teachers* course would be transformative for her because Dr. Rhodes had already begun his own professional transformation.

Kara in College: "I want to figure it out for myself!"

Students work in groups at tables. Each group has a course packet (on paper), blank paper, and a set of AngLegs (plastic pieces that snap together to form flat geometric shapes) to construct different right triangles (see Appendix A for the course packet task). Each group builds squares on each side of the triangles in the task and then records the area of each square in a table provided in their course packets. Several students notice that the area of the square on the hypotenuse is the sum of the areas of the squares on each of the legs. Three people, each in a different group, say out loud some version of: "Oh, that's where $c^2 = a^2 + b^2$ comes from! They never told us that before."

As Dr. Rhodes is talking with a group near the board at the front of the room about their ideas, he turns to draw a triangle on the board and discovers that all of the markers for the whiteboard are dry. Dr. Rhodes leaves the room to get more markers. When he returns five minutes later, Riley explains that the whole class had started discussing whether or not the Pythagorean Theorem was true for triangles that are **not** right triangles. Before Dr. Rhodes has a chance to respond, Kara says, "No, no, no! Don't ask him, I want to figure it out for myself!" Dr. Rhodes asks the class to explore this question and see what they discover.

About ten minutes later, Dr. Rhodes asks the class to stop exploring so they can discuss. He begins with Kara. She tells the other students that "It's not true—it doesn't work for other triangles." Dr. Rhodes asks her how she knows that it's not true, and she responds that she created other triangles that were not right triangles and then computed the areas of the squares on each side. When she added the areas, they were not the same as the area of the square on the longest side. Dr. Rhodes then asks Kara if she found any patterns or relationships among the triangles she tried. For example, did acute, right, or obtuse triangles have squares with larger areas on certain sides? Kara replies that she had not really paid any attention to that aspect so she could not say. Dr. Rhodes then asks the groups to pay attention to acute and obtuse triangles to compare c^2 with $a^2 + b^2$ in each of these cases (see Appendix B).

To examine the mathematics with justice in this vignette, consider the five categories in the Teaching for Robust Understanding (TRU) framework (Schoenfeld et al., 2016). The framework is a minimal set for noticing and weighing justice in instructional decisions.

1. Mathematical Content: Notice that the preservice elementary teachers are engaging in content at a deep and meaningful conceptual level. Yes, they are struggling. One knows that the struggle is productive because students ask questions and connect ideas "Oh, that's where…"

2. Challenge of Cognitive Demand: Both the task and verbal prompts by Dr. Rhodes aim at students doing sense-making. Students have opportunities to work through authentic challenges.

3. Mathematical Identity and Agency: It is apparent that students are in an environment that supports them in building their own mathematical identities and a sense of agency. It is the teacher candidates who initially propose the question about other triangles and they who choose to explore different cases with the AngLegs. Kara is agentic in answering the question at hand and in communicating with classmates about it, expecting she will be supported to do so in this particular class.

4. Equitable Access to Ideas: The vignette demonstrates some signs of what is sometimes termed in the literature as *equity of voice*, different students voicing their ideas, questions, and reasoning.

5. Formative Assessment: The prospective teachers engage in self-assessment during this segment as they try to make sense of what they already know, test the bounds of what they already know, and then consider how to apply what they know to other contexts.

Also present in the excerpt from Kara's college experience, but absent in the TRU framework, are some of Dr. Rhodes early efforts at:

Responsive Instruction: The teacher adapts instruction to support mathematical thinking, following up on Kara's contribution by asking the class to attend to her idea and interrogate her solution.

THE CASE OF DR. RHODES: PART 2

The components described in the TRU framework, along with the extension to responsive instruction, occur through intentional planning. While it may not be reasonable to attend to all components every day, providing regular opportunities for these components to be embedded in class sessions increases the opportunities for future teachers to develop their mathematical knowledge for teaching.

Dr. Rhodes: Process Rather Than Destination

In reflecting on his development as an instructor, Dr. Rhodes commented on coming to a realization about the dynamic nature of learning to teach:

> In my earlier teaching career, the mathematical tasks I used or designed to engage future teachers did not have a rich and textured context, did not include school children's cognitions and experience. And they did not have specific connection to school curriculum or standards or teacher preparation standards (like the MET II, which I read later, and the AMTE standards). When I had learned more from AMTE and was using tasks with rich and textured context, I found the tasks might overwhelm preservice teachers. Most of my future teachers saw *good* teaching as being the *reduction* of the complexities of mathematics. They appreciated using mnemonics (e.g., PEMDAS and "the alligator eats the bigger number") for their own math learning. I rarely heard things like, "I want to figure it out for myself!" I think my future teachers had a hard time distinguishing between themselves as mathematics learners and themselves as professionals learning mathematics to teach it to children. When I started getting specific about switching between their role as learners and as teachers by linking the perspective shift with watching videos of CHILDREN learning, then things started to change.

> For a teacher candidate to go through several mathematics content and methods courses and not be able to distinguish between their own lens of mathematical thoughts as adults and the lens of a child learning the material for the first time demonstrates a failure by the system. Part of our job is to help the teacher candidate move away from self-focus, to attend to noticing and being responsive to differences and similarities across (young) people. Basically, this is the same thing I was talking about in what I had to learn to notice and respond to when students asked the "is this on the test" questions.

> My own lessons learned from this are that teacher candidates need opportunities to interact with people (especially children) who are different from them. Different from who they are now and from who they were as children. That "interaction" can be real or virtual (e.g., through video). They should be able to see classrooms full of children like the 6th grade classrooms I visited a few years ago, where every child was excited to do mathematics and was willing to share their ideas with each other, even when they knew their solutions were not correct. It would be a major step in helping the teacher candidates see teachers in such classrooms, in action. Now, I bombard my prospective teachers with evidence that the vision AMTE has set forth is possible in K–12 classrooms.

Teacher education—as a field—struggles with how to do this effectively. This struggle serves to show that helping teacher candidates to construct and use a mathematical knowledge for teaching that is equitable is a continuing *process* rather than a *destination* that we arrive at with each of the candidates.

Key to Dr. Rhodes' story is his recognition of the importance of focusing on K–12 student thinking as a vehicle to support prospective teachers' emerging content knowledge. He was not attempting to teach his future teachers "methods." Rather, he sought ways that were relevant to a future teacher to scaffold content knowledge development. Importantly, this shift in his teaching supported his students' productive disposition, as captured in Kara's "I want to figure it out for myself!" vignette.

THE CASE OF KARA: PART 3

The final vignette provides a pair of snapshots of Kara's experience in *Geometry for Teachers*. One of the first assignments for Kara was to write about her future classroom. One of the last assignments was to repeat the task and then compare her initial and end-of-semester thoughts. For the comparison, she was asked to reflect on what might have happened during the course to change her perspective or reinforce her initial views.

Kara's Vision of her Future Teacher Self

Initial

In my ideal classroom, I, the teacher, am teaching the meaning of place values. I am asking my 2nd grade students what each place value is. The students are attempting to understand the newly taught material. Within learning about place value, the students are absorbing the material through different color-coded blocks. Each color represents a different place value. A central theme would be to keep trying and to keep growing in patience, which both students and the teacher need.

End-of-Semester

As the teacher, I am encouraging students to think critically without DIRECTLY telling them. The students are using the knowledge they have retained to create problems and answer solutions. The students are learning to work with manipulatives and the MANY possibilities they can conjure to better understand something that they don't know or struggle on. A central theme could be critical thinking for both students and teachers, because as teachers we try our best to allow our students the opportunity to think without telling them exactly what to do. Show guidance. The students are also learning how to explain the knowledge they have and how they get from one point to another: the why is important.

Comparison

Personally, I think [what changed is] the generalization of math and what it can be. At first, I assumed just knowing [the "how"] was enough. But through the semester,

I saw that knowing "why" is crucial to expansion of a child's mind. I've definitely learned that adults don't give students the correct credit they deserve. It's interesting to watch videos on the way one student can figure something out versus what I would do. Kids hold more than anyone could imagine, it's just the chance of allowing them to speak or explain to better understand.

As noted in Kara's reflection, being self-aware and facilitating self-aware learning are valuable (instead of attention and authority being vested largely or solely in the instructor; Teuscher et al., 2016).

NEXT STEPS FOR INSTRUCTORS LEARNING TO TEACH MATHEMATICS WITH JUSTICE

In the process of differentiating learners from one's teacher-self, *decentering* becomes essential. This kind of instructional decentering is, at its most basic, the act of seeing from someone else's point of view and has historical roots in the work of Piaget (1955). It means engaging with other people as a *participant* in interaction, rather than as the center of interaction. In decentering, instructional attention is on uncovering, understanding, and expanding on what students know and do to include novel and non-standard as well as standard mathematical ideas and methods (Hauk & Speer, 2023). In considering the perspectives of future teachers— who may be learning for the first time how to learn and think about mathematics in order to teach it—college instructors are faced with a need for instructional refocusing, one that leverages but is not limited to their skill in advanced mathematics. As teacher educators, they must expand and connect that knowledge to a multi-layered mathematical knowledge for teaching (Ball et al., 2008). Moreover, as illustrated in Dr. Rhodes' journey, to teach mathematics with justice a college instructor will also need to consider the view of future teachers who are in the beginning stages of constructing their own mathematical knowledge for teaching (for their eventual work with children).

Dr. Rhodes case ends with him on the edges of what is commonly referred to in the literature as culturally responsive pedagogy (Gay, 2018). The absence of cultural responsiveness in courses for prospective teachers can have detrimental consequences on both teachers candidates' own learning and that of their future students (Jett, 2013; Ladson-Billings, 2001). The next stage of evolution in the professional growth for those teaching content courses for prospective teachers includes attention to cultural responsiveness. This means that the instruction itself is responsive to the adult students as learners of mathematics, and that the instruction allows opportunities for the prospective teachers to attend to their own (future) responsiveness to learners who are children. This can be achieved by supporting teacher candidates in critical reflection on their future selves and classrooms. A further step, extending culturally responsive pedagogy, is culturally sustaining pedagogy which "seeks to foster to sustain-linguistic, literate, and cultural pluralism as part of schooling for positive social transformation" (Paris &

Alim, 2017, p.1). For example, a future version of Dr. Rhodes could be ready to engage in instruction that would prepare teacher candidates to enact tasks such as those mentioned in Rodriguez et al. (2016) or the "La Lotería" task described by Ramirez and McCollough (2012).

CONCLUSION

In the AMTE *Standards for the Preparation of Teachers of Mathematics* (2017), attention to justice is not only in Standard 4, it is infused in all the standards. Consider the following content-focused indicators and how a justice-oriented classroom can support the journey of becoming well-prepared and effective teachers:

- **Indicator C.1.2. Demonstrate Mathematical Practices and Processes:** Well-prepared beginning teachers of mathematics have solid and flexible knowledge of mathematical processes and practices, recognizing that these are tools used to solve problems and communicate ideas.
- **Indicator C.1.3. Exhibit Productive Mathematical Dispositions:** Well-prepared beginning teachers of mathematics expect mathematics to be sensible, useful, and worthwhile for themselves and others, and they believe that all people are capable of thinking mathematically and are able to solve sophisticated mathematical problems with effort.
- **Indicator C.1.5. Analyze Mathematical Thinking:** Well-prepared beginning teachers of mathematics analyze different approaches to mathematical work and respond appropriately.

Content courses in mathematics for prospective teachers can support the development of mathematical practices and processes (C.1.2). Kara's perception of herself as a learner was shaped by her teachers' views of her as capable (or not) of thinking mathematically and doing productive, effortful, work in problem-solving (C.1.3). Also, Kara shifted her perceptions of mathematics, in part, due to her teachers' responses to her approaches to mathematics (C.1.5). These content standards highlight the critical knowledge and skills well-prepared beginning teachers need to engage students in lively, accessible, and relevant mathematics.

Rochelle Gutiérrez (2013) has noted about teaching mathematics that "all mathematics teachers are identity workers" (p. 16). As can be discerned from Kara's experiences, the mathematics she learned influenced what she thought mathematics was. How she learned that mathematics influenced what she thought was important in the discipline. Collectively, what and how mathematics was taught shaped her view of mathematics and mathematical identity. While the ideas in this chapter are important take-aways for the teaching of *all* mathematics, with preservice teachers the stakes are higher. Prospective teachers taking content courses are beginning to develop their professional identities as teachers of mathematics. It may be particularly critical for future teachers to experience mathematics with justice as learners in college, especially for those who did not have a such an experience in middle or high school.

A potentially useful first step for college faculty who teach preservice elementary teachers is to understand the differences between teaching *with* and *about* justice. A good next step is to examine and enhance their current practices. Tools include readings, such as the Benjamin Banneker Association (2017) position statement regarding mathematics *with*, *about*, and *for* social justice, resources like Berry III et al. (2020), and the application of the TRU framework (Schoenfeld et al., 2016). In the process of reflecting on and enhancing his instruction, Dr. Rhodes reported what research has indicated: attending to disposition along with differentiating and decentering are crucial steps in building flexible knowledge for just mathematics instruction of adult learners who are prospective teachers (Hauk et al., 2014). In a similar vein, Dr. Rhodes demonstrated that skill in multiple discourses (e.g., mathematics, school, teaching) is valuable in a college instructor's mathematical knowledge for teaching future teachers (Jackson et al., 2020). Such knowledge of discourses is far reaching and is "knowledge about the nature of communication, including context and valued forms of inquiry, socio-mathematical norms, and language in, for, and through mathematics in post-secondary educational settings" (Hauk et al., 2017, p. 429).

The goals of this chapter were to offer mathematics faculty who are new to situating justice in their professional work the opportunity for intensive, critical thinking—an opportunity to reach beyond effectiveness or efficiency of instruction to the quality of the human interactions in teaching mathematics. To help accomplish this, the chapter included discussion of the keys of mathematical literacy, mathematics with justice, and examples of what the ideas look like in practice. The examples included information from research and a reference case presented as the accumulation of experiences for Kara Thomas and Dr. Rhodes. The case was a means for exemplifying issues including agency and identity in the mathematics classroom.

In furthering the advocacy called for in AMTE Standard C.4.5, we suggest that culturally responsive and sustaining pedagogies can set the stage for future justice development within mathematics teacher education. Why pursue it? Because justice is like the air, we all have it or none of us has it (Angelou, 1990).

ACKNOWLEDGEMENTS

We would like to acknowledge the prospective teachers who formed the case of Kara. We are grateful, also, to the research and development efforts of many colleagues in the area of mathematics teaching and justice, including the editors, reviewers, and other authors for this volume. This material is based upon work supported by the National Science Foundation under Grant Numbers DUE-1625215 and DUE-1432381.

APPENDIX A: INVESTIGATING THE PYTHAGOREAN THEOREM

Try using these combinations of AngLegs to form triangles:

1. Two orange and one purple
2. Two yellow and one red
3. One orange, one purple, and one green
4. One purple, one blue, and one red

What kind of triangles have you formed?

Now, look on the back of each colored piece of AngLeg. You will see that the length of the piece is recorded in centimeters:

Orange = 5 cm	Purple = 7.07 cm	Green = 8.66 cm
Yellow = 10 cm	Blue = 12.24 cm	Red = 14.14 cm

Here's what I want you to do: form squares on each of the sides of each of the four triangles you created and record the area of each square in the table below.

Triangle #	Length of Shortest Side	Area of Square formed by Shortest Side	Length of Other Short Side	Area of Square formed by Other Short Side	Length of Longest Side	Area of Square formed by Longest Side
(1)						
(2)						
(3)						
(4)						

Once the table is filled in, compare the area of the square formed by the longest side with the area of the squares formed by the two shorter sides. Do you see a relationship between the two? What is it? Describe it.

What relationships might your previous answer suggest about right triangles in general? Describe it here.

If you have not already done so, describe the relationship you just discussed in terms of the side lengths of the original triangle.

APPENDIX B: CONVERSE OF THE PYTHAGOREAN THEOREM

First, complete the following statements.

The Pythagorean Theorem essentially tells us that in right triangles:

That is, if is the hypotenuse (i.e., the longest side), then in terms of the shorter sides and , the Pythagorean Theorem is represented by the relationship:

Now Consider: What if the Triangle is Not a Right Triangle?

Given side lengths of with being the longest side, can we determine if the triangle is acute, right, or obtuse? Use the AngLegs to create some acute, right and obtuse angles. Use these triangles to fill in the table below.

Type of Triangle (Acute, Right, or Obtuse)	$a^2 + b^2$	c^2	Type of Angle Opposite side (Acute, Right, or Obtuse)

What type of triangle do we get when ? What type of angle is opposite the side in this case?

What type of triangle do we get when ? What type of angle is opposite the side in this case?

What type of triangle do we get when ? What type of angle is opposite the side in this case?

So, what does comparing the value of with that of seem to tell us about the triangle in general? What does this tell us about the relationship between sides of the triangle and the angles opposite those sides?

REFERENCES

Abedi, J., & Herman, J. L. (2010). Assessing English language learners' opportunity to learn mathematics: Issues and limitations. *Teachers College Record, 112*(3), 723–746.

Aguirre, J. M., Mayfield-Ingram, K., & Martin, D. B. (Eds.). (2013a). *The impact of identity in K–8 mathematics learning and teaching: Rethinking equity-based practices.* National Council of Teachers of Mathematics.

Aguirre, J. M., Turner, E. E., Bartell, T. G., Kalinec-Craig, C., Foote, M. Q., Roth McDuffie, A., & Drake, C. (2013b). Making connections in practice: How prospective elementary teachers connect to children's mathematical thinking and community funds of knowledge in mathematics instruction. *Journal of Teacher Education, 64*(2), 178–192.

Angelou, M. (1990). Interview. Academy class of 1990. *Academy of Achievement.* https://achievement.org/achiever/maya-angelou/#interview

Association of Mathematics Teacher Educators (AMTE). (2017). *Standards for preparing teachers of mathematics.* https://amte.net/standards

Ball, D. L., Thames, M. H., & Phelps, G. (2008). Content knowledge for teaching: What makes it special? *Journal of Teacher Education, 59*(5), 389–407.

Benjamin Banneker Association (BBA). (2017). *Position statement on implementing a social justice Curriculum: Practices to support the participation and success of African-American students in mathematics.* BBA. http://bbamath.org/wp-content/uploads/2017/11/BBA-Social-Justice-Position-Paper_Final.pdf.

Berryb III, R. Q., Conway IV, B. M., Lawler, B. R., & Staley, J. W. (2020). *High school lessons to explore, understand, and respond to social injustice.* Corwin.

Carter, P. L., & Welner, K. G. (2013). *Closing the opportunity gap: What America must do to give every child an even chance.* Oxford University Press.

Celedón-Pattichis, S., Borden, L. L., Pape, S. J., Clements, D. H., Peters, S. A., Males, J. R., Chapman, O., & Leonard, J. (2018). Asset-based approaches to equitable mathematics education research and practice. *Journal for Research in Mathematics Education, 49*(4), 373–389.

Dean, J. (2013). Living algebra, living wage. In E. Gutstein & B. Peterson (Eds.), *Rethinking mathematics: Teaching social justice by the numbers* (2nd ed., pp. 67–71). Rethinking Schools. https://rethinkingschools.org/books/rethinking-mathematics-second-edition/

Downing, G. A., & Black, B. L. (2020). Measuring the effectiveness of social justice pedagogy on K–8 preservice teachers. In S. S. Karunakaran, Z. Reed, & A. Higgins (Eds.), *Proceedings of the 23rd Annual Conference on Research in Undergraduate Mathematics Education* (pp. 142–150). Boston, MA. http://sigmaa.maa.org/rume/RUME23.pdf

Freire, P. (1970/1993). *Pedagogy of the oppressed.* Continuum.

Gay, G. (2018). *Culturally responsive teaching: Theory, research, and practice* (3rd ed.). Teachers College Press.

Gewertz, C. (2020, December). Teaching math through a social justice lens. *Education Week.* https://www.edweek.org/teaching-learning/teaching-math-through-a-social-justice-lens/2020/12

González, N., Moll, L., & Amanti, C. (2005). *Funds of knowledge.* Erlbaum.

Gutiérrez, R. (2013). Why (urban) mathematics teachers need political knowledge. *Journal of Urban Mathematics Education, 6*(2), 7–19.

Gutstein, E. (2006a). *Reading and writing the world with mathematics: Toward a pedagogy for social justice.* Routledge, Taylor & Francis Group.

Gutstein, E. (2006b). Driving while Black or Brown: The mathematics of racial profiling. In D. Mewborn (Series Ed.), & J. Masingila (Vol. Ed.), *Teachers engaged in research: Inquiry in mathematics classrooms, grades 6–8.* (Vol. 3, pp. 99–118). Information Age Publishing.

Gutstein, E. (2012). Reflections on teaching and learning mathematics for social justice in urban schools. In A. A. Wager & D. W. Stinson (Eds.), *Teaching mathematics for social justice: Conversations with educators* (pp. 63–78). National Council of Teachers of Mathematics.

Gutstein, E., & Peterson, B. (Eds.). (2013). *Rethinking mathematics: Teaching social justice by the numbers* (2nd ed.). Rethinking Schools.

Hauk, S. (2019). Understanding students' perspectives: Mathematical autobiographies of undergraduates who are future K–8 teachers. In S. Hauk, B. Jackson, & J. J. Tsay (Eds.), *Professional resources & inquiry in mathematics education (PRIMED) short-course field-test.* WestEd. https://sfsu.box.com/s/0ikwf3pordvut1vjww1m421yedx9joxf

Hauk, S., Jackson, B., & Tsay, J. J. (2017). Those who teach the teachers: Knowledge growth in teaching for mathematics teacher educators [Conference Long Paper]. In A. Weinberg, C. Rasmussen, J. Rabin, M. Wawro, & S. Brown (Eds.), *Proceedings of the 20th Conference on Research in Undergraduate Mathematics Education* (pp. 428–439). San Diego, CA.

Hauk, S., Khadjavi, L., Kung, D., Piercey, V., & Staley, J. (2022). *Mathematics with, about, and for social justice: A guide for educators.* Carnegie Math Pathways. https://carnegiemathpathways.org/mathematics-with-about-and-for-social-justice-a-new-guide-to-help-you-integrate-social-justice-into-your-math-classroom/

Hauk, S., & Speer, N. (2023). Developing the next generation of change agents in college mathematics instruction. In M. Voigt, J. Hagman, J. Gehrtz, B. Ratliff, N. Alexander, & R. Levy (Eds.), *Justice through the lens of calculus: Framing new possibilities for diversity, equity, and inclusion.* https://doi.org/10.48550/arXiv.2111.11486

Hauk, S., Toney, A., Jackson, B., Nair, R., & Tsay, J. J. (2014). Developing a model of pedagogical content knowledge for secondary and post-secondary mathematics instruction. *Dialogic Pedagogy, 2*(2014), A16–A40.

Jackson, B., Hauk, S., & Tsay, J. J. (2022, May 25). *"I want to figure it out for myself!": Enacting tasks in the courses for future teachers.* Change DIAL Conference, University of Nebraska Lincoln. https://unlcms.unl.edu/cas/csmce/change-dial-conference#workshop6

Jackson, B., Hauk, S., Tsay, J. J., & Ramirez, A. (2020). Professional development for mathematics teacher education faculty: Need and design. *The Mathematics Enthusiast, 17*(2), 537–582. https://scholarworks.umt.edu/tme/vol17/iss2/8/

Jett, C. C. (2013). Culturally responsive collegiate mathematics education: Implications for African American students. *Interdisciplinary Journal of Teaching and Learning, 3*(2), 102–116.

King, Jr., M. L. (1947). The purpose of education. *The Maroon Tiger.* Morehouse Student Newspaper.

Ladson-Billings, G. (1995). Toward a theory of culturally relevant pedagogy. *American Educational Research Journal, 32*(3), 465–491.

Ladson-Billings, G. (2001). *Crossing over to Canaan: The journey of new teachers in diverse classrooms.* Jossey-Bass.

Larnell, G. V., Bullock, E. C., & Jett, C. C. (2016). Rethinking teaching and learning mathematics for social justice from a critical race perspective. *The Journal of Education, 196*(1), 19–29.

Martin, D. B. (2012). Teaching other people's children to teach other people's children: Reflections on integrating equity issues into a mathematics content course for elementary teachers. In L. Jacobsen, J. M. Mistele, & B. Sriraman (Eds.), *Mathematics teacher education in the public interest* (pp. 3–23). IAP.

Moses, R. P. (1994). Remarks on the struggle for citizenship and math/science literacy. *Journal of Mathematical Behavior, 13*(1), 107–111.

Moses, R. P. (2001). *Radical equations: Math literacy and civil rights.* Beacon Press.

National Council of Teachers of Mathematics (NCTM). (2020). *Standards for the preparation of middle level mathematics teachers.* https://www.nctm.org/uploadedFiles/Standards_and_Positions/NCTM_Middle_School_2020_Final.pdf

Okun, T. (2021). *(Divorcing) White supremacy culture: Coming home to who we really are.* https://www.whitesupremacyculture.info

Paris, D. (2012). Culturally sustaining pedagogy: A needed change in stance, terminology, and practice. *Educational Researcher, 41*(3), 93–97.

Paris, D., & Alim, H. S. (Eds.). (2017). *Culturally sustaining pedagogies: Teaching and learning for justice in a changing world.* Teachers College Press.

Piaget, J. (1955). *The language and thought of the child.* Meridian Books.

Ramirez, O. M., & McCollough, C. A. (2012). "La Lotería"-using a culturally relevant mathematics activity with pre-service teachers at a family math learning event. *Teaching for Excellence and Equity in Mathematics, 4*(1), 24–33.

Rodriguez, E. R., Bellanca, J., & Esparza, D. R. (2016). *What is it about me you can't teach?: Culturally responsive instruction in deeper learning classrooms.* Corwin Press.

Schoenfeld, A. H., and the Teaching for Robust Understanding Project. (2016). *An introduction to the teaching for robust understanding (TRU) framework.* Graduate School of Education. http://map.mathshell.org/trumath.php or http://tru.berkeley.edu.

Sensoy, O., & DiAngelo, R. (2017). *Is everyone really equal?: An introduction to key concepts in social justice education.* Teachers College Press.

Strutchens, M. E. (1993). *An exploratory study of the societal and ethnic factors affecting sixth grade African American students' performance in mathematics class.* Unpublished doctoral dissertation, University of Georgia.

Teuscher, D., Moore, K. C., & Carlson, M. P. (2016). Decentering: A construct to analyze and explain teacher actions as they relate to student thinking. *Journal of Mathematics Teacher Education, 19*(5), 433–456. https://doi.org/10.1007/s10857-015-9304-0.

TODOS: Mathematics for ALL and the National Council of Supervisors of Mathematics. (2016). *Mathematics education through the lens of social justice: Acknowledgement, Actions, and Accountability.* Joint position statement. http://www.todos-math.org/socialjustice

Turner, E., Bartell, T. G., Drake, C., Foote, M., McDuffie, A. R., & Aguirre, J. (2019). Prospective teachers learning to connect to multiple mathematical knowledge bases across multiple contexts. In G. M. Lloyd & O. Chapman (Eds), *International handbook of mathematics teacher education* (Volume 3, pp. 289–320). Brill Sense.

White, D. Y., DuCloux, K. K., Carreras-Jusino, A. M., Gonzalez, D. A., & Keels, K. (2016). Preparing preservice teachers for diverse mathematics classrooms through a cultural awareness unit. *Mathematics Teacher Educator*, *4*(2), 164–187.

CHAPTER 2

JUSTICE-ORIENTED MATHEMATICS TEACHER EDUCATION

A Conversation Among Early Childhood and Elementary Mathematics Teacher Educators

Courtney Koestler
Ohio University

Crystal Kalinec-Craig
University of Texas at San Antonio

Eva Thanheiser
Portland State University

Cathery Yeh
University of Texas at Austin

Naomi Jessup
Georgia State University

Anita Wager
Vanderbilt University

As a group of six justice-oriented mathematics teacher educators with diverse social identities and in different positions in mathematics teacher education, we share a window into our process of collaborative professional learning by examining the interplay of our own social identities and the ways in which we prepare prospective early childhood and elementary teachers to address the AMTE Standards for Preparing Teachers of Mathematics (2017). Primary data include letters we wrote to the prospective teachers with whom we work. These letters explore how we define

The AMTE Handbook of Mathematics Teacher Education: Reflection on Past, Present and Future—Paving the Way for the Future of Mathematics Teacher Education, Volume 5
pages 29–43.

justice-oriented mathematics teacher education, how we enact it, and the big ideas we want prospective teachers to come away with from our course, both for them as readers of the letters and for us as a critical friends group. This work serves as one example of how mathematics teacher educators can collaborate to support prospective teachers to be equity- and justice-minded in practice.

INTRODUCTION

In the summer of 2020, we came together as six mathematics teacher educators for a "critical friends" group (Stieha, 2014) to discuss issues of diversity, equity, and justice in mathematics teacher education. Each of us have distinct intersectional social and cultural identities (e.g., ethnicity/race, nationality, religion, gender identity, professional status, areas of research expertise, teaching experiences). We felt the crucial need to continue to grow and deepen our own knowledge and dispositions towards preparing beginning teachers of mathematics to attend to the social, historical, and political contexts of mathematics education. To create a space to support each other as critical mathematics teacher education scholars, we met weekly for over a year to share activities and assignments to implement in our mathematics teacher education courses (methods and content).

During this time, we developed an inquiry into our positionalities as justice-oriented mathematics teacher educators (MTEs) and our individual conceptualizations of justice-oriented mathematics education (and how these might be unique and/or complementary), our goals for our prospective teachers, and how our goals and conceptualizations for justice-oriented work are enacted with our prospective teachers. As part of this inquiry, we wrote "love letters" (Yeh and colleagues, 2022) to the prospective teachers with whom we work, as a way to address these wonderings. In the letters, we attempted to answer the following questions:

- How do we define justice-oriented mathematics teacher education?
- How do we intend to enact justice-oriented mathematics teacher education in our courses and how do our positionalities and particular contexts shape this work?
- What "big ideas" do we want our prospective teachers to take away from our courses?

In this chapter, we share a window into our process of collaborative learning by examining the interplay of our own social identities and the ways we prepare prospective early childhood and elementary teachers to address AMTE's Standards for Preparing Teachers of Mathematics (2017). As context matters, we begin by identifying our positionalities, situating our work within the literature, sharing the method of using letters to prospective teachers as data and findings from analyses, and conclude with implications for mathematics teacher education.

OUR POSITIONALITY

As mathematics teacher education scholars whose work centers on equity and social justice, we offer our analysis prioritizing attention to social identities, the unfixed yet durable histories and trajectories that structure what we know and how we know. Our identities shape who we are professionally, personally, and politically, and this impacts our work as MTEs. Our experiences and expertise shape our different, yet aligned, scholarly work in justice-oriented mathematics education. Here, we briefly share our social identities.

Anita, Professor of Practice at Vanderbilt University, Peabody College, wrote to elementary education students working towards a master's degree. Anita is a white woman whose commitment to justice-oriented mathematics education is driven by the theoretical underpinnings of critical education and her experiences as a 5th grade teacher in a culturally, linguistically, ethnically, and economically diverse elementary school. She endeavors to develop in her students a shared commitment to empowering children to use mathematics to understand the social and political issues in their homes, communities, and world.

Cathery, Assistant Professor of Mathematics Education and Ethnic Studies at the University of Texas at Austin, wrote to prospective teachers working towards both a special education and either an elementary or single subject mathematics teaching credential. Prior to her current role, Cathery was a bilingual teacher for 10+ years visiting student homes while family members and members of community organizations came into class, co-teaching lessons centered on the mathematics that takes place within homes and communities—from braiding hair, doing carpentry and shopping to examining homelessness, gentrification, and affordable housing. Cathery sees teaching as part of movement work; we have to work together in solidarity to challenge the individualism, competition, and complacency that fuels white supremacy.

Courtney, Associate Professor of Curriculum and Instruction at Ohio University, is the Director of the OHIO Center for Equity in Mathematics and Science and wrote letters for prospective teachers in an early childhood-elementary education mathematics methods course (grades PreK–5). They are a white person living in an Appalachian college town located in the poorest county in the state, according to some measures. Courtney draws on their years as an elementary and middle school classroom teacher and K–5 math coach in culturally, linguistically, and economically diverse classrooms and still spends a lot of time working alongside K–5 teacher colleagues and children in their MTE work. They consider critical pedagogy and critical literacy a key feature in their teaching and research.

Crystal, Associate Professor of Curriculum and Instruction at the University of Texas at San Antonio, wrote her letter for her early childhood through grades 6 mathematics methods course. She is a white woman, a fifth generation Texan, and former middle and high school mathematics teacher. She feels most at home when near the US/Mexican border and the Gulf of Mexico. She believes that mathematics is never completely universal, never neutral, and never without context and all children should have Torres' Rights of the Learner (Kalinec-Craig, 2017) to be autonomous, thriving learners. She seeks to find and challenge structures and systems that perpetuate im/explicit bias in mathematics education.

Eva, Professor of Mathematics Education in the mathematics and statistics department of Portland State University, wrote to students in an elementary mathematics content course, a prerequisite for the teacher education program. Her students often have negative prior schooling experiences with mathematics and see mathematics as a collection of rules that they have to memorize to access the next level of education. Eva is an "accidental German" woman raised in Germany by a Jewish Hungarian (single) mother and immigrated to the United States in her twenties; English is her third language. Eva is committed to supporting her students to reconceptualize mathematics as a tool to read and write the world. Justice-oriented work in a mathematics department brings its own challenges, and Eva draws on this group to face those challenges.

Naomi, an Assistant Professor of Mathematics Education in the Early Childhood Elementary Education department at Georgia State University, wrote her letter for early childhood-elementary methods course (grades PreK–5) students working towards a teaching credential which included an urban and ESOL endorsement. Naomi identifies as a Black mother scholar who brings her experiences working alongside culturally, ethnically, linguistically, and neurodiverse children and families as an elementary school teacher, K–8 mathematics instructional coach (school and district), and K–12 formative assessment coach in urban contexts. She is committed to rehumanizing and honoring the voices, knowledge, and contributions of historically marginalized students, parents, and their communities through the interrogation of mathematics education pedagogies and curricula. For Naomi, justice-oriented mathematics provides a mechanism for shifting the utility of mathematics toward understanding, evaluating, creating a more just world.

SITUATING OUR WORK IN THE LITERATURE

We come to this work understanding mathematics (teacher) education is inherently political and non-neutral (Felton-Koestler & Koestler, 2017) and that the future teachers in our classrooms need to understand this in order to "use teaching practices that provide access, support, and challenge in learning rigorous mathematics to advance the learning of every student" (AMTE, 2017, C.2.1). Social justice work is always situated in context, is relational and both political and personal. As such, our views also reflect our social locations and are therefore not a monolith but fluid, intersectional, and evolving, but we, as a collective, agree upon a shared social justice orientation. The following sections briefly outline the scope of the group's views.

Social justice-oriented mathematics builds from the theoretical tradition of critical pedagogy which takes into account the cultural-historical contingency of schooling as a particular form of human activity that not only reproduces knowledge but also social inequities (Frankenstein, 1983; Freire, 2000; Gutstein, 2006; Skovsmose, 1994). The critical theoretical tradition, influenced by the work of Karl Marx, argues that schools are a vehicle for reproduction of existing social structures needed to maintain capitalism (Freire, 2000). Specifically, power operates through everyday material conditions that maintain asymmetry in the social order. Mathematics education is a political project in which power implicitly and explicitly operates through policies, curricula, and practices reproducing dominant ideologies that rationalize and reproduce social and economic hierarchy (Gutiérrez, 2002; Skovsmose & Valero, 2001). Critical mathematics education calls for counterhegemony (Kincheloe & McLaren, 1994). As Kincheloe and McLaren (1994) state, "inquiry that aspires to the name critical must be connected to an attempt to confront the injustice of a particular society or public sphere within the society" (p. 453).

To conceptualize counterhegemony and social justice mathematics, we build on the work of Rico Gutstein. Gutstein (2006) posits that mathematics classrooms should be a space in which students can learn about and from their lives and realities. Gutstein's framework for teaching mathematics for social justice (TMfSJ) applies Freire's framework of reading and writing the world, but with mathematics. In Gustein's model, he defines TMfSJ as having two equally important sets of goals: social justice pedagogical goals and mathematics pedagogical goals. Through the process of using mathematics to read (study and understand) the world, students strengthen their knowledge and application of mathematics. Concurrently, students use mathematics to write (critically analyze and create change) in which students see their agentive role to promote a more democratic society.

Counternarratives in mathematics teacher education curriculum requires historical-political accounts and interpretations that question dominant narratives of neutrality in mathematics. This is what has come to be known as standpoint or sociohistorical positionality, which is tied to how we see the world both as individuals and as a collective in our emergent consciousness. Au (2012) argues that the concept of "curriculum standpoint" enables educators to reflect upon the deeply rooted and contested consciousness and knowledge traditions reflected in any curriculum.

Love Letters To Teacher Candidates

Drawing on the work of Yeh and colleagues (2022), we see the genre of a love letter to students as a means to communicate positionalities and teaching philosophies as a promising practice for teachers and teacher educators. Traditional statements of positionality and philosophies in syllabi are written in a way that are devoid of personal attention to the reader and how they might take up the ideas in the statement. A love letter, on the other hand, communicates a more personal approach to the reader and expresses a concern for how they may (or may not) receive the intention of the ideas as written. From a mathematics education perspective, the love letter also reflects our mathematical identity with a goal toward supporting future teachers to cultivate their own (and their students) positive mathematical identities (AMTE, 2017, C.4.2). For example, a teacher might write a traditional statement of teaching philosophy as positioned within a social justice orientation and cite foundational scholars in the field that support that stance (e.g., "I believe in Freire's resistance to a banking model of education that assumes children come as blank slates to the classroom") (Freire, 1970). A love letter humanizes those same ideas and sentiments from a traditional statement in a way that speaks directly to the students in accessible language that foregrounds the need to emphasize care and compassion about their development and success (e.g., "In our class, you'll learn how no child is a blank slate and how your role as a teacher is to nurture their development in ways so that they see themselves as whole humans first. This is my promise to you as your teacher.").

The notion of love letters connects to the existing literature base of humanizing the practice of teaching and the notion of care. Nel Noddings notion of care in which "in an encounter, the carer is attentive; she or he listens, observes, and is receptive to the expressed needs of the cared-for" (2012, p. 53) can apply to more than just caregivers, but also teachers and teacher educators. The notion of care extends beyond just the simple utterance of the words "I care for you" but also can take the form of visible actions such as extending compassion for prospective teachers' extenuating challenges and giving feedback to prospective teachers with grace and humility by acknowledging how learning to teach is not an easy process. Therefore, the use of love letters pushes back on traditional norms of expressing positionalities and philosophies; it humanizes the ideas, emphasizes a teacher's concern about their students and their successes, and uses accessible language for complex ideas.

METHODS

In deciding how to present our conversation about how identity and experiences shape who we are and how we see ourselves as justice-oriented educators, we drew on dialogic methodology (McCarthy & Moje, 2002) and collaborative autoethnography (Chang et al., 2016). To engage in dialogic and collaborative autoethnography, we wrote several drafts of love letters as a means to explicate our work (Yeh et al., 2022). We engaged in this process as a way to unpack who we are as MTEs and how we ground our social justice work in mathematics teacher education. While the letters serve as the primary data source, a video recorded and transcribed conversation about the letters, regular (often weekly) meeting notes beginning in December of 2020 to discuss the letters, and a shared google drive with teaching resources served as secondary data sources.

Data collection and analysis was ongoing and iterative. After writing the initial letters we shared and discussed them at one of our weekly meetings. Our discussion centered on what we appreciated about the process of writing letters, what we noticed about what we included and what we felt was missing in our own letters in identifying the connections among the enactment of justice-oriented mathematics teacher education, our positionality, and how our particular contexts shape this work. Following this meeting, we rewrote our own letters based on what we learned from others during the discussion. In subsequent (and ongoing) letter revisions, we worked to articulate our embodiment of justice orientations in our practice as MTEs. Therefore, the version of the letters shared here represent a living representation of our practice—the representations that changed and will continue to change over time through praxis.

We then all read all letters separately and individually created themes for what we noticed across the letters. Once we all created themes for each letter, we met to compare our noticings and to decide on common themes and names. For example, we all noticed a theme centered on the political nature of mathematics teaching across many of the letters which was then labeled as *Teaching as Embedded in Social, Political, and Cultural Contexts* (see examples below). The themes included in this paper were selected based on consensus that they are essential across letters.

One final but important note before we present our data analysis and findings: our love letters are not intended to present an exhaustive or complete picture of how we care for our students and/or enact social justice-oriented mathematics teacher education. Limiting letters to two pages, we made certain decisions about what to include and not include. We ask readers to not assume that an omission of an idea in one letter means that the writer does not believe in nor would not enact these ideas in practice. For example, if one writer did not include mention of "funds of knowledge," readers should not assume that she/they excludes this from practice. The writers all agree that the themes identified in the findings collectively honor the letters as living documents that these will change over time, context, and experience.

We analyzed our work to weave together our story by presenting a dialogic unveiling of our collective and individual understanding of the power of our collaboration. Together, these methodologies incorporate individual and interactional analysis, using our autobiographical data to interpret our actions and interactions (Chang et al., 2016).

WHAT WE FOUND

As we set out to write these love letters, the letters were not only a way to articulate to the prospective teachers how we conceptualized justice-oriented mathematics (teacher) education, but they also communicated how we planned to enact our stance through the semester with our prospective teachers and revealed how the commitment connects to a positive mathematical identity (AMTE 2017, C.4.2). The letters were a way to welcome prospective teachers into our classrooms and provided a window into our practices.

All of the letters embedded love and care, but in different ways (some very explicitly, some more generally about teaching). With the explicit notion that these letters are snapshots in time and not static, exhaustive, and/or complete representations of our work given the genre of a letter at the beginning of the semester and at our unique institutions, we realize that this format can be limiting. Next, we share three of the themes that we found across the letters: *teaching as embedded in social, political, and cultural contexts; the role of teachers and practice in creating and enacting positive change;* and *the centering of children and families.*

Teaching as Embedded in Social, Political, and Cultural Contexts

We all come to this work with the view that mathematics and mathematics teaching is political and non-neutral (AMTE 2017, Indicator P.3.3). This theme was evident in all six letters with several of us explicitly stating that mathematics and mathematics teaching are "political." In some of our letters, this perspective was articulated as related to teachers' practice and position; in other letters, it was related to how our prospective teachers and children experience and use mathematics. This theme is also foundational in the AMTE standards (2017). Of the five assumptions about mathematics teacher preparation in these standards, two (1 and 3) explicitly refer to the importance of considering the context of every learner.

Courtney discusses the importance of recognizing one's stance in the classroom. They state:

> … teaching can never be neutral. By this I mean that as a teacher, you have power when you are in front of children. Whether you know it or not- or whether you want to or not- you will be sending messages to children all of the time- about what is important, valued, or valid- about learners, about mathematics, about education.

Courtney continues to share an example:

> … if you give timed tests to children, this will send the message that math is about speed, that quick people are smart (and smart people are quick), math facts are most important, etc. If you pose story problems to work on math facts, this can send a different message to children. And messages are sent all of the time, both implicitly and explicitly.

Cathery illustrates a similar perspective, sharing with her own experience:

> As teachers, we carry such immense power to not only shape students' learning opportunities but students' sense of self….I remember my first day of kindergarten. My family and I came to the United States when I was five years old. I arrived in April and started kindergarten about two months before the school year ended. On the first day of class, the teacher placed me next to Jenny Abo, the only other non-white student in class, with the goal of having her serve as my translator. However, Jenny Abo was Japanese American and only spoke English, and I spoke Mandarin, Chinese. At the end of the school year, I was retained. In California, Kindergarten is optional, but I had to take it twice. That first year represents much of how I felt throughout K–12 schools. The books we read, the problems we solved, the students in my class did not look like me. In my eyes, that did not matter, as I believed it was me that needed to look and be like them. Whiteness taught me to hate the color of my skin. I now know that my story is not unique. Too many of our students feel

the same way. Our work together is to identify how race, racialization, ableism, and smartness operate in tandem to empower some and disempower others. We'll be privileging minoritized voices...

Anita mentions in her letter that "math and teaching math are political but that doesn't mean you have to teach about politics." Eva challenges prospective teachers to consider the traditional view of mathematics as neutral, apolitical, and universal in her letter, specifically calling attention to mathematical content. She states:

Any chart in the newspaper/website is chosen to make a certain point. Being able to deconstruct what is being communicated in the news is an essential part of life and we will work towards the skills to do so. We should always ask ourselves why is this piece of news presented and what is NOT presented? Having a full understanding of why mathematics works will put you into a position to be able to ask such questions.

A subtheme related to the idea that mathematics and mathematics education can never be neutral and that it is always embedded in social, political, and cultural contexts is the way that some of us focused on mathematics as a liberatory tool. Crystal for example states:

First, I believe that mathematics is a civil right...This means that equity and access can be as basic as access to high quality mathematics or as nuanced as seeing that mathematics is a racialized experience. If mathematics is a civil right then mathematics needs to reflect the diversity of all of us, not just those who are the loudest and have the most resources....

She further asserts that mathematics teachers could and should seek "more opportunities to seek and demand justice for all."

Relatedly, Naomi explains in her letter that:

… justice oriented mathematics considers the racialized experiences and generational marginalization that impact learning opportunities and views the learning of mathematics as a civil rights issue....Therefore, access to learning high-quality mathematics is a justice-oriented issue which leads to using mathematics as a liberatory tool.

We know that many prospective teachers who come to us view mathematics teaching as neutral, as apolitical. They do not enter early childhood-elementary teaching as activists. Yet, as MTEs, we see this work as deeply political and see the potential (and need) for all of the prospective teachers with whom we work to see creating and enacting positive change as deeply entwined as part of the job of teaching mathematics well. It is important for us as justice-oriented MTEs to frame our work with our prospective teachers from the outset as non-neutral, as political. It is important that they realize that their work as early childhood and elementary school teachers are shaped by the cultural, social, and political contexts in which they take place.

The Role of Teachers and Practice in Creating and Enacting Positive Change

Across the letters, we articulated different ways that a teacher's role and their practice could serve as a means of creating positive change and the power that teachers hold. The theme described as "role of teachers and their practice" assumes that the work that teachers do, including lesson planning and instruction, are not static events, but can support children, families, communities, and others to embrace our collective humanity. Teachers who help students to read and write the world with mathematics (as inspired by Freire and articulated by Gutstein, 2006, and Stinson & Wager, 2012) is a significant theme that emerges in our letters.

When teachers see their role both in and beyond the classroom, they can find multiple opportunities to help children to use their new knowledge to create positive change. It is through these multiple opportunities that we describe their roles as teachers and aspects of their practice as they move towards justice-oriented mathematics education. An elaboration of AMTE's Standard C.4.1 (UE.6 Ethical Advocates for Students) suggests that teachers "build partnerships with families and communities to work to eliminate institutional and curricular barriers to learning" (2017, p. 91). As Naomi describes:

> I have seen curricular violence occur with schools that use mathematics as a tool of oppression and tracking. In these settings, students are given "survival tactics" for doing mathematics and are not allowed to breathe and ponder, imagine, challenge, and see the potential of what mathematics could do in their everyday lives that would bring joy, solidarity, and wonder.

Naomi also highlights specific aspects of her course that looked at justice-oriented mathematics education through multiple lenses:

> Again, throughout this class, we will constantly examine a) the mathematics content, b) pedagogical practices, c) ways of understanding our students and their complex identities, and d) how our own biases, dispositions, and ways of being socialized into learning mathematics impact our teaching.

It is through these lenses that Naomi sought to counteract the "curricular violence" she witnessed in schools and make space for teachers' roles to enact positive change. To counteract harmful practices, Cathery suggests that her practice is one in which she hopes to create a "mo(ve)ment building to build a coalition of educators, cultural workers, and teacher-activists." Courtney also comments on the damaging experiences children have faced in mathematics classrooms and the role teachers play in providing safer, more welcoming classrooms. As they commented:

> The biggest thing that I want you to learn from my class is the power you have as a teacher. Too many people, *too many children,* have harmful experiences in school, sometimes (oftentimes?) in math. All children should feel important, valued, and valid in our classrooms. Schooling should never be punitive. We should engage in child-

centered practices that support children, not harm them. (Like timed tests! Ah! There is so much written about the detriments of timed tests....but more on those later.)

Consistent across letters is the role of reflection, and how critical reflection must be an ongoing and integral process to teaching. Crystal urges her prospective teachers to take "any opportunities to reflect and pivot your philosophy to something new, something brave, and something bold that opens up more opportunities to seek and demand justice for all." Anita also prompts her prospective teachers to consider their own identities and positionalities saying, "[as] much as I want you to consider your students' personal experiences, you need to also reflect on your own and how they might inform the decisions you make when you teach."

Overall, we acknowledged the inherent and relative power that teachers hold in schools and sought ways to interrupt that power by using their practice and their role as a teacher to redistribute this power to children, families, and communities. Although Eva did not explicitly discuss the topic in her letter, she wholeheartedly agrees with this theme and its importance. Although it is important to recognize the real and perceived constraints teachers have, it is also important to acknowledge the ways in which teachers can act for more equitable and just classrooms, schools, and communities for and with their students and their families.

The Centering of Children and Families

The importance of children and families (AMTE Standard C.2.5) was evident in all letters. Several of us made assertions to prospective teachers about "all children" stating that "all children are smart about mathematics" (Anita), "all children come into school as mathematicians" (Cathery), and "all children should feel important, valued, and valid in our classrooms" (Courtney).

Learners' strengths and children's knowledge as a source of teacher learning was evident in all letters, which is not surprising given our shared backgrounds and expertise in Cognitively Guided Instruction (Carpenter et al., 2014). However, many of us went beyond mentioning the importance of listening to and building on children's thinking.

Crystal notably centered children by framing mathematics as a civil right (e.g., Moses & Cobb, 2002) and described that "equity and access can be as basic as access to high quality mathematics or as nuanced as seeing that mathematics is a racialized experience" for children's learning. Crystal embeds Torres' Rights of the Learner (Kalinec-Craig, 2017) to center the importance of rights of children in mathematics classroom. She states:

When students know that they have rights to exercise as in a democracy, then they can come closer to experiencing mathematics as a civil right. Students who interrupt the teacher to claim a confusion or to claim a mistake are actively participating in a democracy and are co-constructors of their knowledge. It disrupts the notion that the teacher is the only one who can teach and build knowledge. Torres' ROtL are also another place in which we value the diversity of students' ideas, experiences, and reasoning. This is particularly important for students who are immigrants and are balancing between multiple physical spaces, languages, and symbolic notation. Torres' ROTL is a way of thinking that embodies the notion that mathematics is a civil right....

Anita and Cathery both acknowledged the importance of children's lived experiences when considering our mathematics teaching. Anita asserts that teachers must consider the personal, lived experiences of every child in their classroom when designing and enacting lessons in order for teachers to take a justice-oriented mathematics pedagogy that recognizes and builds on children's multiple mathematical knowledge bases. In Cathery's letter, she describes how it is important to listen and understand children's mathematical thinking and their lived experiences, as these provide "the most generative learning opportunities for us as mathematics teachers." She goes on in her letter to describe how prospective teachers will work collaboratively to plan, implement, and reflect on culturally responsive mathematics lessons that build on bilingual children's ways of knowing, being, and communicating in "a Community Math Partnership Program where we'll unlearn and relearn mathematics."

There was also a subtheme of the importance of the work that we are doing now (in our mathematics teacher education courses) and how this will be for the betterment of the children our PTs will work with in the future. For example, when Eva discusses the importance of sense making and problem solving, she says "listening and understanding other students' thinking and making sense of that is the other part and will prepare you to make sense of your students' thinking later on."

Further, Naomi uses historical data to ground the impetus for progressive teaching methods that challenge and, as noted earlier, to disrupt the "curricular violence...[that happens in] schools that use mathematics as [a] tool of oppression and tracking." She urges her prospective teachers to center children and to create "mathematical learning experiences that challenge and question the world around us but also bring joy and liberation."

In all of our letters, and in all of our work as MTEs, children and their contributions play an important role: children's mathematical thinking (obviously), but also the importance of valuing and honoring their agency as valuable members of the mathematical learning community. When advocating for child-centered pedagogies, it is important for us to honor and value children and their brilliance in our work. We want to model learning with and from children so that prospective teachers see this as something integral to mathematics teaching.

IMPLICATIONS AND DISCUSSION

We see the process of letter writing as a vehicle to articulate our mathematics education goals related to equity and justice to prospective teachers and as a tool to learn with and from other MTEs. Just as mathematics is political; mathematics teacher education is political as well. Dismantling historic and ongoing hegemonic structures within mathematics and mathematics education require ongoing reflection.

As we wrote and rewrote these letters in our "critical friends" group, we saw connections between and among our work and ways that we could support and challenge one another in our work as justice-oriented MTEs. In the context of a dialogue with Paulo Freire, Leistyna (2004) noted that critical consciousness "is the ability to analyze, problematize (pose questions) and affect the sociopolitical, economic, and cultural realities that shape our lives" (p. 17). He continues, "For Freire, this process of transformation requires praxis and dialogue." Power, privilege, and positionality are not mentioned, or barely mentioned in mathematics teacher education. Here, we explicitly engaged in a process of praxis and dialogue, posing questions about manifestations of power and privilege in our own lives and how it shapes our actions as justice-oriented MTEs. The letter writing process required each of us, as MTEs, to identify our own social positioning and the relation to power and privilege, to name them, question their existence, and to articulate our own meaning making of what justice-oriented mathematics teacher education means. This process showed how we could honor our individual social identities as immigrants, teachers, women, parents, and the ways we have and can engage in justice-oriented work in our courses.

As bell hooks reminds us, the journey of naming and reclaiming who we are is a space of healing and radical opening: "We are transformed, individually, collectively, as we make radical creative space which affirms and sustains our subjectivity, which gives us a new location from which to articulate our sense of the world" (hooks, 1989, p. 23). We have experienced the progression of ideas, intentionality, and positionality over time with prospective teachers. The discussion and analysis within our "critical friends" group, both theoretically and practically, have been invaluable to our work as MTEs. We hope our process of engaging in praxis and dialogue with students as shared can serve as one model in honoring the subjective, relational and collective nature of learning in mathematics teacher education.

CONCLUSION

In the time that we have been working on this project, more and more state legislators in the U.S. have been advancing "anti-CRT" and "anti-LGBTQ+" legislation, which has pushed us to consider how we would distribute our letters to our prospective teachers. For example, some of us were intending to send these letters via email as an introduction before the semester started. However, a letter about justice-oriented mathematics teaching without context (e.g., before they meet us, without any other sort of readings, discussions, etc.) might be disorienting for prospective teachers. On the other hand, this kind of letter might be helpful in orienting the course, setting the stage, and framing the philosophical underpinnings of the course.

We bring this to a close with a few considerations for other MTEs who might want to engage in similar work. These considerations reflect our own lingering thoughts in our ongoing dialogue with critical friends.

- One of the drawbacks to these letters (or as discussed, the timing of when we send them) is that we might want to reserve some of the ideas in them to be introduced with specific readings at purposeful times during the course. That is, we suggest considering the timing (i.e., what to make explicit at the beginning of the quarter/semester to frame the course versus what to discuss later after trust and relationships are built).
- We find it an ongoing challenge to succinctly articulate explicitly what we value, and to consider what is implicit but still holds true. We suggest making it explicit to students that these brief letters are simply a starting point for a dialogue and that over the course of the quarter/semester you will be delving deeper into the themes of the letter and what they mean for equity- and justice-minded teachers.
- Just as our letters evolved in this process, future letters will continue to evolve, depending on our institutional contexts, social and political contexts, our own positionalities and opportunities to learn and reflect with and from others. Thus, we suggest continually revisiting such efforts to share perspectives and positions that may challenge and support prospective teachers.

Finally, through this process, we are learning how to keep our collective ideas living. For example, in our critical friends group, we considered how to incorporate more ideas into our work (e.g., TMfSJ vs teaching mathematics for liberation); these new ideas brought into question what we needed to add, drop, or revise about our content and pedagogy. We hope by sharing our process and products of inquiry we can challenge other MTEs too. We also hope by sharing our letters with our teacher candidates that they reflect more on our content and pedagogy in community.

REFERENCES

Association of Mathematics Teacher Educators. (2017). *Standards for preparing teachers of mathematics.* Author. https://amte.net/standards.

Au, W. (2012). The long march toward revitalization: Developing standpoint in curriculum studies. *Teachers College Record, 114*(5), 1–30.

Carpenter, T. P., Fennema, E., Franke, M. L., Levi, L., & Empson, S. B. (2014). *Children's mathematics. Cognitively guided instruction.* Heinemann.

Chang, H., Ngunjiri, F., & Hernandez, K.-A. C. (2016). *Collaborative autoethnography.* Routledge.

Felton-Koestler, M. D., & Koestler, C. (2017). Should mathematics teacher education be politically neutral? *Mathematics Teacher Educator, 6*(1), 67–72.

Frankenstein, M. (1983). Critical mathematics education: An application of Paulo Freire's epistemology. *The Journal of Education, 165*(4), 315–339.

Freire, P. (2000). *Pedagogy of freedom: Ethics, democracy and civic courage.* Rowman & Littlefield Publishers.

Gutiérrez, R. (2002). Enabling the practice of mathematics teachers in context: Toward a new equity research agenda. *Mathematical Thinking and Learning, 4*(2–3), 145–187. https://doi.org/10.1207/S15327833MTL04023_4

Gutstein, E. (2006). *Reading and writing the world with mathematics: Toward a pedagogy for social justice.* Routledge.

hooks, b. (1989). Choosing the margin as a space of radical openness. *Framework: The Journal of Cinema and Media, 36*, 15–23.

Kalinec-Craig, C. A. (2017). The rights of the learner: A framework for promoting equity through formative assessment in mathematics education. *Democracy and Education, 25* (2), 1–11.

Kincheloe, J. L., & McLaren, P. L. (1994). Rethinking critical theory and qualitative research. In N. K. Denzin & Y. S. Lincoln (Eds.), *Handbook of qualitative research* (pp. 138–157). Sage.

Leistyna, P. (2004). Presence of mind in the process of learning and knowing. A dialogue with Paulo Freire. *Teacher Education Quarterly, 31*(1), 17–30.

McCarthy, S. J., & Moje, E. B. (2002). Identity matters. *Reading Research Quarterly, 37*(2), 228–238.

Moses, R., & Cobb, C. E. (2002). *Radical equations: Civil rights from Mississippi to the Algebra Project.* Boston, MA: Beacon Press.

Noddings, N. (2012). The language of care ethics. *Knowledge Quest, 40*(5), 52–56.

Skovsmose, O. (1994). Towards a critical mathematics education. *Educational Studies in Mathematics, 27*(1), 35–57.

Skovsmose, O., & Valero, P. (2001). Breaking political neutrality: The critical engagement of mathematics education with democracy. In B. Atweh, H. Forgasz, & B. Nebres (Eds.). *Sociocultural research on mathematics education* (pp. 37–55). Mahwah, NJ: Lawrence Erlbaum.

Stieha, V. (2014). Critical friend. In D. Coghlan & M. Brydon-Miller (Eds.), *The Sage encyclopedia of action research* (pp. 207–208). London, England: Sage Publications.

Stinson, D. W., & Wager, A. A. (2012). A sojourn into the empowering uncertainties of teaching and learning mathematics for social change. In A. A. Wager, & D. W. Stinson (Eds.), *Teaching mathematics for social justice: Conversations with educators* (pp. 3–20). Reston, VA: National Council of Teachers of Mathematics.

Yeh, C., Agarwal-Rangnath, R., Hsieh, B., & Yu, J. (2022). The wisdom in our stories: Asian American motherscholar voices. *International Journal of Qualitative Studies in Education,* 1–14. https://doi.org/10.1080/09518398.2022.2127010

CHAPTER 3

GROWING AND MODELING CULTURALLY RESPONSIVE PEDAGOGIES

A CRP Self-Study Framework for Mathematics Teacher Educators

Lindsay Keazer

Sacred Heart University

Kathleen Nolan

University of Regina

Culturally responsive pedagogies (CRP) are recognized as critically important for the field of mathematics teacher education, yet few studies have focused on mathematics teacher educators (MTEs) growing their own practice as culturally responsive educators. This chapter proposes a self-study framework to support MTEs in growing their CRP. The framework consists of key questions and reflective prompts at the intersections of three dimensions of CRP theory and four components of MTE practice. This framework serves as a pedagogical tool for MTEs to improve upon their pedagogical practice, while also providing a deliberate methodological tool

The AMTE Handbook of Mathematics Teacher Education: Reflection on Past, Present and Future—Paving the Way for the Future of Mathematics Teacher Education, Volume 5
pages 45–65.

with focused questions for MTEs to use in self-study data collection. We suggest that the future of culturally responsive teaching lies in developing the CRP of MTEs, who hold the responsibility to model a living and growing CRP for the teachers with whom they work.

INTRODUCTION

Culturally relevant/responsive pedagogies are increasingly being recognized as critically important for the field of (mathematics) teacher education (Gay 2018; Greer et al., 2009; Ladson-Billings, 2017). Most studies of culturally responsive pedagogies in teacher education have focused on improving cultural responsiveness in K–12 education or with prospective and practicing teachers (PTs) (Nolan & Keazer, 2021a). Few, however, focus on mathematics teacher educators (MTEs) growing their own practice as culturally responsive pedagogues/educators.

Ladson-Billings (1995) developed a theory of culturally relevant pedagogy, which has gained widespread traction. The theory evolved in name over the years with adaptations such as *culturally responsive* (Gay, 2018) and *culturally sustaining* (Alim & Paris, 2017) pedagogies. Ladson-Billings (2014, 2017) has supported these advances in terms, as they represent attempts to respond to issues of misapplication and varied implementation in popular versions of culturally relevant pedagogy. Thus, in seeing these terms as sharing an intended meaning and goal of supporting the mission of responding to and sustaining cultural pluralism, we use the widely-utilized term *culturally responsive pedagogies* (CRP) to encompass the shared meaning amongst all three terms.

We suggest that a critical source for PTs to learn about CRP lies in the opportunities for MTEs such as ourselves to "practice what we preach," and model our own development and growth of CRP. To date, however, no specific tool exists to guide MTEs' self-study of their culturally responsive practice. Thus, the purpose of this chapter is to present such a tool in the form of a self-study framework for MTEs. We propose that the future of culturally responsive teaching lies in developing MTEs' CRP, who in turn can model a living and growing CRP for the PTs with whom they will work. This connects to the AMTE (2017) Standards for developing candidate knowledge and dispositions by encouraging MTEs to model continual growth and reflection of their CRP, in order to equip them to support PTs' development, including commonly overlooked socio-political considerations (Ladson-Billings, 2014), such as indicators C.4.4 focused on power and privilege, and C.4.5 on justice and advocacy. In addition, connections can also be made to the AMTE Standards for program characteristics, which include MTEs with knowledge of CRP (indicator P.3.5), who create opportunities for PTs to explore issues of equity and analyses of power and privilege embedded within mathematics methods courses (indicator P.3.3).

The framework we have constructed and present here is both a pedagogical and methodological tool: pedagogical in that questions are proposed for MTEs to reflect on to improve their culturally responsive practices, and methodological in that the reflection questions are presented in a framework for use in self-study data collection. Both of these functions aid MTEs in aligning teacher education programs with the AMTE (2017) Standards: MTEs who utilize this self-study framework to collect data on growing their CRP will not only model the important work of teachers examining and improving their practice, but ideally will be empowered to explicitly discuss components of CRP being modeled through their pedagogy to provide PTs with much-needed examples of culturally responsive mathematics teaching practices.

As a motivating force for our development of a tool to guide MTE self-study of CRP, we appreciate Han et al.'s (2014) collaborative self-study to explore teacher educator CRP. Theirs was a study which set out to explore and describe "how teacher educators... define, enact, and navigate their roles as culturally responsive educators in a higher education institution" (p. 291). Using general definitions of CRP garnered from the work of Gay and Ladson-Billings, the authors/researchers interviewed each other to ascertain how they each live out their CRP practices. While their study revealed valuable findings related to modeling, relationships, tensions, and opportunities for further growth, our framework offers an opportunity to raise the legitimacy of collaborative CRP self-study in mathematics teacher education, through its robust layering of self-study methodology, CRP theory, and MTE practice. Our self-study framework offers MTEs significant opportunities for rich reflection and growth of their CRP by embracing a crucial dialogue between methodology, theory, and practice.

This chapter is organized into several key sections which lead to the introduction of the framework. Since our framework highlights the dialogue, or interaction, between self-study methodology, CRP theory, and MTE practice, we devote the next three sections of this chapter to elaborating on each of these important layers. First, we describe the choice to draw on self-study methodology as our approach to reflecting on dimensions of CRP and/in mathematics. Following this, we elaborate on the CRP components of our tool, developed from a review of CRP literature in the fields of mathematics teacher education and teacher education (Nolan & Keazer, 2021b). Third, to add specific MTE context to this work, we review research on MTE knowledge, practice and identities in order to identify several key components of MTE practice. After tracing out these three aspects of the dialogue, we present the intersections of CRP theory and MTE practice through a framework of key self-study reflective questions. Subsequently, for each reflective question, we present examples of reflective prompts that could be used as practical stimuli for MTEs' reflections. In other words, by studying the interaction/intersection of each dimension of CRP theory with each component of MTE practice, we have generated a series of reflective questions and corresponding reflective prompts which constitute our self-study framework for growing our practice as culturally responsive MTEs.

ON SELF-STUDY METHODOLOGY IN OUR WORK

As a methodology, self-study is defined as intentional and systematic inquiry into one's own practice (Loughran, 2007). In teacher education, self-study is powerful because of the potential to influence prospective teachers, as well as impact one's own learning and practice as a teacher educator (Nolan, 2015, 2016). The strengths of self-study for studying and improving upon the practice of teacher educators (TEs) are well-documented (Loughran et al., 2004; Williams & Berry, 2016). While self-study has been embraced by individual TEs who focus primarily on improving their own practice (e.g., Nolan, 2015), it has also been drawn on to a great extent recently in collaborations between TEs (Bragg & Lang, 2018; Han et al., 2014; Nolan & Keazer, 2021a), making for a rich and robust research design (Loughran et al., 2004; Samaras, 2011).

While the methodology of self-study is generally defined in consistent ways across the research, the methods of data collection for self-study vary widely. As offered by Vanassche and Kelchtermans (2015) in their review of published self-studies over a period of 22 years (1990–2012), "self-study has no single method inscribed to it. Self-study borrows its repertoire of research methods and strategies from the conventional methods of empirical-analytical and/or qualitative-interpretative research" (p. 515). Hamilton and Pinnegar (1998) suggest self-study researchers use "whatever methods will provide the needed evidence and context for understanding their practice" (p. 240). Across the published self-studies, one finds diverse methods such as interviews with colleagues and students, audio/video recordings of classroom instruction, observations of teaching by critical friends and other peers, meeting notes with critical friends, reflection on student assignments and course evaluations, and, most common, researcher reflective journaling (Gallagher et al., 2011; Vanassche & Kelchtermans, 2015; Williams & Berry, 2016).

Hence, while self-study purports to be a welcome and open methodology with a range of available methods for data collection on any version of self-improvement focused research, we critique its frequently atheoretical stance. In a similar manner, Vanassche and Kelchtermans (2015) critiqued that many self-studies "were mostly descriptive-reflective in nature, started from a broad interest or curiosity and developed largely independent of existing theories, concepts or hypotheses in its operationalization" (p. 520). Self-study research questions aimed at improving or changing practice often lack grounding in a theoretical framework or lens. We advocate for embedding theory into the design of data collection (not merely at the stage of data analysis) to create a purposeful theory-methodology construct (Nolan, 2016). The framework we introduce here frames the research question of how MTEs can grow their CRP by providing specific self-study reflection questions formed at the intersections of CRP theory and MTE practice.

We developed our self-study framework in two levels. In the first level, we present key reflective questions shaped from examining the intersections between dimensions of CRP theory (informed by the work of Ladson-Billings and Gutiérrez) and components of MTE practice (informed by TE and MTE scholarship). In the second level of the framework, we expand upon each of these key reflective questions through a series of corresponding reflective prompts, influenced by a recent review of the work of key scholars in CRP (Nolan & Keazer, 2021b). Our paper unfolds in the next few sections through an elaboration of our methods for arriving at the dimensions of CRP theory, the components of MTE practice, the reflective questions and the reflective prompts—the key aspects that constitute the structure and contents of our framework.

In essence then, our framework provides a pedagogical tool (for MTEs to improve upon their pedagogical practice) while also providing a deliberate methodological tool (focused questions for MTEs to use in self-study data collection). This responds to Vanassche and Kelchtermans' (2015) call "for self-study researchers to engage in systematic dialogue with existing theoretical and conceptual work from the very moment they start framing the issue in their teacher education practice" (p. 522).

THE CRP DIMENSIONS OF OUR FRAMEWORK

We initially sought to develop a tool for MTE self-study of CRP by utilizing existing frameworks designed for MTEs to develop and analyze PTs' culturally responsive practices (e.g., Aguirre & Zavala, 2013; Gallivan, 2017). This initial development work resulted in an early draft of a framework (Keazer & Nolan, 2021) that, while productive for fostering reflection, seemed incomplete at capturing some aspects of CRP that we considered important for our own practices and pedagogies as MTEs. For example, we felt that the sociopolitical consciousness dimension of CRP (Ladson-Billings, 2006) was underdeveloped in that first attempt at a framework. Developing a tool from existing tools runs the risk of diluting aspects of CRP theory, as integral elements may be "lost in translation."

The shortcomings of our first attempt pointed to the need for us to return to the literature and to the roots of CRP theory; to revisit how CRP has been defined and theorized. We conducted a review of CRP-focused literature specific to teacher education and, where available, mathematics teacher education (Nolan & Keazer, 2021b). Through this process, we determined that the theoretical model of culturally relevant pedagogy (Ladson-Billings, 1995) has been most widely-cited, has stood the test of time, and outlines elements of culturally relevant pedagogy that are powerful for assessing and dissecting other approaches to CRP. Ladson-Billings (2006) proposes three elements of culturally relevant pedagogy: academic achievement (i.e., student learning), cultural competence, and sociopolitical consciousness. In using these elements to analyze our own, and other existing work in CRP, we can identify the strengths and better understand the gaps in current practice. Ladson-Billings (2014) warns that the many approaches to CRP tend to neglect or overlook the sociopolitical dimension, since this dimension offers a critical perspective that is missing from the superficial versions of CRP which have gained broad appeal within dominant paradigms. As researchers ourselves, we see this dimension as critically important for guiding the self-study of our CRP, as we apply a critical lens to how our thinking, our theories, and our philosophies manifest in the classroom.

Much like Ladson-Billings' (1995, 2006) efforts to delineate elements of CRP for the field of education, Gutiérrez (2012) contributed four elements of equity within the field of mathematics education. While Gutiérrez's work primarily addresses *equity* in mathematics education, rather than CRP, the four elements are closely connected to Ladson-Billings' work— both sets of elements pay careful attention to achieving balance between dominant approaches to equity/CRP and approaches which highlight the critical and sociopolitical. Moreover, Gutiérrez's four elements were developed with specific attention to the context of mathematics education, a field which has a unique set of political challenges (Bishop, 1990; Burton, 1994; Nolan & Keazer, 2021a). Gutiérrez (2012) advocates for approaches to mathematics education which encompass all four elements in order to achieve an equitable balance: "As researchers concerned with equity, we must keep in mind all four dimensions, even if that means that at times one or two dimensions temporarily shift to the background" (pp. 33–34).

Gutiérrez's four elements of equity are divided into two parts: the dominant axis dimensions (access and achievement), and the critical axis dimensions (identity and power). The dominant elements of equity seek to support and broaden participation and learning in the status quo mathematics that is needed for success in high stakes testing, and "that is involved in making sense of a world that favors the views and perspectives of a relatively elite group" (Gutiérrez, 2007, p. 39). The critical axis, on the other hand, consists of:

> Mathematics that squarely acknowledges the position of students as members of a society rife with issues of power and domination. Critical mathematics takes students' cultural identities and builds mathematics around them in ways that address social and political issues in society, especially highlighting the perspectives of marginalized groups. This is a mathematics that challenges static formalism, as embedded in a tradition that favors the West. (Gutiérrez, 2007, p. 40)

In our literature review (Nolan & Keazer, 2021b), the sets of elements from Ladson-Billings (2006) and Gutiérrez (2012) emerged as crucial in forming the theoretical grounding for the development of CRP dimensions for our framework. By exploring the connections between the four Gutiérrez (2012) elements and the three Ladson-Billings (2006) elements, we merged what we determined to be overlapping or intersecting elements, as shown in Figure 3.1. As a result, we generated three dimensions of CRP theory for use in our framework: a) access and achievement, b) cultural competence/identity, and c) sociopolitical/power. The dimension of *access and achievement* has its roots in Gutiérrez's dominant axis, and was integrated with Ladson-Billings' (2006) element of academic achievement (i.e., student learning). The dimension of access and achievement consists of the resources available to support student participation and success, in addition to student knowledge and learning outcomes, in the dominant approach to mathematics. The dimension of *cultural competence/identity* is an integration of Ladson-Billings' element of cultural competence with Gutiérrez's element of identity. This dimension consists of "understanding mathematics as a cultural practice" (Gutiérrez, 2012, p. 31) and "helping students to recognize and honor their own cultural beliefs and practices while acquiring access to the wider culture, where they are likely to have a chance of improving their socioeconomic status and making informed decisions about the lives they wish to lead" (Ladson-Billings, 2006, p. 36). The third dimension, *sociopolitical/power* was developed from merging the political elements from each theorist. This dimension consists of unpacking the sociopolitical issues of the community and society, and seeking ways to incorporate these issues into the mathematics curriculum, such that mathematics can be reframed as a tool to examine the world and critique issues of power and injustice.

The arrows in Figure 3.1 show connecting elements of CRP and equity which were merged to arrive at three dimensions of CRP for MTEs. These dimensions were used as theoretical grounding in our framework, to focus us on identifying key questions corresponding to each dimension. In doing so, we can use it to identify gaps and work toward balance in ensuring all dimensions are addressed. While we recognize the tensions present in distinguishing between dimensions, reference to these three dimensions provided opportunities for us to construct and categorize the key reflective questions for our framework. Gutiérrez (2012) also acknowledges the inevitable tensions that occur as one tries to balance the critical and dominant axes in mathematics education. She suggests that by accepting and maintaining these tensions, rather than resolving them, we can build new knowledge.

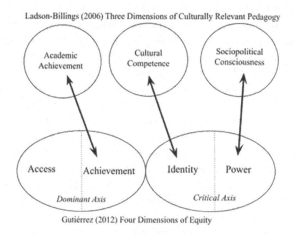

FIGURE 3.1. Merging Theoretical Elements of CRP

THE COMPONENTS OF MTE PRACTICE OF OUR FRAMEWORK

In this section we propose a description of what it is that MTEs should know and do in their work; that is, the key components of MTE practice. Then in the following section, these synthesized components of MTE practice are integrated into our framework, to illustrate how the development and growth of MTEs in CRP can work together with, and for, the development and growth of MTE practices. While scholarship on MTE knowledge and practices has recently grown significantly (Appova & Taylor, 2019; Beswick & Goos, 2018), much of that scholarship draws heavily on existing research that studies the knowledge and practices of TEs in general (e.g., Goos & Beswick, 2021). For this reason, we begin this section by considering the research into the professional knowledge, identities and roles of TEs, before moving into a specific focus on MTE research.

Research in teacher education maintains a consistent focus on the knowledge, practices, and identities of TEs (Davey, 2013; Lunenberg et al., 2014). This includes, according to Berry (2008), how they develop their knowledge of teaching teachers, while serving as role model practitioners who can successfully "articulate their pedagogy in ways that are comprehensible and useful for prospective teachers" (p. 9). In a similar manner, Loughran (2014) discusses the professional development of TEs, proposing that their "knowledge and practice of teaching *and* learning about teaching is intimately tied to: understandings of identity; the challenges and expectations of the teacher education enterprise; and, the place of scholarship as an important marker of knowledge, skill, and ability in the academy" (p. 272). This literature points to the idea, also supported in MTE scholarship, that the knowledge needed for TEs extends beyond that needed by the PTs they teach. Similarly, Beswick and Goos (2018) claim that the forms of knowledge about teaching required by MTEs and mathematics teachers differ considerably, and that MTEs "need to hold this knowledge in a way that is rather different from the way that teachers know it" (p. 418).

TEs have many roles, and hold steep responsibilities, as studied by Lunenberg et al. (2014) who describe six professional roles of teacher educators: teacher of teachers; researcher; coach; curriculum developer; gatekeeper; broker. Also discussing the roles and high standards placed on TEs, Hökkä et al. (2012) offer that TEs "are considered to be academic professionals who are responsible for conducting academic research themselves, keeping up active societal relations, and providing research-based teacher education" (p. 84). Similarly, Goodwin and Kosnik (2013) describe requirements identified across a number of different teacher educator standards, "including model teaching, research and scholarship, leadership in the profession, and ongoing professional development" (p. 337). This literature points to the complex responsibilities of TEs, and the inherent need to continually invest in their own learning and scholarship as they reflect on and refine their knowledge and practice.

As we hone in on examining roles within mathematics teacher education, we think it important to explicitly conceptualize what an MTE is. MTEs have been defined as "professionals who work with practicing and/or prospective teachers to develop and improve the teaching of mathematics" (Jaworski, 2008, p. 1). Since our focus is primarily on MTEs who conduct their work within a university context, we focus the definition of an MTE by specifying that an MTE is an academic who teaches prospective (pre-service) and/or practicing (in-service) teachers (PTs) in the context of a university teacher education program, teaching mathematics content or methods, while also engaged in scholarship in mathematics education and/or related fields.

Outlining the specifics of the knowledge, practices, and identities of MTEs is a growing area of scholarship. In a recent collection edited by Goos and Beswick (2021), chapter authors introduced the complexities involved in describing the kinds of work MTEs do, including how they learn and develop their knowledge and expertise. The book is organized into several themes which contributed to our conceptualization of the components of MTE practice. One theme, focused on the nature of mathematics teacher educator expertise, examined how the types of knowledge theorized for mathematics teachers (e.g., Ball and colleagues) are applicable/transferable to the knowledge of MTEs, as well as how MTE knowledge differs. The research chapters contributing to this theme make clear the importance of MTE knowledge (mathematical, curricular, and pedagogical), as well as the ability to utilize that knowledge for modeling teaching practice. Goos and Beswick (2021) offer that "MTEs require a kind of meta-knowledge which could be described as knowledge for teaching knowledge for teaching mathematics" (p. 3). Contributors writing within the other two themes of the book (focused on how MTEs learn and develop and the methodological challenges in researching MTEs) reflect deeply on the contexts in which MTEs learn, including support for MTEs engaging in critical reflection on their practice (praxis) and staying engaged in research, theory, and scholarship.

Through this examination of the literature on TE knowledge, identities, and practices, and connecting that to the newly developing scholarship of MTEs (e.g., Goos & Beswick, 2021) (see Figure 3.2), we synthesized four general categories of MTE practice being described: mathematical and curricular knowledge; pedagogical knowledge and modeling; research/theory/scholarship; and critical reflection/praxis.

Depicted in the second row of Figure 3.2 (the 5 bubbles) is an overview of the elements of TE practice present in teacher education literature. While the literature draws on different language to describe these elements of TE practice, we synthesized them into four general categories. These four categories, or components, are also fully represented within the themes of the scholarship on MTEs (e.g., Goos & Beswick, 2021). The four components are each briefly described as follows:

1. **Mathematical and curricular knowledge**: extensive knowledge of, and experience with, the mathematics concepts present in K–12 school

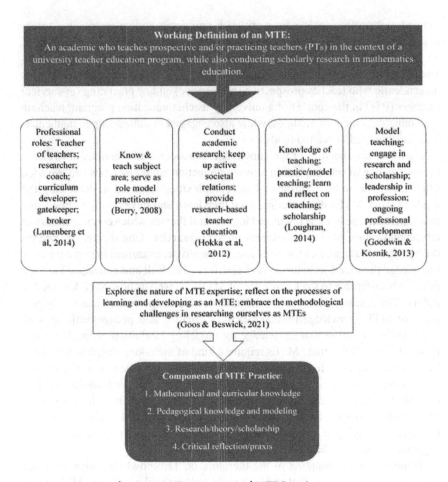

FIGURE 3.2. Developing Key Components of MTE Practice

curricula, in addition to a more comprehensive knowledge of the discipline of mathematics (beyond K–12)— its concepts and relationships between concepts.

2. **Pedagogical knowledge and modeling**: understanding of, and experience with diverse forms of pedagogy, including that which is both general in nature and specific to the content of mathematics, and the ability to model these pedagogies as a strategy for teaching PTs how to teach mathematics.

3. **Research/theory/scholarship**: the theoretical and philosophical perspectives which guide MTEs' work with PTs as well as their research contributions to the wider mathematics education community and society.

4. **Critical reflection/praxis**: reflection by MTEs on their own growth and development within the other three components of practice, their interaction with the tools and theories of the field of mathematics teacher education, and their reflection in/on action with PTs.

In other words, MTEs embrace a praxis approach to addressing questions, tensions, dilemmas, and so forth in teaching, learning, and learning to teach. As Lunenberg et al. (2014) propose: "Dealing with tensions in specific situations requires from teacher educators thorough theoretical knowledge, experience and reflective judgements. By finding the right balance in these tensions at the right moment, teacher educators make sense of their role of teacher of teachers" (p. 28).

In the following section, we examine the intersection points between the above four components of MTE practice and the three dimensions of CRP theory presented earlier, to discuss how they can work together toward the development and growth of CRP within the practices of MTEs.

THE MTE SELF-STUDY FRAMEWORK

In this section, we present our self-study framework for growing our practice as culturally responsive MTEs (see Tables 3.1 and 3.2). The organization of our chapter to this point illustrates the significance of the dialogue between self-study methodology, CRP theory and MTE practice by discussing the individual, and then collective, roles each of these dialogic threads play in the development of our framework.

Table 3.1 illustrates the intersections of CRP theory and MTE practice, where the four components of MTE practices (rows) and the three dimensions of CRP (columns) create 12 points of intersection. At each of these points, a self-study reflective question was designed. The design of these questions was informed by a recent literature review (Nolan & Keazer, 2021b) of the work of key scholars in CRP (e.g., Aguirre & Zavala, 2013; Alim & Paris, 2017; Brown-Jeffy & Cooper, 2011; Chen et al., 2009; Gay, 2018; Gist, 2014; Lingard & Keddie, 2013; Nicol et al., 2020; Sleeter, 2012; White et al., 2016). That literature review, combined with our own self-study reflections on these points of MTE and CRP intersection, provided examples of tensions and challenges that exist at each of these intersections. As an example, consider the intersection of the sociopolitical/power dimension (third column) with the pedagogical knowledge and modeling component of MTE practice (second row); here, we suggest that MTEs pose the reflective question: *To what extent do my pedagogical practices disrupt dominant forms of pedagogy and model the use of mathematics as a tool to critique local and global social issues?* This question was developed from a synthesis of literature (e.g., Aguirre & Zavala, 2013; Alim & Paris, 2017; Gutiérrez, 2012; Willey & Drake, 2013). In Table 3.1, we propose that in most cases, a single reflective question exists for each point of intersection, with two exceptions: in the research/theory/scholarship MTE practice component it seemed more appropriate to construct one question to

TABLE 3.1. Level 1: MTE Self-Study Framework for Growing CRP

Components of MTE Practice	Dimensions of CRP		
	Access/Achievement	Cultural Competence/ Identity	Sociopolitical/Power
Mathematical and curricular knowledge	To what extent am I addressing the need for balance between conceptually deep mathematical knowledge with issues of access, equity, social justice, and language diversity? [M,A]	To what extent am I rethinking mathematics and the curriculum to be responsive to, and build mathematics around, students' cultural identities, while also highlighting the perspectives of marginalized groups? [M,C]	To what extent am I disrupting dominant forms of mathematics and using mathematics to understand and critique local and global social issues? [M,S]
Pedagogical knowledge and modeling	To what extent do my pedagogical practices support students' access to and achievement of mathematics? [P,A]	To what extent do my pedagogical practices draw on my students' mathematical discourse and funds of knowledge? [P,C]	To what extent do my pedagogical practices disrupt dominant forms of pedagogy and model the use of mathematics as a tool to critique local and global social issues? [P,S]
Research/theory/ scholarship	To what extent are my data collection and reflections integrating other MTE perspectives to broaden my awareness of, and research on, my practice? [R,ACS]		
Critical reflection/ praxis	To what extent are my own experiences as a learner interacting/ integrating with what it means to know and learn mathematics? [C,A]	To what extent is my own cultural competence and identity as a teacher interacting/ integrating with the goal to rethink mathematics and build the mathematics around students' cultural identities, while also highlighting the perspectives of marginalized groups? [C,C]	To what extent do I use mathematics to understand, critique, and change inequitable and unjust issues in society and in my life? [C,S]
	What struggles and tensions do I experience (from students, from colleagues, from administrators, and/or from my own experiences) as I attempt to grow my CRP in these areas? [C,ACS]		

facilitate general reflection across all three CRP dimensions and, similarly, in the critical reflection/praxis MTE practice component, a holistic reflective question was added to complement the three key questions and encourage meta-reflection on the struggles and tensions experienced in growing their own CRP practice.

Subsequently, in the second level of the framework, illustrated in Table 3.2, each of the questions from Table 3.1 are expanded upon through a series of corresponding reflective prompts. As noted previously, we gained much insight into developing these reflective prompts from our recent literature review of the work of key scholars in CRP (e.g., Nolan & Keazer, 2021b). In Nolan and Keazer (2021b), we organized the characteristics of CRP described among 25 selected research texts and considered where these characteristics fell among the intersection points of MTE and CRP illustrated in Table 3.1. By synthesizing these characteristics in light of our key reflective questions, we were able to generate a number of example reflective prompts, as shown in Table 3.2. These reflective prompts, among others that MTEs generate themselves, could be used as practical stimuli for MTEs' reflections and self-study data collection. In summary, by studying the interaction of each dimension of CRP theory with components of MTE practice, and informed by our recent review of literature, we generated 11 key reflective questions, along with a corresponding collection of reflective prompts, which constitute our self-study framework for growing our practices as culturally responsive MTEs.

THE SELF-STUDY FRAMEWORK IN ACTION

Here, we provide a glimpse into how one might put this self-study framework into action. As mentioned previously, this framework can serve dual purposes: a) a pedagogical tool to improve teaching and learning, and b) a methodological tool to support self-study data collection. To illustrate the many entry points through which to develop one's culturally responsive teaching, we begin by examining the twelve points of intersection created by the four components of MTE practices (rows) and the three dimensions of CRP (columns).

Reviewing the intersection points captured in Table 3.1, we recommend that an MTE begin by reflecting on areas identified as current strengths in their practice, in addition to identifying others to focus on for further growth. An important consideration is to start small to avoid paralysis; select 1–3 intersection points (and thus, reflection questions) to focus on at a time, in order to give these areas focal attention. A second consideration is to view this work as a necessary part of MTEs honing their ethical mindset and becoming comfortable with the realization that the work of becoming a culturally responsive MTE is never finished. Ladson-Billings (2006) discusses her goal for educators to reconceptualize CRP as less of a thing to do, "and more an ethical position they need to take in order to ensure that students are getting the education to which they are entitled" (p. 40). In the context of MTEs, this means learning to see the growth of our own CRP not as a task to complete, but as a necessary development of our ethical position. In

TABLE 3.2. Level 2: MTE Self-Study Framework for Growing CRP

Key Reflective Question	Reflective Prompts
[M,A] To what extent am I addressing the need for balance between conceptually deep mathematical knowledge with issues of access, equity, social justice, and language diversity?	• I provide opportunities for PTs to engage in high cognitive demand thinking and critical analysis by… • My practices focus on PTs learning key mathematics concepts and their complex relations when I… • That I value social justice and related 'real-world' problems is evident in the way I… • I support English language learners' home languages in learning mathematics by… • Ways that I am supporting and scaffolding academic language development to provide access to English learners is….
[M,C] To what extent am I rethinking mathematics and the curriculum to be responsive to, and build mathematics around, students' cultural identities, while also highlighting the perspectives of marginalized groups?	• My deliberate attempts to increase the mathematical participation of all PTs from diverse backgrounds is evident in… • My aim of fostering sustained and widespread involvement in mathematical activity in the classroom is evident in… • I show that I value curriculum serving as a "mirror" for PTs to see themselves in the curriculum by… • Ways that I have initiated conversations to learn from students and community members in culturally responsive ways are…
[M,S] To what extent am I disrupting dominant forms of mathematics and using mathematics to understand and critique local and global social issues?	• My efforts to present 'non-western' forms of mathematics (e.g., ethnomathematics) to PTs have been mostly un/successful because… • Efforts I have made to better understand local and global social issues of justice as they connect to mathematics are… • Evidence that I look for opportunities in course activities to help students use mathematics to think critically about issues of fairness and equity within local and global issues consists of… • Evidence that I am committed to challenging normative assumptions and sociocultural barriers in mathematics curriculum is… • Evidence that I am mindfully challenging what it means to teach and learn a curriculum for critical literacy is…
[P,A] To what extent do my pedagogical practices support students' access to and achievement of mathematics?	• I draw on PTs' mathematical knowledge in a community of learners by… • I create opportunities for PTs to have input into the pace, direction and/or outcomes of classroom activities by… • Ways that I have rethought my classroom set-up and teaching approach to make participation accessible to all students are… • Ways that I make student thinking and understanding visible and deep are … • It is evident that I value a classroom environment where all PTs feel safe/included/supported by the way I… • Instances when I have modeled ways for responding to bias in the moment are… • Evidence that I am reframing deficit thinking by focusing on strengths is…

TABLE 3.2. Continued

Key Reflective Question	Reflective Prompts
[P,C] To what extent do my pedagogical practices draw on my students' mathematical discourse and funds of knowledge?	• I draw out PTs' funds of knowledge and connect to curriculum when I… • I encourage PTs to be collaborative and responsible for each other's learning through… • Evidence that I create opportunities for students to engage in inquiry around mathematics in authentic ways (for example, the use of mathematical language to explain, reason, debate, or make generalizations) are… • I elicit and incorporate the diverse voices of my students into the co-constructed curriculum by…
[P,S] To what extent do my pedagogical practices disrupt dominant forms of pedagogy and model the use of mathematics as a tool to critique local and global social issues?	• I struggle (or, am productive) with creating openings in my pedagogy for PTs to reach their own interpretations, conclusions, and actions with regard to a sociopolitical issue because… • My strategies to make room for those forms of mathematical knowledge frequently left in the margins of schooling include… • Opportunities that I have identified to decenter whiteness and promote schooling as sustaining cultural plurality are… … • Evidence that I am mindfully reflecting on whose voices are silenced in classroom discussions is…
[R,ACS] To what extent are my data collection and reflections integrating other MTE perspectives to broaden my awareness of, and research on, my practice?	• I have been building relationships with equity-minded colleagues who can offer support and challenge my thinking and actions by… • What I have especially appreciated about other MTEs' perspectives(?) and the self-study community is… • I know that my/our research is contributing toward understanding the impact of CRP on students learning because… • Ways that I will seek out additional collaboration and feedback on my work are…
[C,A] To what extent are my own experiences as a learner interacting/ integrating with what it means to know and learn mathematics?	• As a 'successful' mathematics learner, I am aware of the importance of reflecting on what I think it means to know and learn mathematics because… • I reflect on the ways in which my practice might be further reifying the perception of a rigid/inflexible nature of mathematics by… • Evidence that I am mindfully reflecting on how my own experiences with mathematics influence my interpretation of the ways in which students demonstrate engagement, is …
[C,C] To what extent is my own cultural competence and identity as a teacher interacting/ integrating with the goal to rethink mathematics and build the mathematics around students' cultural identities, while also highlighting the perspectives of marginalized groups?	• My efforts to increase my own knowledge and understanding of the cultures of my students and of society are evident (or not evident) in the way I… • My own identity and experiences of being a teacher are interacting/integrating with key principles of what it means to be culturally responsive by… • Ways that I can use the community and its resources to extend learning are… • Evidence that I am unpacking and investigating my own biases are…

(*continues*)

TABLE 3.2. Continued

Key Reflective Question	Reflective Prompts
[C,S] To what extent do I use mathematics to understand, critique, and change inequitable and unjust issues in society and in my life?	• One way in which I have increased my own awareness of social issues and injustices of the community is... • I continue to work on my own knowledge for mathematizing sociopolitical issues by... • In my practice, I model how mathematics can be used to change inequitable issues in society by...
[C,ACS] What struggles and tensions do I experience (from my students, from my colleagues, from administrators, from my own experiences) as I attempt to grow my CRP in these areas?	• My response to pressures from PTs (and others) to adhere to official curriculum and outcomes is best described as... • If someone challenges me on my focus on CRP in/and mathematics, I explain that... • Evidence that I model the use of learning opportunities when responding to students' regressive practices (e.g., discriminatory comments, biases, etc.) in a productive way is... • Ways that I am advocating for the importance of culturally responsive caring-in-action to my students and to myself as we grow are...

this way, we can model for PTs a living and growing CRP to which they and their students are entitled; we hope they will take on the challenge and responsibility of growing within themselves.

The reflective prompts shown in Table 3.2 take the form of sentence starters to stimulate and scaffold reflection around each key question. We recommend MTEs adopt and adapt these sentence starters, as well as create their own, to record, on a regular basis, reflections and insights in relation to their practice. The repeated practice of reflecting on specific key questions and reflective prompts of the framework will facilitate self-study data collection as well as growth in one's pedagogy.

To illustrate the self-study framework in action, we offer one example from each of the author's reflective research journals as part of their own self-study data collection. In the first example, Nolan reflected on a prompt related to key question [C, C]:

Reflecting on how are "my own identity and experiences of being a teacher are interacting/integrating with key principles of what it means to be culturally responsive?" seems to get a lot of attention in my journal these days. In realizing that I have been well 'trained' as a somewhat traditional mathematics teacher, I don't think I am very comfortable with deliberate efforts to become more knowledgeable about my students' personal lives. While I can (in theory) appreciate the significant impact that my students' own school experiences and their lives outside of university will have on the teachers they become, the idea of tapping into their funds of knowledge and then connecting it to the teaching and learning of mathematics makes me quite uncomfortable at times. I acknowledge that my identity as a mathematics teacher is closely tied, in many ways, to a conscious separation of the personal and the professional in my teaching practice. In fact, I recall reflecting after class one day early in the semester that I didn't even know my students' surnames, let alone where they lived, worked or played. I keep wondering if I can become a culturally responsive pedagogue when I ground my teaching identity in such a separation of the personal and the professional.

This excerpt reflects a dilemma that Nolan faced as she attempted to integrate a key principle of cultural responsiveness—that is, connecting homes and classrooms by eliciting and integrating students' funds of knowledge (Aguirre et al., 2013). She recognizes a conflict with her mathematics teacher identity, which is inadvertently tied to a belief in the separation of home and classroom.

Similarly, author Keazer offers an example of reflection on her practice at the point where the sociopolitical/power dimension of CRP intersects with MTE mathematical and curricular knowledge [M,S]:

I see growth in my willingness to seek opportunities for using mathematics to think critically about issues of fairness or justice, and I see it gradually infiltrating more of my courses. Yet the route in which I am most comfortable traveling is assigning a project that opens up space for students to identify a local or global issue that matters to them, and I encourage their investigation of the issue using mathematics. While this has been a fruitful endeavor, adding joy and mutual learning, I can't shake the feeling that it is just an "add-on" project and fails to uproot the status quo traditional focus of mathematics that is portrayed throughout the curriculum, rather than transforming critical mathematics [to be] an integral part of the everyday practice of using mathematics. What doing mathematics looks like is not yet disrupted enough to allow space for redefining what counts as mathematics and what counts as a mathematics curriculum. I am not yet taking the time for rethinking the everyday mathematics curriculum to embrace the openness of embedding the necessary content within evolving mathematical contexts, connecting to non-western forms of mathematics, or to allow my mathematics curriculum to be moldable and responsive to current events and to the lives and knowledge resources of my ever-changing student population.

This excerpt reflects a dilemma Keazer continues to face. As she finds success taking strides to rethink the mathematics curriculum, she continues to realize the need to further the work of disrupting the traditional mathematics contexts being utilized. "Taking the time" to rethink the mathematics curriculum to align with local knowledge resources and issues of justice goes well beyond just rethinking one's priorities. It involves reflective and listening work that takes substantial time, and taking risks that do not align with the pressures for tenure and accountability through student evaluation surveys. However, the repeated practice of reflecting on an issue allows the dilemma to shift and develop through the cycle of reflection and action, and the use of these key questions and reflective prompts allows us to refocus our lenses to see our practice in new ways.

CONCLUSION

This chapter has unveiled, for the first time, an MTE self-study framework, which we believe makes an innovative and unique contribution to MTE research and practice. This framework grounds self-study research in deliberate and systematic methodological actions, based explicitly in theories of CRP and in the practices of MTEs. In this way, we have highlighted a crucial dialogue between methodology, theory and practice in relation to growing and developing as culturally responsive pedagogues. While the focus in this chapter has been centered on MTEs studying their own CRP, we propose that this crucial methodology-theory-practice dialogue is adaptable to different contexts, such as TEs of other subject areas or even academics working within university mathematics departments. In its capacity to serve as both a methodological tool for research design and a pedagogical tool for becoming stronger culturally responsive pedagogues, the framework holds promise for the wider community of educational pedagogues and researchers.

As mentioned previously, most studies of CRP conducted by teacher educators focus on improving the CRP of prospective and practicing teachers (Nolan & Keazer, 2021a). Few have focused on MTEs growing their own practice as culturally responsive pedagogues/educators. Ladson-Billings (2006) suggests that we should not give ourselves such a "pass" in assuming that the work begins with the teachers themselves, rather than with teacher education programs and ourselves as teacher educators:

> Most discussions of what teachers fail to do give teacher education a pass. We presume that teachers are doing something separate and apart from their preparation. However, I argue that teacher preparation plays a large role in maintaining the status quo. Teacher educators are overwhelmingly White, middle-aged, and monolingual English speakers... This cultural homogeneity of the teacher education profession makes it difficult to persuade convincingly preservice teachers that they should know and do anything different in their classrooms. (p. 38)

The AMTE Standards (2017) also suggests that teacher education programs and MTEs should hold themselves accountable for this work (AMTE, 2017, P.3.3, P.3.5). If MTEs strive for teacher education courses and program to do more than maintain the status quo, efforts aimed at developing CRP in prospective and practicing teachers must begin with MTEs taking up the responsibility of developing their own CRP, to model a living and growing CRP for the teachers with whom we work.

REFERENCES

Aguirre, J., Turner, E. E., Bartell, T. G., Kalinec-Craig, C., Foote, M. Q., Roth McDuffie, A., & Drake, C. (2013). Making connections in practice. *Journal of Teacher Education, 64*(2), 178–192. https://doi.org/10.1177/0022487112466900

Aguirre, J., & Zavala, M. (2013). Making culturally responsive mathematics teaching explicit: A lesson analysis tool. *Pedagogies: An International Journal, 8*(2), 163–190. http://dx.doi.org/10.1080/1554480X.2013.768518

Alim, H. S., & Paris, D. (2017). What is culturally sustaining pedagogy and why does it matter? In D. Paris & H. S. Alim (Eds.), *Culturally sustaining pedagogies: Teaching and learning for justice in a changing world* (pp. 1–21). Teachers College Press.

Appova, A., & Taylor, C. (2019). Expert mathematics teacher educators' purposes and practices for providing prospective teachers with opportunities to develop pedagogical content knowledge in content courses. *Journal of Mathematics Teacher Education, 22*, 179–204. http://dx.doi.org/10.1007/s10857-017-9385-z

Association of Mathematics Teacher Educators (AMTE). (2017). *Standards for preparing teachers of mathematics*. Author. https://amte.net/standards.

Berry, A. (2008). *Tensions in teaching about teaching: Understanding practice as a teacher educator* (1st ed.). Springer.

Beswick, K., & Goos, M. (2018). Mathematics teacher educator knowledge: What do we know and where to from here? *Journal of Mathematics Teacher Education, 21*, 417–427. https://doi.org/10.1007/s10857-018-9416-4

Bishop, A. (1990). Western mathematics: The secret weapon of cultural imperialism. *Race & Class, 32*(2), 51–65. https://doi.org/10.1177/030639689003200204

Bragg, L., & Lang, J. (2018). Collaborative self-study and peer learning in teacher educator reflection as an approach to (re)designing a mathematics education assessment task. *Mathematics Teacher Education and Development, 20*(3), 80–101. https://files.eric.ed.gov/fulltext/EJ1195984.pdf

Brown-Jeffy, S., & Cooper, J. (2011). Toward a conceptual framework of culturally relevant pedagogy: An overview of the conceptual and theoretical literature. *Teacher Education Quarterly, 38*(1), 65–84. https://files.eric.ed.gov/fulltext/EJ914924.pdf

Burton, L. (1994). Clashing epistemologies of mathematics education: Can we see the 'wood' for the 'trees'? *Curriculum Studies, 2*(2), 203–219. https://doi.org/10.1080/0965975940020204

Chen, D., Nimmo, J., & Fraser, H. (2009). Becoming a culturally responsive early childhood educator: A tool to support reflection by teachers embarking on the anti-bias journey. *Multicultural Perspectives, 11*(2), 101–106. http://dx.doi.org/10.1080/15210960903028784

Davey, R. (2013). *The professional identity of teacher educators: Career on the cusp?* Routledge.

Gallagher, T., Griffin, S., Parker, D., Kitchen, J., & Figg, C. (2011). Establishing and sustaining teacher educator professional development in a self-study community of practice: Pre-tenure teacher educators developing professionally. *Teaching and Teacher Education, 27*(5), 880–890. http://dx.doi.org/10.1016/j.tate.2011.02.003

Gallivan, H. (2017). Supporting prospective middle school teachers' learning to revise a high-level mathematics task to be culturally relevant. *Mathematics Teacher Educator, 5*(2), 94–121. https://www.jstor.org/stable/10.5951/mathteaceduc.5.2.0094

Gay, G. (2018). *Culturally responsive teaching: Theory, research, and practice* (3rd edition). Teachers College Press.

Gist, C. D. (2014). The culturally responsive teacher educator. *The Teacher Educator, 49*(4), 265–283. https://doi.org/10.1080/08878730.2014.934129

Goodwin, A. L., & Kosnik, C. (2013). Quality teacher educators = quality teachers? Conceptualizing essential domains of knowledge for those who teach teachers. *Teacher Development, 17*(3), 334–346. https://doi.org/10.1080/13664530.2013.813766

Goos, M., & Beswick, K. (2021). *The learning and development of mathematics teacher educators: International perspectives and challenges.* Springer.

Greer, B., Mukhopadhyay, S., Powell, A. B., & Nelsen-Barber, S. (Eds.). (2009). *Culturally responsive mathematics education.* Routledge.

Gutiérrez, R. (2007). (Re)defining equity: The importance of a critical perspective. In N. S. Nasir & P. Cobb (Eds.), *Improving access to mathematics: Diversity and equity in the classroom* (pp. 37–50). Teachers College Press.

Gutiérrez, R. (2012). Context matters: How should we conceptualize equity in mathematics education? In B. Herbel-Eisenmann, J. Choppin, D. Wagner, & D. Pimm (Eds.), *Equity in discourse for mathematics education: Theories, practices, and policies* (pp. 17–33). Springer.

Hamilton, M. L., & Pinnegar, S. (1998). *Reconceptualizing teaching practice self-study in teacher education.* Falmer Press.

Han, H.S., Vomvoridi-Ivanović, E., Jacobs, J., Karanxha, Z, Lypka, A., Topdemir, C., & Feldman, A. (2014). Culturally responsive pedagogy in higher education: A collaborative self-study. *Studying Teacher Education, 10*(3), 290–312. https://doi.org/10.1080/17425964.2014.958072

Hökkä, P., Eteläpelto, A., & Rasku-Puttonen, H. (2012). The professional agency of teacher educators amid academic discourses. *Journal of Education for Teaching, 38*(1), 83–102. https://doi.org/10.1080/02607476.2012.643659

Jaworski, B. (2008). Mathematics teacher educator learning and development: An introduction. In B. Jaworski & T. Wood (Eds.), *The international handbook of mathematics teacher education* (Vol. 4, pp. 1–13). Sense Publishers.

Keazer, L., & Nolan, K. (2021, Feb. 13). *A tool for reflection: Mathematics teacher educators growing their culturally responsive pedagogies.* [Paper presentation]. 25th Annual Conference of the Association of Mathematics Teacher Educators (AMTE), Virtual.

Ladson-Billings, G. (1995). Toward a theory of culturally relevant pedagogy. *American Educational Research Journal, 32*(3), 465–491. https://doi.org/10.3102%2F00028312032003465

Ladson-Billings, G. (2006). "Yes, but how do we do it?" Practicing culturally relevant pedagogy. In C. Lewis & J. Landsman (Eds.), *White teachers, diverse classrooms: A guide to building inclusive schools, promoting high expectations, and eliminating racism* (pp. 29–43). Stylus Publishing, LLC.

Ladson-Billings, G. (2014). Culturally relevant pedagogy 2.0: a.k.a. The remix. *Harvard Educational Review, 84*(1), 74–84. https://doi.org/10.17763/haer.84.1.p2rj131485484751

Ladson-Billings, G. (2017). The (r)evolution will not be standardized: Teacher education, hip hop pedagogy, and culturally relevant pedagogy 2.0. In D. Paris & H. S. Alim (Eds.), *Culturally sustaining pedagogies: Teaching and learning for justice in a changing world* (pp. 141–156). College Press.

Lingard, B., & Keddie, A. (2013). Redistribution, recognition and representation: Working against pedagogies of indifference. *Pedagogy, Culture & Society, 21*(3), 427–447. https://doi.org/10.1080/14681366.2013.809373

Loughran, J. (2007). Researching teacher education practices: Responding to the challenges, demands, and expectations of self-study. *Journal of Teacher Education, 58*(1), 12–20. http://dx.doi.org/10.1177/0022487106296217

Loughran, J. (2014). Professionally developing as a teacher educator. *Journal of Teacher Education, 65*(4), 271–283. https://doi.org/10.1177%2F0022487114533386

Loughran, J. J., Hamilton, M. L., LaBoskey, V. K., & Russell, T. (Eds.). (2004). *International handbook of self-study of teaching and teacher education practices.* Springer.

Lunenberg, M., Dengerink, J., & Korthagen, F. (2014). *The professional teacher educator: Roles, behaviour, and professional development of teacher educators.* Sense Publishers.

Nicol, C., Archibald, J., Xiiem, Q., & Glanfield, F. (2020). Introduction: Making a difference with/in Indigenous communities. In C. Nicol, A. J. Dawson, J. Archibald, & F. Glanfield (Eds.), *Living culturally responsive mathematics education with/in Indigenous communities* (pp. 1–15). Brill Sense.

Nolan, K. (2015). Beyond tokenism in the field? On the learning of a mathematics teacher educator and faculty supervisor. *Cogent Education, 2*(1), 1065580. https://doi.org/10.1080/2331186X.2015.1065580

Nolan, K. (2016). Stimulating conversations between theory and methodology in mathematics teacher education research: Inviting Bourdieu into self-study research. In M. Murphy & C. Costa (Eds.), *Theory as method in research: On Bourdieu, social theory and education* (pp. 171–190). Routledge.

Nolan, K., & Keazer, L. (2021a). Mathematics teacher educators learn from dilemmas and tensions in teaching about/through culturally relevant pedagogy. In M. Goos & K. Beswick (Eds.), *The learning and development of mathematics teacher educators: International perspectives and challenges* (pp. 301–319). Springer.

Nolan, K., & Keazer, L. (2021b). Developing as culturally responsive mathematics teacher educators: Reviewing and framing perspectives in the research. *International Journal of Humanities, Social Sciences, and Education, 8*(9), 151–163. https://doi.org/10.20431/2349-0381.0809015

Samaras, A. P. (2011). *Self-study teacher research: Improving your practice through collaborative inquiry.* Sage.

Sleeter, C. (2012). Confronting the marginalization of culturally responsive pedagogy. *Urban Education, 47*(3), 562–584. https://doi.org/10.1177%2F0042085911431472

Vanassche, E. & Kelchtermans, G. (2015). The state of the art in self-study of teacher education practices: A systematic literature review. *Journal of Curriculum Studies, 47*(4), 508–528. https://doi.org/10.1080/00220272.2014.995712

White, D., DuCloux, K., Carreras-Jusino, A., González, D., & Keels, K. (2016). Preparing preservice teachers for diverse mathematics classrooms through a cultural awareness unit. *Mathematics Teacher Educator, 4*(2), 164–187. https://doi.org/10.5951/mathteaceduc.4.2.0164

Willey, C., & Drake, C. (2013). Advocating for equitable mathematics education: Supporting novice teachers in navigating the sociopolitical context of schools. *Journal of Urban Mathematics Education, 6*(1), 58–70. https://files.eric.ed.gov/fulltext/EJ1085769.pdf

Williams, J., & Berry, A. (2016). Boundary crossing and the professional learning of teacher educators in new international contexts. *Studying Teacher Education, 12*(2), 135–151. https://doi.org/10.1080/17425964.2016.1192031

CHAPTER 4

CENTERING EQUITY IN MATHEMATICS GRADUATE STUDENT TEACHING PROFESSIONAL DEVELOPMENT

Emily Braley
Johns Hopkins University

Mary E. Pilgrim
San Diego State University

Erica R. Miller

Mathematics Graduate Teachers (MGTs) play an important role in the teaching of undergraduate mathematics, and, as new teachers, often rely on their past experiences as students to inform their teaching. Research and policy call for attending to diversity, equity, and inclusion (DEI), and go further to push for DEI principles to be integrated across all teacher preparation experiences. Current professional development for MGTs varies and does not always integrate DEI principles or often they are a separate add-on training activity. In fact, experts in the field of MGT development

The AMTE Handbook of Mathematics Teacher Education: Reflection on Past, Present and Future—Paving the Way for the Future of Mathematics Teacher Education, Volume 5
pages 67–82.

note that MGTs often make assumptions about their students. Novice MGTs fail to recognize that their students are a diverse group of people with unique experiences and relationships with mathematics that may or may not be positive. We highlight in this chapter an activity we call "the Brown Bag activity" that has been widely used in multiple STEM fields to help MGTs build empathy and better understand their students. We identify critical conversations to have with MGTs early in their teaching that attend to key DEI principles and help deepen their emotional intelligence to be better classroom teachers. Being conscience of and addressing these issues as a community will help move the field of mathematics teacher education forward.

Vignette

The facilitators pass out brown paper bags to each group and tell us to build a mobile. Isn't that one of those things that hangs over a baby's crib? I haven't done that since middle school...but sure. Our group's bag contains a piece of yarn, a brown crayon, a piece of green construction paper, and a wire coat hanger. "Are we sup-

FIGURE 4.1.   I wonder what this has to do with mathematics teaching and learning?

posed to have a theme?" one of my groupmates asks. "I think so." How are we supposed to build a mobile with only one piece of string? "Can we borrow scissors from another group?," I ask one of the activity leaders. Maybe they forgot to give us a pair. "If you think they are finished," the leader responds. Unsure, we carry on and make do without. Our mobile isn't much to look at, but we come up with a theme related to our recent discussions. Later, when all the groups present their mobiles, we see just how limited our bag was. One group got stickers and a feather boa! Other groups had more basic supplies, but we clearly were the most limited. "I wonder what this has to do with mathematics teaching and learning?" I think to myself...

INTRODUCTION

Over the past forty years, faculty within the postsecondary mathematics and mathematics education communities have been working together to better prepare mathematics graduate students for academic careers. Mathematics graduate education has traditionally focused on developing graduate students' content knowledge and research skills. However, the majority of mathematics graduates that pursue a career in academia end up in teaching, not research, intensive positions (Maki et al., 2019). In 1991, the joint Committee on Preparation for College Teaching conducted a survey of PhD-granting mathematics departments in the United States (U.S.) and found that there was a "significant increase in some teaching preparation and monitoring activities" (Case, 1994, p. 22). According to their report, teaching preparation and monitoring included mathematics graduate teacher (MGT)[1] orientation programs, English language training, and faculty observations. In 2005, the College Board of Mathematical Sciences (CBMS) conducted a similar survey and found that mathematics departments had added seminars/classes focused on teaching mathematics and faculty teaching mentors (Ellis, 2014). Surveys of PhD-granting mathematics departments across the U.S. in 2014 and 2021, supported by the Mathematical Association of America (MAA), revealed that a majority of institutions offer some formal preparation to graduate students for teaching mathematics. In fact, in 2021, 96% of responding institutions reported having some formal program for preparing graduate students for their teaching roles (Braley & Bookman, 2022).

As MGT professional development (PD)[2] has become more common, some of these programs have focused on active learning and group work (Haberler & Laursen, 2017a,b,c; Roberson, 2017). We use the CBMS' (2016) characterization of active learning as "classroom practices that engage students in activities, such as reading, writing, discussion, or problem solving, that promote higher-order thinking." We see this as a positive shift away from more traditional, lecture-based pedagogies and towards evidence-based, student-centered approaches to teaching and learning. Some research shows that active learning techniques improve success for all students, reducing failure rates by 55% on average (Freeman et al., 2014). These reformed pedagogies can help address inequities for students who have historically been underrepresented in STEM (e.g., Asera, 2001; Bullock et al., 2015; Laursen et al., 2014).

However, we want to echo the stances of scholars in the field and call for more explicit attention to be given to diversity, equity, and inclusion (Adiredja & Andrews-Larson, 2017; Gutiérrez, 2018; Han et al., 2014). The Association of Mathematics Teacher Educators (AMTE) provides standards for mathematics teacher education which are underpinned by key assumptions which serve as their foundation (2017). One assumption is that a focus on equity should be deeply integrated into all experiences across mathematics teacher preparation programming. Studies have shown that active learning is not enough to combat classroom and societal inequities (Brown, 2018). Without appropriate PD and attention given to issues of equity and inclusion, active learning can still have negative impacts on marginalized populations (Ernest et al., 2019; Johnson et al., 2020; Setren et al., 2019).

By calling for more explicit attention to equity, we are *not* advocating for less attention to active learning. Rather, when MGT PD is centered around equity, active learning can be framed as a *mechanism* for increasing access to mathematics for all students (Abell et al., 2018) and providing "greater opportunities for higher-order thinking" (National Council of Teachers of Mathematics, 2014, p. 63). With graduate students in particular, it is important to emphasize that the students they teach have very different and diverse backgrounds and experiences (Deshler et al., 2015). For example, many graduate students pursue degrees in mathematics because they had positive experiences and enjoyed mathematics growing up, which is not true for many of the students they teach.

In this chapter, we will discuss how scholars have begun to attend to diversity, equity, and inclusion within the field of mathematics education. Then, we will turn our attention to an example of a MGT PD activity that can be used to launch critical conversations that center equity. The purpose of highlighting this activity is to (a) draw a connection between equity work that is being done outside of the domain of mathematics education and (b) provide an example of how an activity can be used to launch different conversations related to equity and inclusion in mathematics classrooms. By highlighting this activity, we are *not* advocating for a single diversity, equity, and inclusion workshop or session to "tick the box." Rather, we use this activity as an *example* of one way that a PD session can be centered around equity.

EQUITY RESEARCH WITHIN MATHEMATICS EDUCATION

Power, Identity, Access and Achievement

Scholars have called on the mathematics education research community to embrace a sociopolitical perspective on equity (Gutiérrez, 2013) and take steps towards advancing equity research in higher education (Adiredja & Andrews-Larson, 2017). The literature base in K–12 mathematics education around equity has a longer history rooted in the social paradigms of the U.S. and informed by the institutions and structures that have perpetuated the privileges of certain students over others for generations. The K–12 mathematics education research community has continued to emphasize that equity should be a responsibility, not a choice (Aguirre et al., 2017).

While the term *equity* has many definitions, we chose to rely on the work of Gutiérrez, since her work is grounded specifically in mathematics education. Gutiérrez (2009) identifies four components of equity (identity, power, access, and achievement). These components interact with one another, often in tension, along two axes: the critical axis of identity and power and the dominant axis of access and achievement. The dominant axis of access and achievement gives us a framework to discuss how well-equipped students are to participate in society, starting with access as a necessary, but insufficient means to achievement. Along the critical axis, identity can be viewed as a precursor to power. When we think about identities that students carry into the classroom, it is crucial to acknowledge the way that one's beliefs about themselves can be changed and influenced within different contexts (Adiredja & Andrews-Larson, 2017). Within the classroom and within the institution, who is perceived to hold power and how students are positioned within hierarchical structures, impact their identities in those contexts (Brown, 2018). Further, students' social identities can impact their participation in mathematical activities both in and out of the classroom (Abell et al., 2018).

The interplay between identity, power, and knowledge arises from social discourse. McHoul and Grace (1993) describe the development of the idea of social discourse, written about extensively by sociologist Michel Foucault, as the ways in which knowledge is constituted within social practices, the power relations that exist within the social practices, and the interplay between them. Gutiérrez (2013) emphasizes that the social discourses are more than the communication but are the paradigms in which we operate. Take for example a classroom where the mode of instruction is lecture based. In this classroom the teacher and textbook hold the authority and the students are positioned as recipients of knowledge (Boaler, 2002a). Brown (2018) draws on the work of Boaler (2002a, 2002b) and Wenger (1998) and describes, in contrast, a classroom where the mode of instruction is discussion-based. In a discussion-based classroom, students have opportunities to interpret, explain, and challenge the status quo (Ladson-Billings, 1995). Discussion-based teaching practices create space to affirm and develop students' identities and present an opportunity to cultivate student agency (Brown, 2018). Ignoring the social discourse and attempting to take a context-free approach to thinking about equity ignores the funds of knowledge that students bring with them to their learning (Barron et al., 2021); a context-free approach continues to privilege identities from the dominant group (Adiredja & Andrews-Larson, 2017; Rubel, 2017). Civil (2016) helps us understand funds of knowledge as the connection among students' everyday lives, mathematics, and mathematics learning. The idea that the social contexts of mathematics teaching and learning are crucial for novice instructors to understand is emphasized in the AMTE's standards (e.g., C4 and P2, 2017) which call for programming to prepare teachers to integrate mathematics, practice teaching, knowledge of student learning, and the social context of mathematics teaching and learning.

In this chapter we highlight an activity originated by Schniedewind and Davidson (1983) that illustrates how a context-free approach to thinking about equity can render students' identities (e.g., race, ethnicity, gender, physical ability, etc.) invisible in the classroom. The original activity was designed "to allow students to experience the frustration and unfairness of a situation where some have an unfair advantage" and "to help students see...what happens in society when people start out with fewer resources but are still expected to achieve equally" (p. 125).

AN EXAMPLE OF HOW TO CENTER EQUITY IN MGT PD

The "Shared Experience"

In 2019, we attended an MAA College Mathematics Instructor Development Source (CoMInDS) workshop at the University of Tennessee, Knoxville, Department of Mathematics (n.d.) on *Improving the Preparation of Graduate Students to Teach Undergraduate Mathematics*. One session at the workshop (co-planned by authors Braley, Pilgrim, and others) focused on ways in which mathematics graduate teachers (MGTs) can be mentored as they work to create inclusive and respectful learning environments[3]. The organizers of this session implemented Schniedewind and Davidson's (1983) activity, which we call the Brown Bag Activity. The activity description in the most recent edition (2014) aims "to help students learn the meaning of 'blame the victim'" and "to enable students to understand that expecting everyone to reach the same goals...and judging them by whether they have reached them is unfair when opportunities are unequal" (p. 176).

At the CoMInDS workshop, groups were provided a coat hanger and materials in a brown paper bag. They were then instructed to work together to build a mobile. The brown paper bags contained varying levels of materials for the groups to work with, ranging from a bag with simply a few crayons, construction paper, and string to a bag that contained the same materials plus tape, scissors, stickers, ribbon, and more. The version of the Brown Bag Activity that we implemented was adapted from Lawrence (1998), which inspired a focus on two questions following the construction of the mobiles:

1.   Did your team ask another team for materials? Why or why not?
2.   Did your team offer other teams materials? Why or why not?

Then a broader question was posed to participants:

3.   What might the "resources bag" represent in terms of how students experience our undergraduate classrooms differently from one another?

We then shifted the focus of the conversation and asked participants to discuss the various goals for delivering a session like this to start a discussion on equity with MGTs. The organizers identified three main learning objectives for this activity when used during MGT PD:

- Help MGTs navigate their new roles as instructors and the power dynamic in their classrooms in a humane way.
- Invite MGTs into a dialogue about equity and inclusion in the classroom.
- Introduce MGTs to the importance of students' identities and sense of belonging.

In the next section we will discuss important conversations for MGTs. Trujillo and Tanner (2014) highlight the importance of moving away from deficit models, where it is assumed that students are lacking in some way, and moving towards dynamic models, where questions are raised about what is lacking in the learning environment that could keep students from learning. The AMTE standards (2017) call for teacher preparation programming to help teacher candidates create new frameworks that support each and every learner, moving away from deficit-thinking models. To examine the deficits of the learning environment Trujillo and Tanner (2014) argue that students' sense of belonging and opportunities to build identities in the classroom can be as important to achievement as "cognitive aspects of learning" (pg. 7). This reinforces the principle from culturally responsive teaching that calls on educators to acknowledge and help preserve student identities and heritage (Barron et al., 2021; Brown, 2018; Gutiérrez, 2018; Han et al., 2014; Ladson-Billings, 1995).

Ways to Frame Critical Conversations that Center Equity

Equity and inclusion should permeate the PD in which MGTs are engaged rather than an optional add-on. The ways in which MGTs think about and support equity and inclusion in their classrooms are not well-understood, and the best practices for how to prepare MGTs in this work are still being examined. The Brown Bag activity provides a way to understand how MGTs' perspectives about equity and inclusion impacts their roles as teachers.

We identify four critical conversations to have with MGTs early in their teaching practice; namely, (a) power dynamics exists in the classroom; (b) the students in your class are not you; (c) empathy is a good place to start—students may be struggling with more than just mathematics; and, (d) equality and equity are not the same. We acknowledge that these are not the *only* conversations that should be had about equity with MGTs, but rather provide these as *examples* of discussions that can be seeded by an activity like the Brown Bag Activity. In this section, we will highlight some of these critical conversation topics and describe how the activity has been used by others to make way for these discussions. The goal of this section is to not focus on the activity, but on the critical conversations that can follow.

Power Dynamics Exist in the Classroom

Lawrence (1998) argues that providing opportunities for participants to explore new ways of viewing themselves in the world is essential. Lawrence used the activity with undergraduates to help reveal to them where they are privileged and bring their assumptions about those less privileged into clearer view. This was specifically done as an exercise to take "a requisite step in unlearning racism" (p. 200). PD providers can engage MGTs about where they are privileged in the mathematics classroom paradigm. Challenging MGTs to frame their understanding of the classroom through the axis of power and identity (Gutiérrez, 2009) can help MGTs confront their positionality in the hierarchical structure of the classroom. This conversation can open the door to talk about how different active learning strategies can center students and shift the power structure in the classroom. Leading with a conversation about equity can help make it clear that attending to equity is essential when implementing active learning, and moreover, that active learning techniques do not automatically produce more equitable classrooms (Brown, 2018).

MGTs may experience tension between the day-to-day complexities of being in the dual role of student and teacher, where they are subordinate to others in positions of power while also having power themselves, namely over undergraduate students. In our experience, conversations about power in the classroom are essential and can also open conversations about where MGTs may have significant challenges teaching and interacting with undergraduate students. For example, MGTs who are in a racial or gender minority may be disproportionately challenged by undergraduate students, receive more complaints, and suffer from lower ratings on teaching evaluations (Lazos, 2012).

Here are examples of guiding questions a facilitator could use after an activity like the Brown Bag Activity to have a conversation about how *power dynamics exist in the classroom:*

1. Did you notice that there were varying levels of resources in the room? What was the communication like between groups?
2. Did you feel that there was a hierarchical structure during this activity? How did it feel?
3. What could have been done differently so that the groups did not have such disparate experiences? How could you implement these ideas in your teaching and learning?

The Students in Your Class are Not You

It is critical to help graduate students understand that *the students in their classes are not necessarily like them*; the students are different from the MGTs and different from each other. (Deshler et al., 2015). Schniedewind and Davidson (2014) elucidate the ways in which students can be different and point to gender, race, age, class background, religion, and able-bodiedness. The classroom experiences of gender minority students can be different from the experience of the dominant gender group in the class. Gutiérrez (2013) points out that researchers used to approach gender inequities in mathematics education from the point of view that studying the successes of boys could help educators understand how to help girls, but this resulted in policies that tried to make girls behave more like boys. This deficit framing makes it seem like there is a "problem" with the girls rather than a problem with the classroom or pedagogies being used to teach the students. Inclusion matters and is emphasized by researchers like Hughes (2018), who found that students who identified as sexual minorities were 7% less likely to be retained in STEM fields than their heterosexual peers. Hughes argues that sexual prejudice persists on college campuses and that STEM fields are more resistant to discussing inequity than counterparts in the arts, humanities, and social sciences. Abell et al. (2018) call on the mathematics community to be aware of how students' social identities can impact their participation when learning mathematics.

San Antonio (2009) has used the Brown Bag Activity with graduate students in education and focuses the debrief on two main questions: (a) How might people respond when they know that their life circumstances are *invisible* to others? (b) How does this lack of awareness affect policies and practices? San Antonio argues that a teacher's readiness to serve as a mentor and advocate for students depends on their ability and willingness to learn more about students' skills, aspirations, and the barriers they face. She emphasizes the importance of teachers not taking a context-free approach and, instead, striving to understand students uniquely and where they come from. This has been reinforced by the work of Barron et al. (2021) who warns against essentializing students and encourages teachers to recognize the funds of knowledge that students bring with them to the classroom. The essential discussion we want to have with new and novice MGTs is that each of their students bring a unique set of identities to the classroom and funds of knowledge that are non-mathematical in nature.

Institutionalized racism plays a role in the broader society in the U.S. but also impacts the institutions and classrooms in which we interact. In her adaptation of the Brown Bag Activity, Lawrence (1998) focuses on how students can be different based on race and ethnicity. She also discusses differences in the financial resources of students. We know that students are not able to perform as well if their basic needs are not met. Are the students in our classrooms able to access food, drink, shelter, sleep, and oxygen and are their basic safety needs being met?

There are many other identities that students value (e.g., engineer, sister, varying physical abilities, athlete, immigrant, trans-border, first-generation, family heritage, etc.). San Antonio (2009) argues that taking the time to get to know students can help teachers understand and leverage the diverse funds of knowledge students bring to the classroom. It may not be possible to represent all student identities or heritages in a mathematics classroom, but attention to opportunities that allow for students to make personally relevant connections to mathematics can demonstrate for students that a diversity of ideas and applications are valued (Tanner, 2013).

Here are examples of guiding questions a facilitator could use after an activity like the Brown Bag Activity to have a conversation about how *the students in your class are not you:*

1. How might students be different from each other and from you? How can their differences be strengths for them in your classroom?
2. What kind of structures and policies can we include that avoids taking a "context-free" approach to our students?
3. What choices can we make in the tasks that we select and ask students to engage with that let them use their diverse funds of knowledge? How can we choose questions that have personal relevance for students? How can we move towards open-ended questions that invite a diverse set of possible answers and problem-solving paths? (Tanner, 2013)

Empathy is a Good Place to Start

When teaching at the undergraduate level, empathy and caring are important ingredients for student success in the mathematics classroom (Uhing, 2020). MGTs need to develop content knowledge and pedagogical content knowledge as well as emotional intelligence in order to be effective in the classroom and address the diverse needs of students (Mortiboys, 2012). Uhing (2020) defines empathy as "the ability to express concern and take the perspective of a student" (pg. 1353) and emphasizes that this includes trying to understand the emotions that a student may experience while doing mathematics. In particular, the kinds of emotions that undergraduate students may experience doing mathematics may differ drastically when compared to the kinds of emotions MGTs experience when doing mathematics (Deshler et al., 2015).

Kjolsing et al. (2016) used an adaptation of the Brown Bag Activity to foster empathy in structural engineering teaching assistants with the goal of improving undergraduate student satisfaction. They wanted to measure whether or not engaging in an activity like the Brown Bag Activity could help graduate students teaching assistants develop empathy for students who lacked resources or needed help accessing resources. As a result of the activity, the graduate students identified that they could "try to be more understanding of the personal circumstances that students may be experiencing" (p. 3) and generally try to learn more about students as people. Kjolsing et al. describe a debrief of the activity where graduate students engage in a critical conversation about resources and emphasize that just because a resource *exists* does not mean that a student will *access* that resource. Importantly this conversation included not just what resources students were able to access, but also how students might feel and how it might impact their state of mind.

Arming MGTs with knowledge of what resources exist, ways in which students might be struggling, and how to connect students to those resources can also help build a MGT's own ability to access resources when needed. Often, when we introduce MGTs to campus resources aimed at undergraduate students, we find that similar resources exist for graduate students. This is especially true of campus health and wellness centers that may provide counseling and mental health services for the entire campus community. We want to help MGTs understand that helping their students and themselves access resources when needed is important for the entire campus community. Ideally this would be a norm for both undergraduate and graduate students, but we know from experience that it can sometimes be hard for students to ask for help. Opening a conversation about empathy and accessing resources centered around undergraduates can provide a segue into helping inform MGTs about the resources available to them.

Here are examples of guiding questions a facilitator could use after an activity like the Brown Bag Activity to have a conversation about how *empathy is a good place to start:*

1. When did you realize that other groups had different resources? How did you feel? Did you ask another group for help or try to ask the facilitator for help or support? Why or why not? (Lawrence, 1998)
2. Describe a way in which a student might have an advantage or disadvantage compared to their peers? What resources might be involved? What feelings might that student be experiencing? (Kjolsing et al., 2016)
3. What kinds of support and resources are available to students on campus? What can we do to help our students access those resources when needed? How might we access those resources ourselves when needed?

Equality and Equity are Not the Same

Schniedewind and Davidson (2014) hone in on the key idea that equality and equity are not the same. They ask the question, "Is the same always fair?" Schniedewind and Davidson argue that using a definition of equality that means people are equally free to act, does not "take into account people's privilege to do so" (pg. 176). In addition, it does not consider a person's personal context and assumes that everyone has access to the same things.

In contrast, if we think about equity along the dominant axis of access and achievement (Gutiérrez, 2013), it emphasizes that equal access is not enough to ensure achievement. Access alone does not address disparities that students may have in their ability to navigate towards achievement; rather, equity can be characterized as the absence of disparities. Tanner (2013) provides a helpful framing of equity for the classroom. Tanner describes equitable teaching as teaching all the students in your classroom, not just the students who are able to fully engage or those who may have already seen the material. This framing calls on teachers to be fair, individually focused, and adaptable, aligning with the AMTE (2017) standard calling on future teachers to commit to the role of advocate for each student. Tanner (2013) further encourages teachers to be explicit that access and equity are key goals in their classroom practice. This demonstrates to students that an instructor wants everyone in the class to be successful and that the instructor and the students are on the same side.

Here are examples of guiding questions a facilitator could use after an activity like the Brown Bag Activity to have a conversation about how *equality and equity are not the same:*

1. What assumptions have people made about you through your education experience? What assumptions have you made about your students before or after the first day of class? Have your assumptions changed? If so, how?
2. Can you think of ways in which colleges and universities might support some people more than others?
3. What happens if we expect all students to do well even when they do not have equal access to resources (Schniedewind & Davidson, 2014)? If students did have equal access to resources would that be enough? What is equity (Gutiérrez, 2009, 2013)?

CONCLUSION

The recent trend to focus more on active learning pedagogies when designing and providing mathematics graduate student professional development is encouraging. However, we join other scholars (e.g., Adiredja & Andrews-Larson, 2017; Boaler, 2002a, 2002b; Brown, 2018; Gutiérrez, 2013, 2018) and call for explicitly centering equity in mathematics education, particularly within MGT PD. Active learning alone is not enough to improve the undergraduate student experience in our mathematics classrooms. While we do *not* promote that PD providers integrate one activity on equity and assume that is enough, we chose to highlight the Brown Bag Activity developed by Schniedewind and Davidson (1983, 2014) as an example of *one* activity that can be used to start conversations about equity. This activity has been modified and adapted by several authors for different audiences and contexts. We highlight four critical conversations, informed by the literature, that could be used in conjunction with the Brown Bag Activity to address issues of power, identity, empathy, and equity versus equality.

We hope to inspire our field to take up these conversations and adapt them to their audience and context. For example, if you are in a context where individuals within your department have expressed the belief that mathematics is "culture free," then perhaps the first step is to engage MGTs in a conversation about how our students differ from us and each other. On the other hand, if the group you are working with already recognizes that we are operating within a mathematics culture that is predominantly white and impacted by institutionalized racism, then perhaps it would be more appropriate to integrate some of the discussion questions related to power dynamics in the classroom. Whatever your situation, we hope that this chapter has provided you with tools, resources, and inspiration for engaging mathematics graduate students in critical conversations related to diversity, equity, and inclusion. Being conscience of and addressing these issues as a community will help move the field of mathematics teacher education forward.

ENDNOTES

1. We decided to use the term "mathematics graduate teacher" (MGT) because we wanted to recognize that they may serve in very different roles (e.g., grader, recitation section leader, instructor of record), depending on their teaching assignment. Two other common terms used in the literature are "graduate teaching assistants" (GTAs) and "graduate student instructors" (GSIs).

2. While the term "professional development" is sometimes used within education to refer to continuing education for practicing teachers, the National Science Board (2008) also uses the term in reference to teacher preparation. Mathematics graduate students fall somewhere between these two communities, since their training and first teaching experiences often coincide. In particular, we want to note that the term "professional development" is used and accepted within the communities of those who facilitate and conduct research on teaching training activities for mathematics graduate students.

3. The vignette and image at the beginning of this chapter highlights an activity from this session.

REFERENCES

Abell, M., Braddy, L., Ensley, D., Ludwig, L., & Soto-Johnson, H. (2018). *MAA Instructional Practices Guide*. The Mathematical Association of America.

Adiredja, A. P., & Andrews-Larson, C. (2017). Taking the sociopolitical turn in postsecondary mathematics education research. *International Journal of Research in Undergraduate Mathematics Education, 3*(3), 444–465. https://doi.org/10.1007/s40753-017-0054-5

Aguirre, J., Herbel-Eisenmann, B., Celedón-Pattichis, S., Civil, M., Wilkerson, T., Stephan, M., Pape, S., & Clements, D. (2017). Equity within mathematics education research as a political act: Moving from choice to intentional collective responsibility. *Journal for Research in Mathematics Education, 48*(2), 124–147.

Asera, R. (2001). *Calculus and community: A history of the emerging scholars program. A report of the national task force on minority high achievement.* College Entrance Examination Board.

Association of Mathematics Teacher Educators. (2017). *Standards for preparing teachers of mathematics: Executive summary.* Author. https://amte.net/sites/amte.net/files/SPTM_ExecSummary.pdf

Barron, H. A., Brown, J. C., & Cotner, S. (2021). The culturally responsive science teaching practices of undergraduate biology teaching assistants. *Journal of Research in Science Teaching, 58*(9), 1320–1358. https://doi.org/10.1002/tea.21711

Boaler, J. (2002a). The development of disciplinary relationships: Knowledge, practice, and identity in mathematics classrooms. *For the Learning of Mathematics, 22*(1), 42–47.

Boaler, J. (2002b). Learning from teaching: exploring the relationship between reform curriculum and equity. *Journal for Research in Mathematics Education, 33*(4), 239–258.

Braley, E., & Bookman, J. (2022, May 14–15). *A survey of programs for preparing graduate students to teach undergraduate mathematics* [Virtual conference presentation]. AMS Spring Western Sectional Meeting.

Brown, S. (2018). E-IBL: An exploration of theoretical relationships between equity-oriented instruction and inquiry-based learning. In A. Weinberg, C. Rasmussen, J. Rabin, M. Wawro, & S. Brown (Eds.), *Proceedings of the 21st Annual Conference on Research in Undergraduate Mathematics Education* (pp. 1–15).San Diego, CA. http://sigmaa.maa.org/rume/RUME21.pdf

Bullock, D., Callahan, J., & Shadle, S. E. (2015, June). *Coherent calculus course design: Creating faculty buy-in for student success* [Conference presentation]. 2015 ASEE Annual Conference & Exposition, Seattle, Washington. https://doi.org/10.18260/p.23694

Case, B. A. (1994). *You're the professor, what next?: Ideas and resources for preparing college teachers.* Mathematical Association of America.

Civil, M. (2016). STEM learning research through a funds of knowledge lens. *Cultural Studies of Science Education, 11*(1), 41–59. https://doi.org/10.1007/s11422-014-9648-2

Conference Board of the Mathematical Sciences. (2016). *Active learning in post-secondary mathematics education.* Author. Retrieved September 23, 2021, from https://www.cbmsweb.org/2016/07/active-learning-in-post-secondary-mathematics-education/

Deshler, J. M., Hauk, S., & Speer, N. (2015). Professional development in teaching for mathematics graduate students. *Notices of the AMS, 62*(6), 638–643.

Ellis, J. (2014). Preparing future professors: Highlighting the importance of graduate student professional development programs in calculus instruction. In S. Oesterle, P. Liljedahl, C. Nicol, & D. Allan (Eds.), *Proceedings of the 38th Conference of the International Group for the Psychology of Mathematics Education and the 36th Conference of the North American Chapter of the Psychology of Mathematics Education* (Vol. 3, pp. 9–16). PME.

Ernest, J. B., Reinholz, D. L., & Shah, N. (2019). Hidden competence: Women's mathematical participation in public and private classroom spaces. *Educational Studies in Mathematics, 102*(2), 153–172. https://doi.org/10.1007/s10649-019-09910-w

Freeman, S., Eddy, S. L., McDonough, M., Smith, M. K., Okoroafor, N., Jordt, H., & Wenderoth, M. P. (2014). Active learning increases student performance in science, engineering, and mathematics. *Proceedings of the National Academy of Sciences, 111*(23), 8410–8415.

Gutiérrez, R. (2009). Framing equity: Helping students "play the game" and "change the game." *Teaching for Excellence and Equity in Mathematics, 1*(1), 4–8.

Gutiérrez, R. (2013). The sociopolitical turn in mathematics education. *Journal for Research in Mathematics Education, 44*(1), 37–68. https://doi.org/10.5951/jresematheduc.44.1.0037

Gutiérrez, R. (2018). The need to rehumanize mathematics. In M. Boston & I. Goffney (Eds.), *Annual perspectives in mathematics education: Rehumanizing mathematics for Black, Indigenous and Latinx students* (pp. 1–10). National Council of Teachers of Mathematics.

Haberler, Z., & Laursen, S. (2017a). CoMInDS program profile: University of Michigan. In *Program profiles, college mathematics instructors development source (CoMInDS).* Mathematics Association of America. http://cominds.maa.org

Haberler, Z., & Laursen, S. (2017b). CoMInDS Program profile: Duke University. In *Program profiles, college mathematics instructors development source (CoMInDS).* Mathematics Association of America. http://cominds.maa.org

Haberler, Z., & Laursen, S. (2017c). CoMInDS program profile: University of Maine. In *Program profiles, college mathematics instructors development source (CoMInDS).* Mathematics Association of America. http://cominds.maa.org

Han, H. S., Vomvoridi-Ivanović, E., Jacobs, J., Karanxha, Z., Lypka, A., Topdemir, C., & Feldman, A. (2014). Culturally responsive pedagogy in higher education: A collaborative self-study. *Studying Teacher Education, 10*(3), 290–312. http://dx.doi.org/10.1080/17425964.2014.958072

Hughes, B. E. (2018). Coming out in STEM: Factors affecting retention of sexual minority STEM students. *Science Advances, 4*(3), eaao6373. DOI: 10.1126/sciadv.aao6373

Johnson, E., Andrews-Larson, C., Keene, K., Melhuish, K., Keller, R., & Fortune, N. (2020). Inquiry and gender inequity in the undergraduate mathematics classroom. *Journal for Research in Mathematics Education, 51*(4), 504–516. https://doi.org/10.5951/jresematheduc-2020-0043

Kjolsing, E., Van Den Einde, L., & Todd, M. (2016). Using a design project to instill empathy in structural engineering teaching assistants. *Journal of Professional Issues in Engineering Education and Practice, 142*(3), 02516001-1–02516001-5. https://doi.org/10.1061/(ASCE)EI.1943-5541.0000279

Ladson-Billings, G. (1995). Toward a theory of culturally relevant pedagogy. *American Educational Research Journal, 32*(3), 465–491. https://doi.org/10.3102/00028312032003465

Laursen, S. L., Hassi, M. L., Kogan, M., & Weston, T. J. (2014). Benefits for women and men of inquiry-based learning in college mathematics: A multi-institution study. *Journal for Research in Mathematics Education, 45*(4), 406–418. https://doi.org/10.5951/jresematheduc.45.4.0406

Lawrence, S. M. (1998). Unveiling positions of privilege: A hands-on approach to understanding racism. *Teaching of Psychology, 25*(3), 198–200. https://doi.org/10.1207/s15328023top2503_8

Lazos, S. R. (2012). Are student teaching evaluations holding back women and minorities?: The perils of "doing" gender and race in the classroom. In G. Gutiérrez y Muhs, Y. F. Niemann, C. G. González, & A. P. Harris (Eds.), *Presumed incompetent: The intersections of race and class for women in academia* (pp. 164–185). University Press of Colorado, Utah State University Press. https://doi.org/10.2307/j.ctt4cgr3k.19

Maki, D., Bookman, J., Jacobson, M., & Speer, N. (2019). Professional preparation of graduate students in the mathematical sciences. In *Mathematics graduate teaching assistant preparation: Instructional activities to prepare graduate students for teaching and research findings in support of those activities* (pp. 3–4). Bloomington, IN.

McHoul, A. W., & Grace, W. (1993). *A Foucault primer: Discourse, power and the subject.* Melbourne University Press.

Mortiboys, A. (2012). *Teaching with emotional intelligence: A step by step guide for higher and further education professionals* (2nd ed.). Routledge.

National Council of Teachers of Mathematics. (2014). *Principles to actions: Ensuring mathematical success for all.* National Council of Teachers of Mathematics.

Roberson, L. (2017). CoMInDS program profile: University of Colorado-Boulder. In *Program profiles, college mathematics instructors development source* (CoMInDS). Mathematics Association of America. http://cominds.maa.org

Rubel, L. H. (2017). Equity-directed instructional practices: Beyond the dominant perspective. *Journal of Urban Mathematics Education, 10*(2), 66–105. https://doi.org/10.21423/jume-v10i2a324

San Antonio, D. M. (2009). Understanding students' strengths and struggles. In M. Scherer (Ed.), *Supporting the whole child: Reflections on best practices in learning, teaching, and leadership* (pp. 210–221). ASCD.

Schniedewind, N., & Davidson, E. (1983). *Open minds to equality: A sourcebook of learning activities to promote race, sex, class, and age equity* (1st ed.). Prentice Hall.

Schniedewind, N., & Davidson, E. (2014). *Open minds to equality: A sourcebook of learning activities to promote race, sex, class, and age equity* (4th ed.). Rethinking Schools.

Setren, E., Greenberg, K., Moore, O., & Yankovich, M. (2019). *Effects of the flipped classroom: Evidence from a randomized trial.* School Effectiveness and Inequality Initiative (SEII).

Tanner, K. D. (2013). Structure matters: Twenty-one teaching strategies to promote student engagement and cultivate classroom equity. *CBE—Life Sciences Education. 12*(3), 322–331. https://doi.org/10.1187/cbe.13-06-0115

Trujillo, G., & Tanner, K. D. (2014). Considering the role of affect in learning: Monitoring students' self-efficacy, sense of belonging, and science identity. *CBE Life Sciences Education, 13*(1), 6–15. https://doi.org/10.1187/cbe.13-12-0241

Uhing, K. (2020). *Exploring pedagogical empathy of mathematics graduate student instructors* (Order No. 27956521). [Doctoral dissertation, The University of Nebraska-Lincoln]. ProQuest.

University of Tennessee, Knoxville Department of Mathematics. (n.d.). *UTK CoMInDS 2019 Workshop.* https://www.math.utk.edu/info/cominds

Wenger, E. (1998). *Communities of practice: Learning, meaning and identity.* Cambridge: Cambridge University Press.

CHAPTER 5

RELATIONAL PRACTICES IN MATHEMATICS TEACHER EDUCATION

Theorizing a Construct to Inform Our Practice

Signe Kastberg
Purdue University

Richard Kitchen
University of Wyoming

Developing productive professional relationships with prospective and practicing teachers is a process and a goal in mathematics education research. As a process, relationships allow mathematics teacher educators (MTEs) to support and challenge teachers. As a goal, relationships encourage teachers to engage in unfamiliar activities and persist when engagement is uncomfortable. Grossman et al. (2009) identified developing relational practice as the single most significant goal in professional practice in general and teacher education specifically. Yet, how MTEs support the development of relational practice is an understudied area. To address how MTEs support the development of relational practice, we draw from research in K–12 mathematics and teacher education to identify areas for theorizing about relational

The AMTE Handbook of Mathematics Teacher Education: Reflection on Past, Present and Future—Paving the Way for the Future of Mathematics Teacher Education, Volume 5 pages 83–100.

practice. We assert that MTE's relational practice serves as an instantiation of practice in work with teachers. We then turn to two examples of relational practice in the form of narratives constructed through dialogic conversations about our practices that describe the context of our work and highlight ways of working. These examples are used to theorize about two areas of relational practice: vulnerability and respect for the dignity and brilliance of teachers. Finally, we argue that MTEs' work as relational practice serves as a foundation for teacher education programs and professional development that provide opportunities for teachers to develop relational practice.

INTRODUCTION

An important goal for Mathematics Teacher Educators (MTEs) is building productive professional relationships with prospective and practicing teachers. The building of such relationships through relational practices has been identified as a process and a goal (Grossman et al., 2009; Noddings, 2003). As a process, relationships afford MTEs opportunities to support and learn with teachers. As a goal, relationships encourage teachers to engage in unfamiliar activities and to persist when engagement is uncomfortable. The construct of relational practice draws from investigations of human interactions as growth fostering (Fletcher, 1998) and depends on the "quality of human relationships" (Grossman et al., 2009, p. 2057). The Association of Mathematics Teacher Educators' (AMTE) Standards for Preparing Teachers of Mathematics (SPTM) (2017) identify the importance of relationships in mathematics teacher education including teachers recognizing "how power relationships effect students' mathematical identities" (p. 7), and "building relationships with families" (p. 18). In this chapter, we take up relational practice as the work of MTEs who seek to support ways of knowing that honor and sustain teachers. We argue that MTEs' work as relational practice serves as a foundation for teacher education programs and professional development that provide opportunities for teachers to develop relational practices.

This chapter contains three central sections. First, we describe a theory of relational practice from K–12 mathematics education research and teacher education including relational competence (Aspelin & Jonsson, 2019). We argue that if the development of teachers' relational practice is the primary goal of teacher education, we should examine the work of MTEs that serves as instantiations of relational practice and informs teachers' relational practice. In mathematics teacher education, relational practice remains undertheorized. To address this gap, we draw from existing research to build a view of MTE's relational practices. Second, to illustrate the potential of this view, we provide examples of and analyze our work as MTEs to illustrate vulnerability and respect for the dignity of teachers as two areas for theorizing about MTEs' relational practices. Third, we argue that such analyses are a crucial first step in theorizing relational practice as a means and an end in teaching about relational practice.

ROOTS FOR THEORIZING ABOUT RELATIONAL PRACTICE

Relationships in K–12 mathematics teacher education have been identified as a critical contextual factor in mathematics learning. Boaler (2008) describes "classrooms that promote relational equity" as a context for learning mathematics in which "students would learn to respect each other's differences, to listen to others who have a different opinion, perspective or experience and to act in equitable ways" (p. 173). Bartell (2011) describes four components in creating a "caring relationship" (p. 65) with students. Building from Noddings (1984, 1992) and Hackenberg (2010a,b), Bartell describes racial, cultural, political, and academic components of relationships that "promote equity in mathematics education" (p. 65). While none of these descriptions refers to mathematics teachers' work as relational practice, all rest on the work of the teacher to interact relationally, build caring relationships, and create contexts with students that promote equity. At the center of these descriptions is a mathematics teacher engaged in relational practice described by Grossman et al. (2009) as depending on human relationships.

Relational practice is a central goal in teachers' education (Grossman et al., 2009). How MTEs should address the development of relational practice remains unclear. Kitchen (2018) identified building professional and personal relationships with students as a priority in mathematics education. Such relationships allow teachers "to press students to work harder" and to encourage students to respond to "academic rigor" (Kitchen, 2018, p. 350). Kitchen (2018) also recommended that mathematics teacher education be structured to provide opportunities for teachers to "personally identify with students" (p. 353), resulting in humanizing, identifying, and empathizing with learners. Such identity informing actions are difficult to create in mathematics teaching that may be informed by views of what it means to do mathematics (Louie, 2018) and hierarchical views of mathematics ability (Louie, 2020).

Aspelin (2021) further describes teaching as "an interactive, mutual, situated process" and a "matter of personal and interpersonal relationships" (p. 594) and highlights relational bonds that transcend social bonds between student and teacher. The development of such bonds involves teachers engaging in "professional vulnerability" (Lasky, 2005, p. 901). Lasky described vulnerability as teachers opening themselves to "the possibility of embarrassment, loss, or emotional pain because they believe that they, another individual, or a situation will benefit from this openness" (p. 901). These scholars highlight the need to support the development of relational practice in mathematics teacher education, but how to support such development is less clear.

Possibilities for promoting relational practice lie in the work Jensen, Aspelin, and colleagues (Aspelin, 2017, 2021; Aspelin & Jonssen, 2019; Jensen et al., 2015) who define "relational competence" (Jensen et al., 2015, p. 206) as "being able to meet students and parents with openness and respect, to show empathy and to be able to take responsibility for one's own part of the relationship as an educator" (p. 206). These authors assert that relational competence can be developed, rather than existing as a natural capacity of teachers. This view aligns with Grossman et al. (2009) who identified the assumption that the capacity to cultivate professional relationships is natural or fixed may act as an obstacle to teachers developing relational practice. Aspelin and Jonssen (2019) explored dimensions of preservice teachers' relational competence and suggest ways teacher educators can support development of such competence.

In the domain of teacher professional development, MTEs seek to support teacher work through relationship development. For example, Bartell (2011) described Battey and Chan's (2010) attention in a multiyear professional development as supporting community and building trusting relationships with teachers that encouraged teachers to "present dilemmas of practice in a safe environment" (p. 68). Over time Battey and Chan describe their relationships with teachers as opening space for discussion of factors that support achievement of students of color. Sztajn, et al. (2007) found that during professional development with teachers, MTEs' "caring relations" supported the "trust-building process" (p. 971). Caring relations eased MTEs' discomfort as they engaged in "the challenge of learning when to push and when to let go" and addressed the explicit and tacit needs of teachers as they care for mathematics learners. Descriptions of MTEs' work provided by Bartell (2011) and Sztajn et al. (2007) illustrate that fostering productive relationships with teachers is central in supporting their relationships with mathematics learners. These examples include descriptions of MTEs' experiences of fostering relationships and building trust, and we assert that these descriptions of MTEs' experience serve as one source of insight in theorizing about MTEs' relational practice.

MATHEMATICS TEACHER EDUCATOR WORK AS RELATIONAL PRACTICE

Research described in the last section begins to fulfill the educational promise of Noddings' (1984) conceptualization of caring relations. Yet this research raises additional questions for MTEs as they seek to understand how to support teachers to develop relational practices that result in relational equity. One category of questions regards how the work of MTEs can be considered as relational. As a first step we assume, as others have (Kastberg et al., 2020; LaBoskey, 2004), that instantiations of practice provide opportunities for teachers to envision and develop mathematics teaching practices. Narratives of MTEs' work with teachers contain evidence of relational practice drawn from "living models" (Steffe & D'Ambrosio, 1995, p. 147) of MTEs' teaching. The development of a theory of relational practice from MTEs' work is an important goal for MTEs support of relational practice in teachers.

To theorize about areas of MTE work that contribute to relational practice we share narratives from our MTE work in the form of life-writing (Eakin, 2020). These narratives are not "objective re-presentations of happenings" (Davis, 1997, p. 357), but stories whose plots, happenings, and action "express a kind of knowledge that uniquely describes human experience in which actions and happenings contribute positively and negatively to attaining goals and fulfilling purposes" (Polkinghorne,1995, p. 8). Our narratives were constructed through an iterative process involving regular, at least monthly, dialogic conversations (Guilfoyle et al., 2004) as an analytical tool in self-based methodology (Chapman et al., 2020) during the pre-pandemic year (2019). Following conversations, we created written descriptions of moments in our work that we identified as contributing to relational practice. As critical friends (LaBoskey, 2004), we probed these descriptions for details about thought, action, and context.

The first author's narrative (Signe) focuses on her work supervising elementary preservice teachers (PTs) in early field experiences. The second author's narrative (Richard) focuses on facilitating inservice teachers' mathematics discussions during professional development (PD). In each narrative we illustrate an area of MTE work that contributes to relational practice: vulnerability and respect for the dignity of teachers. We argue that theorizing about relational practice from MTEs' work is crucial to create teacher education programs and professional development that provide opportunities for teachers to develop relational practices with their students.

Signe's story draws from experience observing PTs hearing and seeing adults engage in verbal assault of students of color in schools. Her story represents ways an MTE's and PTs' vulnerability creates a dialogic space with opportunities for unpacking the complexity of teaching mathematics in school contexts. An outcome is that Signe and her students developed a relational bond (Aspelin, 2021). Richard's story explains how he works to exhibit respect for the dignity of teachers, largely women and people of color, during PD sessions he leads. In these sessions, Richard honors teachers and their mathematical thinking, sending the implicit message that teachers have important mathematical ideas worthy of being expressed, shared, examined, and extended. An outcome is that teachers begin to view themselves as competent learners of mathematics. Participating teachers experienced the joy of doing mathematics that transformed their views of themselves as learners of mathematics and as competent teachers of mathematics as well.

Signe's Story

Since 2010 I have taught mathematics methods with an embedded field experience for preservice elementary teachers. The purpose of my work is to support PTs developing teaching practices and pedagogy. I fulfill that purpose by creating pedagogical situations for PTs in which we can collaborate to consider ways they can build from and support children learning mathematics. The PTs' care for the whole child is the basis of their passion for teaching (Philipp et al., 2007) and we work together to build from wholistic care to mathematical caring (Hackenberg, 2005). I prioritize building care and trust with PTs by sharing resources, providing formative feedback, celebrating successes, and anticipating challenges PTs may face. I am a facilitator and member of a teacher community, as are the PTs. In our community we make explicit our purposes as teachers as we celebrate and address intractable dilemmas in learning about teaching. I share one dilemma in my work that illustrates my practice.

During a field experience visit in the fall of 2018 as I sat observing a lesson, I heard shouting in the hallway. After looking into the hallway, I quickly and awkwardly walked past the White teacher still shouting at a Black child. I was concerned for the child who sat huddled on the floor with the teacher standing over him shouting that he should "grow up." Though I wanted to confront the teacher, I did not. I was afraid–afraid of losing a relationship with the school and opportunities for PTs that my colleagues and I had worked to develop.

As I moved out of the hallway, two PTs and I exchanged knowing glances. We could still hear the teacher shouting in the hallway. Our glances acknowledged that something inappropriate seemed to be taking place in the hallway. Suddenly the teacher appeared in the classroom explaining why he was shouting and how the third-grader had provoked him. He left quickly. I continued talking with the PTs about the mathematics lesson they planned to teach that day.

Back at the university, I confessed to the PTs that I did not know what to do during the shouting incident and when confronted by the teacher with his version of events. I wondered aloud about how often this sort of hallway "dressing down" was happening at the school, particularly with students of color. I described my own behavior as "listening and doing nothing as PTs watched me listen and do nothing." I shared my wonderings, "How can I interrupt such actions in a school? How should I have acted in the moment?" I recognized my potential vulnerability as I planned for this discussion with PTs. Yet, I posed the questions in class in the hope of gathering the PTs' insights about ways to act that aligned with my desire to demonstrate care for children, PTs, and the teacher. PTs quickly responded with their own stories of feeling helpless or seeing helplessness during challenging moments in the schools. The PTs' shared experiences of mentor teachers who did not talk to them or seemed to have "no time for me." They shared the difficulties of negotiating content to be taught, time needed and university requirements with cooperating teachers. Stories about our experiences resulted in discussions of mathematics teaching in challenging contexts and responsibilities of teachers in schools.

And then, the very next semester, with a different class of PTs at a different school, it happened again. PTs were in the midst of teaching a mathematics lesson when I walked in to observe. One PT was assigning the third-graders to small groups for a discussion of problem solving. A White substitute teacher in charge of the class expressed dissatisfaction with a Black child's pace moving to his group. She shouted, "You better go now or else you won't get to be in a group." The boy remained still and responded to the teacher: "I'm not doing anything." The teacher shouted, "I'm calling the principal." The boy responded, "I don't care." The substitute called the principal, and the boy joined his group and began working. The substitute continued to move through the room toward a small group led by one of the PTs. A White girl in the group was drawing a mathematical representation with a pen. The substitute shouted, "We never do math with a pen because you have to erase it! Get a clipboard and a pencil!" Although the PT had asked the children to write in pen and not to erase, she found a pencil for the child and continued facilitating a discussion of a mathematics problem in her small group. I thought: "Maybe she will stop shouting if I walk up to her." Yet, I remained still and silent. I stayed a few more minutes and then left the classroom.

Later, the PTs and I met to discuss their mathematics teaching. Before I could say anything, one PT said "I want to talk to someone about the sub [substitute teacher]." The two PTs recounted statements the substitute made to them about children and to children that they wanted to share with the principal. We agreed I would contact the principal and went on to debrief their teaching. The recording of our conversation contained my hushed tones asserting "Everyone knows yelling is not good for kids. Everyone knows that. So I always wonder how can we set this [transitions between activities in mathematics class] up so that it [yelling] never happens." I worried about the PTs as silent witnesses of a teachers' verbal violence. I wondered, "How am I in this situation again!" I worried that the principal would refuse to meet with us. I was afraid, but requested the meeting anyway. I emailed the PTs' sharing how their courage empowered me to step forward.

> I just wanted you to know that your courage gives me hope and inspiration to take this difficult step. I'm also on the agenda for ElEd to talk with faculty about this issue that has happened in other schools as well. ... I'm reminded of a colleague of mine who I look up to... she said secrets about hurting others must not be kept. This should be our mantra. You are showing me how to live that powerful message. Thank you! Signe

The principal promptly responded and met with the PTs and me the next week. She shared challenges the school faced in staffing and the value of having a local community member as a substitute whom many of the children enjoyed. The principal described the complexity of the schooling landscape, but by meeting with us, she showed her care for us, the students, and the substitute.

In both incidents, when the PTs and I talked about how we responded with silence, we expressed a desire to learn how to act in difficult moments. At the university and in the school setting, the PTs were learning to teach mathematics. While mathematics content was not the focus of this story, the contexts for learning to teach mathematics, much like the context in which mathematics is learned, are identity forming and contribute to what is knowable about mathematics teaching (Bibby, 2009). Actions of MTEs in such communities have been shown to be significant factors in building trust (Sztajn et al., 2007). In the story described, the PTs and I wrestled with the complexity of power relationships involved between the actors in the situations and how they effected not only students' mathematical identities, but also relationships with families described in SPTM (AMTE, 2017). As a White woman, I learned to remain silent to belong. I wonder if it was White fragility (DiAngelo, 2018), "conflict of caring… that could not be resolved and must simply be lived" (Noddings, 2003, p. 55), or my personal experiences hearing yelling growing up that kept me silent.

Together, the PTs' stories and my own revealed the complexity of being in a school to contribute to mathematics teaching and yet not being part of the school. Each of us understood our need to be in the school setting as learners and teachers of mathematics teaching, but also to respect that the school teachers and students formed a community without us. I was ashamed of my example as a silent bystander and I wanted to build a path of action for future situations. The PTs and I admitted our present human limitations wrestling with the "conflict of caring" (Noddings, 2003, p. 55) involved in our silence. We told stories of our purposes in teaching in general and mathematics teaching specifically and how to address these purposes and build strength for overcoming our silences. Emailing and talking with the principal helped me take a step toward relationships based on shared engagement in the school community rather than maintenance of a "guest" status. Importantly, it was my students who empowered my action and the principal who illustrated that we were valued stakeholders in the school community.

Richard's Story

For every academic year since 2004-05 with the exception of one, I have led monthly full-day PD sessions for primary and secondary-level teachers of mathematics who teach in culturally and linguistically diverse, rural schools in northern New Mexico, USA. The vast majority of the participating teachers were pre-tertiary teachers who are women and people of color, primarily of Hispanic[1] and Native American descent. In each "Institute" session offered through the Los Alamos National Laboratory Math and Science Academy (MSA),[2] teachers regularly engage in problem solving activities, share their solution strategies with one another in small group settings and with the whole group, and examine other teachers' mathematical solutions. The primary goals for the Institute include: (1) teachers engaging in problem solving as a means to impact their understanding of elementary mathematics, (2) teachers gaining insight into alternative methods of representing problem solutions, and (3) teachers learning first-hand through experience about pedagogy that inherently demonstrates respect for the dignity of participants.

One of my primary goals as the Institute instructor is to leverage teachers' mathematical ideas (LTMI) during our sessions as a means to support them developing deeper mathematical knowledge and stronger mathematical agency (Kitch-

---

[1] The term "Hispanic" is commonly used in New Mexico to denote individuals and communities of Spanish descent.

[2] The Laboratory's Math and Science Academy provides quality STEM education professional development for teachers in northern New Mexico (Los Alamos National Laboratory, n.d.). In addition to the Institute, MSA provides job-embedded professional learning for teachers in their classrooms. The professional learning includes lesson observations with debriefings, student-centered coaching, and instructional modeling.

en, 2020). This involves noticing teachers' mathematical ideas and having them share these ideas with their peers (Jacobs et al., 2010; Sherin & van Es, 2009). More than just noticing their mathematical ideas, in LTMI the focus of instruction *is* the teachers' ideas since much of my instruction is essentially informed by the participating teachers' ideas. This entails working actively to both highlight and build on teachers' ideas during instruction as a means to support teachers reflecting on and potentially revising their mathematical thinking (Schoenfeld, 1985).

The vast majority of Institute participants, mostly women and people of color, experienced mathematics in impoverished ways when they attended school. Most of the teachers are from rural communities who rarely had mathematics teachers who supported them to construct deeper mathematical understandings (Gravemeijer et al., 2016). Instead, they were primarily subjected to rote instruction in which they were expected to memorize math facts, algorithms, procedures, and mathematical rules (Davis & Martin, 2008; Kitchen, 2003; Martin, 2013). Many Institute participants viewed schools and schooling with skepticism based upon their experiences attending poorly funded schools in Northern New Mexico. As students, they experienced intellectual, psychological, and emotional violence (Joseph & Smith, 2018) in schools where their cultures and languages were essentially ignored. Some were not allowed to speak their native indigenous language or Spanish in school (Mondragón & Stapleton, 2005). Many believe they are "not good" in mathematics. When this notion is unpacked further, teachers come to realize that they were not the problem, the problem was the poor quality of mathematics instruction they received. Historically, mathematics has served as a gatekeeper, privileging White and Asian middle class and upper-middle class students (Kitchen et al., 2007; Tate, 2008). From a Vygotskian perspective (1929/1993), the fact that traditional instruction impeded the teachers learning mathematics via exploration, discussion, and through exposure to a plethora of mathematical representations underscores how they were denied access to important mediation tools to learn mathematics with meaning.

My primary strategy for building relationships with Institute participants is through listening to and learning about their past experiences in mathematics and through the mathematics that we study. To exemplify the latter, I offer the following example that emerged as teachers solved the following task that approximates a fifth-grade Common Core State Standard in Mathematics (CCSSM) Operations & Algebraic Thinking Standard (National Governors Association Center for Best Practices [NGA] & Council of Chief State School Officers [CCSSO], 2010).[3]

There are initially T tons of sand in a pile. Half of a ton of sand is dumped onto the pile and then 1/4 of the sand in the new, larger pile, is removed. Write two equivalent expressions in terms of T for the number of tons of sand that are in the pile now.

---

[3] As of this writing, 41 of 50 states in the United States, the District of Columbia, four U.S. territories, and the Department of Defense Education Activity had voluntarily adopted the Common Core State Standards (Common Core State Standards Initiative, 2019).

As teachers worked on the problem individually and then with group members, I walked around observing and listening to how participants explained their solutions to one another. It quickly became clear that many of the teachers were struggling to interpret T in the problem. In response, we collectively discussed a problem solving strategy that is emphasized throughout the Ir-Rational Number Institutes; use numbers whenever possible to make a situation more concrete and less abstract. After making this recommendation, I noticed that one of the teachers, substituted 8 for T and wrote the following in her notebook:

8 tons + 1/2 ton
8 1/2 tons – 1/4(8 1/2) tons =

This teacher, a woman who had recently immigrated to the United States from Mexico, was not sure how to proceed to discover how many tons of sand were now in the pile after starting with 8 tons of sand, adding 1/2 ton to her pile, and then removing 1/4 of the 8 1/2 tons of sand. We collaborated to find a solution together, deriving the following in writing:

8 1/2 tons – 1/4(8 1/2) tons = 8 1/2 tons – 1/4 · 8 – 1/4 · 1/2 tons =
8 1/2 tons – 2 tons – 1/8 tons = (8 1/2 – 2 – 1/8) tons =
(6 1/2 – 1/8) tons = (6 4/8 – 1/8) tons = 6 3/8 tons

At this point, I asked her if she would be willing to share our work with the entire group as a means to help the others make meaning of the problem. She accepted and proceeded to show our work with the aid of a document camera on a screen above a whiteboard in the front of the room. After her presentation, I highlighted aspects of the work, making an effort to emphasize in particular how we had used the distributive property of multiplication over subtraction to help us derive a solution. Our work served as a gateway to exploring this property, a powerful property that I could now highlight and refer back to as a means to support the teachers' learning. I made it clear that because the presenting teacher had helped us make meaning of the sand problem, we were now well positioned to derive an expression in terms of T for the number of tons of sand remaining in the pile. Next, working collaboratively with the entire group, I led the class in deriving the following on the whiteboard. I wrote this derivation directly below the teacher's work which remained projected on the screen above the whiteboard:

T + 1/2 – 1/4(T + 1/2) = T + 1/2 – 1/4T – 1/8 =
T – 1/4T + 1/2 – 1/8 = 3/4T + 4/8 –1/8 =
3/4T + 3/8

After deriving this expression, we collectively discussed whether this expression would give us 6 3/8 tons if we substituted 8 tons for T. Participants checked that this worked. We also had a conversation about how we could use the commutative property of addition to commute 1/2 – 1/4T to create the equivalent expression – 1/4T + 1/2. Lastly, we talked about expressions that were equivalent to 3/4T + 3/8, namely some of the expressions that preceded this expression in the derivation above.

FUTURE: THEORIZING RELATIONAL PRACTICE

In this section, we turn back to the narratives and consider vulnerability and re-spect for the dignity of participants by intentionally noticing and showcasing teachers' mathematical brilliance (Kitchen, 2020) as two areas of MTE work that contribute to the development of relational practice. We discuss these two areas of MTEs' work using research literature in mathematics education and teacher education and provide insights about future research.

Vulnerability

Signe's story highlights the uncertainty (Simpson, 2019) and vulnerability MTEs describe as essential in the development of relationships and trust (e.g. Sztajn et al., 2007). In her uncertainty, Signe called into question her own silence and the "authority of position" (Munby & Russell, 1994, p. 92) of teachers. In so doing, Signe attempted to interrupt the idea that her own silence and authority of position to teach about mathematics teaching should be considered normal prac-tice. Instead, she drew attention to "authority of experience" (p. 92), highlighting ways experience can inform future practice. Signe sought to refocus the PTs' at-tention on the context of teaching mathematics with implications for relationships learners have with teachers and mathematics. In that refocusing, Signe sought to build her own and PTs' resistance to the notion that mathematics teaching and learning and learning to teach mathematics is apolitical and impersonal.

Signe's story contains vulnerability as a trigger for pain and relational growth. Lasky (2005) described vulnerability as "a multidimensional, multifaceted emo-tional experience" (p. 901) that is triggered in situations where one feels powerless or afraid or in situations where a person opens themselves to possible embarrass-ment, pain, or loss. The lack of control Signe felt in seeing children experience verbal violence resulted in her withdrawing emotionally in the moment. In retro-spect, Signe was not powerless in these moments of risk. As Signe opened herself up to possible embarrassment with the PTs, sharing her sense of powerlessness at the school, Signe was in control of the risk she took. Most importantly, it was the PTs' response to Signe's vulnerability with their own experience of risk, that created a dialogic space of safety in the community of teachers with whom she worked. Through this situation, Signe experienced a relational bond (Aspelin, 2017) with the PTs, seeing herself as more than socially and professionally bond-ed to PTs, but as one learning to teach about mathematics teaching relationally.

Signe's story does not focus on mathematics but instead takes up the relational context in which PTs' learn about mathematics teaching. Bibby (2009) asserts that learners' experiences of learning mathematics are "generated and held in-side relationships" (p. 128). Similarly, Battey (2013) illustrates ways teachers use of power to address behavior informs relationships. We extend these claims to hypothesize about experiences of learning to teach mathematics that are gener-ated and held inside relationship. All mathematics learning and learning about mathematics teaching occurs in context. Teachers described as "identity workers" (AMTE, 2017, **Indicator P.3.3.–Address the Social Contexts of Teaching and Learning**), need the sort of emotional support described as emerging from relationships with MTEs. Much like the relationships that teachers build with mathematics learners highlighted in the SPTM (AMTE, 2017, p. 7), MTEs' relationships with PTs informs PTs' identities as mathematics teachers. MTEs' instantiations of relational practice involve the unconscious and go beyond supporting the development of rationality to ways "relationships and subjectivities frame and constitute what is it known and knowable" (Bibby, 2009, p. 123) about mathematics teaching. Such instantiations become contexts for learning to teach mathematics and are a critical part of what teachers come to know regarding learning and teaching mathematics. Sztajn et al. (2007) note that "teachers who are learning and changing their practices are in a potentially delicate position because they are vulnerable to their peers' opinions, the professional developers' perceptions, and their administrators' expectations" (p. 973). Recasting this idea in the context of MTEs and PTs learning to teach, illustrates vulnerability they are likely to feel at the university as well as in the school context. Signe demonstrates awareness of this vulnerability in teaching about mathematics teaching. Her movement to share her experiences with the PTs and the school principal within two different communities as Sztajn et al. suggested, helped her to "feel less vulnerable" (2007, p. 973) than her first encounter.

Embracing vulnerability in work with PTs involves much more than confronting PTs with dilemmas of teaching. Creating communities in which PTs and MTEs engage in and share ways they are learning about teaching builds relationships. Of importance in the work of MTEs is becoming aware of contradictions (Whitehead, 1989) in practice (Kastberg et al., 2020; Lovin et al., 2012) that reveal expectations of PTs that are not reflected in the MTE's practice. In Signe's case, she expected PTs to be advocates for children and support their agency as mathematics learners, yet in examining her own practice, she wondered if she was an advocate for children and for PTs. Consciousness of this contradiction initiated movement toward transformation of her practice. Each MTE's lived experience, contexts, and goals are different resulting in differences in how MTEs develop instantiations of relational practice, yet a critical first step may be looking for contradictions in one's own practice.

Respect for Dignity

Richard's story reflects one of his primary goals–to model instruction that demonstrates respect for the dignity of participants by intentionally noticing and showcasing teachers' mathematical brilliance (Kitchen, 2020). Through LTMI, teachers' knowledge is leveraged to extend their mathematical thinking largely by building on their ideas, particularly the ideas of women and people of color. The use of LTMI intentionally challenges the sort of instruction that Institute participants experienced in the past; instruction in which they were constructed as mathematically deficient. Employing LTMI also contests instruction that positions those from the dominant culture (i.e., male, White, heterosexual) as the most mathematically capable. Through rendering teachers' ideas as the center piece of instruction, many teachers realize, some for the first time in their lives, that they are competent in mathematics.

Richards's story highlights one MTE's ideology in practice that challenges deficit views of teachers as mathematical knowers. Moreover, through centering teachers' mathematical ideas as a central focus of instruction, Richard argues that his relational practices with teachers have developed largely through mathematics. By continually demonstrating that he takes teachers' mathematical ideas seriously, Richard's relationships with teachers have been positively enhanced. Louie (2020) argued that for teachers to attend to their students' mathematical brilliance, PTs need to learn how to notice and discuss students' brilliance. Battey and Chan (2010) remind MTEs that cultivating teachers' respect for the dignity of learners cannot be achieved simply by telling teachers they should respect learners. Instead, demonstrating respect for teachers' dignity and brilliance models for teachers how to develop relational practices with their students through the mathematics that they teach.

MTEs can demonstrate respect for the dignity of teachers, both preservice and practicing teachers, by actively seeking to listen to and build on teachers' mathematical ideas in their work. For this to happen, MTEs must incorporate mathematical problem solving as a central part of their work with teachers. From a generative perspective, solving problems involves generating ideas rather than merely applying procedural knowledge to derive solutions (English & Gainsburg, 2016). Hence, to solve problems teachers must use intuition, prior knowledge, and experiences to resolve problem situations (Lester & Kehle, 2003). As teachers solve problems, they generate alternative methods to solve and represent their problem solutions (Kitchen, 2020). MTEs can take advantage of these varied solutions and representations to help teachers gain not only mathematical insights from their peers, but also insights about pedagogy. Specifically, teachers can learn firsthand as problem-solvers about the power of pedagogy that inherently values their ideas, representations, and mathematical ways of thinking (Kitchen et al., 2021). Ultimately, after experiencing pedagogy that respects the dignity of learners, teachers want the same for their students. Specifically, teachers want to learn how to incorporate instruction in their classrooms that is respectful toward and embraces the diverse and brilliant ideas of their students (Kitchen et al., 2021).

CONCLUSION

In this chapter we have focused on the construct of relational practice as drawing from investigations of human interactions as growth fostering (Fletcher, 1998). Described as depending on "quality of human relationships" (Grossman et al., 2009, p. 2057), relational practice in the work of MTEs serves as an instantiation that provides opportunities for teachers' relational practices. We have addressed the undertheorized construct of relational practice by identifying two areas of

MTEs' work that we assert belong to relational practice: vulnerability and respect for dignity. These two areas of work have been identified as supporting or informing relationships. Existing mathematics education literature focused on K–12 mathematics learners and their teachers (e.g. Battey, 2013) as well as MTEs and teachers (e.g. Sztajn et al., 2007) suggest that the identified areas and others such as trust and care (Sztajn et al., 2007) warrant additional scholarly inquiry (Lee & Mewborn, 2009). Studies of the work of MTEs are likely to reveal growth fostering actions and behaviors that allow for theorizing about relational practice.

We have illustrated how vulnerability in MTE interactions with teachers can serve as an opportunity for teachers and MTEs to learn about relational practice. While we hypothesize that making one's self vulnerable with teachers is part of a relational practice, descriptions of teachers' responses to such situations are needed. We anticipate that MTE and teacher descriptions of moments involving vulnerability will contain evidence of relational interactions (Battey, 2013) and relational competence (Aspelin & Jonsson, 2019).

We have also illustrated respect for the dignity and mathematical brilliance of teachers within one MTE's work. While we hypothesize that respect for dignity and mathematical brilliance is part of relational practice, we anticipate that, as in mathematics teachers' work, MTE ideology and culture play a role (Louie, 2018). MTEs' respect for dignity and the mathematical brilliance of teachers informs the construction of counternarratives to those focused on hierarchies of ability (Louie, 2020), yet like vulnerability, descriptions of teachers' responses to such situations are needed. We anticipate that MTE and teacher descriptions of moments involving respect for dignity will contain evidence of caring relationships (Bartell, 2011) and relational interactions (Battey, 2013).

Moving forward, the work of MTEs examined through self-based methodologies (Chapman et al., 2020) has potential for contributing insights about relational practice. We caution that such studies, to be useful in contributing to theorizing about relational practice, need to include the voices of teachers. The narratives we constructed contain these voices to illustrate ways vulnerability and respect for dignity were responded to by teachers, yet more can be done to gather evidence of the ways MTE work is experienced by teachers. As with explorations of teacher practice, MTE relational practice can be understood as complex, involving cognition, affect, identity, ideology, and culture and happening in the moments of teaching. Existing work in mathematics teacher education such as Sztajn et al. (2007) illustrates the complexity of theorizing about relational practice where trust, care, and vulnerability are involved in relationships. It also demonstrates the importance of engaging in teachers' communities and understanding their views on school culture as a means to engage in difficult discussions about deficit views they may hold about learners (Battey & Chan, 2010).

In terms of understanding how MTEs' practices support the development of "identity workers" by providing emotional support for such development called for by AMTE (2017, **Indicator P.3.3.–Address the Social Contexts of Teaching**

and Learning), there is a need to understand how MTEs' work can be viewed as an instantiation of relational practice. What we can say is that there is evidence that relational practice can be taught (Aspelin & Jonsson, 2019). Yet our understanding of how to do so is certainly in its infancy. Most promising are studies that unpack the intricacies of teachers' work from the perspective of MTEs as we have done, but also include the perspectives of teachers. Building productive relationships with teachers that are both a means and an end should be a goal for teacher education programs. Fostering relational practices in mathematics teacher education is a vital response to racial and class-based injustices (Flores, 2008; Kitchen et al., 2007; Leonard & Martin, 2013) in the United States. While we are not so naïve as to believe that relational practices will eliminate these injustices, we do believe that integrating relational practices into our work as MTEs will provide teachers with vital emotional support as they counteract the legacy of mistrust, racial division, and hatred in efforts to heal the many social, cultural and political divides that currently exist in the United States.

REFERENCES

Aspelin, J. (2017). In the heart of teaching: A two-dimensional conception of teachers' relational competence. *Educational Practice and Theory, 39*(2), 39–56.

Aspelin, J. (2021). Teaching as a way of bonding: A contribution to the relational theory of teaching. *Education Philosophy and Theory, 53*(6), 588–596.

Aspelin, J., & Jonsson, A. (2019). Relational competence in teacher education. Concept analysis and report from a pilot study. *Teacher Development, 23*(2), 264–283,

Association of Mathematics Teacher Educators. (2017). *Standards for preparing teachers of mathematics*. https://amte.net/standards

Bartell, T. G. (2011). Caring, race, culture, and power: A research synthesis toward supporting mathematics teachers in caring with awareness. *Journal of Urban Mathematics Education, 4*(1), 50–74.

Battey, D. (2013). "Good" mathematics teaching for students of color and those in poverty: The importance of relational interactions within instruction. *Educational Studies in Mathematics, 82*(1), 125–144.

Battey, D., & Chan A. (2010). Building community and relationships that support critical conversations on race: The case of cognitively guided instruction. In M. Q. Foote (Ed.), *Mathematics teaching and learning in K–12* (pp. 137–149). Palgrave Macmillan. https://doi.org/10.1057/9780230109889_10

Bibby, T. (2009). How do pedagogic practices impact on learner identities in mathematics? In L. Black, H. Mendick, & Y. Solomon (Eds.), *Mathematical relationships in education: Identities and participation* (pp. 123–135). Routledge.

Boaler, J. (2008). Promoting 'relational equity' and high mathematics achievement through an innovative mixed-ability approach. *British Educational Research Journal, 34*(2), 167–194.

Chapman, O., Kastberg, S., Suazo-Flores, E., Cox, D., & Ward, J. (2020). Mathematics teacher educators' learning through self-based methodologies. In K. Beswick & O. Chapman (Eds.), *The mathematics teacher educator as a developing professional* (pp. 157–187). Brill.

Common Core State Standards Initiative. (2019). *Standards in your state*. www.corestandards.org/standards-in-your-state/

Davis, B. (1997). Listening for differences: An evolving conception of mathematics teaching. *Journal for Research in Mathematics Education, 28*(3), 355–376.

Davis, J., & Martin, D. B. (2008). Racism, assessment, and instructional practices: Implications for mathematics teachers of African American students. *Journal of Urban Mathematics Education, 1*(1), 10–34.

DiAngelo, R. (2018). *White fragility: Why it's so hard for white people to talk about racism*. Beacon Press.

Eakin, P. J. (2020). *Writing life writing: Narrative, history, autobiography*. Routledge.

English, L., & Gainsburg, J. (2016). Problem solving in a 21st-century mathematics curriculum. In L. D. English & D. Kirshner (Eds.), *Handbook of international research in mathematics education* (pp. 313–335). Routledge.

Fletcher, J. (1998). Relational practice a feminist reconstruction of work. *Journal of Management Inquiry, 7*(2), 163–186.

Gravemeijer, K., Bruin-Muurling, G., Kraemer, J. M., & van Stiphout, I. (2016). Shortcomings of mathematics education reform in the Netherlands: A paradigm case? *Mathematical Thinking and Learning, 18*(1), 25–44.

Grossman, P., Compton, C., Igra, D., Ronfeldt, M., Shahan, E., & Williamson, P. (2009). Teaching practice: A cross-professional perspective. *Teachers College Record, 111*(9), 2055–2100.

Guilfoyle, K., Hamilton, M., Pinnegar, S., & Placier, P. (2007). The epistemological dimensions and dynamics of professional dialogue in self-study. In J. J. Loughran, M. L. Hamilton, V. K. LaBoskey, & T. Russel (Eds.), *International handbook of self-study of teaching and teacher education practices* (pp. 1109–1167). Springer.

Hackenberg, A. J. (2005). A model of mathematical learning and caring relations. *For the Learning of Mathematics, 25*(1), 45–51.

Hackenberg, A. J. (2010a). Mathematical caring relations: A challenging case. *Mathematics Education Research Journal, 22*(3), 57–83.

Hackenberg, A. J. (2010b). Mathematical caring relations in action. *Journal for Research in Mathematics Education, 41*(3), 236–273.

Jacobs, V. R., Lamb, L. L. C., & Philipp, R. A. (2010). Professional noticing of children's mathematical thinking. *Journal for Research in Mathematics Education, 41*(2), 169–202. https://doi.org/10.5951/jresematheduc.41.2.0169

Jensen, E., Skibsted, E. B., & Christensen, M. V. (2015). Educating teachers focusing on the development of reflective and relational competences. *Educational Research for Policy and Practice, 14*(3), 201–212.

Joseph, N. M., & Smith, A. (2018). *Black woman first, mathematician second: Toward an understanding of black women's constructions of mathematics identity* [Paper presentation]. American Educational Research Association 2018 Annual Meeting, New York.

Kastberg, S. E., Lischka, A. E., & Hillman, S. L. (2020). Characterizing mathematics teacher educators' written feedback to prospective teachers. *Journal of Mathematics Teacher Education, 23*(2), 131–152. https://doi.org/10.1007/s10857-018-9414-6

Kitchen, R. S. (2003). Getting real about mathematics education reform in high poverty communities. *For the Learning of Mathematics, 23*(3), 16–22.

Kitchen, R. S. (2018). A commentary with urgency: Looking across theoretical perspectives to put relationship building with underserved students at the forefront of our work. In S. E. Kastberg, A. M. Tyminski, A. E. Lischka, & W. B. Sanchez (Eds.), *Building support for scholarly practices in mathematics methods* (pp. 343–358). Association of Mathematics Teacher Educators Professional Book Series, Volume 3. Information Age Publishing.

Kitchen, R. (2020). The power of their ideas: Leveraging teachers' mathematical ideas in professional development. *International Journal of Mathematical Education in Science and Technology, 53*(7), 1835–1858. https://doi.org/10.1080/002073 9X.2020.1847337

Kitchen, R. S., DePree, J., Celedón-Pattichis, S., & Brinkerhoff, J. (2007). *Mathematics education at highly effective schools that serve the poor: Strategies for change.* Lawrence Erlbaum Associates.

Kitchen, R., Martinez-Archuleta, M., Gonzales, L., & Bicer, A. (2021). Actualizing change after experiencing significant mathematics PD: Hearing from teachers of color about their practice and mathematical identities. *Education Sciences, 11*(11), 710. https://doi.org/10.3390/educsci11110710

LaBoskey, V. K. (2004). The methodology of self-study and its theoretical underpinnings. In J. J. Loughran, M. L. Hamilton, V. K. LaBoskey, & T. Russell (Eds.), *International handbook of self-study of teaching and teacher education practices* (pp. 817–869). Springer.

Lasky, S. (2005). A sociocultural approach to understanding teacher identity, agency and professional vulnerability in a context of secondary school reform. *Teaching and Teacher Education, 21*(8), 899–916.

Leonard, J., & Martin, D. B. (2013). *The brilliance of black children in mathematics: Beyond the numbers and toward new discourse.* Information Age Publishing.

Lee, H. S., & Mewborn, D. S. (2009). Mathematics teacher educators engaging in scholarly practices and inquiry. In D. S. Mewborn & H. S. Lee (Eds.), *Scholarly practices and inquiry in the preparation of mathematics teachers* (pp. 1–6). Association of Mathematics Teacher Educators (AMTE) Monograph Series, Volume 6. AMTE.

Lester, F. K., & Kehle, P. (2003). From problem solving to modeling: The evolution of thinking about research on complex mathematical activity. In R. Lesh & H. M. Doerr (Eds.), *Beyond constructivism: Models and modeling perspectives on mathematics problem solving, learning, and teaching* (pp. 501–517). Erlbaum.

Los Alamos National Laboratory. (n.d.). *Math & science academy for teachers.* https://www.lanl.gov/community/education/teacher-resources/math-science-academy.php

Louie, N. L. (2018). Culture and ideology in mathematics teacher noticing. *Educational Studies in Mathematics, 97*(1), 55–69.

Louie, N. (2020). Agency discourse and the reproduction of hierarchy in mathematics instruction. *Cognition and Instruction, 38*(1), 1–26.

Lovin, L. H., Sanchez, W. B., Leatham, K. R., Chauvot, J. B., Kastberg, S. E., & Norton, A. H. (2012). Examining beliefs and practices of self and others: Pivotal points for change and growth for mathematics teacher educators. *Studying Teacher Education, 8*(1), 51–68.

Martin, D. B. (2013). Race, racial projects, and mathematics education. *Journal for Research in Mathematics Education, 44*(1), 316–333. https://doi.org/10.5951/jresematheduc.44.1.0316

Mondragón, J. B., & Stapleton, E. S. (2005). *Public education in New Mexico.* University of New Mexico Press.

Munby, H., & Russell, T. (1994). The authority of experience in learning to teach: Messages from a physics methods class. *Journal of Teacher Education, 45*(2), 86–95.

National Governors Association Center for Best Practices & Council of Chief State School Officers. (2010). *Common core state standards for mathematics.* http://www.corestandards.org/assets/CCSSI_Math%20Standards.pdf.

Noddings, N. (1984). *Caring: A feminine approach to ethics and moral education.* University of California Press.

Noddings, N. (1992). *The challenge to care in schools: An alternative approach to education.* Teachers College Press.

Noddings, N. (2003). Is teaching a practice? *Journal of Philosophy of Education, 37*(2), 241–251.

Philipp, R. A., Ambrose, R., Lamb, L. L., Sowder, J. T., Schappelle, B. P., Sowder, L., Thanheiser, E., & Chauvot, J. (2007). Effects of early field experiences on the mathematical content knowledge and beliefs of prospective elementary school teachers: An experimental study. *Journal for Research in Mathematics Education, 38*(5), 438–476.

Polkinghorne, D. (1995). Narrative configuration in qualitative analysis. *International Journal of Qualitative Studies in Education, 8*(1), 5–23. DOI:10.1080/0951839950080103.

Schoenfeld, A. (1985). Metacognitive and epistemological issues in mathematical understanding. In E. A. Silver (Ed.), *Teaching and learning mathematical problem solving: Multiple research perspectives* (pp. 361–380). Erlbaum.

Sherin, M. G., & van Es, E. A. (2009). Effects of video club participation on teachers' professional vision. *Journal of Teacher Education, 60*(1), 20–37. https://doi.org/10.1177/0022487108328155

Simpson, A. (2019). Being "challenged" and masking my own uncertainty: My parallel journey with elementary prospective teachers. *Studying Teacher Education, 15*(2), 217–234. DOI: 10.1080/17425964.2019.1587608

Steffe, L., & D'Ambrosio, B. (1995). Toward a working model of constructivist teaching: A reaction to Simon. *Journal for Research in Mathematics Education, 26*(2), 146–159.

Sztajn, P., Hackenberg, A. J., White, D. Y., & Allexsaht-Snider, M. (2007). Mathematics professional development for elementary teachers: Building trust within a school-based mathematics education community. *Teaching and Teacher Education, 23*(6), 970–984.

Vygotsky, L. S. (1929/1993). Introduction: Fundamental problems of defectology. In R. W. Rieber, & A. S. Carton (Eds.), *The collected works of L. S. Vygotsky, volume 2: The fundamentals of defectology.* Plenum Press. (Original work published 1929).

Whitehead, J. (1989). Creating a living educational theory from questions of the kind, "How do I improve my practice?" *Cambridge Journal of Education, 19*(1), 41–52.

CHAPTER 6

DOCTORAL PREPARATION IN MATHEMATICS TEACHER EDUCATION

Starlie Chinen
University of Washington

Saba Din
McGill University

There has been an increased focus on preparing and supporting teachers to teach mathematics and what research might move the field of mathematics teacher education forward. While this is relevant to the work of both in-service and preservice teachers, an often-forgotten mechanism in the cycle of mathematics teacher education involves the preparation of doctoral students as future mathematics teacher educators. This chapter provides a summary of the current literature on doctoral student learning in mathematics education by answering the following questions: (1) What knowledges and skills should doctoral students in mathematics education learn through their doctoral studies? (2) What structures and initiatives have supported this learning? We identified six categories of knowledge and skills necessary for doctorates in mathematics teacher education. Structures that support learning across doctoral programs are similar, however the specific aspects within them vary.

The AMTE Handbook of Mathematics Teacher Education: Reflection on Past, Present and Future—Paving the Way for the Future of Mathematics Teacher Education, Volume 5
pages 101–120.

Based on our findings, we offer direction for further investigation regarding doctoral student learning in mathematics education to improve the field of mathematics teacher education more broadly.

INTRODUCTION

The 2017 AMTE Standards for Preparing Teacher of Mathematics identify the knowledge, skills, and dispositions necessary for being a mathematics educator. The 2017 AMTE document was written for a teacher preparation audience and offers examples for how these various knowledge, skills, and dispositions are cultivated in our mathematics teaching force. An oft-forgotten mechanism in the cycle of mathematics teacher education, however, is the preparation of doctoral students as future mathematics teacher educators. In a survey of doctoral graduates in mathematics education from 1997–2014, Shih et al., (2020) found approximately 70% of doctoral graduates in mathematics education in the United States (U.S.) hold teaching positions in higher education. The high percentage of doctoral graduates working in higher education suggests this population has a significant influence on the experiences and learning opportunities of future K–12 students through the teachers they help to prepare. While the majority of doctoral graduates in mathematics education go on to support new teacher learning, the learning opportunities and requirements of doctoral programs across the U.S. vary greatly in terms mathematics content coursework, research opportunities, courses in education more broadly, and access to engage in teacher education work. These differences are driven in part, by the expertise and experiences of faculty members, quantity of faculty in mathematics education, and budget limitations. This variability is problematic because the variation in programs and lack of consensus about what it means to be a high-quality doctoral program in mathematics education makes it challenging to identify aspects for programs to improve and grow (Reys, 2018).

When we use the term "mathematics education" throughout this chapter, we are referring to the broader field of mathematics education that also includes mathematics teacher education. Doctoral students' learning in mathematics education, including mathematics teacher education, is additionally important because in the last two decades there has been a reported need for individuals with doctorates in mathematics education in higher education (Reys, 2000; Reys et al., 2019; Reys & Reys, 2016; Shih et al., 2018). The literature of doctoral learning in mathematics education in the U.S. is limited, and even less is readily available about doctoral learning in mathematics education in non-U.S. contexts. A common reason for this is a widespread assumption that teaching experience is sufficient preparation for being a teacher educator (e.g., Hollins et al., 2014). Yet, scholars who do research focused on teacher educator learning argue that while teaching experience is necessary for becoming a teacher educator, it is insufficient because supporting preservice teacher learning requires knowledge and practices that go beyond the scope of what is typically part of K–12 teaching (Abell et al., 2009; Dunn, 2016;

Goodwin et al., 2014; Hollins et al., 2014; Jacobs et al., 2015; Zeichner, 2005). A small, but growing body of literature identifies the particular knowledges, skills, and dispositions necessary for future mathematics teacher educators (Jaworski, 2008; Reys, 2018, 2017; Reys & Dossey, 2008; Shih et al., 2016; Zaslavsky & Leikin, 2004).

To address this issue there have been three national conferences (Reys & Dossey, 2008; Reys & Kilpatrick, 2001) focused on the state of doctoral education in mathematics education in the U.S. which includes topics such as: identifying common challenges across programs, increasing the production of high-quality doctorates in mathematics education, and outlining the various knowledge and skills doctoral students should acquire during their PhD experience (Reys & Dossey, 2008; Reys & Kilpatrick, 2001). Offerings in this chapter do not include an analysis of the topics and conversations held at the 2021 conference as materials from this meeting have yet to be disseminated. Furthermore, there have been several initiatives that focused on systemic improvement in mathematics education. The Centers for Learning & Teaching and MetroMath, for example, have focused on a multi-tiered approach to improving mathematics education through work with students, teachers, community and graduate programs. Despite these efforts, there continues to be a shortage of doctorates in mathematics education (Kilpatrick, 2008; Shih et al., 2020).

A few scholars have repeatedly called for increased attention to the preparation of doctoral students in mathematics education (Shih et al., 2020; Reys et al., 2019a; Reys, 2018; Reys & Reys, 2016). This chapter echoes this call and provides a summary of the literature on doctoral student learning in mathematics education over the past 20 years, relying heavily on perspectives of U.S.-based mathematics education researchers. Our investigation of the literature on doctoral student learning in mathematics education focused on the following two questions: *What knowledges and skills should doctoral students in mathematics education learn through their doctoral studies? What structures and initiatives have supported this learning?*

This chapter also offers direction for what the field might seek to investigate, including (1) a clarification of the various knowledges, skills, and dispositions deemed important for doctoral students in mathematics education to learn, (2) studies on the relationship between theory and the various knowledges, skills, and dispositions with a particular investigation into the underpinning values of these particular knowledges, skills, and dispositions, and (3) given more recent calls to diversify the field of mathematics teachers, attention to the experiences and preparation of doctoral students of color in mathematics education.

METHODS FOR LITERATURE REVIEW

We identified the literature included in this review through searches of ERIC, Education Source, and the library databases at our institutions. Given the limitedness of this field, we chose to select any article or report that addressed issues of doctoral student learning in mathematics education that were written in English. We searched for a combination of the following terms: "doctorate," "mathematics education PhD," "preparation," "doctoral learning," and "mathematics teacher education." As we read articles and reports we also relied heavily on the citation of articles and included relevant studies into our collection of articles. The earliest literature we were able to find was from 2001. While we sought to include the experiences and perspectives of doctoral student preparation in mathematics education in non-U.S. contexts, we found few reports or studies in non-U.S. settings. Therefore, we did not include citations from non-U.S. contexts.

Our search yielded two compilations of non-peer reviewed papers on doctoral student learning in mathematics education that were developed in preparation for or response to the two national conferences on doctoral student learning in the U.S. (Reys & Dossey, 2008; Reys & Kilpatrick, 2001). In addition, our search led us to one set of guiding principles set out by the Association of Mathematics Teacher Education (AMTE) (2002) that outlines necessary aspects of doctoral programs in mathematics education, and 16 other journal articles or reports that address aspects of doctoral student learning in mathematics education, the history of doctoral student learning in mathematics education, or report on the number of doctoral students in mathematics education. Of these reports and articles (see Appendix A), only six studies used empirical data (Chazan & Lewis, 2008; Reys & Reys, 2016; Shih et al., 2016, 2018; Shih et al., 2019; Shih et al., 2020) while all other reports drew on the authors' perspectives. Chazan and Lewis (2008) surveyed participants at the 2008 national conference on doctoral student learning in mathematics education to learn about expectations related to the mathematics content knowledge necessary for doctoral preparation in mathematics teacher education. Reys & Reys (2016) look at data across surveys of earned doctorates from 1962–2014. Shih et al. (2016, 2018, 2019, 2020) describe a profile of mathematics education doctoral graduates' backgrounds and preparation in the U.S. by analyzing survey results from 500 mathematics education doctorates from 1997 to 2014 who attended 23 different institutions that were a part of the Centers for Learning and Teaching (described in greater detail below).

It is important to note that the claims and stances presented in the articles and reports in this literature are derived mainly from the authors' experiences, survey data from faculty across the U.S. in mathematics education, and in one study, the experiences of doctoral students in "novel" and "conventional" doctoral coursework (Chazan & Lewis, 2008). Notably, these reports and survey-based studies have focused on mapping the range of experiences doctoral students bring into PhD programs and the various learning opportunities they have (e.g., types of coursework, opportunities for research, etc.). In what follows, we highlight themes from our analysis of the literature on doctoral student learning in mathematics education.

REVIEW OF LITERATURE—KNOWLEDGE AND SKILLS

While much of the scholarship around doctoral student learning in mathematics education has focused on mapping the field, that is, describing the history and current context of doctoral programs in mathematics education, some of the conversation revolves around the particular knowledge and experiences doctoral students need to acquire. Based on the literature, we identified six categories of knowledge and skills necessary for doctorates in mathematics education that are important for supporting the development of mathematics teacher educators who are able to support mathematics teacher preparation in both content, practice, and research. They include: *mathematics content knowledge, research skills, teaching and learning mathematics, experience with mathematics curricula, equity,* and *policy*. Each of these categories is described in more detail below.

Mathematics Content Knowledge

Given that doctorates in mathematics education go on to support teachers of mathematics, an uncontended set of knowledge they must acquire or have already acquired is mathematics content knowledge. While all scholars agree that mathematics content is a cornerstone of the knowledge necessary to do the work of mathematics teacher education, there is less consensus on how much mathematics is necessary. The "plus six" rule dictates doctoral students should have expertise in school mathematics content six years beyond the highest level of mathematics education they teach (Dossey & Lappan, 2001). That is, a high school mathematics teacher educator should have mastery of mathematics content through the second year of a masters degree program in mathematics. At the elementary level, this allows mathematics teacher educators to "engage in conversations with teachers over the full K–12 range" (Dossey & Lappan, 2001, p. 68). At the secondary level, it provides mathematics teacher educators the foundation they need to "discuss articulation issues across the boundaries of elementary, secondary, and post-secondary education" (Dossey & Lappan, 2001, p. 68). Thus, some scholars argue graduate level mathematics courses should be a requirement in doctoral programs in mathematics education while other programs set these as requirements for admission into the PhD. While only a few scholars take on the "plus six" language to quantify the necessary mathematics content knowledge doctoral students must acquire, no scholars argue against the importance of doctorates in mathematics education having a strong mathematics content background.

As stated earlier, all scholars seem to agree that mathematics content knowledge is an important part of developing mathematics educators. What is less established is agreement on whether the "plus six" rule is necessary. In a study of 500 PhD graduates in mathematics education from 1997 to 2014 from institutions that were Centers for Learning and Teaching (NSF, 2004), approximately 23% of doctoral students reported taking no mathematics courses as part of their doctoral studies. Of those involved in the study, 31% reported entering the doctoral program after completing masters degrees in mathematics and another 31% of respondents reported having a masters degree in mathematics education (which may or may not have required mathematics coursework), potentially signifying a large population of doctoral students in mathematics education do not need additional mathematics coursework. In addition, 40% of survey participants indicated they enrolled in 1–4 mathematics courses as part of their doctoral program. Despite the relatively low number of mathematics coursework most doctoral students take during their doctoral programs, 60–70% of doctoral students reported feeling prepared to teach a wide variety of mathematics education course, 90% reported feeling their preparation for engaging in qualitative research was adequate or very well addressed, and 80% rated their preparation for engaging in quantitative research as adequate or very well addressed (Shih et al., 2019, 2016). The discrepancy in the percentage of doctoral students who do not take many mathematics courses and still report feeling well prepared for their work in university settings leads to increased questions about the validity of the "plus six" rule.

Others have suggested that programs should instead focus on developing doctoral students' stances as "life-long learner[s] of mathematics" (Chazan & Lewis, 2008, p. 81) through their coursework. Little, however, is described about the quality of graduate level mathematics courses that promote a sense of a life-long learning of mathematics. AMTE, the leading collective of mathematics teacher educators, takes a similar stance as Chazan and Lewis (2008), describing the necessary mathematics knowledge for doctoral students in mathematics education to acquire as "broad and deep mathematical knowledge both to identify the big ideas in the pre-K–14 mathematics curriculum and to examine how those ideas develop throughout the curriculum" (AMTE, 2002, p. 4).

Research Skills

The most cited set of knowledges and skills described in this literature is the importance of cultivating scholars who can produce meaningful and rigorous research. Research, although not always directly tied to the work of preparing future mathematics teachers, is often a major part of the job responsibilities of faculty positions. In addition, engaging in teacher education related research can be useful in supporting mathematics teacher learning as a teacher educator may draw on their research to improve their practice. As such, one's ability to continue working as a mathematics teacher educator in a university setting depends, in part, on a scholar's ability to produce research. Scholars seem to unanimously agree that doctoral students are best prepared when they are exposed to learning opportunities related to multiple traditions of research (Rico et al., 2008; Thornton et al., 2001). This category of knowledge and skills is cited by current scholars as one of the areas that many doctoral students have the least amount of previous experience with given that many doctoral students have rich histories as K–12 educators or in mathematics and little experience engaging in research (Ferrini-Mundy, 2008). This, some scholars argue, should be the focus of doctoral student learning because being a contributing member of the academic community is a major part of the work of being an academic and often requires doctoral students to go through an "epistemological shift" from the work doctoral students previously engaged in (Liljedahl, 2018; Monaghan, 2019; Nardi, 2015). Shih et al. (2016) reported that 95% of graduates enrolled in less than four courses in educational statistics. This suggests that while there is unanimous agreement that attaining research skills and knowledges from different traditions is important, there is a lack of agreement across programs about what qualifies as sufficient.

In addition to the quantity of courses in quantitative methods Shih et al. (2016) asked doctoral graduates in mathematics teacher educator positions in university settings how well prepared they felt about their qualitative and quantitative research preparation. 90% of doctoral graduates reported their preparation in qualitative methods as "adequate" or "very well addressed" and 80% of all faculty surveyed rated their preparation in quantitative methods as "adequate" or "very well addressed" in their programs. These statistics seem to further raise questions about what necessary and adequate levels of preparation needed to engage in mathematics education research.

There is variation across programs on the amount of research related experience or coursework doctoral students in mathematics education should receive. Golde (2008) offers an apprentice model for supporting doctoral students to learn from a more expert mentor who is able to guide and cultivate their knowledge and skill acquisition. The idea of the importance of apprenticing into the work of mathematics education research is highly supported across the literature (Lambdin & Wilson, 2001; Lester & Carpenter, 2001; Reys et al., 2001; Wolff, 2001).

Teaching and Learning Mathematics

Given that 40% of the work of doctorates in mathematics education is teaching (most commonly mathematics content or methods courses), a large focus of the core knowledge and skills doctoral students need to learn is about teaching and learning mathematics at various levels (K–12, undergraduate, post-baccalaureate). Wilson and Franke (2008) suggest, "Teaching mathematics to teachers involves an additional layer of complexity [than teaching mathematics to K–12 students]. Preparing teachers requires an enhanced set of knowledge and skills—the learner is a teacher of mathematics as well as a student of mathematics" (p. 103). This involves an understanding of the "fundamental theories of learning mathematics" (AMTE, 2002, p. 8), also known as specialized knowledge of mathematics needed for teaching (Wilson & Franke, 2008). Other scholars name the importance of fluency in using technology to promote mathematics learning (Heid & Lee, 2008), knowledge of learning theories (Reys et al., 2008), and the ability to communicate effectively with teachers, administrations, and teacher candidates (Fennell et al., 2001). For many institutions this requires doctoral students have previous K–12 teaching experiences (Fennell et al., 2001; Lambdin & Wilson, 2001; Ragan, 2001; Reys, 2017; Reys et al., 2019b; Thornton et al., 2001) and experiences teaching at the university level during the PhD programs (AMTE, 2002; Blume, 2001; Fennell et al., 2001; Lambdin & Wilson, 2001; Ragan, 2001). In addition to the experience of teaching at the K–12 and university level, many scholars agree that doctoral students must learn to be "critically reflective" about their own practice and develop dispositions towards improving their teaching craft (AMTE, 2002).

Experience with Mathematics Curricula

One aspect of supporting mathematics teacher learning is unpacking and assessing mathematics curricula and making adaptations for a teacher's instruction for a particular group of students. Thus, it is crucial doctoral students obtain experience with mathematics curricula as part of their PhD programs. Curriculum, as described in the literature, attends to the sequence of mathematical ideas students might learn in a particular grade or course and the activities that support the learning of those mathematical ideas. AMTE (2002) suggests doctoral students should be able to "design effective curricula and learning environments to facilitate the development of deep and connected mathematical understanding" (p. 7). This requires knowledge "of current theories and research about human learning, how to connect different areas of mathematics, and how students come to appreciate mathematics as a discipline" (p. 7). In addition, Long (2001) and Silver and Walker (2008) describe the importance of understanding the past and current political context around mathematics curricula at the local (school/district), state, and national levels. For example, learning about how national standards set by National Council of Teachers of Mathematics or National Assessment of Educational Progress influence the textbook adoption lists developed in some states (Silver & Walker, 2008).

Equity

Despite Shih et al.'s (2019) findings that 80% of doctoral students feel satisfied with their preparation, the most under addressed topics identified by survey participants were topics related to diversity and equity. While coded language is often used to describe teachings within doctoral education, it is not always clear what this language refers to. For example, Presmeg and Wagner (2001) name "political awareness" (p. 75) as one of several bulleted items that are important for doctoral students to learn, although what is meant by this term is not elaborated. In another example, the title "historical, social, political, and economic contexts of education" (AMTE, 2002, p. 5) is used as a heading that largely focuses on doctoral students learning about the history of various aspects of mathematics education such as teaching, learning, assessment, and technology. In its description, there is a single sentence that attends specifically to issues of equity that focuses on the potential importance of doctoral students learning about the "historical evolutions of the equity movement" as it equips mathematics educators to "understand and respond to current efforts to address inequities" (p. 5). This treatment of the "equity movement" as something finite and singular adds to the uncertainty of what is meant by "equity."

Taylor and Kitchens (2008) are the only scholars who center equity as a foundational point of learning for doctoral students in mathematics education. In particular, they name topics such as: how diverse learners learn, strategies for supporting struggling learners; the role of parents in mathematics education; and strategies for preparing teacher candidates to work with diverse learners. In addition to these necessary knowledges, Taylor and Kitchens (2008) urge institutions to infuse issues of equity across all aspects of the doctoral program. These authors suggest all doctoral students be required to have experiences working in a diversity of settings that are unfamiliar to them and that they cultivate an "appreciation of diversity/equity issues" and theories. For example, immersing themselves in schools with racially diverse student populations. They also suggest doctoral students need to learn more about how to integrate mathematics curricula with issues of social justice. This might include learning about effective curricula for diverse learners to better address the needs of students with learning differences. These topics and experiences, according to Taylor and Kitchens (2008), are important because they address the need for doctoral programs to focus on issues of equity that can promote more equitable mathematics teaching.

Policy

Policy is a less described area of expertise to which doctoral students in mathematics teacher education need exposure. Policy issues might include, as Fey (2001) suggests, understanding the context in which research is situated because it will allow scholars to produce more meaningful and impactful research. While many of the issues related to policy in mathematics education can be learned about through coursework, scholars who write about the importance of policy in doctoral education suggest active or observational engagement in real policy work (e.g., attending school board meetings) is likely better to support doctoral student learning (Long, 2001; Middleton & Dougherty, 2008).

SUPPORTS FOR DOCTORAL STUDENT LEARNING

The categories described above represent the major content and skills doctoral students in mathematics education should learn. Among this literature, there is also a push to consider the types and quality of experiences doctoral students should have that promote these knowledges and skills. Golde (2008), for example, offers the idea of *intellectual communities* that are broadly inclusive of different status and expertise, purposeful and knowledge-centered, flexible, respectful and generous, and deliberately tended to meet the learning needs of doctoral students. Teuscher et al., (2008), doctoral students at the time their paper was published, comment on the positive impact of *intellectual communities* that involved shared authority and a high level of interaction on their learning.

Despite differences in how faculty support doctoral student learning, one quality of doctoral programs that seems to be consistent across all contexts is a dissertation requirement. In all settings, the multiple-chapter or three-paper dissertation has largely been used as a final milestone in the PhD program. Stiff (2001), however, raised the question of whether other forms of the doctoral dissertation might be a better indication of doctoral preparation to engage in the work of mathematics education. Some possibilities include curriculum development projects or historical reviews. While Stiff (2001) proposed expanding the form of the PhD dissertation, he also acknowledged the limitations of what changes might be currently possible in academic settings.

Over the past two decades conversations about doctoral preparation in mathematics education have been ongoing. While the categories listed above are commonly agreed upon knowledges and skills doctoral students in mathematics education should learn, the nuances of what content and the level of expertise doctoral students should obtain in each of these categories remains particular to each institution. Reys (2018) advocates that the field identify qualities of high-quality U.S. doctoral programs in mathematics education and create an accreditation process for institutions with doctorate programs in mathematics education. These qualities, Reys (2018) suggests, will support programs to identify areas to improve. This sentiment is shared more broadly: Lappan et al. (2008) reported that 47% of participants at the 2008 national conference on doctoral programs in mathematics education agreed national standards and an accreditation process for doctoral programs in mathematics education should exist.

INITIATIVES FOCUSED ON SUPPORTING DOCTORAL STUDENTS IN MATHEMATICS EDUCATION

Over the course of the last two decades, a few initiatives focused on doctoral preparation in mathematics education have surfaced. In what follows, we describe, based on limited reports and articles, the purpose and function of this work. All of these initiatives were government funded and their work has since ended, although members of these initiatives continue to apply their learning in academic settings.

Centers for Learning and Teaching

The Centers for Learning and Teaching (CLT) were developed to support the "preparation of science, technology, engineering, and mathematics educators, as well as the establishment of meaningful partnerships among education stakeholders, especially PhD-granting institutions, school systems, and informal education performers" (NSF, 2004). To be considered a "center," programs (that may be made up of numerous institutions) needed to address three components: the diversification of STEM leaders and professionals, enhancement of content knowledge and pedagogical skills of the STEM teaching force, and developing research in STEM education that could support national improvement. In particular "each CLT institution was engaged in improving its doctoral preparation (in quality and quantity)" (Reys & Reys, 2017, p. 386). Institutions across the CLTs made improvements to their programs in relation to the number of doctoral students recruited, the number of mathematics education courses offered, more specific pathways within a mathematics PhD (e.g. urban education, rural education, social justice/equity/diversity, curriculum, or teaching), opportunities for engaging in supported or collaborative research projects, opportunities to work as TAs for more experienced faculty, and opportunities to participate in grant writing (Reys & Reys, 2017).

MetroMath

MetroMath was an initiative developed by a team of three universities and four school districts to address the issue of poor mathematics learning outcomes for students by focusing on "strong teacher professional development, rigorous research, and a graduate program" (Penn GSE News, 2010). By integrating the graduate program within this initiative, it allowed for strengthened collaboration between teachers and doctoral students to explore research questions about urban education, teaching, and student learning (EurekAlert!, 2003).

MOVING THE FIELD FORWARD—FUTURE RESEARCH

In this section, we provide ideas for future research based on our analysis of the literature by identifying opportunities for improvement in our understanding of doctoral student learning to become mathematics teacher educators. First, given that the previously described major categories of doctoral student learning in mathematics education are not equally represented or explored across the literature, a future line of research might focus on further clarifying the core knowledges, skills, and dispositions necessary for doctoral student learning in mathematics education (Reys et al., 2019b), perhaps in relation to the 2017 AMTE Standards for Preparing Teachers of Mathematics. This research focus is informed by the lack of clarity about exactly what doctoral students in mathematics education need to learn. For example, if we look at the mathematics curriculum category, what particular theories of teaching and learning mathematics should doctoral students be fluent in and be able to apply in the creation and assessment of various mathematics curricula? Are there particular stances about how students develop an appreciation of mathematics that one should take? Or should doctoral students be aware of a range of stances about developing an appreciation of mathematics and be able to articulate and defend their particular stance? Reys et al. (2019a) offer a series of future lines of research to investigate this issue that include looking at program requirements, particular course syllabi, statistics on admissions, acceptance, graduation rates, and attrition, looking for alignment and dissonance between programs and the AMTE Standards (2017) to *Guide the Design and Implementation of Doctoral Programs in Mathematics Education*, among other topics. This and studies like this might clarify the core knowledges, skills, and dispositions that are critical for supporting the development of future doctorates in mathematics educators. This would require further investigation into what is considered "effective" in relation to these knowledges, skills, and dispositions.

Additionally, there is a gap in understanding how the core knowledges, skills, and dispositions support teacher educators in being effective in their work with teacher candidates in mathematics. Therefore, we call the field to conduct research that attends to the impact of the core knowledges, skills, and dispositions on the effective preparation of mathematics teachers. For example, scholars identify mathematical content knowledge as a cornerstone of the core knowledges doctoral students in mathematics teacher education must obtain. Some scholars attempt to quantify the amount of mathematical knowledge necessary with a "plus six" requirement (Dossey & Lappan, 2001; Reys, 2017). However, the plus six requirement seems to be arbitrarily defined, as there is no research that indicates how this standard of knowledge benefits and contributes to the effectiveness of the teacher educator. Thus, a future line of research might investigate the mathematical content that expert mathematics teacher educators at the elementary, middle, and high school levels draw on. This research may help us better understand (1) what mathematical content knowledge is relevant for their work and (2) how particular mathematical content knowledge impacts the work of effective mathematics teacher educators. In addition, this research may shed light on the importance of deeply understanding the mathematical content knowledge that precedes the grade levels a mathematics teacher educator focuses on, and in particular, how students come to learn these concepts (e.g., how fractions are taught in elementary is crucial for high school mathematics).

Future research might also take a more conceptual approach, considering the ways these various core knowledges interact in theory and practice. That is, how the assumptions underpinning each of these knowledges is either aligned with or in contradiction to one another. For example, *mathematical content knowledge* and *historical, social, political, and economic contexts in education* impact what we see as important mathematical content knowledge for students to learn. If we come to understand the hegemonic history of what came to count as "school mathematics," we might expand what is currently described as mathematical content knowledge necessary for doctoral students (and teachers) to learn (e.g., ethnomathematics) or support future mathematics teacher educators to identify the assumptions in the decisions they make related to what is included in the canon of mathematics.

Looking across this literature, there is also a clear need for more studies that focus on the experience of the learner. The majority of available literature draws on the perspectives of the papers' authors. It is problematic to make unilateral decisions about what is meaningful for doctoral student learning to become mathematics teacher educators on the basis of such few perspectives and experiences. This is not to suggest that the experiences and expertise of practicing doctorates in mathematics education is not valuable, but rather that assumptions about what is generalizable to the population of PhD experiences in mathematics education from an individual's perspective is inappropriate. Thus, we urge future research to focus on how these critical core knowledges, skills, and dispositions are facilitated, enacted, and taught to doctoral students throughout their programs. In particular, research should focus on how this learning occurs for doctoral students across programs with vastly different requirements. Research on the various ways doctoral students learn these core knowledges, skills, and dispositions will provide insight into possible learning trajectories for doctoral students across programs and support programs to consider the mechanisms through which doctoral students learn about the core knowledges, skills, and dispositions. This type of work has the potential to further inform doctoral programs in devising structures and supports for the development of effective future mathematics educators. This approach might also shed light on an unknown demographic of doctoral students in mathematics teacher education, namely those who do not complete the programs. Understanding their experiences, challenges, and what pushes them to leave doctoral programs may provide the field with information on how to better support doctoral students more generally.

Finally, an area we deem critical for future research focuses on the experiences of doctoral students who have been historically marginalized in educational settings through institutional racism, classism, sexism, homophobia, etc. The 2007 status update on doctorates in mathematics education in the U.S. reports 20% of all doctorates in mathematics education are earned by people of color (Reys et al., 2008). Given calls to diversify the teaching field as an important mechanism for improving educational and systemic justice (Ahmad & Boser, 2014; AMTE, 2017) it is not hard to see a parallel call to diversify the field of doctorates in mathematics education who go on to support teacher learning. Thus, attending to the experiences of this population of doctoral students may shed light on how a majority White field might support these doctoral students. By conducting such studies, we might highlight how institutions and individuals within institutions either promote or harm the diversification of the field of mathematics education. With a deeper understanding, we can begin to shift the educational system in ways that negate and remove institutional biases that negatively impact many students in our education systems. One might investigate the long-term impact of the Centers for Learning and Teaching and the work in which graduates of these initiatives have engaged.

SUMMARY

The central aim of this chapter was to provide an overview of the available literature on doctoral student learning in mathematics education in the U.S. and provide avenues for future research that may support the field as a way to consider how to improve the preparation of mathematics teachers.

Many programs are similar in their emphasis of the mathematical content knowledge, research skills, understandings of theories of teaching and learning of mathematics, and experience with the mathematics curricula. However, the details of what constitutes each of these content areas and how they might be learned vary greatly amongst institutions within the U.S. These differences in coursework and experiences available to doctoral students vary based on the expertise and practice of the limited number of faculty at each institution. In addition, content focused on policy and equity is lacking across doctoral programs, and although educational, cultural, and political contexts differ from nation to nation, such issues are critical for all doctoral students to develop expertise, not only for improving the mathematics learning experiences and opportunities for students and future mathematics teachers, but to shape the field in progressive ways.

Given this literature base, we argued for three major areas of research. First, clarifying the core knowledges, skills, and dispositions necessary for doctoral student learning in mathematics teacher education. This includes an interrogation of the underlying assumptions and values that these knowledges, skills, and dispositions represent and the impact of these knowledges, skills, and dispositions on practicing mathematics teacher educator work. More specifically, this might look at how these knowledge, skills, and dispositions support doctoral students in supporting mathematics teacher learning. Second, we advocate for work that centers around the learners' perspectives. This includes taking into account the experiences of doctoral students in mathematics teacher education who represent nondominant identity groups. Third, we suggest studies focused on the experiences of underrepresented doctoral students to promote diversity in the field.

APPENDIX A: TABLE OF REPORTS AND ARTICLES USED IN LITERATURE REVIEW

Citation	Empirical
AMTE. (2002). *Principles to guide the design and implementation of doctoral programs in mathematics education.* Author.	
Jaworski, B. (2008). Development of mathematics teacher educators and its relation to teaching development. In B. Jaworski & T. Wood (Eds.), *The international handbook of mathematics teacher education: The mathematics teacher educator as a developing professional* (Vol. 4, 335–361). Rotterdam, The Netherlands: Sense.	
Liljedahl, P. (2018). Mathematics education graduate students' thoughts about becoming researchers. *Canadian Journal of Science, Mathematics, and Technology Education, 18*(1), 42–57.	
Monaghan, J. (2019). The practices of mathematics education doctoral students. *Educação Matemática Pesquisa, 21*(4), 53–62.	

Citation	Empirical
Nardi, E. (2015). "Not like a big gap, something we could handle": Facilitating shifts in paradigms in the supervision of mathematics graduates upon entry into mathematics education. *International Journal of Research in Undergraduate Mathematics Education, 1*, 135–156.	
National Science Foundation. (2004, February 20). *Centers for learning and teaching (CLT)*. https://www.nsf.gov/pubs/2004/nsf04501/nsf04501.htm	
Reys, R. E. (2000). Doctorates in mathematics education: an acute shortage. *Notices of the American Mathematical Society, 47*(10), 1267–1270.	
Reys, R. (2018). The preparation of a mathematics education: The case of Carey. *Canadian Journal of Science, Mathematics and Technology Education, 18*(1), 58–67.	
Reys, R. E. (2017). Doctorates in mathematics education: How they have evolved, what constitutes a high-quality program, and what might lie ahead. In In J. Cai (Ed.), *Compendium for Research in Mathematics Education* (pp. 934–948). NCTM.	
Reys, R. (2016). Some thoughts on doctoral preparation in mathematics education. *Journal of Mathematics Education at Teachers College, 7*(2), 22–35.	
Reys, R. & Dossey, J. (2008). *US doctorates in mathematics education: Developing stewards of the discipline.* Providence, RI: American Mathematical Society. (contains 27 reports/commentaries)	X (one paper in this report)
Reys, R. & Kilpatrick, J. (2001). *One field, many paths: US doctoral programs in mathematics education.* Providence, RI: American Mathematical Society.	
(Contains 23 reports/commentaries)	
Reys, B. & Reys, R. (2017). Strengthening doctoral programs in mathematics education: A continuous process. *Notices of the American Mathematical Society, 64*(4), 386–389.	
Reys, R. & Reys, B. (2016). A recent history (5 decades) of the production of doctorates in mathematics education, *Notices of the American Mathematical Society, 63*(8), 936–939.	X
Reys, R., Reys, B., & Shih, J. (2019). Doctoral preparation in mathematics education—Time for research and a widespread professional conversation. *Journal of Mathematics Education at Teachers College, 10*(2), 21–27.	
Reys, R., Reys, B., Shih, J., & Safi, F. (2019). Doctoral programs in mathematics education: A status report of size, origin of program leadership, and recommended institutions. *Notices of the American Mathematical Society, 66*(2), 212–217.	

Citation	Empirical
Shih, J. Reys, R., & Engledowl, C. (2018). Issues of validity in reporting the number of doctorates in mathematics education. *Investigations in Mathematics Learning, 10*(1), 1–8.	X
Shih, J., Reys, R., & Engledowl, C. (2016). Profile of research preparation of doctorates in mathematics education in the United States. *Far East Journal of Mathematical Education, 16*(2), 135–148.	X
Shih, J. Reys, R., Reys, B. & Engledowl, C. (2019). A profile of mathematics education doctoral graduates' background and preparation in the United States. *Investigations in Mathematics Learning, 11*(1), 16–28.	X
Shih, J., Reys, R., Reys, B. & Engledowl, C. (2020). Examining the career paths of doctorates in mathematics education working in institutions of higher education. *Investigations in Mathematics Learning, 12*(1), 1–9.	X

REFERENCES

Abell, S., Rogers, M., Hanuscin, D., Lee, M., & Gagnon, M. (2009). Preparing the next generation of science teacher educators: A model of developing PCK for teaching science teachers. *Journal of Science Teacher Education, 20*(1), 77–93.

Ahmad, F., & Boser, U. (2014). *America's leaky pipeline for teachers of color: Getting more teachers of color into the classroom.* Center for American Progress. https://cdn.americanprogress.org/wp-content/uploads/2014/05/TeachersOfColor-report.pdf

AMTE. (2002). *Principles to guide the design and implementation of doctoral programs in mathematics education.* Author.

AMTE. (2017). *Standards for preparing teachers of mathematics.* Author.

Blume, G. (2001). Beyond course experiences: the role of non-course experiences in mathematics education doctoral programs. In R. Reys & J. Kilpatrick (Eds.), *One field, many paths: US doctoral programs in mathematics education* (pp. 87–94). American Mathematical Society.

Chazan, D., & Lewis, J. (2008). The mathematical education of doctorates in mathematics education. In R. Reys & J. A. Dossey (Eds.), *US doctorates in mathematics education: Developing stewards of the discipline* (pp. 75–86). American Mathematical Society.

Dossey, J., & Lappan, G. (2001). The mathematical education of mathematics educators in doctoral programs in math education. In R. Reys & J. Kilpatrick (Eds.), *One field, many paths: US doctoral programs in mathematics education* (pp. 67–72). American Mathematical Society.

Dunn, A. (2016). "It's dangerous to be a scholar-activist these days": Becoming a teacher educator amidst the hydra of teacher education. *Teacher Education Quarterly, 43*(4), 3–29.

EurekAlert! (2003, October 9). *Rutgers leads on $10m NSF grant for urban math instruction.* https://www.eurekalert.org/news-releases/638051

Fennell, F., Briars, D., Crites, T., Gay, S., & Tunis, H. (2001). Reflections on the match between jobs and doctoral programs in mathematics education. In R. Reys & J. Kilpatrick (Eds.), *One field, many paths: US doctoral programs in mathematics education.* American Mathematical Society.

Ferrini-Mundy, J. (2008). What core knowledge do doctoral students in mathematics education need to know? In R. Reys & J. A. Dossey (Eds.), *US doctorates in mathematics education: Developing stewards of the discipline* (pp. 63–74). American Mathematical Society.

Fey, J. (2001). Doctoral programs in mathematics education: Features, options, and challenges. In R. Reys & J. Kilpatrick (Eds.), *One field, many paths: US doctoral programs in mathematics education* (pp. 55–62). American Mathematical Society.

Golde, C. (2008). Creating a broader vision of doctoral education: Lessons from the Carnegie initiative on the doctorate. In R. Reys & J. A. Dossey (Eds.), *US doctorates in mathematics education: Developing stewards of the discipline* (pp. 53–62). American Mathematical Society.

Goodwin, A. L., Smith, L., Souto-Manning, M., Cheruvu, R., Tan, M. Y., Reed, R., & Taveras, L. (2014). What should teacher educators know and be able to do? Perspectives from practicing teacher educators. *Journal of Teacher Education, 65*(4), 284–302.

Heid, M. K., & Lee, H. S. (2008). Using technology in teaching and learning mathematics: What should doctoral students in mathematics education know? In R. Reys & J. A. Dossey (Eds.), *US doctorates in mathematics education: Developing stewards of the discipline.* (pp. 117–128). American Mathematical Society.

Hollins, E., Luna, C., & Lopez, S. (2014). Learning to teach teachers. *Teacher Education, 25*(1), 99–124.

Jacobs, J., Yendol-Hoppey, D., & Dana, N. (2015). Preparing the next generation of teacher educators: The role of practitioner inquiry. *Action in Teacher Education, 37*(4), 373–396.

Jaworski, B. (2008). Development of mathematics teacher educators and its relation to teaching development. In B. Jaworski & T. Wood (Eds.), *The international handbook of mathematics teacher education: The mathematics teacher educator as a developing professional* (Vol. 4, pp. 335–361). Sense.

Kilpatrick, J. (2008). Doctoral programs in mathematics education: An international perspective. In R. Reys & J. A. Dossey (Eds.), *US doctorates in mathematics education: Developing stewards of the discipline* (pp. 177–180). American Mathematical Society.

Lambdin, D., & Wilson, D. (2001). The teaching preparation of mathematics educators in doctoral programs in mathematics education. In R. Reys & J. Kilpatrick (Eds.), *One field, many paths: US doctoral programs in mathematics education* (pp. 77–84). American Mathematical Society.

Lappan, G., Newton, J., & Teuscher, D. (2008). Accreditation of doctoral programs: A lack of consensus. In R. Reys & J. A. Dossey (Eds.), *US doctorates in mathematics education: Developing stewards of the discipline* (pp. 215–222). American Mathematical Society.

Lester, F., & Carpenter, T. (2001). The research preparation of doctoral students in mathematics education. In R. Reys & J. Kilpatrick (Eds.), *One field, many paths: US doctoral programs in mathematics education* (pp. 63–66). American Mathematical Society.

Liljedahl, P. (2018). Mathematics education graduate students' thoughts about becoming researchers. *Canadian Journal of Science, Mathematics, and Technology Education, 18*(1), 42–57.

Long, V. (2001). Policy-A missing but important element in preparing doctoral students. In R. Reys & J. Kilpatrick (Eds.), *One field, many paths: US doctoral programs in mathematics education* (pp. 141–144). American Mathematical Society.

Middleton, J., & Dougherty, B. (2008). Doctoral preparation of researchers. In R. Reys & J. A. Dossey (Eds.), *US doctorates in mathematics education: Developing stewards of the discipline* (pp. 139–146). American Mathematical Society.

Monaghan, J. (2019). The practices of mathematics education doctoral students. *Educação Matemática Pesquisa, 21*(4), 53–62.

Nardi, E. (2015). "Not like a big gap, something we could handle": Facilitating shifts in paradigms in the supervision of mathematics graduates upon entry into mathematics education. *International Journal of Research in Undergraduate Mathematics Education, 1*, 135–156.

National Science Foundation. (2004, February 20). *Centers for learning and teaching (CLT)*. https://www.nsf.gov/pubs/2004/nsf04501/nsf04501.htm

Penn GSE News. (2010, February 12). *Creating young urban mathematicians*. https://www.gse.upenn.edu/content/creating-young-urban-mathematicians

Presmeg, N., & Wagner, S. (2001). Preparation in mathematics education: Is there a basic core for everyone? In R. Reys & J. Kilpatrick (Eds.), *One field, many paths: US doctoral programs in matematics education*. American Mathematical Society.

Ragan, G. (2001). My doctoral program in mathematics education. In R. Reys & J. Kilpatrick (Eds.), *One field, many paths: US doctoral programs in mathematics education* (pp. 145–152). American Mathematical Society.

Reys, R. E. (2000). Doctorates in mathematics education: an acute shortage. *Notices of the American Mathematical Society, 47*(10), 1267–1270.

Reys, R. (2018). The preparation of a mathematics education: The case of Carey. *Canadian Journal of Science, Mathematics and Technology Education, 18*(1), 58–67.

Reys, R. E. (2017). Doctorates in mathematics education: How they have evolved, what constitutes a high-quality program, and what might lie ahead. In J. Cai (Ed.), *Compendium for research in mathematics education* (pp. 934–948). NCTM.

Reys, R. (2016). Some thoughts on doctoral preparation in mathematics education. *Journal of Mathematics Education at Teachers College, 7*(2), 22–35.

Reys, R., & Dossey, J. (2008). *US doctorates in mathematics education: Developing stewards of the discipline*. American Mathematical Society.

Reys, R., Glasgow, B., Ragan, G., & Simms, K. (2001). Doctoral programs in mathematics Education in the United States: A status report. In R. Reys & J. Kilpatrick (Eds.), *One field, many paths: US doctoral programs in mathematics education* (pp. 19–40). American Mathematical Society.

Reys, R., Glasgow, B., Teuscher, D., & Nevels, N. (2008). Doctoral programs in mathematics Education in the United States: A status report. In R. Reys & J. Dossey (Eds.), *US doctorates in mathematics education: Developing stewards of the discipline* (pp. 3–18). American Mathematical Society.

Reys, R. & Kilpatrick, J. (2001). *One field, many paths: US doctoral programs in mathematics education*. American Mathematical Society.

Reys, B., & Reys, R. (2017). Strengthening doctoral programs in mathematics education: A continuous process. *Notices of the American Mathematical Society, 64*(4), 386–389.

Reys, R. & Reys, B. (2016). A recent history (5 decades) of the production of doctorates in mathematics education, *Notices of the American Mathematical Society, 63*(8), 936–939.

Reys, R., Reys, B., & Shih, J. (2019a). Doctoral preparation in mathematics education— Time for research and a widespread professional conversation. *Journal of Mathematics Education at Teachers College, 10*(2), 21–27.

Reys, R., Reys, B., Shih, J., & Safi, F. (2019b). Doctoral programs in mathematics education: A status report of size, origin of program leadership, and recommended institutions. *Notices of the American Mathematical Society, 66*(2), 212–217.

Rico, L., Fernandez-Cano, Castro, E., & Torralbo, M. (2008). Post-graduate study program in mathematics education at the university of Granada (Spain). In R. Reys & J. Dossey (Eds.), *US doctorates in mathematics education: Developing stewards of the discipline* (pp. 203–214). American Mathematical Society.

Shih, J., Reys, R., & Engledowl, C. (2011). Profile of research preparation of doctorates in mathematics education in the United States. *Far East Journal of Mathematical Education, 16*(2), 135–148.

Shih, J., Reys, R., & Engledowl, C. (2018). Issues of validity in reporting the number of doctorates in mathematics education. *Investigations in Mathematics Learning, 10*(1), 1–8.

Shih, J., Reys, R., Reys, B., & Engledowl, C. (2019). A profile of mathematics education doctoral graduates' background and preparation in the United States. *Investigations in Mathematics Learning, 11*(1), 16–28.

Shih, J., Reys, R., Reys, B., & Engledowl, C. (2020). Examining the career paths of doctorates in mathematics education working in institutions of higher education. *Investigations in Mathematics Learning, 12*(1), 1–9.

Silver, E. & Walker, E. (2008). Making policy issues visible in the doctoral preparation of mathematics educators. In R. Reys & J. Dossey (Eds.), *US doctorates in mathematics education: Developing stewards of the discipline* (pp. 97–102). American Mathematical Society.

Stiff, L. (2001). Discussions on different forms of doctoral dissertations. In R. Reys & J. Kilpatrick (Eds.), *One field, many paths: US doctoral programs in mathematics education* (pp. 85–86). American Mathematical Society.

Taylor, E., & Kitchens, R. (2008). Doctoral programs in mathematics education: Diversity and equity. In R. Reys & J. Dossey (Eds.), *US doctorates in mathematics education: Developing stewards of the discipline* (pp. 111–116). American Mathematical Society.

Teuscher, D., Marshall, A., Newton, J., & Ulrich, C. (2008). Intellectual communities: Promoting collaboration within and across doctoral programs in mathematics education. In R. Reys & J. Dossey (Eds.), *US doctorates in mathematics education: Developing stewards of the discipline* (pp. 233–243). American Mathematical Society.

Thornton, C., Hunting, R., Shaughnessy, J., Sowder, J., & Wolff, K. (2001). Organizing a new doctoral program in mathematics education. In R. Reys & J. Kilpatrick (Eds.), *One field, many paths: US doctoral programs in mathematics education* (pp. 95–100). American Mathematical Society.

Wilson, P., & Franke, M. (2008). Preparing teachers in mathematics education doctoral Programs: Tension and strategies. In R. Reys & J. Dossey (Eds.), *US doctorates in mathematics education: Developing stewards of the discipline* (pp. 103–110). American Mathematical Society.

Wolff, K. (2001). Recruiting and funding doctoral students. In R. Reys & J. Kilpatrick (Eds.), *One field, many paths: US doctoral programs in mathematics education* (pp. 107–1140). American Mathematical Society.

Zaslavsky, O., & Leikin, R. (2004). Professional development of mathematics teacher educators: Growth through practice. *Journal of Mathematics Teacher Education, 7*(1), 5–32.

Zeichner, K. (2005). Becoming a teacher educator: A personal perspective. *Teaching and Teacher Education, 21*(2), 117–124.

SECTION II

PREPARING TEACHERS TO TEACH SPECIFIC AREAS OF
MATHEMATICS AND APPLIED MATHEMATICS

CHAPTER 7

ENGAGING AND PREPARING EDUCATORS TO TEACH STATISTICS AND DATA SCIENCE

Susan A. Peters
University of Louisville

Anna Bargagliotti
Loyola Marymount University

Christine Franklin
American Statistical Association

The omnipresence of data in today's world has prompted calls for increased focus on developing PreK–12 students' data literacy. Teachers of mathematics will bear responsibility for answering these calls, and mathematics teacher educators will need to ensure that teachers are prepared to teach the statistics and data science content required for data literacy. In this chapter, we review empirical and expository literature to summarize current knowledge about statistics and data science education. We highlight the insights that extant literature offers for teacher education, describe the promise of current innovations in data science, and identify research needed to advance teacher education in data science.

The AMTE Handbook of Mathematics Teacher Education: Reflection on Past, Present and Future—Paving the Way for the Future of Mathematics Teacher Education, Volume 5
pages 123–150.

INTRODUCTION

We live in a world where we are surrounded by data. From data presented in news on topics such as elections and public health, to data analytics reported from social media and fitness apps, data permeate many aspects of daily life. And, the amount of data is growing and predicted to continue growing exponentially well into the future (e.g., Finzer, 2013; Khvoynitskaya, 2020). Recognition of current and growing data proliferation prompted observation about an "urgency [for] teaching and learning about the skills and concepts of data science" (Dorsey & Finzer, 2017, p. 4). Although there is no agreed-upon definition of data science (ASA, 2015; NASEM, 2020), there is consensus about its multidisciplinary nature and its focus on extracting insights from large and complex raw data typically produced incidentally from technological developments such as social networking (NASEM, 2018). Data science draws from disciplines such as computer science and information technology to manage data storage and access and statistics to visualize and describe data and model relationships among variables (e.g., NASEM, 2018). Although data science education draws from statistics education, the additional importance afforded to data provenance, data cleaning, data moves, and data management shifts the emphasis of statistics education to include greater consideration of data prior to analyses.

Calls for data science education in PreK–12 focus on the need for all students to become data literate—for students to "identify, collect, evaluate, analyze, interpret, present, and protect data" (ODI, 2016, p. 2)—by developing data habits of mind from data investigations (Finzer, 2013). PreK–12 statistics education is changing to address elements of data science education due to the amount and types of data now available and the need to differently manage and handle that data; however, reasoning with the data largely remains unchanged.

Several organizations and states are paving the way to develop a data literate citizenry and prepare students for further study in data science. Co-sponsored by the National Council of Teachers of Mathematics (NCTM) and the American Statistical Association (ASA), the *Pre-K–12 Guidelines for Assessment and Instruction in Statistics Education II* (GAISE II) (Bargagliotti et al., 2020) provide a framework to guide the statistical and data science education of PreK–12 students. Of note are foci added since publication of the original GAISE report (Franklin et al., 2007)—additions that specifically identify knowledge and skills for managing and handling data. Newly emphasized content foci include questioning throughout the investigative process, considering data types beyond categorical and quantitative (e.g., pictures, geospatial), cleaning and interrogating secondary data in addition to collecting and analyzing data, developing multivariate thinking, and increasingly using technology (beyond graphing calculators) for data investigations.

Additionally, some states are proposing more statistics and data science content in their PreK–12 standards to prepare students for a future workforce that will require data literacy (e.g., Levitt, 2019). For example, Oregon's mathematics content standards now include a data reasoning domain in grades K–12 (ODE, 2021) that places additional emphasis on formulating statistical investigative questions and considering secondary data by interrogating, for example, how data were collected and measured. They proposed that high school students complete 25% of their core mathematics content credits in data science and statistics, and additional credit in one of three pathways including a data science pathway (ODE, 2022). California highlighted the increasing importance of data science by including a separate data science chapter to explain data-science related standards in the most recent draft of their framework available at the time of writing, the 2022 Revision of the Mathematics Framework (CDE, 2022), which draws extensively on GAISE II. Georgia proposed integrating a framework for statistical reasoning in all K–8 and high school math courses and a statistical reasoning course as an option for secondary students (GDE, 2021). With the ongoing proliferation of data, even more states are likely to acknowledge the importance of and need to address elements of data science in PreK–12 education.

Although data science is multidisciplinary, PreK–12 data science and statistics standards tend to be housed under mathematics and thus will most likely be taught by teachers of mathematics. Research connecting teacher knowledge with student achievement in mathematics (e.g., Charalambous et al., 2020) and statistics (Callingham et al., 2016) suggests the importance of teachers knowing both the content they teach and the pedagogy to teach it for students to achieve proficiency. The relative newness of data science elements in PreK–12 standards, research that suggests teachers' statistical understandings are similar to those of students (e.g., Batanero et al., 2011) and not very strong (Lovett & Lee, 2018), and continuing perceptions of teachers' poor preparation in statistics (e.g., Franklin et al., 2015) suggest that as mathematics teacher educators (MTEs), we need to give additional consideration to how to best prepare teachers to teach statistics and to teach elements of data science. For example, to address the need for increased attention to questioning, Arnold and Franklin (2021) show how questioning takes on different forms throughout statistical investigations. As MTEs begin to prepare teachers for data science education, they must grapple with how to use questions to guide investigations within a classroom. Additionally, due to the computing skills necessary to work with some large secondary data sets, MTEs must decide how much technology, programming, and coding is necessary for teachers to meet growing demands in data science education.

In this chapter, we use empirical and expository literature to summarize current knowledge about statistics education in general and data science education in particular. Throughout, we highlight students' and teachers' learning barriers and interventions that offer insights for teacher education. Through awareness of learners' typical struggles and knowledge about and inclusion of interventions found to be successful in reducing learners' struggles, MTEs can help teachers to not only develop content knowledge but also develop pedagogical content knowledge (PCK) by learning in ways that parallel recommendations for teaching statistics and data science (e.g., Heaton & Mickelson, 2002). We also identify research gaps and offer suggestions to advance teacher education in data science.

STATISTICS AND DATA SCIENCE EDUCATION

Research provides considerable insight into the teaching and learning of statistics content detailed in the original GAISE report (Franklin et al., 2007), which is still relevant for teaching much of the statistics and data science content advocated for inclusion in PreK–12 curricula. Much of the research is summarized in the *International Handbook of Research in Statistics Education* (Ben-Zvi et al., 2018) and focuses on learning and teaching statistics in relation to components of statistical problem solving. The statistical problem solving (investigative) process consists of the four components of formulating statistical investigative questions, collecting or considering the data needed to answer the questions, analyzing the data, and interpreting results in the context of the data (Bargagliotti et al., 2020; cf. Wild & Pfannkuch, 1999). Although data analysis and interpretation have typically been included in PreK–12 statistics education, the importance of the first two components becomes more evident with the use of complex secondary data sets in data science—data that require particular attention to formulating questions and considering data. We begin by elucidating some of the insights that statistics education literature provides for MTEs to identify typical barriers to learning statistics and to design activities for prospective and inservice teachers before reviewing research on learning and teaching elements of data science.

Statistics Education Research: Formulating Statistical Investigative Questions

Although investigative questions are important for "doing statistics," relatively little research focuses on this aspect of statistical problem solving. Existing research suggests that students and teachers have difficulty posing nondeterministic questions appropriate for statistical investigation. For example, after 14–15 year olds participated in a unit on the entire investigative cycle, including two lessons focused on posing investigative questions, fewer than half of their questions were suitable for investigation (Arnold, 2008). Similarly, prospective elementary teachers struggled to pose investigative questions that could be addressed quantitatively; their questions tended to be "all very one-dimensional with one right answer" (Heaton & Mickelson, 2002, p. 46) or could be answered with categorical data (Leavy, 2010).

Supporting students (and teachers) in posing investigative questions might begin with introducing criteria that emerged from teaching experiments: a good investigative question is one that "has both the variable(s) and population(s) clear; has the intention clear; is able to be answered with the data; is about the whole group; and is interesting" (Arnold, 2013, p. 135). Follow-up activities might resemble those experienced by nine-year old students who participated in a unit on posing investigative questions in which they generated characteristics of investigative questions, classified class-generated questions as investigative and non-investigative, considered data required to address investigative questions, and collaboratively refined their own investigative questions (Allmond & Makar, 2010). The percentage of students who were able to pose good questions after instruction increased by 56%. Also important for success was knowledge of context, which complements Lehrer and Schauble's (2002) suggestion that learners need familiarity with phenomena under investigation to pose good investigative questions. Teachers might benefit from similar activity in contexts of interest to them to develop their statistics content knowledge. Considerable numbers of researchers suggest that teachers, as learners, should undergo the same learning experiences as their students (e.g., Burgess, 2008; Pfannkuch, 2008) and engage with projects that can be adapted for school use (MacGillivray & Pereira-Mendoza, 2011).

Researchers also recognize that teachers "need particular knowledge around students and the interrogative cycle" (Burgess, 2008) to facilitate students' developing capacities to pose good investigative questions. Thus, teachers should be made aware of the obstacles students face in their learning as well as strategies to help students overcome the obstacles. Although statistics education research offers insights for teacher education, researchers continue to call for "a focus on effective pedagogy in the statistical education of teachers" (MacGillivray & Pereira-Mendoza, 2011, p. 119) to better inform the endeavors of MTEs. The field would benefit from additional research and development such as design experiments (Cobb, Confrey et al., 2003) to facilitate teachers' development of knowledge about statistics and students (KSS) and knowledge about statistics and teaching (KST) (Groth, 2007), particularly for formulating questions.

Statistics Education Research: Collecting/Considering Data

Multiple knowledgeable teachers acknowledge that the problem-solving component of collecting data—design—was the component they felt least knowledgeable about and least prepared to teach (Peters, 2014). Much of the research related to data collection focuses on learners' conceptions of sampling and their understanding of concepts such as random and representative. A potential barrier for developing robust understandings of sampling is the difference between every day and technical use of terms such as random and sample (e.g., Kaplan, Rogness et al., 2014; Watson & Moritz, 2000) and common notions of fairness. For example, the fifth graders observed by Jacobs (1999) did not associate fairness with equiprobable selection but with the selection process. They desired sampling methods that would ensure even small population diversities were represented. A second barrier relates to the complexity of learning to reason proportionally (e.g., Behr et al., 1992) and developing multiplicative conceptions of sample and sampling (Saldanha & Thompson, 2002)—viewing samples as "quasi-proportional" subsets of populations. Implicit in multiplicative conceptions are notions of sampling representativeness—samples will have characteristics similar to those of their respective populations—and sampling variability—samples are not all identical and thus do not match their respective populations exactly (e.g., Rubin et al., 1990). Overreliance on representativeness leads to deterministic beliefs that samples tell everything about populations, whereas overreliance on variability leads to deterministic beliefs that samples tell nothing about populations. As individuals with considerable mathematics but somewhat limited statistics experiences, teachers may have a propensity for deterministic reasoning (Meletiou-Mavrotheris, 2007) and struggle to balance notions of sampling representativeness and sampling variation.

Several researched interventions help learners navigate past barriers associated with sampling. In one study, seventh graders critiqued newspaper claims made from sample data and used mock articles to report sample and survey results set within familiar contexts to make and defend claims and progress from personal opinion to using data and sampling concepts to support claims (Osana et al., 2004). Sixth graders who engaged with a sequence of activities focused on the meaning and role of samples, the role of sampling methods and sample size, and conducting studies using the investigative cycle showed improved reasoning about the role of sample size and methods such as simple random sampling and stratified sampling (Meletiou-Mavrotheris & Paparistodemou, 2015). Fifth graders began developing robust informal conceptions of sampling variability and sampling representativeness by using technology to select repeated random samples from a known finite population; many of the fifth graders were able to hypothesize a reasonable population proportion by considering the sample statistics (Watson & English, 2016). Ninth graders deepened their understandings of sampling variability and sampling representativeness by comparing a random sample selected from each of two populations, drawing additional random samples from the populations, and comparing sample medians for each population and between populations (Saldanha & McAllister, 2014). These latter three interventions offer promise for developing understandings not only of samples but also for ideas such as inference.

Randomization is an important concept not only for sampling in observational studies but also for group assignment in experimental design. Designing experiments requires consideration of both sources of variation and ways to control variation from those sources. Students at the elementary level can recognize sources of error (that cause variation) in data (Masnick & Klahr, 2003), but few learners at any level consider random assignment as a strategy to control variation, even when enrolled in an introductory course that emphasizes the role of randomization (Derry et al., 2000). Consideration of methods to control systematic and random variability involves reasoning rarely seen in Groth's (2003) study to investigate secondary students' understanding of experimental design. The importance of design, however, cannot be understated. Teachers who exhibited robust understandings of variation credited design with helping them to make connections among different statistical concepts and areas of statistics and for helping them to see the "big picture" of statistics, which they equated with the statistical problem-solving process (Peters, 2014).

Carefully designed interventions in experimental design can offer learning benefits. In a teaching experiment designed to consider error from multiple sources, fourth graders compared heights of rockets with different physical features and explored whether differences could be attributed to random error or were indicative of systematic error (Petrosino et al., 2003). Students used data they collected to argue that observed differences were systematic. In general, learners who have experiences with designing two or more experiments as part of completing the entire problem-solving process tend to realize considerable learning by recognizing shortcomings in initial designs and planning improvements to reduce error (e.g., Anderson-Cook & Dorai-Raj, 2001; Mackisack, 1994).

Learning about experimental design and sampling remains under-researched, particularly in teacher education, and would benefit from increased focus. Despite the dearth of research, insights into teachers' struggles and how instructors and MTEs can sequence activities to deepen students' (and teachers') understandings of sampling and experimental design can be gleaned from this body of research with students, particularly because teachers' learning often parallels students' learning in statistics (e.g., Batanero et al., 2011). Engaging teachers in their own investigations that attend to all components of statistical problem solving is critical for developing teachers' content knowledge (Burgess, 2011), and having teachers reflect on their and their instructors' work at the conclusion of such investigations might help to develop teachers' KST. To develop teachers' KSS, some researchers suggest that teachers might learn best by observing and interacting with students as students engage in investigations (Burgess, 2011). The field would benefit from research to facilitate teachers' development of KSS and KST as well as video cases of students engaging in investigations, the barriers they encounter such as those detailed above, and students' means of overcoming the barriers.

Statistics Education Research: Analyzing Univariate Data

Much existing statistics education research focuses on students' and teachers' reasoning with and understanding of statistical measures and graphical representations. For example, research investigating elementary-aged through college-aged students' conceptions of average and mean reveals that many are able to calculate numerical summary values without understanding the meaning of their results (e.g., Mokros & Russell, 1995; Pollatsek et al., 1981). Like students, many elementary and secondary teachers struggle to view measures of center as more than procedures (e.g., Jacobbe, 2012; Jacobbe & Carvalho, 2011; Leavy & O'Loughlin, 2006). They also struggle to view measures of central tendency as representative (or "typical") values (e.g., Groth & Bergner, 2006), to conceive of the mean in multiple ways (Gfeller et al., 1999), to apply the mean to higher-level problems (Gfeller et al., 1999), to estimate values for the mean from graphical representations of data (Callingham, 1997; Sorto, 2004), and to use the mean to compare groups (Hammerman & Rubin, 2004; Leavy & O'Loughlin, 2006). Research also reveals that school students have intuitive conceptions of variability and are able to reason about the spread and range of data relative to a center (e.g., Reading & Shaughnessy, 2004), but they struggle to move beyond intuition and reason with formal measures of variation even after engaging with activities designed to advance their reasoning with formal measures (Garfield et al., 2007). For example, university students looked for rules to compare standard deviations for multiple pairs of distributions (delMas & Liu, 2005); very few students employed a conceptual approach to dynamically coordinate the estimated location of a mean with deviations from the mean to reason about standard deviation. Teachers' conceptions of variability are similar to those of students (Mooney et al., 2014), and their understandings of formal measures of variation tend to be procedural (Silva & Coutinho, 2006, 2008).

Effective strategies for developing understandings of statistical measures include engaging learners with describing and inventing statistical measures of center and spread and comparing distributions (Bakker & Gravemeijer, 2004; Konold & Pollatsek, 2002; Petrosino et al., 2003). For example, Ben-Zvi (2004) found that middle grades students moved towards considering and integrating multiple characteristics of data sets, including measures of center and spread, shape, and outliers, when comparing distributions. Another strategy to support students' and teachers' developing understandings for univariate data is what has become known as a growing samples activity (e.g., Konold & Pollatsek, 2002). Middle grades students participating in growing samples activities deepened their understandings of distribution by changing their focus from individual datum to the aggregate of data (e.g., Bakker, 2004). A third set of strategies comes from Peters (2014), who interviewed teachers with robust understandings of variation (Peters, 2011) to uncover their perceptions and recollections of learning factors. She then used the results to develop professional learning experiences for middle and high school teachers (Peters, 2018; Peters & Stokes-Levine, 2019; Peters et al., 2014). Elements effective for broadening teachers' perspectives on measures of center and variation included teachers' engagement with activities designed to prompt cognitive conflict, opportunities to consider multiple perspectives through multiple representations and rational discourse, and opportunities to examine the premises underlying measures and procedures.

As observed by Peters et al. (2014, 2019), graphical representations can be used to build learners' understanding of central tendency and variability, particularly in the context of repeated measurements of objects such as table lengths (e.g., Konold & Harradine, 2014; Lehrer et al., 2011). Prior to formal instruction on representing data, students gravitate toward individual views of data by constructing pictures/pictographs (Leavy & Hourigan, 2018) and case value plots— bar graphs in which each datum is represented by a bar with length corresponding to magnitude (e.g., Lehrer & Schauble, 2004). After being introduced to standard representations, middle school, high school, and university students struggle to interpret the information conveyed in histograms (e.g., Kaplan, Gabrosek et al., 2014) and boxplots (e.g., Bakker et al., 2004). Similarly, preservice and practicing teachers have limited graphical comprehension (Jacobbe & Horton, 2010; Pierce & Chick, 2013) as evidenced by their separating histogram rectangles or inadequately labeling axes with real numbers in graph constructions (Bruno & Espinel, 2009) or their difficulty "in seeing graphs as tools for establishing conclusions" (González et al., 2011).

Research suggests that invention and comparison of displays can promote students' (and teachers') development of graphical comprehension (Friel et al., 2001) and facilitate their development of aggregate reasoning (e.g., Lehrer et al., 2007). For example, in one study, fifth-graders invented data displays that resulted in different shapes for the same data (Lehrer & Schauble, 2004); comparing and contrasting displays prompted students to move toward aggregate perspectives. Similarly, seventh-graders began discussing the aggregate shape of a distribution as they invented representations for a collection of student weight measures (Bakker & Gravemeijer, 2004). Although teachers should develop knowledge about typical student errors and obstacles with interpreting graphical displays and with appropriate instructional approaches to address those difficulties (e.g., González et al., 2011), preservice and practicing elementary teachers tend to underestimate the complexities associated with learning and teaching graphical representations (Leavy, 2015). Yet, practicing primary and secondary teachers are more confident about teaching graphical representations than other data and chance topics (Watson, 2001). Researchers recommend that MTEs expose teachers to using graphs as part of the statistical problem-solving process (González et al., 2011), media graphs that are technically inaccurate or incomplete (Watson, 1997), and explore data using representations afforded by technology tools such as TinkerPlots (Lee & Hollebrands, 2011) to develop teachers' graphical competence.

Statistics Education Research: Analyzing Bivariate and Multivariate Data

A first step for reasoning with bivariate data, particularly covariational reasoning, is moving past prior beliefs about relationships between variables to interpret relationships suggested from data (Batanero et al., 1997; Estepa & Batanero, 1996; Estepa et al., 1999). Also important is viewing bivariate distributions in three dimensions, with the third dimension being the relative frequency of data values at the orthogonal intersection of explanatory and response variable values (Cobb, McClain, & Gravemeijer, 2003).

To facilitate such reasoning, Konold (2002) suggested using "sliced scatterplots" for students to see "each vertical slice of data in this plot as a distribution of a discrete group, [and for students to] apply skills they have learned in comparing two distributions to visually compare the centers of the distributions in the sliced scatterplot" (p. 3). Insights about sequencing activities to support students' (and teachers') covariational reasoning can be gleaned from the teaching experiments of Cobb, McClain, and Gravemeijer (2003), whose eighth-grade students were able to view a scatterplot as a series of distributions of univariate data with prompting from the researchers, and of Gil and Gibbs (2017). Gil and Gibbs found that twelfth-grade students began to reason covariationally in a multivariate setting, saw nonlinear models, and modeled covariation after engaging with activities and tools such as Gapminder (https://www.gapminder.org/) and iNZight (https://inzight.nz/).

Teachers need considerable understanding to teach using methods similar to those mentioned. For example, just to teach the correlation coefficient, teachers need to know how to compute and interpret correlation coefficient along with the rationale behind the computation, understand properties of the correlation coefficient, and know definitions, differences, and relationships among correlation, association, and regression (Casey, 2010). Secondary teachers may have appropriate content knowledge to informally find lines of best fit (e.g., Casey & Wasserman, 2015), but they have difficulty supporting students' analyses and interpretations of the correlation coefficient and in reasoning about the regression line (Quintas et al., 2014). Some teachers might benefit from engagement with regression activities designed specifically for teachers and with research in mind (e.g., Bargagliotti et al., 2014) to further develop their PCK.

Statistics Education Research: Interpreting Results

Recently, researchers have begun exploring how students build and interact with statistical models to draw conclusions from data. They focus on how students "integrate sample data, probabilistic models, context, and inference in a technology-enhanced learning environment" (Pfannkuch et al., 2018, p. 1113). In the context of data modeling, students can develop understandings of measures as aggregate descriptors of distributions that guide the inferences they can make from data (Aridor & Ben-Zvi, 2019; Makar, 2014) in conjunction with their knowledge of context (Dierdorp et al., 2011; Langrall et al., 2011). Reasoning about and drawing conclusions from formal inference has proven to be more problematic than informal inferential reasoning (e.g., Aquilonius & Brenner, 2015), even for secondary teachers (Liu & Thompson, 2009). Relatively recent technology developments prompted many in statistics education to teach inference using nonparametric methods such as randomization tests (e.g., Biehler et al., 2015). Doing so, however, does not remove difficulties associated with interpreting results of confidence intervals or significance tests, such as for experimental results.

Pfannkuch and colleagues (2015) identified twelve elements within uncertainty that "we think need to be addressed in instruction to enable students to appreciate and grasp more fully the thinking and argumentation underpinning the designed experiment and the randomization test" (p. 117), and Lee and colleagues (2016) detailed key conceptualizations and capabilities that are important for learners to successfully make inferences using repeated sampling. In two courses designed specifically for prospective primary and secondary teachers—one focused on the investigative process and informal inferential reasoning and the other on simulations—Biehler and colleagues (2015) used a framework for randomization testing to develop activities and support materials for conducting randomization tests and interpreting results. Many of the teachers were able to use technology to carry out randomization tests and make inferences about group comparisons but had difficulty generating adequate null hypotheses and interpreting p-values. Preservice primary teachers are able to reason informally about inferences but might have difficulty selecting sufficiently complex data for drawing inferences and "developing pedagogical contexts which would advance the informal inferential reasoning of primary level students" (Leavy, 2010, p. 62) and need further development of their PCK for advancing children's learning. As with other components of statistical problem solving, the field would benefit from research that investigates how to facilitate teachers' development of KSS and KST for drawing inferences from data.

Statistics Education Research: Statistical Problem Solving and Lingering Questions

Relatively few studies report on students engaging in the entire statistical problem-solving process but instead focus on components of the process; even fewer studies focus on teachers engaging in the process or how they engage students in statistical problem solving. Several studies suggest benefit from engaging teachers with the entire investigative process. Middle and secondary teachers who attended a professional development (PD) program designed to capitalize on research on student learning (e.g., incorporating a growing samples activity) and on the statistical problem-solving process showed quantitatively and qualitatively significant improvement in statistics content knowledge after completing the program (Peters, 2018; Peters & Stokes-Levine, 2019; Peters et al., 2014). Similarly, a PD program focused on comparing distributions provided quantitative and qualitative evidence of teachers' increased knowledge for most aspects of statistical problem solving (Madden, 2008), and a graduate course designed around the principles of active learning through technology-rich interventions offered evidence of teachers' technological pedagogical statistical knowledge growth (Madden, 2019). Although these studies provide insights into how teachers might gain content knowledge and even PCK, they provide few insights into if or how that knowledge translates into designing and implementing instruction.

The absence of studies focused on the entire statistical problem-solving process led Watson and colleagues (2018) to pose the following unanswered questions for researchers.

> Are some stages of the practice more difficult than others? If so, which are they and why are they more difficult? Does difficulty depend on the context, implying that the most difficult stages are different for different contexts? Is long-term retention greater for the stages of the practice if they have been embedded in a meaningful context than if they have been taught in an isolated manner? Does the exposure to complete investigations and drawing informal inferences at school build the appropriate foundation for formal inference when introduced at the college level? (p. 124)

To their questions about doing statistics, we add questions about teaching that remain unanswered. Are some stages of statistics problem solving more difficult to teach than others? What statistical, pedagogical, technological, intersectional, and specialized knowledge do teachers need to teach statistical problem solving? How is that knowledge best developed? How are different types of teacher knowledge related, and in what order do they develop? How is statistical knowledge for teaching similar to or different from mathematical knowledge for teaching (see Groth, 2007)? Although the number of studies that investigate teachers' learning of statistical content and pedagogy is increasing, much remains unknown. Considerably more research is needed to offer a solid research base for teacher education in statistics.

DATA SCIENCE EDUCATION IN PRE-K–12

Some statisticians suggest that an overhaul of primary and secondary statistics education may be needed for students to achieve data literacy (Wild et al., 2018). GAISE II (Bargagliotti et al., 2020) provides one vision to guide PreK–12 statistics and data science education. The shift in discussion from statistics education to data science education is particularly highlighted within the problem-solving component of Collect/Consider data because in today's world, the data we analyze often are secondary and thus need to be considered rather than collected.

A considerable body of expository literature complements GAISE II and offers suggestions for appropriate content and instruction in PreK–12 data science education (e.g., Biehler et al., 2018; Davies & Sheldon, 2021) as well as guidance for teacher education in data science. There are curriculum frameworks developed by an international team of computer scientists and statisticians that "provide the basis for development of courses in Introductory Data Science for students in their final two years of secondary school, and of courses to teach teachers how to teach Introductory Data Science" (IDSSP Curriculum Team, 2019, p. 3). Innovative curricular materials for data science at the middle and high school levels also are available to assist teachers and MTEs with developing learners' data literacy and offer insights for teachers' development of content knowledge in data science. These materials include Data Clubs materials for students in grades 6–8 (TERC, 2021), youcubed data science lessons for grades 6–10 students (youcubed, n.d.a) and high school data science course materials (youcubed, n.d.b), the Introduction to Data Science Curriculum for high school students (Gould et al., 2019), the Bootstrap: Data Science module that addresses middle and high school mathematics and computer science standards (Schanzer et al., 2017), and An Introduction to Data Science with CODAP materials that align with using the Common Online Data Analysis Platform (CODAP; https://codap.concord.org/app/static/dg/en/cert/index.html) for analyses (Erickson, 2021). These curriculum materials differ in both their lengths and their content foci. Implementation times range from several hours to a full school year, and the extent to which they treat elements of data science that are not part of traditional statistics education (e.g., coding, data ethics) differs. For the most part, these materials encourage students to use technology to explore large- (or at least medium-) sized multivariate data sets, to visualize data in many forms, and to formulate models and use them to make predictions, often with respect to issues of social responsibility.

Research into teaching and learning data science content beyond traditional statistics is limited, but insights nonetheless can be gleaned from related research and novel interventions. Researchers note that students (and teachers) in traditional settings typically work with "well-behaved" data that facilitates computation or illustration of particular methods (e.g., Rubin, 2021). Data science, however, often requires working with messy secondary data that requires interrogating data and attending to data provenance. Arnold and colleagues (2021) illustrate how to utilize a large rich data set in the classroom for students to consider data provenance. Data science also requires different ways of dealing with data, including attending to issues of data retrieval, data cleaning, data wrangling, and data munging (Rubin & Erickson, 2018). These data handling skills, respectively, relate to searching for and downloading public data, cleaning data, altering data for analyses, and finalizing data for analyses by following stringent processes that Meng (2021) refers to as minding data. Erickson and colleagues (2019) describe data moves such as sorting, filtering, grouping, and calculating that students can perform with CODAP to alter the contents, structure, or values of data for analyses. The data moves available in CODAP align with coding verbs in the R tidyverse (https://www.tidyverse.org/) to offer students basic data-wrangling needs while setting them up for coding to enact more powerful moves (Erickson & Chen, 2020). With little introduction to data moves or CODAP, high school students can enact data moves and reason about potential causes for variation while using data moves (Wilkerson et al., 2021). For these students, sorting within groups highlighted natural variation; filtering focused attention on potential sources of variability by reducing expected variation; and calculating revealed new or clarified existing patterns of variation such as by converting kilograms to pounds for relatability.

For students (and teachers) to extract meaningful results from complex data, they should use technology tools to reason about relationships among multiple variables. High-achieving students as young as nine can reason about data with one dependent and two independent variables (Ridgway & McCusker, 2003), and students aged 12–15 perform only slightly worse on assessments that require reasoning with multivariate data (Ridgway et al., 2007). In both of these studies conducted by Ridgway and colleagues, students' multivariate reasoning was supported by interactions with dynamic technology. Biehler and colleagues (2013) describe the affordances that dynamic technology tools such as TinkerPlots (and CODAP) allow for students to reason with data, including multivariate data, through the use of color gradients to compare three variables simultaneously. A growing body of work investigates students' explorations with data using dynamic technologies (e.g., Haldar et al., 2018; Rubin & Mokros, 2018) and provides insights into how technology supports development of important data science skills, including creating and analyzing multiple representations of data for large data sets with multiple attributes. For example, third-grade students' informal inferential reasoning was supported by dynamically linked multiple representations on which they could operate quickly with immediate feedback (Paparistodemou & Meletiou-Mavrotheris, 2008). Similarly, elementary students quickly became independent users of dynamic technology and created multiple data plots to explore relationships among multiple variables and make sense of data to investigate hypotheses (Fitzallen & Watson, 2010).

Beyond traditional data representations, students (and teachers) should explore data with nontraditional representations. Engel and colleagues (2020) provide an overview of historic data visualizations, modern approaches to visualizing statistics, and interactive visualizations available to today's students. Recent and developing innovations in supporting middle school (Rubin & Mokros, 2019) and high school students' (Gould et al., 2016) explorations with data offer additional promise for informing and advancing data science education, including designing data science interventions for teachers.

TEACHER EDUCATION IN DATA SCIENCE

Many current recommendations for teachers' statistical preparation were released prior to calls for elements of data science to be included in PreK–12 education. However, these recommendations provide a baseline from which to build teacher preparation in elements of data science. The *Statistical Education of Teachers* (SET) (Franklin et al., 2015) expands on recommendations from the *Mathematical Education of Teachers II* (MET II) (CBMS, 2012) for middle level teachers to complete a teacher-specific statistics course and for high school teachers to complete statistics courses focused on statistical problem solving. The authors recommend additional coursework for PreK–12 teachers to be prepared to teach statistics. Recent standards by mathematics education organizations (AMTE, 2017; NCTM, 2020a,b) do not explicitly mention data science as part of teachers' mathematical preparation, yet they cite the foundational importance of teachers developing robust understandings of both the content they will teach and the ways to teach that content. In order to meet AMTE and NCTM standards, teacher preparation programs will need to respond to increased demands in PreK–12 for statistics and data science content. The SET report provides a starting point for needed coursework and content, but additional consideration must be given to preparing teachers to teach the new content and foci articulated in GAISE II, potentially by integrating data science into current statistics preparation, the development of new courses, or ongoing PD.

Research that explores teacher education in data science is sparse. In one of only a few such studies, nine secondary teachers reasoned with participatory sensing data—complex and multivariate opportunistic data collected by teachers (Gould et al., 2017). These data provided teachers with meaningful data that shared many features of big data in a context of interest to them. In this novel setting, the researchers tracked teachers' statistical problem solving and found that the more successful teachers were stronger in their question-posing and exhibited stronger links between questioning and considering data. These teachers had the option of analyzing data using a dashboard visualizer or by coding in RStudio. Whether teachers of mathematics in PreK–12 will need to teach data science via coding at some point is an open question, but statistics education researchers are conducting investigations in this area. For example, Fergusson and Pfannkuch (2021) implemented a teaching experiment to create tasks with three design considerations in a six-phase approach to facilitate teachers' transition from using the VIT Online tool to a code-driven tool for randomization tests. Upon implementation, they found that learning tasks reflective of the design principles and considerations supported teachers' usage of code-driven tools and their development of integrated statistical and computational thinking. These two studies offer some beginning insights for teacher preparation in data science, but much remains unknown about developing teachers' content knowledge and PCK for teaching data science.

Until further research is conducted, unique teacher-specific courses and professional learning experiences can offer additional insights for teacher education in data science. For example, Loyola Marymount University offers a "Statistics and Data Science for Teachers" course that could inform how data science is incorporated into teachers' learning and teaching. Insights also can be gleaned from professional learning to advance inservice teachers' knowledge for teaching statistics and data science, such as from the InSTEP project and MOOCs (Friday Institute, 2021a,b,c,d) or from professional learning programs designed for teachers to meet with success implementing data science curricula (e.g., Gould et al., 2016). Insights gained from instructors teaching these courses and programs and research conducted with some of the innovations can inform the design of future courses and professional learning experiences that include data science.

Although research on teacher education in data science might be sparse, opinions about data science teacher education are not. There is general consensus that teachers currently do not have the necessary content knowledge, much less the PCK, to teach the data science content articulated in GAISE II and IDSSP, particularly if coding is required (e.g., Gould et al., 2016; Horton, 2015; Sentance, 2018). There also is some consensus on how teachers can develop content knowledge needed for teaching data science. First, teachers need opportunities to engage in the entire statistical problem-solving process by using technology to explore complex, multivariate data (e.g., Goode, 2018; Zapata-Cardona & Martinez-Castro, 2023). Work with secondary students suggests that initial explorations might need to be semi-structured in order to focus investigations on important concepts (Kazak et al., 2023). MTEs could benefit from repositories of large data sets from interesting contexts and accompanying lessons for teachers to consider these data as part of engaging with the entire statistical problem-solving process. Relatedly, teachers should explore data as learners and reflect on their experiences as teachers to develop both content knowledge and PCK (e.g., Lee & Hollebrands, 2008; Ubrilla & Gorgorió, 2020) as well as a stronger sense of self efficacy (e.g., Estrada et al., 2011). Those explorations should be designed for teachers to become computationally nimble and comfortable with data moves using technology such as CODAP. Lastly, preparing teachers to teach data science will take considerable time (Davies & Sheldon, 2021; Gould et al., 2016). The uniqueness and magnitude of learning data science and learning to teach data science led Davies and Sheldon (2021) to suggest that teacher educators "should consider establishing, for example, a Post Graduate Certification in Education (PGCE) in D[ata]S[cience]" (p. S66).

The absence of research on the learning and teaching of data science means that data science education is wide open for study. It also means, however, that the number of research-informed strategies for teaching the data science content that has not traditionally been part of PreK–12 statistics education is minimal. Little information exists to suggest how students (and teachers) acquire and prepare data for analysis, how they reason with nontraditional visualizations of data, how they learn best to reason with multivariate data sets consisting of multiple types of data, what supports they need to accomplish the preceding, and what teachers need to know to accomplish the preceding with students and how they come to know it. If, indeed, it is true that "there is little doubt that teachers able to deliver a course in D[ata]S[cience] are in short supply" (Davies & Sheldon, 2021, p. S64), then it may also be reasonable to assume that MTEs capable of delivering a teacher education course or PD program in data science are in short supply. We leave you with one last question that might need to be answered before any recommendations on the teaching and learning of data science can be implemented successfully: How can we support MTEs to prepare teachers for teaching data science? As a field, we need to find answers to these questions sooner rather than later. We hope that this chapter provided you with some resources and strategies to consider for your work with teachers.

CONCLUDING THOUGHTS

Considerable research exists to inform PreK–12 statistics education. However, considerable gaps exist, particularly with research on teacher education in statistics. Further, although assumptions might be made that increased emphasis on PreK–12 statistics means prospective teachers enter college with statistical savviness, evidence suggests that well after the release of Common Core State Standards in Mathematics (NGACBP & CCSSO, 2010), prospective teachers are still not well prepared to teach the statistics content outlined in the standards (Lovett & Lee, 2017). Continued concerns about teachers' statistical knowledge coupled with increased demands in PreK–12 for statistics and elements of data science require that we reconsider current teacher preparation in statistics. As a field, we can benefit from renewed commitment to increasing research not only in statistics education but also in data science education and revisiting and redesigning statistics and data science elements of teacher preparation to ensure alignment with extant research and innovations for teachers to be proficient with facilitating students' data literacy development.

REFERENCES

Allmond, S., & Makar, K. (2010). Developing primary students' ability to pose questions in statistical investigations. In C. Reading (Ed.), *Data and context in statistics education: Towards an evidence-based society. Proceedings of the eighth International Conference on Teaching Statistics. Ljubljana Slovenia, 11–16 July 2010.* International Statistical Institute.

American Statistical Association (ASA). (2015, August 8). *ASA statement on the role of statistics in data science.* https://www.amstat.org/asa/files/pdfs/POL-DataScienceStatement.pdf

Anderson-Cook, C. M., & Dorai-Raj, S. (2001). An active learning in-class demonstration of good experimental design. *Journal of Statistics Education, 9*(1). https://doi.org/10.1080/10691898.2001.11910645

Aquilonius, B. C., & Brenner, M. E. (2015). Students' reasoning about *p*-values. *Statistics Education Research Journal, 14*(2), 7–27. https://doi.org/10.52041/serj.v14i2.259

Aridor, K., & Ben-Zvi, D. (2019). Students' aggregate reasoning with covariation. In G. Burrill & D. Ben-Zvi (Eds.), *Topics and trends in current statistics education research: International perspectives* (pp. 71–94). Springer. https://doi.org/10.1007/978-3-030-03472-6_4

Arnold, P. (2008). What about the P in the PPDAC cycle? An initial look at posing questions for statistical investigation. *Proceedings of the 11th International Congress of Mathematics Education (ICME-11), Monterrey, Mexico, 6–13 July 2008.*

Arnold, P. (2013). *Statistical investigative questions: An enquiry into posing and answering investigative questions from existing data.* [Doctoral dissertation] The University of Auckland. https://researchspace.auckland.ac.nz/handle/2292/21305

Arnold, P., & Franklin, C. (2021). What makes a good statistical question? *Journal of Statistics and Data Science Education, 29*(1), 122–130. https://doi.org/10.1080/26939169.2021.1877582

Arnold, P., Perez, L., & Johnson, S. (2021). Using photographs as data sources to tell stories. *Harvard Data Science Review, 3*(4). https://doi.org/10.1162/99608f92.f0a7df71

Association of Mathematics Teacher Educators (AMTE). (2017). *Standards for preparing teachers of mathematics.* https://amte.net/standards

Bakker, A. (2004). Reasoning about shape as a pattern in variability. *Statistics Education Research Journal, 3*(2), 64–83. https://doi.org/10.52041/serj.v3i2.552

Bakker, A., Biehler, R., & Konold, C. (2004). Should young students learn about box plots? In G. Burrill & M. Camden (Eds.), *Curricular development in statistics education. International Association for Statistical Education (IASE) Roundtable. Lund, Sweden* (pp. 163–173). International Statistical Institute. https://iase-web.org/documents/papers/rt2004/1_Frontmatter.pdf?1402524986

Bakker, A., & Gravemeijer, K. P. E. (2004). Learning to reason about distribution. In D. Ben-Zvi & J. Garfield (Eds.), *The challenge of developing statistical literacy, reasoning, and thinking* (pp. 327–352). Kluwer. https://doi.org/10.1007/1-4020-2278-6_7

Bargagliotti, A., Anderson, C., Casey, S., Everson, M., Franklin, C.,Gould, R., Haddock, J., & Watkins, A. (2014). Project-SET materials for the teaching and learning of sampling variability and regression. In K. Makar, B. de Sousa, & R. Gould (Eds.), *Sustainability in statistics education. Proceedings of the 9th International Conference on Teaching Statistics (ICOTS9, July, 2014), Flagstaff, AZ, USA.* International Statistical Institute.

Bargagliotti, A., Franklin, C., Arnold, P., Gould, R., Johnson, S., Perez, L., & Spangler, D. (2020). *Pre-K–12 guidelines for assessment and instruction in statistics education II (GAISE II): A framework for statistics and data science education.* American Statistical Association. https://www.amstat.org/asa/files/pdfs/GAISE/GAISEI-IPreK-12_Full.pdf

Batanero, C., Burrill, G., & Reading, C. (Eds.). (2011). *Teaching statistics in school mathematics—Challenges for teaching and teacher education.* Springer. https://doi.org/10.1007/978-94-007-1131-0

Batanero, C., Estepa, A., & Godino, J. D. (1997). Evolution of students' understanding of statistical association in a computer-based teaching environment. In J. B. Garfield & G. Burrill (Eds.), *Research on the role of technology in teaching and learning statistics: Proceedings of the 1996 IASE Round Table Conference* (pp. 191–205). International Statistical Institute.

Behr, M., Harel, G., Post, T., & Lesh, R. A. (1992). Rational number, ratio, and proportion. In D. A. Grouws (Ed.), *Handbook of research on mathematics teaching and learning* (pp. 296–333). National Council of Teachers of Mathematics.

Ben-Zvi, D. (2004). Reasoning about variability in comparing distributions. *Statistics Education Research Journal, 3*(2), 42–63. https://doi.org/10.52041/serj.v3i2.547

Ben-Zvi, D., Makar, K., & Garfield, J. (Eds.). (2018). *International handbook of research in statistics education.* Springer. https://doi.org/10.1007/978-3-319-66195-7

Biehler, R., Ben-Zvi, D., Bakker, A., & Makar, K. (2013). Technology for enhancing statistical reasoning at the school level. In M. A. Clements, A. M. Bishop, C. Keitel, J. Kilpatrick, & F. K. S. Leung (Eds.), *Third international handbook of mathematics education* (pp. 643–689). Springer. https://doi.org/10.1007/978-1-4614-4684-2_21

Biehler, R., Budde, L., Frischemeier, D., Heinemann, B., Podworny, S., Schulte, C., & Wassong, T. (Eds.) (2018). *Paderborn symposium on data science education at school level 2017: The collected extended abstracts.* Universitätsbibliothek Paderborn. https://doi.org/10.17619/UNIPB/1-374

Biehler, R., Frischemeier, D., & Podworny, S. (2015). Preservice teachers' reasoning about uncertainty in the context of randomization tests. In A. Zieffler & E. Fry (Eds.), *Reasoning about uncertainty: Learning and teaching informal inferential reasoning* (pp. 129–162). Catalyst Press.

Bruno, A., & Espinel, M. C. (2009). Construction and evaluation of histograms in teacher training. *International Journal of Mathematical Education in Science and Technology, 40*(4), 473–493. https://doi.org/10.1080/00207390902759584

Burgess, T. A. (2008). Teacher knowledge for teaching statistics through investigations. In C. Batanero, G. Burrill, C. Reading, & A. Rossman (Eds.), *Joint ICMI/IASE Study: Teaching statistics in school mathematics. Challenges for teaching and teacher education. Proceedings of the ICMI Study 18 and 2008 IASE Round Table Conference.* International Commission on Mathematical Instruction and International Association for Statistical Education.

Burgess, T. A. (2011). Teacher knowledge of and for statistical investigations. In C. Batanero, G. Burrill, & C. Reading (Eds.), *Teaching statistics in school mathematics: Challenges for teaching and teacher education. New ICMI study series* (Vol. 14, pp. 259–270). Springer. https://doi.org/10.1007/978-94-007-1131-0_26

California Department of Education (CDE). (2022, March). *Mathematics framework chapter 5 data science, TK–12: Second field review draft.* https://www.cde.ca.gov/ci/ma/cf/

Callingham, R. A. (1997). Teachers' multimodal functioning in relation to the concept of average. *Mathematics Education Research Journal, 9*(2), 205–224. https://doi.org/10.1007/BF03217311

Callingham, R., Carmichael, C., & Watson, J. M. (2016). Explaining student achievement: The influence of teachers' pedagogical content knowledge in statistics. *International Journal of Science and Mathematics Education, 14*(7), 1339–1357. https://doi.org/10.1007/s10763-015-9653-2

Casey, S. A. (2010). Subject matter knowledge for teaching statistical association. *Statistics Education Research Journal, 9*(2), 50–68. https://doi.org/10.52041/serj.v9i2.375

Casey, S. A., & Wasserman, N. H. (2015). Teachers' knowledge about informal line of best fit. *Statistics Education Research Journal, 14*(1), 8–35. https://doi.org/10.52041/serj.v14i1.267

Charalambous, C. Y., Hill, H. C., Chin, M. J., & McGinn, D. (2020). Mathematical content knowledge and knowledge for teaching: Exploring their distinguishability and contribution to student learning. *Journal of Mathematics Teacher Education, 23*(6), 579–613. https://doi.org/10.1007/s10857-019-09443-2

Cobb, P., Confrey, J., diSessa, A. A., Lehrer, R., & Schauble, L. (2003). Design experiments in educational research. *Educational Researcher, 32*(1), 9–13. https://doi.org/10.3102/0013189X032001009

Cobb, P., McClain, K., & Gravemeijer, K. (2003). Learning about statistical covariation. *Cognition and Instruction, 21*(1), 1–78. https://doi.org/10.1207/S1532690X-CI2101_1

Conference Board of the Mathematical Sciences (CBMS). (2012). *The mathematical education of teachers II.* American Mathematical Society and Mathematical Association of America.

Davies, N., & Sheldon, N. (2021). Teaching statistics and data science in England's schools. *Teaching Statistics, 43*(S1), S52–S70. https://doi.org/10.1111/test.12276

delMas, R., & Liu, Y. (2005). Exploring students' conceptions of the standard deviation. *Statistics Education Research Journal, 4*(1), 55–82. https://doi.org/10.52041/serj.v4i1.525

Derry, S. J., Levin, J. R., Osana, H. P., Jones, M. S., & Peterson, M. (2000). Fostering students' statistical and scientific thinking: Lessons learned from an innovative college course. *American Educational Research Journal, 37*(3), 747–773. https://doi.org/10.3102/00028312037003747

Dierdorp, A., Bakker, A., Eijkelhof, H., & van Maanen, J. (2011). Authentic practices as contexts for learning to draw inferences beyond correlated data. *Mathematical Thinking and Learning, 13*(1), 132–151. https://doi.org/10.1080/10986065.2011.538294

Dorsey, C., & Finzer, W. (2017, Fall). The data science education revolution. *@Concord, 21*(2), 4–6.

Engel, J., Campos, P., Nicholson, J., Ridgway, J., & Teixeira, S. (2020). Visualization multivariate data: Graphs that tell stories. In P. Arnold (Ed.), *New skills in the changing world of statistics education: Proceedings of the roundtable conference of the International Association for Statistical Education (IASE), July 2020.* International Statistical Institute.

Erickson, T. (2021). *An introduction to data science with CODAP.* Concord Consortium. https://codap.xyz/awash/

Erickson, T., & Chen, E. (2020). Introducing data science with data moves and CODAP. *Teaching Statistics, 43*(S1), S124–S132. https://doi.org/10.1111/test.12240

Erickson, T., Wilkerson, M. H., Finzer, W., & Reichsman, F. (2019). Data moves. *Technology Innovations in Statistics Education, 12*(1). https://doi.org/10.5070/T5121038001

Estepa, A., & Batanero, C. (1996). Judgments of correlation in scatterplots: Students' intuitive strategies and preconceptions. *Hiroshima Journal of Mathematics Education, 4,* 25–41.

Estepa, A., Batanero, C., & Sanchez, F. T. (1999). Students' intuitive strategies in judging association when comparing two samples. *Hiroshima Journal of Mathematics Education, 7,* 17–30.

Estrada, A., Batanero, C., & Lancaster, S. (2011). Teachers' attitudes towards statistics. In C. Batanero, G. Burrill, & C. Reading (Eds.), *Teaching statistics in school mathematics: Challenges for teaching and teacher education. New ICMI study series* (Vol. 14, pp. 163–174). Springer. https://doi.org/10.1007/978-94-007-1131-0_18

Fergusson, A., & Pfannkuch, M. (2021). Introducing teachers who use GUI-driven tools for the randomization test to code-driven tools. *Mathematical Thinking and Learning, 24*(4), 336–356. https://doi.org/10.1080/10986065.2021.1922856

Finzer, W. (2013). The data science education dilemma. *Technology Innovations in Statistics Education, 7*(2). https://doi.org/10.5070/T572013891

Fitzallen, N., & Watson, J. (2010). Developing statistical reasoning facilitated by TinkerPlots. In C. Reading (Ed.), *Data and context in statistics education: Towards an evidence-based society. Proceedings of the Eighth International Conference on Teaching Statistics (ICOTS8, July, 2010), Ljubljana, Slovenia.* International Statistical Institute.

Franklin, C. Bargagliotti, A., Case, C., Kader, G., Scheaffer, R., & Spangler, D. A. (2015). *The statistical education of teachers.* American Statistical Association.

Franklin, C., Kader, G., Mewborn, D., Moreno, J., Peck, R., Perry, M., & Scheaffer, R. (2007). *Guidelines for Assessment and Instruction in Statistics Education (GAISE) report: A Pre-K–12 curriculum framework.* American Statistical Association.

Friday Institute for Educational Innovation. (2021a). *InSTEP: Invigorating statistics teacher education through professional online learning.* Retrieved January 12, 2021 from https://www.fi.ncsu.edu/projects/instep/

Friday Institute for Educational Innovation. (2021b). *Teaching statistics through data investigations online professional development course.* Retrieved January 12, 2021 from https://www.fi.ncsu.edu/projects/teaching-statistics-through-data-investigations-mooc-ed/

Friday Institute for Educational Innovation. (2021c). *Teaching statistics through inferential reasoning online professional learning course.* Retrieved January 12, 2021 from https://www.fi.ncsu.edu/projects/teaching-statistics-through-inferential-reasoning-mooc-ed/

Friday Institute for Educational Innovation. (2021d). *Amplifying statistics and data science in classrooms.* Retrieved August 23, 2021 from https://place.fi.ncsu.edu/local/catalog/course.php?id=33&ref=1

Friel, S. N., Curcio, R. F., & Bright, G. W. (2001). Making sense of graphs: Critical factors influencing comprehension and instructional implications. *Journal for Research in Mathematics Education, 32*(3), 124–158. https://doi.org/10.2307/749671

Garfield, J., delMas, R., & Chance, B. (2007). Using students' informal notions of variability to develop an understanding of formal measures of variability. In M. C. Lovett & P. Shah (Eds.), *Thinking with data* (pp. 117–148). Erlbaum.

Georgia Department of Education. (2021). *Georgia's K–12 mathematics standards.* Retrieved June 8, 2022 from https://www.gadoe.org/Curriculum-Instruction-and-Assessment/Curriculum-and-Instruction/Pages/GA-K12-Math-Standards.aspx

Gfeller, M. K., Niess, M. L., & Lederman, N. G. (1999). Preservice teachers' use of multiple representations in solving arithmetic mean problems. *School Science and Mathematics, 99*(5), 250–257. https://doi.org/10.1111/j.1949-8594.1999.tb17483.x

Gil, E., & Gibbs, A. L. (2017). Promoting modeling and covariational reasoning among secondary school students in the context of big data. *Statistics Education Research Journal, 16*(2), 163–190. https://doi.org/10.52041/serj.v16i2.189

González, M. T., Espinel, M. C., & Ainley, J. (2011). Teachers' graphical competence. In C. Batanero, G. Burrill, & C. Reading (Eds.), *Teaching statistics in school mathematics: Challenges for teaching and teacher education. New ICMI study series* (Vol. 14, pp. 187–197). Springer. https://doi.org/10.1007/978-94-007-1131-0_20

Goode, J. (2018). Data science in computer science classrooms: A United States perspective. In R. Biehler, L. Budde, D. Frischemeier, B. Heinemann, S. Podworny, C. Schulte, & T. Wassong (Eds.), *Paderborn symposium on data science education at school level 2017: The collected extended abstracts* (pp. 70–72). Universitätsbibliothek Paderborn. https://doi.org/10.17619/UNIPB/1-374

Gould, R., Bargagliotti, A., & Johnson, T. (2017). An analysis of secondary teachers' reasoning with participatory sensing data. *Statistics Education Research Journal, 16*(2), 305–334. https://doi.org/10.52041/serj.v16i2.194

Gould, R., Machado, S., Johnson, T. A., & Molyneux, J. (2019). *Introduction to data science.* https://curriculum.idsucla.org/

Gould, R., Machado, S., Ong, C., Johnson, T., Molyneux, J., Nolen, S., Tangmunarunkit, H., Trusela, L., & Zanontian, L. (2016). Teaching data science to secondary students: The Mobilize introduction to data science curriculum. In J. Engel (Ed.), *Promoting understanding of statistics about society. Proceedings of the Roundtable Conference of the International Association of Statistics Educators (IASE), July 2016.* International Statistical Institute.

Groth, R. (2003). High school students' levels of thinking in regard to statistical study design. *Mathematics Education Research Journal, 15*(3), 252–269. https://doi.org/10.1007/BF03217382

Groth, R. E. (2007). Toward a conceptualization of statistical knowledge for teaching. *Journal for Research in Mathematics Education, 38*(5), 427–437. https://doi.org/10.2307/30034960

Groth, R. E., & Bergner, J. A. (2006). Preservice elementary teachers' conceptual and procedural knowledge of mean, median, and mode. *Mathematical Thinking and Learning, 8*(1), 37–63. https://doi.org/10.1207/s15327833mtl0801_3

Haldar, L. C., Wong, N., Heller, J. I., & Konold, C. (2018). Students making sense of multilevel data. *Technology Innovations in Statistics Education, 11*(1). https://doi.org/10.5070/T5111031358

Hammerman, J. K., & Rubin, A. (2004). Strategies for managing statistical complexity with new software tools. *Statistics Education Research Journal, 3*(2), 17–41. https://doi.org/10.52041/serj.v3i2.546

Heaton, R. M., & Mickelson, W. T. (2002). The learning and teaching of statistical investigation in teaching and teacher education. *Journal of Mathematics Teacher Education, 5*(1), 35–59. https://doi.org/10.1023/A:1013886730487

Horton, N. J. (2015). Challenges and opportunities for statistics and statistical education: Looking back, looking forward. *The American Statistician, 69*(2), 138–145. https://doi.org/10.1080/00031305.2015.1032435

International Data Science in Schools Project Curriculum Team (IDSSP). (2019). *Curriculum frameworks for introductory data science.* http://idssp.org/files/IDSSP_Frameworks_1.0.pdf

Jacobbe, T. (2012). Elementary school teachers' understanding of the mean and median. *International Journal of Science and Mathematics Education, 10,* 1143–1161. https://doi.org/10.1007/s10763-011-9321-0

Jacobbe, T., & Carvalho, C. (2011). Teachers' understanding of averages. In C. Batanero, G. Burrill, & C. Reading (Eds.), *Teaching statistics in school mathematics: Challenges for teaching and teacher education. New ICMI study series* (Vol. 14, pp. 199–209). Springer. https://doi.org/10.1007/978-94-007-1131-0_21

Jacobbe, T., & Horton, R. M. (2010). Elementary school teachers' comprehension of data displays. *Statistics Education Research Journal, 9*(1), 27–45. https://doi.org/10.52041/serj.v9i1.386

Jacobs, V. R. (1999). How do students think about statistical sampling before instruction? *Mathematics Teaching in the Middle School, 5*(4), 240–46, 263. https://doi.org/10.5951/MTMS.5.4.0240

Kaplan, J. J., Gabrosek, J. G., Curtiss, P., & Malone, C. (2014). Investigating student understanding of histograms. *Journal of Statistics Education, 22*(2). https://doi.org/10.1080/10691898.2014.11889701

Kaplan, J. J., Rogness, N. T., & Fisher, D. G. (2014). Exploiting lexical ambiguity to help students understand the meaning of *random. Statistics Education Research Journal, 13*(1), 9–24. https://doi.org/10.52041/serj.v13i1.296

Kazak, S., Fujita, T., & Turmo, M. P. (2023). Students' informal statistical inferences through data modeling with a large multivariate dataset. *Mathematical Thinking and Learning, 25*(1), 23–43. https://doi.org/10.1080/10986065.2021.1922857

Khvoynitskaya, S. (2020, January 30). *The future of big data: 5 predictions from experts for 2020–2025.* iTransition. https://www.itransition.com/blog/the-future-of-big-data

Konold, C. (2002). Alternatives to scatterplots. In B. Phillips (Ed.), *Developing a statistically literate society. Proceedings of the Sixth International Conference on Teaching Statistics, Cape Town, South Africa* [CD-ROM]. International Statistical Institute.

Konold, C., & Harradine, A. (2014). Contexts for highlighting signal and noise. In T. Wassong, D. Frischemeier, P. R. Fischer, R. Hochmuth, & P. Bender (Eds.), *Using tools for learning mathematics and statistics* (pp. 237–250). Springer. https://doi.org/10.1007/978-3-658-03104-6_18

Konold, C., & Pollatsek, A. (2002). Data analysis as the search for signals in noisy processes. *Journal for Research in Mathematics Education, 33*(4), 259–289. https://doi.org/10.2307/749741

Langrall, C., Nisbet, S., Mooney, E., & Jansem, S. (2011). The role of context expertise when comparing data. *Mathematical Thinking and Learning, 13*(1), 47–67. https://doi.org/10.1080/10986065.2011.538620

Leavy, A. M. (2010). The challenge of preparing preservice teachers to teach informal inferential reasoning. *Statistics Education Research Journal, 9*(1), 46–67. https://doi.org/10.52041/serj.v9i1.387

Leavy, A. M. (2015). Looking at practice: Revealing the knowledge demands of teaching data handling in the primary classroom. *Mathematics Education Research Journal, 27*(3), 283–309. https://doi.org/10.1007/s13394-014-0138-3

Leavy, A. M., & Hourigan, M. (2018). Inscriptional capacities and representations of young children engaged in data collection during a statistical investigation. In A. Leavy, M. Meletiou-Mavrotheris, & E. Paparistodemou (Eds.), *Statistics in early childhood and primary education: Supporting early statistical and probabilistic thinking* (pp. 89–107). Springer. https://doi.org/10.1007/978-981-13-1044-7_6

Leavy, A., & O'Loughlin, N. (2006). Preservice teachers' understanding of the mean: Moving beyond the arithmetic average. *Journal of Mathematics Teacher Education, 9*(1), 53–90. https://doi.org/10.1007/s10857-006-9003-y

Lee, H. S., Doerr, H. M., Tran, D., & Lovett, J. N. (2016). The role of probability in developing learners' models of simulation approaches to inference. *Statistics Education Research Journal, 15*(2), 216–238. https://doi.org/10.52041/serj.v15i2.249

Lee, H., & Hollebrands, K. (2008). Preparing to teach mathematics with technology: An integrated approach to developing technological pedagogical content knowledge. *Contemporary Issues in Technology and Teacher Education, 8*(4), 326–341. https://citejournal.org/volume-8/issue-4-08/mathematics/preparing-to-teach-mathematics-with-technology-an-integrated-approach-to-developing-technological-pedagogical-content-knowledge/

Lee, H. S., & Hollebrands, K. F. (2011). Characterising and developing teachers' knowledge for teaching statistics with technology. In C. Batanero, G. Burrill, & C. Reading (Eds.), *Teaching statistics in school mathematics: Challenges for teaching and teacher education. New ICMI study series* (Vol. 14, pp. 359–370). Springer. https://doi.org/10.1007/978-94-007-1131-0_34

Lehrer, R., Kim, M.-J., & Jones, R. S. (2011). Developing conceptions of statistics by designing measures of distribution. *ZDM Mathematics Education, 43*(5), 723–736. https://doi.org/10.1007/s11858-011-0347-0

Lehrer, R., Kim, M.-J., & Schauble, L. (2007). Supporting the development of conceptions of statistics by engaging students in measuring and modeling variability. *International Journal of Computers for Mathematical Learning, 12*(3), 195–222. https://doi.org/10.1007/s10758-007-9122-2

Lehrer, R., & Schauble, L. (Eds.). (2002). *Investigating real data in the classroom: Expanding children's understanding of math and science.* Teachers College Press.

Lehrer, R., & Schauble, L. (2004). Modeling natural variation through distribution. *American Educational Research Journal, 41*(3), 635–679. https://doi.org/10.3102/00028312041003635

Levitt, S. (2019, October 2). *America's math curriculum doesn't add up.* http://freakonomics.com/podcast/math-curriculum/

Liu, Y., & Thompson, P. W. (2009). Mathematics teachers' understandings of proto-hypothesis testing. *Pedagogies, 4*(2), 126–138. https://doi.org/10.1080/15544800902741564

Lovett, J., & Lee, H. (2017). New standards require teaching more statistics: Are preservice secondary mathematics teachers ready? *Journal of Teacher Education, 68*(3), 299–311. https://doi.org/10.1177/0022487117697918

Lovett, J. N., & Lee, H. S. (2018). Preservice secondary mathematics teachers' statistical knowledge: A snapshot of strengths and weaknesses. *Journal of Statistics Education, 26*(3), 214–222. https://doi.org/10.1080/10691898.2018.1496806

MacGillivray, H., & Pereira-Mendoza, L. (2011). Teaching statistical thinking through investigative projects. In C. Batanero, G. Burrill, & C. Reading (Eds.), *Teaching statistics in school mathematics: Challenges for teaching and teacher education. New ICMI study series* (Vol. 14, pp. 109–120). Springer. https://doi.org/10.1007/978-94-007-1131-0_14

Mackisack, M. (1994). What is the use of experiments conducted by statistics students? *Journal of Statistics Education, 2*(1). https://doi.org/10.1080/10691898.1994.11910461

Madden, S. R. (2008). *High school mathematics teachers' evolving knowledge of comparing distributions* [Unpublished doctoral dissertation]. Western Michigan University. http://iase-web.org/documents/dissertations/08.Madden.Dissertation.pdf

Madden, S. (2019). Exploring secondary teacher statistical learning: Professional learning in a blended format statistics and modeling course. In G. Burrill & D. Ben-Zvi (Eds.), *Topics and trends in current statistics education research: International perspectives* (pp. 265–282). Springer. https://doi.org/10.1007/978-3-030-03472-6

Makar, K. (2014). Young children's explorations of average through informal inferential reasoning. *Educational Studies in Mathematics, 86*(1), 61–78. https://doi.org/10.1007/s10649-013-9526-y

Masnick, A. M., & Klahr, D. (2003). Error matters: An initial exploration of elementary school children's understanding of experimental error. *Journal of Cognition and Development, 4*(1), 67–98.

Meletiou-Mavrotheris, M. (2007). The formalist mathematical tradition as an obstacle to stochastical reasoning. In K. François & J. P. Van Bendegem (Eds.), *Philosophical dimensions in mathematics education* (pp. 131–156). Springer. https://doi.org/10.1007/978-0-387-71575-9_7

Meletiou-Mavrotheris, M., & Paparistodemou, E. (2015). Developing students' reasoning about samples and sampling in the context of informal inference. *Educational Studies in Mathematics, 88*(3), 385–404. https://doi.org/10.1007/s10649-014-9551-5

Meng, X.-L. (2021). Enhancing (publications on) data quality: Deeper data minding and fuller data confession. *Journal of the Royal Statistical Society: Series A (Statistics in Society), 184*(4), 1161–1175. https://doi.org/10.1111/rssa.12762

Mokros, J., & Russell, S. J. (1995). Children's concepts of average and representativeness. *Journal for Research in Mathematics Education, 26*(1), 20–39. https://doi.org/10.2307/749226

Mooney, E., Duni, D., van Meenen, E., & Langrall, C. (2014). Preservice teachers' awareness of variability. In K. Makar, B. de Sousa, & R. Gould (Eds.), *Sustainability in statistics education. Proceedings of the 9th International Conference on Teaching Statistics, Flagstaff, AZ, USA.* International Statistical Institute.

National Academies of Sciences, Engineering, and Medicine (NASEM). (2018). *Data science for undergraduates: Opportunities and options.* The National Academies Press. https://doi.org/10.17226/25104

National Academies of Sciences, Engineering, and Medicine (NASEM). (2020). *Roundtable on data science postsecondary education: A compilation of meeting highlights.* The National Academies Press. https://doi.org/10.17226/25804

National Council of Teachers of Mathematics (NCTM). (2020a). *Standards for the preparation of middle level mathematics teachers.* Author.

National Council of Teachers of Mathematics (NCTM). (2020b). *Standards for the preparation of secondary mathematics teachers.* Author.

National Governors Association Center for Best Practices & Council of Chief State School Officers. (2010). *Common core state standards for mathematics.* Author.

Oceans of Data Institute. (2016). *Building global interest in data literacy: A dialogue. Workshop report.* Education Development Center. http://oceansofdata.org/our-work/building-global-interest-data-literacy-dialogue-workshop-report

Oregon Department of Education. (2021, October). *2021 Oregon mathematics content standards adopted version 5.2.2.* Retrieved June 6, 2022 from https://www.oregon.gov/ode/educator-resources/standards/mathematics/Documents/2021OregonMathStandards.pdf

Oregon Department of Education (ODE). (2022, March). *Oregon mathematics guidance document—guidance version 5.2.5.* Retrieved June 6, 2022 from https://www.oregon.gov/ode/educator-resources/standards/mathematics/Documents/K12FullVersionwithGuidance.pdf

Osana, H. P., Leath, E. P., & Thompson, S. E. (2004). Improving evidential argumentation through statistical sampling: Evaluating the effects of a classroom intervention for at-risk 7th-graders. *Journal of Mathematical Behavior, 23*(3), 351–370. https://doi.org/10.1016/j.jmathb.2004.06.005

Paparistodemou, E., & Meletiou-Mavrotheris, M. (2008). Developing young students' informal inference skills in data analysis. *Statistics Education Research Journal, 7*(2), 83–106. https://doi.org/10.52041/serj.v7i2.471

Peters, S. A. (2011). Robust understandings of variation. *Statistics Education Research Journal, 10*(1), 52–88. https://doi.org/10.52041/serj.v10i1.367

Peters, S. A. (2014). Developing understanding of statistical variation: Secondary statistics teachers' perceptions and recollections of learning factors. *Journal of Mathematics Teacher Education, 17,* 539–582. https://doi.org/10.1007/s10857-013-9242-7

Peters, S. A. (2018). Professional development to transform middle and high school teachers' knowledge about distribution. In M. A. Sorto, A. White, & L. Guyot (Eds.), *Looking back, looking forward. Proceedings of the tenth International Conference on Teaching Statistics (ICOTS-10, July, 2018)*, Kyoto, Japan. International Statistical Institute.

Peters, S. A., & Stokes-Levine, A. (2019). Teacher learning: Measures of variation. In G. Burrill & D. Ben-Zvi (Eds.), *Topics and trends in current statistics education research: International perspectives* (pp. 245–264). Springer. https://doi.org/10.1007/978-3-030-03472-6

Peters, S. A., Watkins, J. D. , & Bennett, V. M. (2014). Middle and high school teachers' transformative learning of center. In. K. Makar, B. de Sousa, & R. Gould (Eds.), *Sustainability in statistics education. Proceedings of the 9th International Conference on Teaching Statistics (ICOTS-9), Flagstaff, AZ*. International Statistical Institute.

Petrosino, A. J., Lehrer, R., & Schauble, L. (2003). Structuring error and experimental variation as distribution in the fourth grade. *Mathematical Thinking and Learning, 5*(2–3), 131–156. https://doi.org/10.1080/10986065.2003.9679997

Pfannkuch, M. (2008). Training teachers to develop statistical thinking. In C. Batanero, G. Burrill, C. Reading, & A. Rossman (Eds.), *Joint ICMI/IASE Study: Teaching statistics in school mathematics. Challenges for teaching and teacher education. Proceedings of the ICMI Study 18 and 2008 IASE Round Table Conference*. International Commission on Mathematical Instruction and International Association for Statistical Education.

Pfannkuch, M., Ben-Zvi, D., & Budgett, S. (2018). Innovations in statistical modeling to connect data, chance and context. *ZDM, 50*(7), 1113–1123. https://doi.org/10.1007/s11858-018-0989-2

Pfannkuch, M., Budgett, S., & Arnold, P. (2015). Experiment-to-causation inference: Understanding causality in a probabilistic setting. In A. Zieffler & E. Fry (Eds.), *Reasoning about uncertainty: Learning and teaching informal inferential reasoning* (pp. 95–127). Catalyst Press.

Pierce, R., & Chick, H. (2013). Workplace statistical literacy for teachers: Interpreting box plots. *Mathematics Education Research Journal, 25*(2), 189–205. https://doi.org/10.1007/s13394-012-0046-3

Pollatsek, A., Lima, S., & Well, A. D. (1981). Concept or computation: Students' understanding of the mean. *Educational Studies in Mathematics, 12*(2), 191–204. https://doi.org/10.1007/BF00305621

Quintas, S., Ferreira, R. T., & Oliveira, H. (2014). Attending to students' thinking on bivariate statistical data at secondary level: Two teachers' pedagogical content knowledge. In K. Makar, B. de Sousa, & R. Gould (Eds.), *Sustainability in statistics education. Proceedings of the 9th International Conference on Teaching Statistics, Flagstaff, AZ, USA*. International Statistical Institute.

Reading, C., & Shaughnessy, J. M. (2004). Reasoning about variation. In D. Ben-Zvi & J. Garfield (Eds.), *The challenge of developing statistical literacy, reasoning, and thinking* (pp. 201–226). Kluwer. https://doi.org/10.1007/1-4020-2278-6_9

Ridgway, J., & McCusker, S. (2003). Using computers to assess new educational goals. *Assessment in Education, 10*(3), 309–328. https://doi.org/10.1080/0969594032000148163

Ridgway, J., Nicholson, J., & McCusker, S. (2007). Reasoning with multivariate evidence. *International Electronic Journal of Mathematics Education, 2*(3), 245–269. https://www.iejme.com/download/reasoning-with-multivariate-evidence.pdf

Rubin, A. (2021). What to consider when we consider data. *Teaching Statistics, 43*(S1), S23–S33. https://doi.org/10.1111/test.12275

Rubin, A., Bruce, B., & Tenney, Y. (1990). *Learning about sampling: Trouble at the core of statistics*. In D. Vere-Jones (Ed.), *Proceedings of the third International Conference on Teaching Statistics* (Vol. 1, pp. 314–319). International Statistical Institute.

Rubin, A., & Erickson, T. (2018). Tools, best practices, and research-based reminders. In R. Biehler, L. Budde, D. Frischemeier, B. Heinemann, S. Podworny, C. Schulte,

& T. Wassong (Eds.), *Paderborn symposium on data science education at school level 2017: The collected extended abstracts* (pp. 103–106). Universitätsbibliothek Paderborn. https://doi.org/10.17619/UNIPB/1-374

Rubin, A., & Mokros, J. (2018). Data clubs for middle school youth: Engaging young people in data science. In M. A. Sorto, A. White, & L. Guyot (Eds.), *Looking back, looking forward. Proceedings of the tenth International Conference on Teaching Statistics (ICOTS10, July, 2018), Kyoto, Japan.* International Statistical Institute.

Rubin, A., & Mokros, J. (2019). *Designing and exploring a model for data science learning for middle school youth. One-page project description for NSF.* Retrieved January 12, 2021 from https://drive.google.com/file/d/1cLhoPIuHHJ_7AbQgqalQ4U07 ns2qcTGb/view

Saldanha, L., & McAllister, M. (2014). Using re-sampling and sampling variability in an applied context as a basis for making statistical inferences with confidence. In K. Makar, B. deSousa, & R. Gould (Eds.), *Sustainability in statistics education (Proceedings of the 9th International Conference on the Teaching of Statistics, Flagstaff, Arizona, July 13–18).* International Statistical Institute.

Saldanha, L., & Thompson, P. W. (2002). Conceptions of sample and their relationship to statistical inference. *Educational Studies in Mathematics, 51*(3), 257–270. https://doi.org/10.1023/A:1023692604014

Schanzer, E., Fisler, K., & Krishnamurthi, S. (2017). *Bootstrap: Data science.* https://bootstrapworld.org/materials/spring2021/en-us/courses/data-science/

Sentance, S. (2018). Data science and data literacy in schools: Opportunities and challenges. In R. Biehler, L. Budde, D. Frischemeier, B. Heinemann, S. Podworny, C. Schulte, & T. Wassong (Eds.), *Paderborn symposium on data science education at school level 2017: The collected extended abstracts* (pp. 84–89). Universitätsbibliothek Paderborn. https://doi.org/10.17619/UNIPB/1-374

Silva, C. B., & Coutinho, C. Q. S. (2006). The variation concept: A study with secondary school mathematics teachers. In A. Rossman & B. Chance (Eds.), *Proceedings of the seventh International Conference on Teaching Statistics (ICOTS-7), Salvador, Brazil* [CD-ROM]. International Statistical Institute.

Silva, C. B., & Coutinho, C. Q. S. (2008). Reasoning about variation of a univariate distribution: A study with secondary math teachers. In C. Batanero, G. Burrill, C. Reading, & A. Rossman (Eds.), *Joint ICMI/IASE study: Teaching statistics in school mathematics. Challenges for teaching and teacher education. Proceedings of the ICMI Study 18 and 2008 IASE Round Table Conference.* International Commission on Mathematical Instruction and the International Association for Statistical Education.

Sorto, M. A. (2004). *Prospective middle school teachers' knowledge about data analysis and its applications to teaching* [Unpublished doctoral dissertation]. Michigan State University.

TERC. (2021, June 24). *Modules.* DataClubs. https://www.terc.edu/dataclubs/modules/

Ubrilla, F. M., & Gorgorió, N. (2020). From a source of real data to a brief news report: Introducing first-year preservice teachers to the basic cycle of learning from data. *Teaching Statistics, 43*(S1), S110–S123. https://doi.org/10.1111/test.12246

Watson, J. M. (1997). Assessing statistical literacy through the use of media surveys. In I. Gal & J. Garfield (Eds.), *The assessment challenge in statistics* (pp. 197–201). IOS Press.

Watson, J. M. (2001). Profiling teachers' competence and confidence to teach particular mathematics topics: The case of chance and data. *Journal of Mathematics Teacher Education, 4*(1), 305–337. https://doi.org/10.1023/A:1013383110860

Watson, J. M., & English, L. D. (2016). Repeated random sampling in year 5. *Journal of Statistics Education, 24*(1). https://doi.org/10.1080/10691898.2016.1158026

Watson, J. M., Fitzallen, N. Fielding-Wells, J., & Madden, S. (2018). The practice of statistics. In D. Ben-Zvi, K. Makar, & J. Garfield (Eds.), *International handbook of research in statistics education* (pp. 105–138). Springer. https://doi.org/10.1007/978-3-319-66195-7

Watson, J. M., & Moritz, J. B. (2000). Developing concepts of sampling. *Journal for Research in Mathematics Education, 31*(1), 44–70. https://doi.org/10.2307/749819

Wild, C. J., & Pfannkuch, M. (1999). Statistical thinking in empirical enquiry. *International Statistical Review, 67*(3), 223–265. https://doi.org/10.1111/j.1751-5823.1999.tb00442.x

Wild, C. J., Utts, J. M., & Horton, N. J. (2018). What is statistics? In D. Ben-Zvi., K. Makar, & J. Garfield (Eds.), *International handbook of research in statistics education* (pp. 5–36). Springer. https://doi.org/10.1007/978-3-319-66195-7_1

Wilkerson, M. H., Lanouettte, K., & Shareff, R. L. (2021). Exploring variability during data preparation: A way to connect data, chance, and context when working with complex public datasets. *Mathematical Thinking and Learning, 24*(4), 312–330. https://doi.org/10.1080/10986065.2021.1922838

youcubed. (n.d.-a). *Data science lessons*. Youcubed. https://www.youcubed.org/data-science-lessons/

youcubed. (n.d.-b). *Explorations in data science: Adaptable, project-based curriculum for high school*. Youcubed. https://hsdatascience.youcubed.org/

Zapata-Cardona, L., & Martinez-Castro, C. A. (2023). Statistical modeling in teacher education. *Mathematical Thinking and Learning, 25*(1), 64–78. https://doi.org/10.1080/10986065.2021.1922859

CHAPTER 8

PREPARING TEACHERS OF STATISTICS

A Critical Read of Standards, Review of Past Research, and Future Directions

Travis Weiland
University of Houston

Christopher Engledowl
University of Florida

Susan O. Cannon
University of Georgia

In this chapter we consider the past, present, and future of the statistics education of K–12 mathematics teachers. We consider the past and present in terms of what we know about the preparation and support of K–12 mathematics teachers for teaching statistics from research and current policies. We consider the future in terms of important directions of research and mathematics teacher education aimed at improving the preparation and support of K–12 mathematics teachers for teaching statistics. Based on our review of the literature, we found a dearth of research on how to prepare teachers of statistics. In particular, we documented a lack of research

The AMTE Handbook of Mathematics Teacher Education: Reflection on Past, Present and Future—Paving the Way for the Future of Mathematics Teacher Education, Volume 5
pages 151–174.

on the social contexts of statistics teaching and learning and students as learners of statistics. Based on these findings, we provide questions to drive future research in this area. We further implore mathematics teacher educators (MTEs), mathematics and statistics education researchers, and teacher education organizations to prioritize attention, professional development, and scholarship in this area.

INTRODUCTION

The current post-truth discourses undermining the democratic ethics and pillars of our society highlight the dire need for all citizens to be statistically literate (i.e., able to critically engage in, construct, and critique arguments with data). Furthermore, data collection and analysis are inherent in daily life, such as in popular devices like fitbits or smart watches that collect data on your movement, websites that collect information on browsing activity, or the reading of data visualizations in social media posts. We are constantly a participant in data investigations (Raz, 2016). Because a primary goal of public education is to prepare students to be critical citizens in our democratic society (Labaree, 1997), developing students' statistical literacy is crucial for K–12 education (Bargagliotti et al., 2020; Franklin et al., 2007; Weiland, 2017). Statistics has been a part of the K–12 mathematics curriculum for many decades now (Scheaffer & Jacobbe, 2014), yet serious consideration of how to prepare and support teachers of mathematics to teach statistics has really only begun in the 21st century (Franklin et al., 2015).

In the *Standards for Preparing Teachers of Mathematics (SPTM)*, the Association of Mathematics Teacher Educators (AMTE, 2017) acknowledged the importance of including statistics concepts and practices in preparing teachers of mathematics with statements such as, "Well-prepared beginning teachers of mathematics possess robust knowledge of mathematical and statistical concepts that underlie what they encounter in teaching. They engage in appropriate mathematical and statistical practices and support their students in doing the same" (p.6). Such statements make clear the importance of mathematical AND statistical concepts and practices that demarcate the disciplines as distinct. Despite such positioning, there is little mention of research about how to prepare and support mathematics teachers' preparation to teach statistics in relevant handbooks and handbook chapters (e.g. Ben-Zvi et al., 2018; Langrall et al., 2017; Shaughnessy, 2007). The lack of clarity in the research on *how* to support mathematics teachers' preparation to teach statistics is further exacerbated by the explosion of data science and its infusion into the K–12 curriculum.

OBJECTIVES AND GOALS

In this chapter our goal is to consider the **past and present** in terms of what we know about the preparation and support of K–12 mathematics teachers for teaching statistics from research and current policies. We consider the **future** in terms of important directions of research and mathematics teacher education aimed at improving the preparation and support of K–12 mathematics teachers for teaching statistics. Based on these goals from this point forward, when we discuss teaching statistics, teachers of statistics (ToS), or supporting teachers, we are referring to K–12 mathematics teachers. Towards our goals, we address the following questions:

1. What do policies relative to mathematics teacher education recommend about the preparation and support of ToS?
2. What do we know from past research about how to prepare and support ToS?
3. What future research is needed in order to better prepare ToS?
4. What are possible future directions for mathematics teacher education in regard to preparing and supporting ToS?

We will begin by analyzing the current policy documents that provide recommendations for preparing and supporting ToS. We will use a synthesis of the findings of our policy analysis to frame a systematic review of the research literature on preparing and supporting ToS. Based on our review of the literature and analysis of the policy documents, we will identify areas of need to propose future directions, taking into consideration current trends and issues.

POLICY RECOMMENDATIONS

Our goal in this section is to address our first question: *What do policies relative to mathematics teacher education recommend about the preparation and support of ToS?* Policy recommendations differ by location. In this section, we present some of the major policy recommendations made by professional organizations such as the Conference Board of Mathematical Sciences (CBMS), American Statistical Association (ASA), and the Association of Mathematics Teacher Educators (AMTE), which are organizations with international reach and impact, but are still firmly situated in the K–12 education context of the United States (U.S.). The issue of preparing teachers to teach statistics was not included in policy documents until 2012 when the Mathematical Education of Teachers II (MET II) report (Conference Board of the Mathematical Sciences, 2012) released recommendations on how to prepare mathematics teachers to teach statistics. In the MET II, recommendations were made as to the main understandings a mathematics teacher should have in measurement and data for elementary (K–5) pre-service teachers, and statistics and probability for middle grades (6–8) and high school (9–12) pre-service teachers. The MET II also provided some course recommendations for pre-service teacher programs and professional development, but they provided little discussion of scope or sequence of those courses.

Building from the momentum created by the first Guidelines for Assessment and Instruction in Statistics Education (GAISE) Framework (Franklin et al., 2007) and the MET II (CBMS, 2012), the ASA sponsored the Statistical Education of Teachers (SET) report provided detailed recommendations (for summary see

TABLE 8.1. Summary of Six Major Recommendations for the Statistical Education of Teachers.

Recommendation 1	"Prospective teachers need to learn statistics in ways that enable them to develop a deep conceptual understanding of the statistics they will teach." (p.5)
Recommendation 2	"Prospective teachers should engage in the statistical problem-solving process— formulate statistical questions, collect data, analyze data, and interpret results—regularly in their courses." (p.5)
Recommendation 3	"Because many currently practicing teachers did not have an opportunity to learn statistics during their pre-service preparation programs, robust professional development opportunities need to be developed for advancing in-service teachers' understanding of statistics." (p.6)
Recommendation 4	"All courses and professional development experiences for statistics teachers should allow them to develop the habits of mind of a statistical thinker and problem-solver, such as reasoning, explaining, modeling, seeing structure, and generalizing." (p.6)
Recommendation 5	"At institutions that prepare teachers or offer professional development, statistics teacher education must be recognized as an important part of a department's mission and should be undertaken in collaboration with faculty from statistics education, mathematics education, statistics, and Mathematics." (p.6)
Recommendation 6	"Statisticians should recognize the need for improving statistics teaching at all levels." (p. 6)

Table 8.1) on preparing and supporting ToS. For preparing elementary teachers, the SET recommended that institutions provide a special section of introductory statistics for future teachers, a course specifically for teachers, or relevant statistics content infused into existing mathematics content courses. For middle grades, the SET recommends two courses: one, a modern introduction to statistics course with an emphasis on simulation-based approaches, and two, a course focused on the middle grades, similar to the content in *Essential Understandings* (Kader & Jacobbe, 2013). Three statistics courses are suggested for high school teachers: 1) a modern introduction to statistics course, 2) a course building from the introduction course to include randomization and classical procedures for comparing two parameters, basic principles of the design and analysis of sample surveys and experiments, inference in simple linear regression, and tests of independence/homogeneity for categorical data, and 3) a course on statistical modeling with multiple regression techniques, exponential and power models, models for analyzing designed experiments, and logistic regression models. The SET report also provides a detailed list of key understandings for teachers at each grade level, including discussion of the key difference between mathematics and statistics. Furthermore, the report provides a synthesis of prior research on the teaching and learning of statistics in K–12 schools including a review of research on teacher preparation, which is scant. It is of note that there is some overlap of this literature review and

our own in terms of time frame, however, the review in the SET report is not described as systematic like ours. We highly recommend MTEs review the SET report and incorporate its recommendations into their teacher education programs.

The SPTM (AMTE, 2017) provided a much-needed update for the field around what type of preparation K–12 teachers need after the large shifts to state standards that took place after CCSS-M (NGAC & CCSSO, 2010). In particular, the SPTM laid out four major standard clusters with an associated broad list of indicators that provided, for the first time, inclusion of the social contexts of teaching and learning. Within the scope of these four standard clusters there is specific mention of probability and statistics as an important domain for teacher content and pedagogical content knowledge.

Specifically, for early childhood (Pre-K–2), AMTE (2017) calls for beginning teachers to "understand the foundations of statistical reasoning" (p. 52), especially focusing on the investigative statistical problem-solving process—as described in the GAISE reports (Bargagliotti et al., 2020; Franklin et al., 2007) and the SET report (Franklin et al., 2015)—which includes formulating a statistical question, collecting data, analyzing data, and interpreting results. Moreover, the SPTM connects these statistical ideas to number and measurement concepts that are focal across Pre-K–2. Similarly, across grades 3–5, the SPTM leans on the MET II report, recommending use of "data displays to ask and answer questions about data" (p. 80), and experience with measures of center and spread to summarize data.

For preparing middle grades ToS, the SPTM relies heavily upon the SET report, MET II report, and the types of experiences recommended in the GAISE report (Franklin et al., 2007), especially the statistical problem-solving process. Moreover, echoing the expectations described in CCSS-M, the SPTM describes a focus on variability, appropriate selection and use of data displays, comparing samples, drawing informal inferences about a population, and the importance of simulations and experiments to support student learning.

At the high school level, the SPTM provides more detail about the important experiences that teachers should have, including further explanation of the importance of recognizing the distinct differences between mathematics and statistics—an understanding that should be a focus in elementary and middle grades as well. As with the middle grades, the SPTM relies heavily upon the SET report, including a major focus on simulations and statistical modeling using statistics-based technological tools to answer real-world questions with relevant implications for students.

CURRENT RESEARCH ON PREPARING MATHEMATICS TEACHERS TO TEACH STATISTICS & PROBABILITY

In this section we will address our second question: *What do we know about how to prepare and support ToS?* To do so, we conducted a systematic literature review, and in the next section, we describe the method followed by our findings.

Method

Systematic reviews of literature can "shape a view or vision of the state of a field of research" (Murphey et al., 2017, p. 3). The field of statistics teacher education is in the process of finding its place within mathematics education research. In order to address the critical questions (Alexander, 2020) we raised above and to consider the past and future directions of the field of statistics teacher education, we conducted a systematic review of peer-reviewed articles on the statistics education of teachers using the ERIC search database (https://eric.ed.gov/) in July of 2021. We limited our search to articles published in the last ten years—the most recent collection of research on statistics teacher education was compiled in a book edited by Batanero et al. (2011). Our initial search terms and results can be seen in Table 8.2.

From the initial search results, we each took a set of results and did an initial filtering of the results delimiting them to include only mathematics and statistics *education* journals in order to retain a focus on the research on statistics *within* the fields of mathematics and statistics education. In particular, we only kept publications in journals that would reasonably be seen by MTEs, drawing from past research on journal quality in mathematics education (Williams & Leatham, 2017). We also read the abstracts of these articles and excluded any articles that clearly had no relation to the education of mathematics teachers. As a note, no articles that came up in our search for data science in teacher education or preparation made it through this filtering of the data pointing to an area of need in terms of future research. We recorded all the pertinent reference information on each article that met the criterion discussed as well as the abstract and keywords provided by ERIC. Each of us then examined the remaining articles and collectively discussed and determined whether to keep each article based on our previously agreed upon inclusion criteria. After two rounds of filtering, 44 articles remained. Additionally, we added one article from *Mathematics Teacher Educator* that did not come up in our search but was important to include as it is published in the joint AMTE and NCTM flagship journal. Our final analysis then consisted of 45 articles (41 are included in Tables 8.6 and 8.7). We then divided these articles among us and extracted the following: research questions or aims, major findings/contributions,

TABLE 8.2.   Initial Search Terms and Number of Results.

Search Terms	Number of Results
"Statistics" and "Teacher Education"	859
"Statistics" and "Teacher Preparation"	77
"Probability" and "Teacher Education"	92
"Probability" and "Teacher Preparation"	12
"Data science" and "teacher education"	26
"Data science" and "teacher preparations"	3

grade band (K–5, 6–8, 9–12), and preservice or inservice teacher education. We then coded each article for the SPMT (AMTE, 2017) standard it addressed: C.1. Mathematics Concepts, Practices, and Curriculum; C.2. Pedagogical Knowledge and Practices for Teaching Mathematics; C.3. Students as Learners of Mathematics; and C.4. Social Contexts of Mathematics Teaching and Learning

Findings

To begin our findings section, we start by briefly describing where the articles were published before delving into their substance in terms of what audiences they focused on (i.e., preservice or inservice), grade bands, and which SPTM(s) they addressed. Next, we synthesize the major findings in the research, grouped by the SPMTs. Table 8.3 shows the journals the 45 articles were published in that met our final inclusion criteria. Not surprisingly, the most frequent journals were the two top research journals for statistics education. The Journal of Mathematics Teacher Education (JMTE) was also present towards the top as one of the top journals for research in mathematics teacher education. Notably, the *Mathematics Teacher Educator* (MTE) journal only had one article related to statistics education even though the SPTM include the preparation of teachers to teach statistics. However, there

TABLE 8.3. Number of Publications per Journal

Journal Name	Number of Publications
Statistics Education Research Journal (SERJ)	12
Journal of Statistics Education (JSE)	7
Journal of Mathematics Teacher Education (JMTE)	4
International Journal of Mathematical Education in Science and Technology (IJMEST)	4
International Journal of Science and Mathematics Education (IJSME)	4
Contemporary Issues in Technology and Teacher Education (CITE)	3
Proceedings of the Annual Meeting of the North American Chapter of the International Group for the Psychology of Mathematics Education (PME-NA)	2
Mathematics Education Research Journal (MERJ)	2
ZDM	2
Canadian Journal of Science, Mathematics and Technology Education (CJSMTE)	1
Educational Studies in Mathematics (ESM)	1
Investigations in Mathematics Learning (IML)	1
Problems, Resources, and Issues in Mathematics Undergraduate Studies (PRIMUS)	1
Mathematics Teacher Educator	1

TABLE 8.4. Number of Publications per Year

2012	2013	2014	2015	2016	2017	2018	2019	2020	2021
5	7	9	1	2	5	3	6	5	2

TABLE 8.5. Counts and Proportions of Publications Addressing Each of the Standards for Preparation of Teachers of Mathematics

Standard	Number of Publications	Proportion of Publications
C.1. Mathematics Concepts, Practices, and Curriculum	34	76%
C.2. Pedagogical Knowledge and Practices for Teaching Mathematics	14	31%
C.3. Students as Learners of Mathematics	5	11%
C.4. Social Contexts of Mathematics Teaching and Learning	1	2%
Mathematics Teacher Educator Specific	7	16%

were several articles in the journal CITE-Math, which is sponsored by AMTE, and was included despite not making Williams and Leatham's (2017) list because of AMTE's role in its publication. Table 8.4 displays when the articles that were included in our review were published. There is a notable peak during 2013–2014 that has not been repeated since. Overall, the trend is that only a handful of manuscripts addressing the statistical education of teachers were published each year in journals MTEs would likely read. We also found a predominant focus on studying preservice teachers, with 25 (57%) articles focused on that audience, 17 (39%) focused on inservice teachers, and two (5%) focused on both.

To dig deeper into which aspects of the preparation of ToS the articles investigated, we categorized each by which SPTM (AMTE, 2017) they addressed. Table 8.5 shows an overview of those results. The vast majority of articles focus on standard C.1. with statistics content knowledge at the core of most of those studies. A strong second was standard C.2., which addresses the pedagogical considerations of teaching statistics, with a substantial number of those articles addressing both C.1. and C.2. What is notably absent is research around standards C.3. or C.4., which address students as learners of mathematics and social contexts of mathematics teaching and learning, respectively. In the sections that follow, we elaborate on the major findings from the research in our review, grouped by the SPMTs to make direct connections to policy.

Synthesis of Research on Mathematics Concepts, Practices, and Curriculum

Across all the studies included in our review, 33 had at least one component that was coded as related to AMTE (2017) standard C.1. Of the 6 sub-standards of C.1., only *C.1.1. Know Relevant Mathematical Content* and *C.1.3. Exhibit Productive Mathematical Dispositions* were coded. Three studies were coded to be in both standards (de Souza, 2014; Hannigan et al., 2013; Pierce & Chick, 2013). Including these three in both counts, there were 8 studies coded as C.1.3. and 28 coded as C.1.1. (see Table 8.6).

Looking deeper within C.1.1, empirical investigations of knowledge cut across preservice teacher and inservice teacher populations (n = 14 and 12 respectively, with two overlapping) and one textbook analysis (Jones & Jacobbe, 2014). Methodologically, there were both qualitative and quantitative studies. Quantitative studies included between about 100 to 300 participants and included methods of ANOVA, multiple linear regression, t-tests, chi-square tests, and descriptives. Qualitative studies included between 4 and 290 participants and included methods of document analysis, interviews, and case studies. A few mixed studies were also conducted, drawing on regression, chi-square, descriptives, document analysis, and interviews, with between about 80 and 2000 participants. The majority of studies examined topics of center, spread, shape and visualizations (n = 8), broader assessments of statistics knowledge (e.g., using the Comprehensive Assessment of Outcomes in a First Statistics course (CAOS) or the Levels of Conceptual Understanding of Statistics (LOCUS); n = 4), and using TinkerPlots to examine the investigative cycle or random sampling (n = 3), with other topics involving the examination of z-scores and p-value interpretations, informal statistical inference, asking statistical questions, probabilistic fairness, nominal categorical data analysis, association between knowledge and attitude, the statistical learning environment, broadly looking at preservice teacher coursework as preparation for

TABLE 8.6. References Coded for Sub-standards of C.1: Mathematics Concepts, Practices, and Curriculum

C.1.1.: Content		C.1.3.: Dispositions
Bansilal (2014)	Hourigan (2020)	de Souza (2014)*
Biehler et al. (2018)	Jacobbe (2012)	Hannigan et al. (2013)*
Casey et al. (2018)	Jones (2014)	Harrell-Williams et al. (2015)
Casey et al. (2020)	Leavy (2014)	Leavy (2021)
Chick & Pierce (2011)	Lee et al. (2012)	Martins et al. (2012)
de Souza (2014)*	Lee et al. (2013)	Pierce & Chick (2013)*
de Vetten et al. (2018)	Lee et al. (2014)	Whitaker (2016)
Engledowl & Tarr (2020)	Lovett & Lee (2018)	Yang (2014)
Frischemeier & Biehler (2018)	Magalhaes & Magalhaes (2014)	
Green et al. (2018)	North et al. (2014)	
Groth (2012)	Odom & Bell (2017)	
Groth (2013)	Olson (2016)	
Groth (2014)	Peters (2013)	
Hannigan et al. (2013)*	Pierce & Chick (2013)*	
Hobden (2014)		

*Note: * indicates inclusion in both standards

teaching statistics, and building capacity to teach statistics. A few others were theoretical or reflections/commentaries.

Looking deeper within C.1.3, empirical investigations of beliefs, attitudes, and dispositions cut across preservice and inservice teacher populations ($n = 3$ and 5, respectively). Only two qualitative studies were conducted, one with 12 participants (Whitaker, 2016), which drew on interviews, and one with 175 participants (Martins et al., 2012), which drew upon open-ended survey responses. Quantitative studies included between about 130 and about 300 students and used methods such as ANOVA, t-tests, and Rasch Analysis. One mixed methods study with about 2000 students drew upon chi-square and focus group interviews (and was also included in C.1.1). These studies examined self-efficacy, disposition towards boxplots, statistics teacher identity development, change in attitudes across a course, teacher reasons for responses on an attitude survey, and the relationship between knowledge and attitudes.

Synthesis of Research on Pedagogical Knowledge and Practices for Teaching Mathematics

As the standards document states, "well-prepared beginners realize that the teaching of mathematics has its own nuances and complexities specific to the discipline" (https://amte.net/node/2257, para 2), so too does the teaching of statistics. Within our review, there were 12 manuscripts related to pedagogical knowledge and practices for teaching statistics (see Table 8.7). The majority of these studies were evenly distributed between C2.2., planning for effective instruction, and C.2.4, analyzing teaching practice. Notably, there were no investigations that directly explored C.2.1, priming equitable teaching. All of the investigations in this section used qualitative methods, including narrative inquiry, lesson study, video analysis and document analysis. There were 7 studies involving primary settings and four in middle or secondary settings, with one study spanning K–12.

TABLE 8.7. References Coded for Sub-standards of C.2.: Pedagogical Knowledge and Practices for Teaching Mathematics

C.2.2. Plan for Effective Instruction	C.2.3. Implement Effective Instruction	C.2.4. Analyze Teaching Practice	C.2.5. Enhance Teaching Through Collaboration
Chick et al., (2011)*	Casey et. al. (2018)	Chick et al. (2011)*	de Souza (2014)*
Casey et al. (2020)	Groth (2014)	de Souza (2014)*	Estrella (2019)*
Groth (2012)	Hourigan et al. (2020)	Estrella (2019)*	Grando et al. (2020)*
Groth et al. (2013)		Grando et al. (2020)*	Leavy (2014)*
Huey et al. (2018)		Groth (2017)	
Leavy (2014)*		Leavy (2014)*	
Lee et al. (2012)			

*Note: * indicates inclusion in more than one standard

The standard for planning for effective instruction includes a focus on clear understandings of content, and in particular learning goals, as the starting point of instructional design. Huey et al. (2018), in a study of the lesson plans of 16 teachers across 4 U.S institutions, found that the teachers' plans lacked important conceptual understandings about variation. In particular, related to the importance of learning goals, the teachers struggled with defining the purpose of teaching standard deviation. Chick and Pierce (2012), investigated Australian preservice teachers' lesson planning as well and considered how capable teachers were of incorporating real-world data into their lesson plans to build statistical literacy. They found that the majority of the teachers in the study designed lessons that were at low levels in the statistical literacy hierarchy and were more focused on data reading and graph making. In contrast to these examples, Casey et al. (2020) found promise in the use of ESTEEM materials to prepare preservice teachers to teach statistics. In their study of technology-enhanced task design three fourths of the 75 PSTs across 5 U.S. universities designed tasks that required students to engage in multiple phases of the statistical investigation cycle and to relate the data to the context from which it was generated multiple times in analysis. In another promising study, Leavy (2015) engaged 73 preservice primary teachers in Japanese lesson study of 3 years. She found that the teachers' attention to the pedagogical knowledge led to development of pure content knowledge. In developing and refining their lessons, teachers sought out additional pure statistical content knowledge.

The standard for implementing effective instruction focuses on teaching practices drawn from *Principles to Actions* (National Council of Teachers of Mathematics, 2014). Groth (2014) describes the use of a NAEP-based (National Assessment of Educational Progress) item on median and mean to assess prospective teachers' statistical knowledge for teaching. Groth considered the ways in which teachers maintained cognitive demand, encouraged metacognition, and provided opportunities for students to construct knowledge. Further, Casey et al. (2019) found that engaging teachers in constructing graphs resulted in the teachers improving their analyses of student work, noting the lack of frequencies of outcomes and missing axes labels in student work examples. In a more fine-grained consideration of teaching practices, Hourigan and Levy (2019) investigated the use of materials such as dice and spinners to construct activities exploring fairness. Through the study, they found that preservice teachers' (PSTs) understandings of fairness did not align with formal definitions to construct appropriate tasks for students to explore the concepts.

Several studies considered the ways in which preservice and inservice teachers analyze and reflect on their practice (C.2.4), and within this group, there were 4 studies that also addressed C.2.5 in that the participants analyzed practice through collaboration. Two studies engaged teachers in video analysis. Estrella et al. (2020) worked with teachers to consider the admittance (or not) of cognitive demand in lessons, and Groth (2017) worked to develop and refine learning trajectories. Leavy (2015) and de Souza et al. (2014) engaged teachers in designing lessons or instructional sequences and then refining them within small groups. Both found the refinement of lessons and resulting discussions led to increases in conceptual understanding and statistical content knowledge of teachers. Grando and Lopes (2020) engaged primary inservice teachers in Brazil in reflection on moments of creative insubordination within their practice and connected these moments to problematizing within statistical investigations.

Synthesis of Research on Students as Learners of Mathematics

Across the studies in our review, we found 3 articles that were coded as involving AMTE (2017) Standard *C.3 Students as Learners of Mathematics*. All 3 articles were coded as the sub-standard *C.3.1. Anticipate and Attend to Students' Thinking About Mathematics Content*, 2 were empirical qualitative investigations (Park & Lee, 2019; Shin, 2020), with Park and Lee (2019) focusing on elementary preservice teachers and Shin (2020) focusing on secondary preservice mathematics teachers, and Shin (2021) presented a framework for use in research.

In their investigation of the ways preservice elementary teachers in a Korean higher education institution identify whether students are exhibiting equiprobability bias and the strategies they use to correct it, Park and Lee (2019) reported that some of the PSTs offered strategies that included direct interaction with students "based on students' current understanding" while others offered strategies that included direct explanations and asking closed questions (p. 1601). In a similar study, Shin (2020) examined secondary level PSTs' noticing of students' understanding of mean and variability. The PSTs were enrolled in a statistics course for mathematics education students that was aligned to the GAISE report (Franklin et al., 2007) and included both content and pedagogy. Based on qualitative analysis of PSTs' work and interviews, Shin (2020) found that they struggled to notice differences in the sophistication of students' reasoning about variability, especially when there were unequal group sizes, but they were successful when it came to noticing differences in sophistication of students' reasoning about the mean.

Shin (2021) reported on the same overall project from Shin (2020) but proposed a framework "for understanding how secondary [preservice mathematics teachers] attend to salient statistical elements used by students and how they make sense of and decide how to respond to students' reasoning about variability" (p. 700). Given the paucity of studies identified in this domain that were included in our review, there is much work left to do in order to understand how to support preservice and inservice teachers in developing "foundational understandings of students' mathematical knowledge, skills, and dispositions" (AMTE, 2017, p. 6).

Research Synthesis on the Social Contexts of Mathematics Teaching and Learning

Only one article was found in our review that addressed standard C.4. related to the social contexts of statistics teaching and learning. In particular, the study focused on forms of creative insubordination taken up by two practicing teachers of mathematics in how they taught statistics to their students and the types of activities they implemented. The authors operationalized the term 'creative insubordination' to refer to "professionals acting counter to rules and directives when seeking to protect those to whom they provide services and in order to provide better conditions" (Grando & Lopes, 2020, p. 621). Working in the context of Brazil the authors found that the teachers they worked with were exhibiting actions of creative insubordination by, "proposing problems in statistics and probability, listening to the children and encouraging them to problematize and create strategies and procedures for solving statistical problems, in a perspective that encompasses their social practices" (p. 621). It is exciting to see such work however we need significantly more in this area to begin to understand how to prepare teachers.

Synthesis of Research on Mathematics Teacher Education

In addition to manuscripts that focused directly on knowledge and pedagogies of preservice or inservice teachers, our search included 7 manuscripts reporting on the work of statistics teacher education. This category of manuscripts included descriptions of curriculum for statistics courses involving: collaboration with mathematicians (Sears et al. 2019); implementation and redesign of curriculum across four U.S. institutions (Lee et al., 2013); focus on collaborative learning in Brazil (Magalhaes & Magalhaes, 2014); lessons learned through the design and implementation of a course for inservice teachers through their masters' program (Schmid et al., 2014); the design and review of statistical literacy modules for primary teachers (Bilgin et al., 2017); and the effects of a reform-oriented statistics course on the attitudes of PSTs toward statistics (Leavy et al., 2021). In addition Groth (2014) shared a theoretical framework to mapping pedagogical knowledge for teaching statistics to PSTs performance on assessment items, while Green et al. (2018) considered the ways that inservice middle grades teachers were able to incorporate statistical concepts learned in a methods course into their subsequent action research proposals. These collaborations and considerations of the design of statistics curriculum for teachers give hope for continued growth in statistics education research.

FUTURE RESEARCH NEEDED!

In this section we work to address our third question: *What future research is needed in order to better prepare ToS?* To begin, data science continues to gain momentum in K–12 education. Yet, from our search, there seems to be a dearth of research around how to prepare mathematics teachers to teach data science.

It could be argued that this work is more within the purview of computer science (CS) in the K–12 curriculum because of the amount of programming and data visualization commonly associated with CS. However, it is gaining support in states like California (see California Department of Education (CDE) et al., 2022) to make courses in data science count as mathematics courses on students' transcripts firmly placing the responsibility of data science curriculum on teachers of mathematics and, by extension, on MTEs. This brings up many important questions about teachers' preparation, especially when only 5% of high school teachers in the U.S. report feeling very well prepared to teach CS and only 31% to teach statistics and probability (Banilower et al., 2018). Although we commend the forward thinking of California in including data science in their mathematics curriculum, we wonder: Is California "putting the cart before the horse?" Perhaps first we should define, describe, and/or build a framework of data science in K–12 educational settings and carefully consider the connections to the K–12 mathematics standards. Arguably, data science still does not have a clear identity making it even more challenging to then prepare teachers to teach it.

Additionally, there needs to be significant conversation about where in the K–12 mathematics curriculum should data science go. GAISE II (Bargagliotti et al., 2020) and the California Frameworks (California Department of Education (CDE), Instructional Quality Commission (IQC), & State Board of Education (SBE); 2022) suggest threaded throughout, with the Framework also suggesting a course in high school. Oregon has taken steps to incorporate GAISE II recommendations but does not go so far as to deeply engage in incorporating data science though their unpacking documents mention creating a pathway to data science or creating a core course that would be half algebra/geometry and half data science/statistics. Georgia has created CS courses in data science, but it is still not present in the mathematics curriculum. Given the lack of research on this topic we have little understanding about how this works or looks in action. However, given that many states are jumping into such initiatives, there are significant issues to explore in research and practice around teacher education to meet these new demands on mathematics teachers. In an effort to jump-start research in data science teacher education, we pose the following questions for consideration:

- What concepts and practices do mathematics teachers need to understand to be prepared to teach data science in K–12 educational settings?
- What technological tools are pedagogically effective at supporting students' learning of data science concepts and practices?
- What technological and technological pedagogical content knowledge do mathematics teachers need to teach data science?
- How do teachers learn data science concepts and practices and how to teach them?
- How can MTEs prepare mathematics teachers to teach concepts and practices in effective and meaningful ways?

To effectively incorporate data science into mathematics teacher education curriculum, we also need to consider how these efforts integrate with and extend from our current efforts at preparing ToS so that this is not just another thing added onto an already overwhelming list of learning objectives for MTEs to incorporate into teacher education programs and experiences.

Though data science seems to be part of the future trajectory of the school mathematics curriculum, it is incumbent on us to point out that although statistics has been part of the mathematics curriculum for decades, there is still very little research on the statistical education of teachers—as shown in our systematic review of the literature over the past 10 years. In other words, there is a significant need for research on the preparation and support of ToS to teach the data related concepts and practices that have been in the official curriculum of many states for decades. Special issues in journals and calls at conferences for research associated with the preparation and support of ToS could help to draw more researchers into this area. Although there is a need for research across many areas in statistics teacher education, we point to specific areas of greatest need and make some suggestions for potentially fruitful directions in hopes of providing some focus for the field.

In returning to the SPTM, based on our review of the literature, a clear area of need is research related to the social contexts of mathematics teaching and learning (Standard C.4.). Study of how to prepare and support ToS in understanding the social, historical, and institutional contexts of statistics and how that affects teaching and learning was absent from the research included in our systematic review. Mathematics education began taking a sociopolitical turn a decade ago (Gutiérrez, 2013). However, statistics education has not made such a turn, particularly in the context of teacher education.

Because statistics is an overlapping, but distinct, discipline with mathematics (Cobb & Moore, 1997), there are differences in preparing teachers to teach statistics (Groth, 2007, 2015) and in the social, historical, and instructional contexts of statistics that teachers need to understand and incorporate into their teaching. For example, modern statistics has strong roots in the eugenics movement from the work of Francis Galton, Karl Pearson, and Ronald Fisher (Clayton, 2020). These men were quite prolific white supremacists and their work was frequently taken up to support Nazi justification of the systematic extermination of anyone deemed impure (Clayton, 2020; Parrinder, 1997). However, it is ubiquitous to see textbooks discussing Pearson's correlation coefficient without challenging why we still use Pearson's name in relation to this measure or include his picture prominently in sections on statistics. To help seed scholarship in researching the social contexts of statistics teaching and learning in teacher preparation, we pose the following questions for consideration:

- How should we prepare ToS to challenge disciplinary roots in white supremacy?

- How do we prepare ToS to problematize common disciplinary methods of measuring social constructs like race, gender, and ethnicity?
- How do we cultivate positive statistical identities in students and ToS?
- How can we draw upon students' statistical strengths?
- How can students' funds of knowledge be drawn upon to support the teaching and learning of statistics?
- How can ToS use statistics to advocate for, and with, their students?
- How can MTEs support teachers in providing opportunities for students to learn statistics in the mathematics curriculum?

We would also like to note that there are important connections to considering data science relative to this standard as big data, machine learning, and algorithmic predictive models are increasingly a part of how governments govern, and corporations sell products and have been shown time and again to have racist and sexist assumptions programed into them (O'Neil, 2016). This means considering the ethics of data science should be an important component of discussions within teacher education, along with ethical issues in statistics.

Regarding research on preparing and supporting ToS in understanding students as learners of mathematics AND statistics (Standard C.3.), there is more research in this area; however, there are still substantial gaps. This could be connected to the relative dearth of research on how students think and reason statistically—a point raised by others (Langrall et al., 2017; Shaughnessy, 2007). For example, statistical practices differ from mathematical ones, yet such differences have not been highlighted or investigated in research on the education of ToS. The SET report (Franklin et al., 2015) attempts to unpack such differences, but research is quite thin in regard to students' development of statistical practices—especially in the context of the first two stages of the statistical problem-solving process: formulating statistical questions and collecting/selecting data (Watson et al., 2018). Much of statistics education research does not even consider practices but focuses more on knowledge, beliefs, and attitudes. Another highly noticeable gap is that although teacher noticing research has become quite prevalent in mathematics education, there is a lack of this type of research in statistics education. To spark research in this area, we pose the following questions:

- What statistical practices should be incorporated into K–12 statistics education, and how do they differ from mathematical practices?
- What statistical dispositions are important for students to develop and how might ToS foster them?
- How do ToS notice students' statistical thinking and reasoning?
- How can MTEs develop and support ToS' noticing of students' statistical thinking and reasoning?

Regarding research on pedagogical knowledge and practices for teaching mathematics AND statistics (Standard C.2), this area has been studied more-so than the previous two, however, there is still no research on implementing equitable teaching practices, and very little on implementing effective instruction. It is of note that both Leavy (2015) and de Souza et al. (2014) found that attention to pedagogical knowledge and practices, in their studies, also built statistical content knowledge. There is also promising research (de Souza et al., 2014; Estrella et al., 2020; Grando & Lopez, 2020; Leavy, 2015) on connections between collaborative lesson or task study and increased pedagogical knowledge for teaching statistics. Given the limited research in statistics education for teachers and the remarkable gaps in all areas, these opportunities for connections across standards are important to explore in future research. To spark research in this area, we pose the following questions:

- How do the disciplinary differences between mathematics and statistics inform differences in the pedagogical knowledge and practices needed to teach statistics?
- How might statistics education build on the body of literature on pedagogical knowledge for teaching mathematics while acknowledging the distinct disciplines?
- How can equitable teaching be promoted in the teaching of statistics?
- What opportunities might statistics as a discipline offer in regard to equitable teaching practices that are unique from the ways that equitable teaching practices are being studied in mathematics education research?

Research on mathematics AND statistics concepts, practices, and curriculum was by far the most researched area in the past ten years. However, all of that work was focused either on knowing relevant content knowledge or exhibiting productive dispositions, which collectively only account for a third of the indicators for this standard. In our examination of the literature around *C.1. Know Relevant Mathematical Content* (AMTE, 2017), despite finding the largest number of studies coded as involving this standard ($n = 33$), there is still a lot of work needed in every aspect of this standard. Moreover, under the C.1. standard, there are 6 sub-standards, but we only found studies that related to C.1.1.—about mathematical and statistical knowledge—and C.1.3.—about mathematical and statistical dispositions. Even when looking at C.1.1., arguably the most researched area across the history of statistics education, the coverage of topics is quite thin and largely focused on measures of center, spread, shape of distribution, and common data visualizations such as dotplots, histograms, and boxplots, and some of these also involved use of dynamic software like TinkerPlots (Learn Troop, 2022) or CODAP (The Concord Consortium, 2020).

Quite surprisingly, only one study was found that involved informal statistical inference (ISI), only two appeared to have involved simulations, and only three appeared to have explicitly involved engagement in the statistical problem-solving process—all of which are essential for teachers to be prepared to teach and students to experience across all of K–12 (Bargagliotti et al., 2020; Franklin et al.,

2007, 2015). Another shocking finding was the lack of studies examining teacher beliefs, identities, dispositions, and self-efficacy (standard C.1.3.), with only 8 studies measuring it, and only one of which attempted to examine it in tandem with ToS' knowledge. Moreover, it appears as though only 4 of the 8 studies made use of an instrument with validity evidence to support its use—a point others have called for the field to give more attention to (Whitaker et al., 2021). We pose the following questions:

- What types of instruments does the field need to create and provide validity evidence for, to support investigations into preservice/inservice teachers' knowledge, engagement in statistical practices, critical analysis of curriculum materials, and of student work?
- What supports do preservice/inservice teachers need to fully engage in the statistical problem-solving process?
- What supports do preservice/inservice teachers need when making sense of, and making use of, simulations for understanding statistics and representing and analyzing data?
- What experiences support the development of positive and productive beliefs, attitudes, dispositions, self-efficacy, and identities toward statistics, and how are these related to the development of sophisticated forms of statistical knowledge?

FUTURE DIRECTIONS FOR MATHEMATICS TEACHER EDUCATION

In our final section, we draw upon our discussion from all the previous sections to respond to our fourth question: *What are possible future directions for mathematics teacher education in regard to preparing and supporting ToS?* To begin, the GAISE College Report (GAISE College Report ASA Revision Committee, 2016) and GAISE II Report (Bargagliotti et al., 2020)—which collectively make suggestions for what students should learn in Pre-K–16—were published after the writing of the SET report (Franklin et al., 2015) and were not consulted in the writing of the SPTMs (AMTE, 2017), and thus have new implications for the concepts and practices teachers will need to know to teach statistics. We also note that in our anecdotal experiences, many MTEs are not familiar with the GAISE or SET reports, which points to an area of need in terms of MTE professional development. Furthermore, the emergence of data science—and now data science education—have significant implications for the K–12 curriculum (Engel, 2017; Gould, 2010; Martinez & LaLonde, 2020) and speak to future directions of teacher education. For example, California has put in their revised curriculum framework a course in data science to count as a high school mathematics course (CDE, IQC, & SBE; 2022). This change is pending but provokes many questions: Who will teach such courses? How will mathematics teachers and educators prepare themselves and support other teachers to teach such courses? What skills must teachers master to teach such courses (e.g., Do they need to know how to code in R or Python (Muenchen, 2017)?) What concepts and practices must they become attuned to?)

As a field, statistics teacher education is small and has much room to grow. We encourage MTEs to partner with statistics teacher educators and researchers to begin to address the systemic issues in the preparation and support of ToS.

REFERENCES

Alexander, P. (2020). Methodological guidance paper: The art and science of quality systematic reviews. *Review of Educational Research, 90*(1), 6–23.

Association of Mathematics Teacher Educators. (2017). *Standards for the preparation of teachers of mathematics.* https://amte.net/standards

Bansilal, S. (2014). Using an APOS framework to understand teachers' responses to questions on the normal distribution. *Statistics Education Research Journal, 13*(2), 42–57. https://doi.org/10.52041/serj.v13i2.279

Banilower, E. R., Smith, P. S., Malzahn, K. A., Plumley, C. L., Gordon, E. M., & Hayes, M. L. (2018). *Report of the 2018 NSSME+.* Horizon Research, Inc. https://horizon-research.com/NSSME/wp-content/uploads/2020/04/Report_of_the_2018_NSSME.pdf

Bargagliotti, A., Franklin, C., Arnold, P., Gould, R., Johnson, S., Perez, L., & Spangler, D. (2020). *Pre-K–12 guidelines for assessment and instruction in statistics education II* (GAISE II) (2nd ed.). American Statistical Association.

Batanero, C., Burrill, G., & Reading, C. (Eds.). (2011). *Teaching statistics in school mathematics—Challenges for teaching and teacher education.* Springer.

Ben-Zvi, D., Makar, K., & Garfield, J. (Eds.). (2018). *International handbook of research in statistics education.* Springer.

Biehler, R., Frischemeier, D., & Podworny, S. (2018). Elementary preservice teachers' reasoning about statistical modeling in a civic statistics context. *ZDM, 50*, 1237–1251. https://doi.org/10.1007/s11858-018-1001-x

Bilgin, A. A. B., Date-Huxtable, E., Coady, C., Geiger, V., Cavanagh, M., Mulligan, J., & Petocz, P. (2017). Opening real science: Evaluation of an online module on statistical literacy for pre-service primary teachers. *Statistics Education Research Journal, 16*(1), 120–138. https://doi.org/10.52041/serj.v16i1.220

California Department of Education (CDE), Instructional Quality Commission (IQC), & State Board of Education (SBE). (2022). *Mathematics framework chapter 5 data science, TK–12: Second field review draft.* https://www.cde.ca.gov/ci/ma/cf/documents/mathfwchapter5sfr.docx

Casey, S. A., Albert, J., & Ross, A. (2019). Developing knowledge for teaching graphing of bivariate categorical data. *Journal of Statistics Education, 26*(3), 197–213. https://doi.org/10.1080/10691898.2018.1540915

Casey, S. A., Hudson, R., Harrison, T., Barker, H., & Draper, J. (2020). Preservice teachers' design of technology-enhanced statistical tasks. *Contemporary Issues in Technology and Teacher Education, 20*(2). https://citejournal.org/volume-20/issue-2-20/mathematics/preservice-teachers-design-of-technology-enhanced-statistical-tasks

Clayton, A. (2020). How eugenics shaped statistics: Exposing the damned lies of three science pioneers. *Nautilus*, 92. https://nautil.us/issue/92/frontiers/how-eugenics-shaped-statistics

The Concord Consortium. (2020). *Common online data analysis platform (CODAP)*. https://codap.concord.org/

Chick, H. L., & Pierce, R. (2012). Teaching for statistical literacy: Utilising affordances in real-world data. *International Journal of Science and Mathematics Education, 10*, 339–362. https://doi.org/10.1007/s10763-011-9303-2

Cobb, G. W., & Moore, D. S. (1997). Mathematics, statistics, and teaching. *The American Mathematical Monthly, 104*(9), 801–823.

Conference Board of the Mathematical Sciences. (2012). *The mathematical education of teachers II*. American Mathematical Society and Mathematical Association of America.

de Souza, A. C., Lopes, C. E., & de Oliveira, D. (2014). Stochastic education in childhood: Examining the learning of teachers and students. *Statistics Education Research Journal, 13*(2), 58–71. https://doi.org/10.52041/serj.v13i2.282

de Vetten, A., Schoonenboom, J., Keijzer, R., & van Oers, B. (2019). Pre-service primary school teachers' knowledge of informal statistical inference. *Journal of Mathematics Teacher Education, 22*, 639–661. https://doi.org/10.1007/s10857-018-9403-9

Engel, J. (2017). Statistical literacy for active citizenship: A call for data science education. Statistics Education Research Journal, 16(1), 44–49.

Engledowl, C., & Tarr, J. E. (2020). Secondary teachers' knowledge structures for measures of center, spread & shape of distribution supporting their statistical reasoning. *International Journal of Education in Mathematics, Science and Technology, 8*(2), 146–167. https://doi.org/10.46328/ijemst.v8i2.810

Estrella, S., Zakaryan, D., Olfos, R., & Espinoza, G. (2020). How teachers learn to maintain the cognitive demand of tasks through Lesson Study. *Journal of Mathematics Teacher Education, 23*, 293–310. https://doi.org/10.1007/s10857-018-09423-y

Franklin, C., Bargagliotti, A., Case, C., Kader, G., Scheaffer, R., & Spangler, D. (2015). Statistical education of teachers. American Statistical Association. http://www.amstat.org/education/SET/SET.pdf

Franklin, C., Kader, G., Mewborn, D., Moreno, J., Peck, R., Perry, M., & Scheaffer, R. (2007). Guidelines for assessment and instruction in statistics education (GAISE) report: A pre-K–12 curriculum framework. American Statistical Association.

Frischemeier, D., & Biehler, R. (2018). Preservice teachers comparing groups with Tinkerplots—An exploratory laboratory study. *Statistics Education Research Journal, 17*(1), 35–60. https://doi.org/10.52041/serj.v17i1.175

GAISE College Report ASA Revision Committee. (2016). Guidelines for assessment and instruction in statistics education college report 2016. American Statistical Association. http://www.amstat.org/education/gaise

Gould, R. (2010). Statistics and the modern student. International Statistical Review, 78(2), 297–315.

Grando, R. C., & Lopes, C. E. (2020). Creative insubordination of teachers proposing statistics and probability problems to children. *ZDM, 52*, 621–635. https://doi.org/10.1007/s11858-020-01166-6

Green, J. L., & Blankenship, E. E. (2013). Primarily statistics: Developing an introductory statistics course for pre-service elementary teachers. *Journal of Statistics Education, 21*(3), 1–20. https://doi.org/10.1080/10691898.2013.11889683

Green, J. L., Smith, W. M., Kerby, A. T., Blankenship, E. E., Schmid, K. K., & Carlson, M. A. (2018). Introductory statistics: Preparing in-service middle-level mathematics teachers for classroom research. *Statistics Education Research Journal, 17*(2), 216–238. https://doi.org/10.52041/serj.v17i2.167

Groth, R. E. (2007). Toward a conceptualization of statistical knowledge for teaching. *Journal for Research in Mathematics Education, 38*(5), 427–437.

Groth, R. E. (2012). The role of writing prompts in a statistical knowledge for teaching course. *Mathematics Teacher Educator, 1*(1), 23–40. https://doi.org/10.5951/mathteaceduc.1.1.0023

Groth, R. E. (2014). Using work samples from the National Assessment of Educational Progress (NAEP) to design tasks that assess statistical knowledge for teaching. *Journal of Statistics Education, 22*(3), 1–28. https://doi.org/10.1080/10691898.2014.11889712

Groth, R. E. (2015). Working at the boundaries of mathematics education and statistics education communities of practice. *Journal for Research in Mathematics Education, 46*(1), 4–16.

Groth, R. E. (2017). Developing statistical knowledge for teaching during design-based research. *Statistics Education Research Journal, 16*(2), 376–396. https://doi.org/10.52041/serj.v16i2.197

Groth, R. E., & Bergner, J. A. (2013). Mapping the structure of knowledge for teaching nominal categorical data analysis. *Educational Studies in Mathematics, 83*, 247–265. https://doi.org/10.1007/s10649-012-9452-4

Gutiérrez, R. (2013). The sociopolitical turn in mathematics education. *Journal for Research in Mathematics Education, 44*(1), 37–68.

Hannigan, A., Gill, O., & Leavy, A. M. (2013). An investigation of prospective secondary mathematics teachers' conceptual knowledge of and attitudes towards statistics. *Journal of Mathematics Teacher Education, 16*(6), 427–449. https://doi.org/10.1007/s10857-013-9246-3

Hobden, S. (2014). When statistical literacy really matters: Understanding published information about the HIV/AIDS epidemic in South Africa. *Statistics Education Research Journal, 13*(2), 72–82. https://doi.org/10.52041/serj.v13i2.281

Hourigan, M., & Leavy, A. M. (2019). Pre-service teachers' understanding of probabilistic fairness: Analysis of decisions around task design. *International Journal of Mathematical Education in Science and Technology, 51*(7), 997–1019. https://doi.org/10.1080/0020739X.2019.1648891

Huey, M. E., Champion, J., Casey, S. A., & Wasserman, N. H. (2018). Secondary mathematics teachers' planned approaches for teaching standard deviation. *Statistics Education Research Journal, 17*(1), 61–84. https://doi.org/10.52041/serj.v17i1.176

Jacobbe, T. (2012). Elementary school teachers' understanding of the mean and median. *International Journal of Science and Mathematics Education, 10*(5), 1143–1161. https://doi.org/10.1007/s10763-011-9321-0

Jones, D. L., & Jacobbe, T. (2014). An analysis of the statistical content in textbooks for prospective elementary teachers. *Journal of Statistics Education, 22*(3). https://doi.org/10.1080/10691898.2014.11889713

Kader, G. D., & Jacobbe, T. (2013). *Developing essential understandings of statistics: Grades 6–8.* National Council of Teachers of Mathematics.

Labaree, D. F. (1997). Public goods, private goods: The American struggle over educational goals. *American Educational Research Journal, 34*(1), 39–81.

Langrall, C., Makar, K., Nilsson, P., & Shaughnessy, M. (2017). Teaching and learning probability and statistics: An integrated approach. In *Compendium of research in mathematics education* (pp. 490–525). National Council of Teachers of Mathematics.

Learn Troop. (2022). *TinkerPlots.* https://www.tinkerplots.com/

Leavy, A. M. (2015). Looking at practice: Revealing the knowledge demands of teaching data handling in the primary classroom. *Mathematics Education Research Journal, 27*, 283–309. https://doi.org/10.1007/s13394-014-0138-3

Leavy, A. M., Hourigan, M., Murphy, B., & Yilmaz, N. (2021). Malleable or fixed? Exploring pre-service primary teachers' attitudes towards statistics. *International Journal of Mathematical Education in Science and Technology, 52*(3), 427–451. https://doi.org/10.1080/0020739X.2019.1688405

Lee, H. S., Doerr, H., Ärlebäck, J., & Pulis, T. (2013). Collaborative design work of teacher educators: A case from statistics. In M. Martinez & A. Castro Superfine (Eds.), *Proceedings of the 35th annual meeting of the North American Chapter of the International Group for the Psychology of Mathematics Education.* (pp. 357–364). University of Illinois Chicago.

Lee, H. S., Kersaint, G., Harper, S. R., Driskell, S. O., Jones, D. L., Leatham, K., Angotti, R. L., & Adu-Gyamfi, K. (2014). Teachers' use of transnumeration in solving statistical tasks with dynamic statistical software. *Statistics Education Research Journal, 13*(1), 25–52. https://doi.org/10.52041/serj.v13i1.297

Lee, H. S., Kersaint, G., Harper, S. R., Driskell, S. O., & Leatham, K. R. (2012). Teachers' statistical problem solving with dynamic technology: Research results across multiple institutions. *Contemporary Issues in Technology and Teacher Education, 12*(3). https://citejournal.org/volume-12/issue-3-12/mathematics/teachers-statistical-problem-solving-with-dynamic-technology-research-results-across-multiple-institutions/

Lovett, J. N., & Lee, H. S. (2017a). Incorporating multiple technologies into teacher education: A case of developing preservice teachers' understandings in teaching statistics with technology. *Contemporary Issues in Technology and Teacher Education, 17*(4). https://citejournal.org/volume-17/issue-4-17/mathematics/incorporating-multiple-technologies-into-teacher-education-a-case-of-developing-preservice-teachers-understandings-in-teaching-statistics-with-technology

Lovett, J. N., & Lee, H. S. (2017b). Preservice secondary mathematics teachers' statistical knowledge: A snapshot of strengths and weaknesses. In E. Galindo & J. Newton (Eds.), *Proceedings of the 39th annual meeting of the North American Chapter of the International Group for the Psychology of Mathematics Education.* Hoosier Association of Mathematics Teacher Educators.

Lovett, J. N., & Lee, H. S. (2019). Preservice secondary mathematics teachers' statistical knowledge: A snapshot of strengths and weaknesses. *Journal of Statistics Education, 26*(3), 214–222. https://doi.org/10.1080/10691898.2018.1496806

Magalhaes, M. N., & Magalhaes, M. C. C. (2014). A critical understanding and transformation of an introductory statistics course. *Statistics Education Research Journal, 13*(2), 28–41. https://doi.org/10.52041/serj.v13i2.278

Martinez, W., & LaLonde, D. (2020). Data science for everyone starts in kindergarten: Strategies and initiatives from the American Statistical Association. *Harvard Data Science Review, 2.3*. https://doi.org/10.1162/99608f92.7a9f2f4d

Martins, J. A., Nascimento, M., & Estrada, A. (2012). Looking back over their shoulders: A qualitative analysis of Portuguese teachers' attitudes towards statistics. *Statistics Education Research Journal, 11*(2). https://doi.org/10.52041/serj.v11i2.327

Muenchen, R. A. (2017). *The popularity of data science software*. R4stats.com. http://r4stats.com/articles/popularity/

Murphey, P. K., Knight, S. L., & Dowd, A. C. (2017). Familiar paths and new directions: Inaugural call for manuscripts. *Review of Educational Research, 87*(1), 3–6.

National Council of Teachers of Mathematics. (2014). *Principles to action: Ensuring mathematics success for all*. Authors.

National Governors Association Center for Best Practices & Council of Chief State School Officers. (2010). *Common core state standards for mathematics*. Authors.

North, D., Gal, I., & Zewotir, T. (2014). Building capacity for developing statistical literacy in a developing country: Lesson learned from an intervention. *Statistics Education Research Journal, 13*(2), 15–27. https://doi.org/10.52041/serj.v13i2.276

Odom, A. L., & Bell, C. V. (2017). Developing PK–12 preservice teachers' skills for understanding data-driven instruction through inquiry learning. *Journal of Statistics Education, 25*(1), 29–37. https://doi.org/10.1080/10691898.2017.1288557

Olsen, T. (2016). Preservice secondary teachers perceptions of college-level mathematics content connections with the common core state standards for mathematics. *Investigations in Mathematics Learning, 8*(3), 1–15. https://doi.org/10.1080/24727466.2016.11790351

O'Neil, C. (2016). *Weapons of math festruction*. Crown.

Park, M., & Lee, E. J. (2019). Korean preservice elementary teachers' abilities to identify equiprobability bias and teaching strategies. *International Journal of Science and Mathematics Education, 17*(8), 1585–1603. https://doi.org/10.1007/s10763-018-9933-8

Parrinder, P. (1997). Eugenics and utopia: Sexual selection from Galton to Morris. *Utopian Studies, 8*(2), 1–12.

Peters, S. (2013). Developing understanding of statistical variation: Secondary statistics teachers' perceptions and recollections of learning factors. *Journal of Mathematics Teacher Education, 17*(6), 1–44. https://doi.org/10.1007/s10857-013-9242-7

Pierce, R., & Chick, H. (2013). Workplace statistical literacy for teachers: Interpreting box plots. *Mathematics Education Research Journal, 25*, 189–205. https://doi.org/10.1007/s13394-012-0046-3

Raz, G. (2016, September 9). *Big data revolution*. TED Radio Hour Podcast. https://www.npr.org/programs/ted-radio-hour/492296605/big-data-revolution

Scheaffer, R. L., & Jacobbe, T. (2014). Statistics Education in the K–12 Schools of the United States: A Brief History. *Journal of Statistics Education, 22*(2). https://doi.org/10.1080/10691898.2014.11889705

Schmid, K. K., Blankenship, E. E., Kerby, A. T., Green, J. L., & Smith, W. M. (2014). The development and evolution of an introductory statistics course for in-service middle-level mathematics teachers. *Journal of Statistics Education, 22*(3), 1–22. https://doi.org/10.1080/10691898.2014.11889715

Sears, R., Kersaint, G., Burgos, F., & Wooten, R. (2019). Collaborative effort to develop middle school preservice teachers' mathematical knowledge. *Problems, Resources, and Issues in Mathematics Undergraduate Studies, 29*(9), 965–981. https://doi.org/10.1080/10511970.2018.1532936

Shaughnessy, M. (2007). Research on statistics learning and reasoning. In F. K. Lester (Ed.), *Second handbook of research on mathematics teaching and learning* (pp. 957–1009). Information Age Publishing.

Shin, D. (2020). Prospective mathematics teachers' professional noticing of students' reasoning about mean and variability. *Canadian Journal of Science, Mathematics and Technology Education, 20*, 423–440. https://doi.org/10.1007/s42330-020-00091-w

Shin, D. (2021). A framework for understanding how preservice teachers notice students' statistical reasoning about comparing groups. *International Journal of Mathematical Education in Science and Technology, 52*(5), 699–720. https://doi.org/10.1080/0020739X.2019.1699968

Watson, J., Fitzallen, N., Fielding-Wells, J., & Madden, S. (2018). The practice of statistics. In D. Ben-Zvi, K. Makar, & J. Garfield (Eds.), *International handbook of research in statistics education* (pp. 105–137). Springer International Publishing. https://doi.org/10.1007/978-3-319-66195-7_4

Weiland, T. (2017). Problematizing statistical literacy: An intersection of critical and statistical literacies. *Educational Studies in Mathematics, 96*(1), 33–47. https://doi.org/10.1007/s10649-017-9764-5

Whitaker, D. (2016). The development of a professional statistics teaching identity. *Proceedings of the 38th Annual Meeting of the North American Chapter of the International Group for the Psychology of Mathematics Education, 992–999.*

Whitaker, D., Bolch, C., Harrell-Williams, L., Casey, S., Huggins-Manley, C., Engledowl, C., Tjoe, H. (2021). The search for validity evidence for instruments in statistics education: Preliminary findings. In R. Helenius, E. Falck (Eds.), *Statistics education in the era of data science, Proceedings of the International Association for Statistics Education (IASE) 2021 Satellite Conference*, Aug-Sept 2021, Online conference. https://iase-web.org/documents/papers/sat2021/IASE2021%20Satellite%20170_WHITAKER.pdf?1649974217

Williams, S. R., & Leatham, K. R. (2017). Journal quality in mathematics education. *Journal for Research in Mathematics Education, 48*(4), 369–396.

Yang, K.-L. (2014). An exploratory study of Taiwanese mathematics teachers' conceptions of school mathematics, school statistics, and their differences. *International Journal of Science and Mathematics Education, 12*, 1497–1518. https://doi.org/10.1007/s10763-014-9519-z

CHAPTER 9

(TOWARD) ESSENTIAL STUDENT LEARNING OBJECTIVES FOR TEACHING GEOMETRY TO SECONDARY PRE-SERVICE TEACHERS

Tuyin An
Georgia Southern University[1,2]

Amanda Brown
University of Michigan

Steve Cohen
Roosevelt University

Henry Escuadro
Juniata College

Mike Ion
University of Michigan

Nathaniel Miller
University of Northern Colorado

Ruthmae Sears
University of South Florida

Stephen Szydlik
University of Wisconsin Oshkosh

Steven Boyce
Portland State University

Orly Buchbinder
University of New Hampshire

Dorin Dumitrascu
Adrian College

Patricio Herbst
University of Michigan

Erin Krupa
North Carolina State University

Laura J. Pyzdrowski
West Virginia University

Julia St. Goar
Merrimack College

Sharon Vestal
South Dakota State University

The AMTE Handbook of Mathematics Teacher Education: Reflection on Past, Present and Future—Paving the Way for the Future of Mathematics Teacher Education, Volume 5
pages 175–197.
Copyright © 2024 by Information Age Publishing
www.infoagepub.com

In this chapter, we describe the process that led to the development of a set of essential Student Learning Objectives (SLO) for undergraduate Geometry for Teachers courses. We briefly summarize each SLO, connect these SLOs to the AMTE Standards for Preparing Teachers of Mathematics (SPTM) and Mathematical Association of America's Committee on the Undergraduate Program in Mathematics (CUPM) Curriculum Guide, and provide recommendations for implementation and future stewardship of the SLOs.

INTRODUCTION

Secondary (Grades 6–12) geometry has historically played an important role in preparing students for their undergraduate mathematics learning because it offers students unique opportunities to study and engage in mathematical reasoning and proof (González & Herbst, 2006; National Council of Teachers of Mathematics [NCTM], 2000; National Governors Association Center for Best Practices [NGA Center] & Council of Chief State School Officers [CCSSO], 2010). Proving is "an essential component of doing, communicating, and recording mathematics" (Schoenfeld, 1994, p. 27). For secondary geometry courses to realize this potential, it is crucial for those assigned to teach it to understand the content, know the best practices for teaching the content, and be able to reflect on their teaching (Schoenfeld, 1988). Many secondary teacher preparation programs require a Geometry for Teachers course (GeT course, hereafter) which is expected to cultivate pre-service teachers' (PSTs') knowledge for teaching secondary geometry. However, GeT courses at various institutions in the United States have had highly varied content coverage. For example, from their survey of 108 undergraduate geometry courses, Grover and Connor (2000) catalogued five broad types of courses with substantially different foci. Similarly, the Mathematical Association of America's Committee on the Undergraduate Program in Mathematics (CUPM, hereafter) lists nine possible types of undergraduate geometry courses in its Geometry Course Report (Venema et al., 2015). On the one hand, some variability is to be expected as a consequence of institutional needs and academic freedom; on the other hand, it is important that all GeT courses ensure that future teachers develop the knowledge they need to teach high school geometry.

The Association of Mathematics Teacher Educators' *Standards for Preparing Teachers of Mathematics* (AMTE-SPTM, hereafter) emphasizes that preparation of effective mathematics teachers requires content-specific knowledge, skills, and dispositions (AMTE, 2017). The AMTE-SPTM tacitly opens the possibility to develop a process to identify and steward the development of consensus on such knowledge, skills, and dispositions in geometry. Several lines of thought contribute to a need for consensus. For example, mathematics educators continue to ponder how the adoption of the secondary level Common Core State Standards for Mathematics (CCSS-M, hereafter) (NGA & CCSSO, 2010) should impact the content and teaching of both high school and college level geometry courses (Harel, 2014; Nirode, 2013). Following the adoption of CCSS-M, the Conference

Board of the Mathematical Sciences updated their document, *Mathematical Education of Teachers* (MET II, hereafter), to better reflect what mathematics teachers should know in order to align their teaching with CCSS-M. Additionally, the MAA's CUPM developed a curriculum guide for mathematics departments (Zorn, 2015), including recommendations for what coursework should be offered for prospective high school mathematics teachers (Tucker et al., 2015). The CUPM recommendations include a variety of possible geometry courses that departments could offer (Venema et al., 2015).

Facing this possible diversity in offerings, teacher education programs need to ensure quality and relative consistency of outcomes across institutions. A working group from the GeT: A Pencil community of instructors proposed that one way to contribute to reducing the variability in the outcomes in the preparation of secondary geometry teachers could be to formulate and steward a set of student learning objectives (SLOs) that could be utilized by instructors of GeT courses. *GeT: A Pencil* is a faculty online learning community (FOLC, hereafter, see An et al., 2023) of GeT instructors formed as part of an NSF-funded project that seeks to develop an inter-institutional network for supporting undergraduate GeT instructors to work collectively towards the goal of improving the capacity of secondary geometry teaching. A commitment to a process of stewarding these SLOs across programs and over time might make the variability of the knowledge PSTs bring to practice more predictable (Bryk et al., 2015).

In this chapter, we describe the process that led to the development of this set of essential SLOs for GeT courses, briefly summarize each SLO, connect these SLOs to the AMTE-SPTM and MAA CUPM Curriculum Guide, and provide recommendations for implementation and stewarding of the SLOs.

THEORETICAL FRAMEWORK

An important problem often described in the literature on educational improvement is that of balancing implementation fidelity with adaptation to context (Datnow & Park, 2012). While traditional reform approaches rely on an implementation model that controls variability to achieve main effects of interventions, the instructional improvement approach (Bryk et al., 2015) embraces the notion of local variability and balances attention to main effects with an interest in reducing the variability of outcomes. This applies to teacher education too, as programs and courses vary across universities, yet schools and districts need to be able to trust that the teachers they hire have the necessary qualifications. Just as the AMTE-SPTM provides guidance that may support reducing variability in the outcomes of programs, SLOs for specific mathematical domains (e.g., geometry) can support the same goal of reducing variability in the outcomes of courses that cover such domains.

Importance of K–12 Geometry

Geometry is a critical component of the school mathematics curriculum across grades K–12. Students are expected to start learning attributes and relationships of geometric shapes in kindergarten and should be able to develop proofs for geometric theorems by the end of their high school education (NCTM, 2000; NGA & CCSSO, 2010). Teachers must support students connecting their "developing spatial awareness, and their ability to visualize, to their developing knowledge and understanding of, and ability to use, geometric properties and theorems" (Jones, 2000, p. 84).

High school geometry plays an important role in preparing students for their undergraduate mathematics learning because it offers students unique opportunities to study mathematical reasoning and proof. Additionally, given the literature that suggests students need additional instructional time to master the visuospatial reasoning demanded by school geometry, digital tools have become more prevalent in the K–16 geometry curriculum to facilitate the achievement of this mastery (Sinclair & Bruce, 2015; Sinclair et al., 2016). Thus, it is crucial to cultivate well-prepared high school geometry teachers who are well prepared to facilitate geometry instruction in several ways.

Subject Matter Knowledge for Secondary Geometry Teachers

A fundamental role of the GeT course is to provide opportunities for PSTs to develop robust knowledge for teaching secondary geometry. But what should constitute this knowledge base, and how should it be determined? One resource is the CBMS (2012) MET II, which provides recommendations for potential topics in a course focused on high school geometry from an advanced mathematics viewpoint. These recommendations vary in their specificity; their commonality is building domain-specific content knowledge related to the work of teaching a secondary geometry curriculum.

Other educational researchers have sought to identify the knowledge PSTs require by developing theoretical frameworks and instruments for measuring necessary content knowledge (e.g., Ball et al., 2008; Blömeke et al., 2016; Copur-Gencturk & Lubienski, 2013; Herbst & Kosko, 2014; Hill et al., 2004; Krauss et al., 2008; McCrory et al., 2012; Mohr-Schroeder et al., 2017; Thompson, 2016). Such work has helped to illustrate ways that variations in teachers' subject matter knowledge can inform teachers' instructional actions such as choosing the appropriate givens for a problem or in accurately understanding students' work (Ko & Herbst, 2020). This research on teacher knowledge has begun to have an influence on how teacher education programs prepare PSTs (Ball & Forzani, 2009; Clay et al., 2012) in geometry and other content courses (Lai, 2019; Lischka et al., 2020; Murray & Star, 2013; Steele, 2013). Given that many variations of the GeT course exist (Grover & Connor, 2000; Venema et al., 2015), it may be difficult to draw firm conclusions about what PSTs can be expected to learn from the different types of GeT courses (Wu, 2011). Nevertheless, reducing the variability in outcomes among these various GeT courses remains a laudable goal. We suggest such a goal can be supported by a community of instructors stewarding a set of student learning objectives (SLOs) with the aim of improving the GeT course and thereby improving the capacity for the teaching of secondary geometry.

DEVELOPING THE STUDENT LEARNING OUTCOMES (SLOS)

The *GeT: A Pencil* FOLC is made up of about 30 people who identify either as mathematicians or as mathematics educators, drawn from differing institutions—ranging from those that grant only baccalaureate degrees to doctoral-granting institutions (An et al., 2023). The experiences of these faculty are also diverse as all faculty ranks are represented. The diversity of viewpoints represented in the larger GeT: A Pencil community is an important resource that has aided in the ongoing development and stewardship of the SLOs.

One of the primary activity structures within GeT: A Pencil is that of working groups, in which groups of instructors collaborate towards particular aspects of improvement of GeT courses. In Fall 2018, a working group of fifteen instructors (WG–1, hereafter) began meeting to discuss what knowledge outcomes should be expected from students completing GeT courses. In those discussions, the WG–1 encountered a fundamental problem: unlike many college mathematics courses, there is not a broad consensus about what should be included in a GeT course, as there are many different types of GeT courses (Grover & Connor, 2000; Venema et al., 2015). Despite this lack of consensus, the WG–1's ongoing discussions revealed substantial overlap in courses' learning objectives. By Spring 2019, WG–1 had come to the realization that to make progress on the goal of improving secondary geometry instruction, the overlap of learning objectives across GeT courses could and should be identified to help inform other GeT course instructors, particularly those new to teaching the course.

In response to this realization, a group of GeT instructors was formed in Fall 2019 to work toward the goal of identifying a list of essential SLOs for inclusion in GeT courses—where *essential* meant the identification of content knowledge that *all* prospective secondary geometry teachers should have the opportunity to learn. This group (SLO-WG, hereafter) met consistently from Fall of 2019 through Summer of 2022 (monthly for the first year and a half and bi-weekly for the subsequent year)—working on the development, elaboration, and stewardship of the GeT SLOs. Building on the prior work of WG–1, the SLO-WG proceeded on the assumption that it was not practical to propose a unique GeT course due to the high variability in institutions, program structures and student populations. Instead, the SLO-WG set out to identify a core set of SLOs that should be included in any GeT course, regardless of type.

Members of the SLO-WG elected to use a winnowing strategy to create a list of essential SLOs—with each group member having the opportunity to contribute a set of student learning objectives they thought were essential to one master list.

Their work was guided by discussions and reflections of learning objectives drafted as part of earlier work in WG–1. Instructors drew from or were informed by the following sources in developing learning objectives: (1) instructors' own previous course syllabi and materials, (2) standards documents for secondary geometry (e.g., CCSS-M, NCTM's Principles to Standards), (3) curricular guidelines and recommendations specific to college geometry (e.g., MET II, MAA Curriculum Guide Geometry Course Report), and (4) descriptive research on undergraduate geometry courses (e.g., Grover & Connor, 2000). After this initial drafting took place, the SLO-WG culled the master list based on common themes. These themes became the focus of subsequent meetings in which the collective worked toward the construction of common statements that all participants agreed were essential learning objectives. These later discussions, some of which are ongoing at the time of the writing of this chapter, were guided by the further reflection and discussion regarding the interpretations of standards and guidance documentation (e.g., the CCSS-M, the MET II, the MAA CUPM Curriculum Guide, AMTE-SPTM).

In Spring 2020, the SLO-WG was invited by the larger GeT: A Pencil community to write elaborations of the SLOs for inclusion in the community's newsletter GeT: The News! (e.g., An et al., 2021). Based on feedback obtained on those initial newsletter elaborations, the SLO-WG further developed the SLOs—including making revisions to the SLOs themselves as well as writing longer elaborations (i.e., 1–2-page narratives about each one) that aimed to describe aspects of the SLOs that would be most important for consideration by GeT course instructors. For each SLO, a subgroup of 2–3 people wrote an initial draft of a narrative. The whole SLO-WG discussed the draft, and then the subgroup revised it. This process of revision followed by discussion was iterated until the SLO-WG was satisfied with the narrative.

Through these discussions the SLO-WG has become more aware of the ways and reasons why GeT courses, including both the objectives and enactments, align and diverge across institutions. While some of these divergences are the result of instructor preferences (i.e., academic freedom), many others find their source in larger institutional particularities, structures, and constraints of which an instructor has little control (e.g., the percentage of PSTs that populate the course, prerequisite and corequisite courses, the state mandates for curricular topics). This awareness has led to a realization amongst members of the SLO-WG that it may not be feasible for a given instructor in a given semester to address every one of the SLOs in their institution's particular GeT course (more details about this are provided in "Implementing the SLOs"). Members of the SLO-WG have accepted that the content identified in the SLOs may have to be considered aspirational. Despite these realities, the SLO-WG has maintained its commitment to a shared understanding of the SLOs as essential for PSTs—identifying a set of minimal learning objectives to be included within any program responsible for preparing secondary teachers to teach geometry. We explore the implications of these discussions later in this chapter.

THE SLOS

The SLOs include ten broad categories of geometry content recommended for PSTs to learn. The ten SLOs are further elaborated in Table 9.1.

- SLO 1 **[Proofs]:** Derive and explain geometric arguments and proofs.
- SLO 2 **[Critique Reasoning]:** Evaluate geometric arguments and approaches to solving problems.
- SLO 3 **[Secondary Geometry Understanding]:** Understand the ideas underlying current secondary geometry content standards and use them to inform their own teaching.
- SLO 4 **[Axiomatic Systems]:** Understand the relationship between axioms, theorems, and different geometric models in which they hold.
- SLO 5 **[Definitions]:** Understand the role of definitions in mathematical discourse.
- SLO 6 **[Technologies]:** Effectively use technologies to explore geometry and develop understanding of geometric relationships.
- SLO 7 **[Euclid]:** Demonstrate knowledge of Euclidean geometry, including the history and basics of Euclid's *Elements*, and its influence on mathematics as a discipline.
- SLO 8 **[Constructions]:** Carry out basic Euclidean constructions and justify their correctness.
- SLO 9 **[Non-Euclidean Geometries]:** Compare Euclidean geometry to other geometries such as hyperbolic or spherical geometry.
- SLO 10 **[Transformations]:** Use transformations to explore definitions and theorems about congruence, similarity, and symmetry.

Process Objectives: All GeT courses should give PSTs many chances to experience and develop proficiency with the mathematical process skills of problem solving, oral and written communication of mathematical ideas, and productive collaboration. They should also have opportunities to engage with the progression of geometric exploration, conjecturing, and construction of arguments.

In addition to the one-paragraph executive summaries found in Table 9.1, the SLO-WG has created longer 1–2-page narratives for each of the SLOs which can be found at www.getapencil.org. This site is purposed to disseminate the SLOs and narratives while still allowing for further development. The SLOs have evolved over time, and we envision them as compiled into a living document that will continue to change as more people engage with it.

TABLE 9.1. Executive Summaries of the Essential GeT SLOs

After completing a Geometry for Teachers course, students should (be able to):

SLO 1 [Proofs] Derive and explain geometric arguments and proofs.	Proof is a cornerstone of mathematics, and a GeT course should enhance a student's ability to read and write proofs of theorems, apply them, and explain them to others. Geometry offers one of the best opportunities for students to take ownership of proof in both the undergraduate and secondary curriculum. Because geometric proofs come with natural visualizations, they provide a rich environment for students to think deeply about the arguments that are being made and to make sense of each statement. Students should understand that proof is the means by which we demonstrate deductively whether a statement is true or false, and that geometric arguments may take many forms. They should understand that in some cases one type of proof may provide a more accessible or understandable argument than another.
SLO 2 [Critique Reasoning] Evaluate geometric arguments and approaches to solving problems.	Students should have opportunities not only to write proofs, but also to evaluate proofs and other types of reasoning. Geometry courses provide a natural setting for students to reflect on their own reasoning, share their reasoning with one another, and then critique the reasoning of their peers. If students only ever see correct arguments given by teachers and textbooks, they may not carefully evaluate the arguments because they assume they will always be correct. The ability to evaluate other people's arguments is an important real-world skill that is related to, but separate from, the skill of proof writing. Critiquing reasoning is a competency that needs to be practiced in order to improve it and is an essential skill for future geometry teachers.
SLO 3 [Secondary Geometry Understanding] Understand the ideas underlying current secondary geometry content standards and use them to inform their own teaching.	Future secondary geometry teachers must deeply understand specialized content that is aligned to national and state secondary standards, know the best practices for teaching the content, and be able to reflect on their teaching. Due to limited time and instructors' varied preferences in content selection, it is not practical to suggest a list of geometry topics to be covered in a GeT course. Thus, the GeT course should focus on helping students understand essential ideas emphasized in secondary geometry standards and use them to inform their future teaching. GeT instructors should be able to incorporate teacher preparation standards into their course designs in a way that fits their teaching agenda and introduce the national and state curriculum standards to future teachers. In addition, a GeT course should foster the construction of pedagogical content knowledge by sharing teaching techniques and by engaging students in conversations about teaching geometry content.
SLO 4 [Axiomatic Systems] Understand the relationships between axioms, theorems, and different geometric models in which they hold.	Geometry courses are one of the few places where students have opportunities to engage explicitly with axiomatic systems. Mathematical theories can be developed from a small set of axioms with theorems proven from those axioms. However, GeT students may have limited prior experience working with axiomatic systems. Students in GeT courses should gradually develop the ability to: (a) recognize and communicate the distinction between axioms, definitions, and theorems, and describe how mathematical theories arise from them, (b) construct logical arguments within the constraints of an axiomatic system, and (c) understand the roles of geometric models such as the plane, the sphere, the hyperbolic plane, etc., in identifying which theorems can or cannot be proven from a given set of assumptions.

SLO 5 [Definitions] Understand the role of definitions in mathematical discourse.

In a geometry class, definitions can be a fruitful area for students to explore. Students can propose their own definitions for geometric objects and relationships, they can engage in class discussions about mathematical definitions versus vague descriptions, and they can compare and contrast definitions that refer to different properties. Determining whether and when definitions have equivalent meanings prepares prospective teachers for the varieties of geometric definitions they may encounter in teaching secondary geometry. Prospective geometry teachers should understand the role of precision in definitions for geometric terms and relationships, including understanding that some geometric terms and relationships must remain undefined. They should also understand that there are a variety of acceptable choices for some geometric definitions, and that these choices can influence the structure of proofs. For example, proving that two lines are parallel because they do not intersect can be very different from proving that they are parallel because they are everywhere equidistant.

SLO 6 [Technologies] Effectively use technologies to explore geometry and develop understanding of geometric relationships.

Geometry courses need to utilize technology to help develop geometric reasoning and deepen students' understanding of geometric concepts and relationships. Two types of technologies that are beneficial for teaching geometry are dynamic geometry environments (DGEs) and digital proof tools (DPTs). DGEs support student understanding by allowing students to explore properties of geometric figures dynamically, which provides advantages over using paper and pencil. These explorations help students generate their own conjectures, test their conjectures, and provide justification and understanding for theorems. DGEs have been an important tool for teaching geometry since the 1990s. DPTs are an emerging technology that provide students with interactive figures to manipulate and opportunities to practice writing proofs with immediate feedback.

SLO 7 [Euclid] Demonstrate knowledge of Euclidean geometry, including the history and basics of Euclid's Elements and its influence on math as a discipline.

The study of many mathematical subjects can be illuminated by looking at their histories, but this is especially true of geometry. Euclidean geometry is named after Euclid, the Greek mathematician who lived in Alexandria around 300 BC. Euclid systematized the knowledge of geometry and included it in thirteen books called The Elements. In The Elements, Euclid set out a sequence of Definitions, Postulates, Common Notions (axioms for geometry), and Propositions (theorems derived logically from the preceding materials). For most of the 2400 years since it was written, it was considered an essential text and the gold standard of mathematical rigor. Students need to know this history to place modern ideas about proof into context and to understand mathematics as a human endeavor.

SLO 8 [Constructions] Carry out basic Euclidean constructions and to justify their correctness.

Traditional geometric constructions are those done exclusively with a compass and straightedge. However, the term can be used more generally to include other tools and manipulatives such as paper-folding (e.g., using origami or patty paper), dynamic geometry environments, or transparent mirrors (e.g., MIRAs). Constructions remain an essential part of Euclidean geometry and therefore of a GeT course. Constructions support the development of mathematical thinking in several essential ways: they provide a natural opportunity for making mathematical arguments, encourage the use of precise mathematical language in communication, impart a sense of where assumptions in building mathematical systems come from, open discussions of the historical development of geometry, especially the work of Euclid, and give future teachers experience with the curriculum they will be expected to teach.

(continues)

TABLE 9.1.   Continued

SLO 9 [Non-Euclidean Geometries] Compare Euclidean geometry to other geometries such as hyperbolic or spherical geometry.

Just as visiting another country can offer one a richer perspective on their own culture, the study of non-Euclidean geometries can help students to develop a deeper understanding of Euclidean geometry. The term "non-Euclidean geometry" is interpreted broadly here, referring to any geometry different from Euclidean, including spherical, hyperbolic, incidence, and taxicab geometries, among others. The choice of which non-Euclidean geometries to consider might depend on the demands of a particular GeT course, but all non-Euclidean geometries offer rich opportunities to explore and visualize novel and engaging worlds. Learning the properties of non-Euclidean geometries puts preservice teachers in the position of their students who may be learning Euclidean geometry for the first time. Moreover, non-Euclidean geometries challenge student assumptions about what is "true" or "obvious" in Euclidean geometry (e.g., the shape of a parabola or the angle sum in a triangle). In this way, exploring non-Euclidean geometries naturally encourages questions of "Why?" and "How?" and supports the development of mathematical thinking, especially the need for justification.

SLO 10 [Transformations] Use transformations to explore definitions and theorems about congruence, similarity, and symmetry.

Two main types of transformations arise in GeT courses: isometries (also known as congruence transformations) and similarity transformations. Isometries include reflections, rotations, translations, and glide reflections; a similarity transformation is the composition of an isometry and a dilation. A GeT course can contain a dedicated unit on transformations, or transformational concepts can be integrated throughout the course (even to the extent of a purely axiomatic treatment). To enhance and facilitate prospective teacher learning, familiar function notation can be incorporated in an introduction to transformations. This may lead to a better understanding of a sequence of transformations as a composition. While some GeT courses may begin with an informal approach to the understanding of reflections and rotations and extend the concepts of symmetry to study geometric shapes using transformations, others may instead begin solely with sequences of reflections, which can be used to generate all other isometries of the plane.

Process Objectives

All GeT courses should give PSTs many chances to experience and develop proficiency with the mathematical process skills of problem solving, oral and written communication of mathematical ideas, and productive collaboration. They should also have opportunities to engage with the progression of geometric exploration followed by conjecturing and construction of arguments.

ALIGNMENT WITH AMTE STANDARDS FOR PREPARING
TEACHERS OF MATHEMATICS

The AMTE-SPTM (2017) is a set of aspirational standards to help mathematics educators within teacher preparation programs prepare high-quality Pre-K–12 beginning teachers of mathematics. These standards are intended to:

> Guide the improvement of individual teacher preparation programs, inform the accreditation process of such programs, influence policies related to preparation of teachers of mathematics, and to promote national dialogue and action related to preparation of teachers of mathematics. (AMTE, 2017, p. xiv)

The creation of SLOs for GeT courses certainly addresses assisting with improving the GeT course in individual preparation programs and promoting national dialogue related to the preparation of teachers of mathematics, since the group consists of people from various universities. Another intent of these standards is to "engage the mathematics teacher education community in continued research and discussion about what candidates must learn during their initial preparation as teachers of mathematics" (AMTE, 2017, p. 1). The SLO-WG is an example of a community of mathematicians and mathematics educators attempting to articulate what mathematics teachers should know about secondary geometry. As noted above, we view these SLOs as essential, in the sense that we agree that all pre-service secondary mathematics teachers should have the opportunity to learn them, ideally in the context of their GeT course. However, we also acknowledge that for some instructors it might not be feasible to fit all ten SLOs into a single course. Later in this chapter, we provide more discussion regarding the various ways we can envision these SLOs being utilized by instructors to inform both the design of a GeT course and the larger teacher education program that includes the course.

While there are many connections between the SLOs and the AMTE-SPTM, one of the most direct connections we see is with Standard C.1. Mathematics Concepts, Practices, and Curriculum (see Table 9.2).

Here, we illustrate some of the ways the SLOs and Process Objectives make connections to various aspects of that standard. SLOs 1 [Proof], 4 [Axiomatic Systems], 5 [Definitions], 7 [Euclid], 8 [Constructions], 9 [Non-Euclidean Geometries], and 10 [Transformations] provide a geometry-specific perspective on what is involved in PSTs knowing the mathematical content (C.1.1). The Process Objectives along with SLO 3 [Secondary Geometry Understanding] provide further elaboration of what it means for future geometry teachers to demonstrate mathematical practices and processes, exhibit productive mathematical dispositions, and analyze the geometry content within the curriculum (C.1.2, C.1.3, C.1.4). Being able to explore geometry using dynamic geometry software (SLO 6 [Technologies]) and performing constructions (SLO 8 [Constructions]) provide more subject-specific articulation on the AMTE-SPTM's focus on PSTs' ability to use mathematics tools and technology (C.1.6). The analysis of mathematical

TABLE 9.2. Connections Between the SLOs, AMTE SPTM, and the MAA CUPM Curriculum Guide

Student Learning Objective	AMTE Standards for Preparing Teachers of Mathematics	MAA CUPM Curriculum Guide
SLO 1 [Proofs] Derive and explain geometric arguments and proofs.	C.1.1: Know relevant mathematical content HS.1. Essential Understandings of Mathematics Concepts and Practices in High School Mathematics HS.7. Mathematical Content Preparation of Teachers of Mathematics at the High School Level ML.1. Essential Understandings of Mathematics Concepts and Practices in Middle School Mathematics	Cognitive Recommendation 1: Students should develop effective thinking and communication skills. (CogR1) Cognitive Recommendation 4: Students should develop mathematical independence and experience open-ended inquiry. (CogR4) Content Recommendation 2: Students majoring in the mathematical sciences should learn to read, understand, analyze, and produce proofs at increasing depth as they progress through a major. (ConR2) Content Recommendation 4: Mathematical sciences major programs present key ideas and concepts from a variety of perspectives to demonstrate the breadth of mathematics. (ConR4)
SLO 2 [Critique Reasoning] Evaluate geometric arguments and approaches to solving problems.	C.1.5: Analyze mathematical thinking HS.1; HS.7; ML.1.	CogR1; CogR4 ConR2; ConR4
SLO 3 [Secondary Geometry Understanding] Understand the ideas underlying current secondary geometry content standards and use them to inform their own teaching.	C.1.2: Demonstrate mathematical practices and processes C.1.3: Exhibit productive mathematical dispositions C.1.4: Analyze the mathematical content of curriculum HS.1; HS.7; ML.1.	Cognitive Recommendation 2: Students should learn to link applications and theory. (CogR2) CogR4 ConR4 Content Recommendation 5: Students majoring in the mathematical sciences should experience mathematics from the perspective of another discipline. (ConR5) Content Recommendation 6: Mathematical sciences major programs should present key ideas from complementary points of view: continuous and discrete; algebraic and geometric; deterministic and stochastic; exact and approximate. (ConR6)
SLO 4 [Axiomatic Systems] Understand the relationships between axioms, theorems, and different geometric models in which they hold.	C.1.1 HS.1; HS.7	CogR1; CogR4 ConR2; ConR4;

Student Learning Outcome	Standards	Recommendations
SLO 5 [**Definitions**] Understand the role of definitions in mathematical discourse.	C.1.1 HS.1; HS.7	CogR1; CogR4 ConR2; ConR4
SLO 6 [**Technologies**] Effectively use technologies to explore geometry and develop understanding of geometric relationships.	C.1.6: Use mathematical tools and technology HS.2: Use of Tools and Technology to Teach High School Mathematics HS.1; HS.7; ML.1.	Cognitive Recommendation 3: Students should learn to use technological tools. (CogR3) CogR4; ConR4; ConR6
SLO 7 [**Euclid**] Demonstrate knowledge of Euclidean geometry, including the history and basics of Euclid's *Elements* and its influence on math as a discipline.	C.1.1 HS.1; HS.7	CogR4 ConR4
SLO 8 [**Constructions**] Be able to carry out basic Euclidean constructions and to justify their correctness.	C.1.1; C.1.6 HS.1; HS.2; HS.7; ML.1.	CogR4 ConR4;
SLO 9 [**Non-Euclidean Geometries**] Compare Euclidean geometry to other geometries such as hyperbolic or spherical geometry.	C.1.1 HS.2; HS.7	CogR4 ConR4
SLO 10 [**Transformations**] Use transformations to explore definitions and theorems about congruence, similarity, and symmetry.	C.1.1 HS.2; HS.7; ML.1.	CogR2; CogR4 ConR4; ConR6
Process Objectives	C.1.2: Demonstrate mathematical practices and processes C.1.3: Exhibit Productive Mathematical Dispositions	CogR1; CogR2; CogR3; CogR4

thinking (C.1.5) is seen in SLO 2 [Critique Reasoning] (AMTE, 2017; An et al., 2022). Our process objectives align with AMTE-SPTM's focus on mathematical processes (C.1.2, C.1.3).

In addition to the connections with the AMTE-SPTM's Standard C.1, we see connections to the AMTE-SPTM's elaboration regarding the preparation of PSTs for grade levels 6–8 as outlined in the Chapter 6 of the AMTE-SPTM. The GeT course SLOs align with ML.1. The SLOs that are most relevant for teaching middle school mathematics content are SLO 10 [Transformations],1 [Proofs] and 2 [Critique Reasoning]. These are particularly related to the knowledge of the Pythagorean Theorem and its many proofs, some of which rely on transformations, and SLO 3 [Secondary Geometry Understanding] emphasizes exposing PSTs to their state content standards and the CCSS-M Mathematical Practices (NGA & CCSSO, 2010). In addition, ML.1. stipulates that PSTs need to be able to select and use appropriate tools, such as dynamic geometry and physical manipulatives, to solidify geometry concepts, which directly relates to SLO 6 [Technologies] and SLO 8 [Constructions].

There are also numerous connections between the SLOs and those portions of AMTE-SPTM which focus on preparing high school mathematics teachers (outlined in Chapter 7 of AMTE-SPTM); we find that the SLOs align with HS.1, HS.2, and HS.7—which are also stated in Figure 9.2. In particular, HS.1 outlines the essential content and practices for PSTs and relates crucially to secondary geometry understanding provided in all ten SLOs. The focus on using tools and technology in HS.2 is aligned with SLO 6 [Technologies] and SLO 8 [Constructions], with particular attention on the geometry-specific activities which demand such tools as dynamic geometry environments. For example, SLO 8 [Constructions] provides guidance regarding the need for PSTs to gain facility over the tools and technologies specific to the situation of doing constructions (e.g., compass and straightedge, patty paper, MIRA™). In this way, SLOs 6 [Technologies] and 8 [Constructions] provide support for the AMTE-SPTM recommendation that beginning high school mathematics teachers should be familiar with manipulatives and know how to use them. Moreover, the SLOs provide subject-specific guidance regarding the situations and tools for which PSTs need to gain expertise. While we only carefully describe connections to the high school mathematics teacher preparation from AMTE-SPTM Chapter 7, the SLOs certainly address similar standards for preparing middle-level teachers of mathematics in Chapter 6 of AMTE-SPTM, particularly ML.1.

In the description of HS.7, the AMTE-SPTM document mentions MET II and the CUPM curriculum guide suggestions. Specifically, the AMTE-SPTM recommends that PSTs take "at least three content courses relevant to teaching high school mathematics" (AMTE-SPTM, p. 136) We maintain that a GeT course is particularly relevant and critically important since most high school students take at least one semester devoted to Euclidean geometry. The narrative for HS.7 also discusses the importance of PSTs being exposed to "mathematical practices and processes" in their mathematics coursework. This idea is addressed in all ten SLOs as well as the Process Objectives.

Chapter 9 in the AMTE-SPTM is devoted to *Enacting Effective Preparation of Teachers of Mathematics*, which is essentially what our group has attempted to do through the SLOs for the GeT course. The process that we followed in creating the SLOs was built upon the following assumptions of the AMTE-SPTM:

- **Assumption 3:** Learning to teach mathematics requires a central focus on mathematics.
- **Assumption 4:** Multiple stakeholders must be responsible for and invested in preparing teachers of mathematics.
- **Assumption 5:** Those involved in mathematics teacher preparation must be committed to improving their effectiveness in preparing future teachers of mathematics. (AMTE, 2017, pp. 1–2)

We assembled a group made up of multiple stakeholders involved in mathematics teacher education to identify the central mathematical ideas essential to any GeT course, in order to try to improve the future preparedness of mathematics teacher educators. Our FOLC members have been actively engaging in the improvement of their GeT courses under the guidance of the SLOs.

IMPLEMENTING THE SLOS

Prospects and Challenges of Implementing the SLOs within a Single GeT Course

It can be a daunting challenge to adequately address all the SLOs in a single GeT course. There are many reasons for this. For one, some GeT courses are not intended solely for PSTs, which makes it difficult to justify the inclusion of some SLOs, such as SLO 3 [Secondary Geometry Understanding]. Even when populated by all or mostly PSTs, there are sometimes institutional constraints that may challenge an instructor's sense of agency for designing the GeT course. One common constraint for instructors' agency is previously-established course descriptions—some of which promise exploration of topics quite different from those one might reach with a focus on the SLOs (e.g., a focus on differential geometry). Another constraint instructors face is abbreviated course formats (e.g., summer offerings of the GeT course) which can make comprehensive inclusion difficult.

All of these constraints point to an important challenge in implementing the SLOs into a single GeT course—finding ways to include all the SLOs within the larger set of institutional expectations for the course. One strategy for including all of the SLOs in a single course is to integrate objectives around carefully chosen activities and concepts. For example, in an activity that has been used by several of the authors in their own GeT courses (Miller, 2010a, Chapter 3), students read Euclid's Proposition 1 on constructing an equilateral triangle; construct equilateral triangles and squares in GeoGebra; use GeoGebra to find the symmetries of these shapes; identify all the triangles and quadrilaterals that have some of these symmetries; and make arguments that they have found all of the possibilities. This single activity touches on almost all the SLOs, perhaps excluding non-Euclidean geometries. It provides a historical context around which students can learn to read, write, and explain proofs of material from the secondary curricula, this time engaging in the content in a more rigorous way. As another example, students can explore the geometry of the hyperbolic plane using dynamic geometry software to try to see how familiar definitions, constructions, postulates, theorems, and proofs change in that context (Miller, 2010b, Chapter 4; Szydlik, 2001, Teaching Examples section). This kind of activity also touches on all the SLOs except for SLO 10 on transformations.

Even if one designs learning opportunities to address multiple SLOs simultaneously, there will undoubtedly be some SLOs that are emphasized more than others. This can be due to instructor preference or expertise as well as students' preparation, experiences, and goals. For instance, in some programs, students enter a GeT course without having (recently) taken a high school geometry course. Accordingly, providing opportunities for GeT students to master secondary geometry content should be included in the GeT course to support equitable course and program outcomes. A fuller inclusion of secondary geometry content will likely mean that some geometry topics will not be addressed to the depth of knowledge an instructor might wish. Though the level of treatment of SLOs will vary to some extent based on the context of a given GeT course, it is important to establish consistent expectations for the depth and scope of SLOs incorporated within a GeT course from year to year to maintain its purpose and value as part of a teacher education program.

Prospects and Challenges of Implementing the SLOs Across a Program

Though it may be possible to address all the SLOs in a single GeT course, another option would be to distribute the SLOs across multiple courses in a teacher education program. For example, certain SLOs may fit naturally into a history of mathematics course, a proofs course, or a teaching methods course. Reasons to consider this option might include: accommodating multiple faculty members' strengths and interests when making teaching assignments, strategically integrating content SLOs and other types of objectives (such as those related to pedagogy) through key courses in the program or allowing for a gradual building of a depth of knowledge throughout a sequence of courses. A faculty member with a strong interest and background in the history of mathematics could easily address several of the objectives such as SLO 7 [Euclid] and 9 [Non-Euclidean Geometries] in the history of mathematics course. If both the history of mathematics and the GeT courses are required of all students in a program, addressing some objectives in history of mathematics could allow a GeT course instructor to spend more time on other objectives or allow the leading of students toward a deeper understanding of particular SLOs. For example, an instructor could use additional time when addressing SLO 10 [Transformations] by making meaningful connections to abstract algebra.

While we see the implementation of the SLOs across a program as a viable alternative to the implementation of the SLOs within a single course, we are somewhat cautious about this form of implementation. Special care would need to be taken when using a programmatic distribution of objectives as there are situations and circumstances that can conspire against efficacy in teacher preparation. Changes in teaching assignments can lead to misunderstandings of what is to be taught in specific courses and the lens through which the content is explored. Even when content objectives are well documented in institutionally-approved syllabi, the academic flexibility with respect to percentage of objectives addressed and depth of coverage with respect to those objectives invoked by an instructor can vary. On the other hand, more global teacher education program changes (such as those experienced when a program becomes involved in multi-institutional projects) may make local decisions with respect to course objectives difficult or unlikely.

For a programmatic approach to SLO implementation to be viable, it is necessary to have good collaboration and management not only within a department but likely, in some cases, across colleges. Such collaboration and integration would likely benefit preparation for program accreditation and lead to a cohesive pathway of courses for students.

FUTURE DIRECTIONS: ARTICULATING A PROCESS FOR STEWARDING THE SLOS

In figuring out how to improve the geometric preparation of secondary mathematics teachers, we have deliberately avoided the extremes of leaving each instructor on their own or attempting to prescribe a curriculum or pedagogy. Instead, we have found strength in being part of an inter-institutional community of instructors where we can sharpen our ideas and work to develop commonalities. The idea of drafting a set of SLOs was a rallying point for us, and the task of writing the SLOs and their elaborations enabled us to learn from each other while clarifying the SLOs and the ways they can be implemented. In all of this, inclusivity and collaborative work have been more important than deadlines or reaching a final text. In fact, as we have indicated above, the SLOs are meant to be a living document, to be continuously improved under the stewardship of the community of instructors.

The choice to treat the SLOs as a living rather than final document is an intentional one, moving beyond simply inviting other instructors to use the SLOs toward a more inclusive goal of inviting other instructors to actively contribute to the ongoing development and stewardship work that is ahead of us. Our vision for the work is to continue in the manner we began—using a more democratic model of course development—by inviting new instructors to join us and have a voice in the ongoing revision, refinement, and stewardship of these and other related SLO artifacts.

The group that has thus far stewarded the development of the SLOs includes the authors of this chapter along with other members of our community of GeT instructors who have lent their tacit support while they worked on other projects within GeT: A Pencil. But across the United States, there are other instructors we have yet to connect with as well as individuals recently added to the ranks of those teaching a college geometry course. We would like to reach out to them and invite them to join our community. As the community grows by incorporating new members and as existing members' thinking evolves, the text can be enriched in many ways. A process for stewarding the SLOs would include providing leadership and organization of events and artifacts that could produce the motivation for reading, instructionally applying, and contributing to documenting the SLOs. Whereas not everybody may be able to join a working group that meets biweekly like ours, there may be other activities they could join. Writings like this chapter and the elaborations of the SLOs we have written (e.g., An et al., 2021, 2022) offer opportunities for engagement, and we would like to hear from more geometry instructors in response to these writings. A survey is currently being administered that will provide university geometry instructors the opportunity to express how much they value each SLO in comparison with other possible learning objectives; the survey is a way for us to reach out and another way for instructors to make their voices heard.[3]

Based on the SLOs, we have created assessment items and collected initial student responses that inform the development of a set of artifacts for all instructors to use. The assessment items will be shared with instructors along with ideas for how to use them in the classroom for formative purposes. We are keen on getting feedback from instructors as well as samples of students' work that could eventually be a resource for us to document and benchmark the attainment of the SLOs.

To build on the assessment items, we expect to continue collecting, reviewing, revising, and documenting tasks or other materials that instructors can later reference to support their teaching of the SLOs. These artifacts, like the SLOs themselves, will also be living documents that instructors could not only use but actively contribute to by offering ideas for how they can be used in GeT courses (Herbst, 2021; Szydlik, 2021). This process of using and documenting teaching and assessment materials has the potential to engage GeT instructors in a different kind of collective practice—one which engages the larger community of instructors into the work of continuously discussing the meaning of the SLOs and thus supporting consensus building around these living documents. Meanwhile, members of the community will also seek ways to share the SLOs with other geometry instructors, be they mathematicians or mathematics educators, through conference presentations and workshops.

CONCLUDING THOUGHTS

In this chapter, we have shared the ongoing work of a community of GeT instructors to articulate a set of essential student learning objectives for the GeT course. This work has been guided by the twin imperatives of (1) improving the GeT course towards the goal of (2) increasing the capacity for the teaching of secondary geometry. This chapter makes two important contributions of significance for the AMTE audience. The first, and perhaps most obvious, is the presentation of the SLOs for use by other GeT instructors. We have been careful to present the SLOs in a way that provides some crucial guidance, based on our collective experiences, about the various ways that the SLOs might be taken up in practice. In this way, the content presented in this chapter has the capacity to move the field forward. Related to this first contribution, we have also aimed to clearly outline our visions for the SLOs going forward; namely that we conceive of the SLOs as a living document—open to the voices of other GeT instructors and in need of ongoing stewardship.

While the presentation of the SLOs is an important contribution to the field, a second contribution of this chapter lies in the way it outlines a process that animates the ideals outlined in AMTE's vision for the improvement of university mathematics courses. Specifically, this chapter describes a process by which instructors of undergraduate mathematics courses can be collectively engaged in the kind of work envisioned by AMTE. For one, the effort is taken up by a group of instructors (rather than an individual) gathered across institutional boundaries. This kind of cross-institutional collaboration affords the work numerous advantages not available from similarly purposed work carried out by individuals drawn from a single program/institution. This aspect of the work responds directly to Call to Action #4 outlined by AMTE (2017):

> Faculty in programs preparing teachers of mathematics must build collaborations with faculty in other programs preparing teachers of mathematics. Learning from and with colleagues from other institutions and providers can accelerate progress in their improvement efforts, with faculty benefitting from experiences and results of each site. The networked improvement community model proposed by Bryk et al. (2015) may be particularly useful in building knowledge across programs (cf. Martin & Gobstein, 2015, p. 166)

Also, our work illustrates the notion expressed by AMTE, namely that improvement of teacher preparation is ongoing and cyclical, rather than linear and complete (see Figure 9.1). The ongoing, cyclical nature of our work is perceivable not only from the processes we have used so far toward the development of the SLOs but is also evident in our articulation of the work that is ahead. Specifically, we strategically framed this chapter in ways that not only encourage the dissemination of the SLOs, but also seek the opportunity to build a common vision with the larger community of GeT instructors, anticipating the need to reconsider the SLOs in light of new and differing contexts that may help further refine the SLOs.

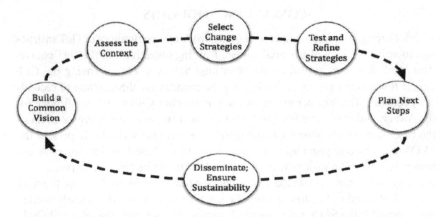

FIGURE 9.1. AMTE-SPTM's Model (AMTE, 2017, p. 165) of the ongoing and cyclical nature of the work of improving mathematics teacher preparation programs.

Perhaps the most fundamental aspect of our collective work is the way in which our group positions itself as open and growing, committed from its inception to a model of working collegially with one another and other instructors, rather than presuming a top-down approach to improvement. This commitment has meant that from the beginning the work has drawn from a diversity of voices. One of the ways this diversity has shown up is through the inclusion of both mathematicians and mathematics teacher educators who often come with differing ideas about what the improvement of the geometry content preparation of beginning secondary teachers might look like. In this way, the work we have described reflects the ideals outlined by AMTE in Call to Action #1:

> Mathematics and statistics teacher educators must collaborate with mathematicians and statisticians to enact the vision for the content preparation of beginning teachers presented in these standards, building on related documents offered by other professional organizations (e.g., CBMS, 2012; Franklin et al., 2016; Tucker et al., 2015). (AMTE, 2017, p. 166)

Another way this diversity has been present is with the drawing of individuals from different kinds of institutions across the country, who come with differing kinds of institutional constraints which necessarily shape their perspectives and approaches to the GeT course. These differences have meant we have sometimes had to have difficult conversations in which agreement is not easily found.

Despite, or perhaps even because of, these differing perspectives, we have made substantial gains towards the creation of a common vision regarding the geometry-specific content preparation of beginning mathematics teachers. Through our work on the SLOs, we have learned to hear one another through our differences—developing a greater appreciation for each other, an understanding of the complexity of our work as GeT instructors, and the possibilities for progress and improvement through collaboration.

ENDNOTES

1. *The names of the authors are in alphabetical order. All authors contributed to the paper equally.
2. Work reported in this chapter has been done with support of NSF grant DUE- 1725837 (Patricio Herbst, PI). All opinions are those of the authors and do not necessarily represent the views of the Foundation.
3. As of December 2022, the GeT Instructor survey had gathered nearly 140 responses (Herbst, unpublished raw data).

REFERENCES

An, T., Berzina-Pitcher, I., Bigelow, V., Brown, A., Buchbinder, O., Herbst, P., Miller, N., Prasa, P.V., Pyzdrowski, L.J., St Goar, J., Sears, R., Szydlik, S., & Vestal, S. (2023). A cross-institutional faculty online learning community: Community-guided faculty development in teaching college geometry for teachers. In S. Linder, C. Lee, & K. High (Eds.), *The handbook of STEM faculty development* (pp. 325–336). Information Age Publishing. https://www.infoagepub.com/products/Handbook-of-STEM-Faculty-Development

An, T., Boyce, S., Buchbinder, O., Cohen, S., Dumitrascu, D. C., Escuadro, H., Krupa, E., Miller, N., Pyzdrowski, L., Szydlik, S., & Vestal, S. (2022). GeT course Student Learning Outcome #2. *GeT: The News!, 3*(3). https://www.gripumich.org/v3-i3-sp2022/#get-course-student-learning-outcome-2

An, T., Boyce, S., Cohen, S., Escuadro, H., Krupa, E., Miller, N., Pyzdrowski, L., Szydlik, S., & Vestal, S. (2021). GeT course Student Learning Outcome #4. *GeT: The News!, 2*(2). https://gripumich.org/v2-i2-w2021/#get-course-student-learning-outcome-4

Association of Mathematics Teacher Educators (AMTE). (2017). *Standards for preparing teachers of mathematics.* Online available at: https://amte.net/standards

Ball, D. L., & Forzani, F. M. (2009). The work of teaching and the challenge for teacher education. *Journal of Teacher Education, 60*(5), 497–511.

Ball, D. L., Thames, M. H., & Phelps, G. (2008). Content knowledge for teaching: What makes it special. *Journal of Teacher Education, 59*(5), 389–407.

Blömeke, S., Busse, A., Kaiser, G., König, J., & Suhl, U. (2016). The relation between content-specific and general teacher knowledge and skills. *Teaching and Teacher Education, 56*(2016), 35–46.

Bryk, A. S., Gomez, L. M., Grunow, A., & LeMahieu, P. G. (2015). *Learning to improve: How America's schools can get better at getting better.* Harvard Ed. Press.

Clay, E., Silverman, J., & Fischer, D. J. (2012). Unpacking online asynchronous collaboration in mathematics teacher education. *ZDM, 44*(6), 761–773.

Conference Board of the Mathematical Sciences (CBMS) (Ed.). (2012). *The mathematical education of teachers II* (Vol. 17, Issues in Mathematics Education). American Mathematical Society. https://www.cbmsweb.org/archive/MET2/met2.pdf

Copur-Gencturk, Y., & Lubienski, S. T. (2013). Measuring mathematical knowledge for teaching: A longitudinal study using two measures. *Journal of Mathematics Teacher Education, 16*(3), 211–236.

Datnow, A., & Park, V. (2012). Conceptualizing policy implementation: Large-scale reform in an era of complexity. In G. Sykes, B. Schneider, & D. Plank (Eds.), *Handbook of education policy research* (pp. 348–361). Routledge.

González, G., & Herbst, P. (2006). Competing arguments for the geometry course: Why were American high school students supposed to study geometry in the twentieth century? *International Journal for the History of Mathematics Education, 1*(1), 7–33.

Grover, B., & Connor, J. (2000). Characteristics of the college geometry course. *Journal of Mathematics Teacher Education, 3*(1), 47–67.

Harel, G. (2014). Common Core State Standards for geometry: An alternative approach. *Notices of the AMS, 61*(1), 24–35.

Herbst, P. (2021). Illustrating a modeling approach to high school geometry: The pool problem. *GeT: The News!, (2)*3. https://www.gripumich.org/v2-i3-s2021/#illustrating-a-modeling-approach-to-high-school-geometry-the-pool-problem

Herbst, P., & Kosko, K. (2014). Mathematical knowledge for teaching and its specificity to high school geometry instruction. In J. Lo, K. R. Leatham, & L. R. Van Zoest (Eds.), *Research trends in mathematics teacher education* (pp. 23–45). Springer.

Hill, H. C., Schilling, S. G., & Ball, D. L. (2004). Developing measures of teachers' mathematics knowledge for teaching. *Elementary School Journal, 105*(1), 11–30.

Jones, K. (2000). Critical issues in the design of the school geometry curriculum. In B. Barton (Ed.), *Readings in mathematics education* (pp. 75–90). University of Auckland.

Ko, I., & Herbst, P. (2020). Subject matter knowledge of geometry needed in tasks of teaching: relationship to prior geometry teaching experience. *Journal for Research in Mathematics Education, 51*(5), 600–630.

Krauss, S., Brunner, M., Kunter, M., Baumert, J., Blum, W., Neubrand, M., & Jordan, A. (2008). Pedagogical content knowledge and content knowledge of secondary mathematics teachers. *Journal of Educational Psychology, 100*(3), 716.

Lai, Y. (2019). Accounting for mathematicians' priorities in mathematics courses for secondary teachers. *The Journal of Mathematical Behavior, 53*(4), 164–178.

Lischka, A. E., Lai, Y., Strayer, J. F., & Anhalt, C. (2020). Developing mathematical knowledge in and for teaching in content courses. *The Mathematics Teacher Education Partnership: The power of a networked improvement community to transform secondary mathematics teacher preparation* (Volume 4, pp. 119–141). IAP.

Martin, W. G., & Gobstein, H. (2015). Generating a networked improvement community to improve secondary mathematics teacher preparation: Network leadership, organization, and operation. *Journal of Teacher Education, 66*(5), 482–493. https://doi.org/10.1177/0022487115602312

McCrory, R., Floden, R., Ferrini-Mundy, J., Reckase, M. D., & Senk, S. L. (2012). Knowledge of algebra for teaching: A framework of knowledge and practices. *Journal for Research in Mathematics Education, 43*(5), 584–615.

Miller, N. (2010a). Modern Geometry I. *Journal of Inquiry-Based Learning in Mathematics, 17*. https://jiblm.org/downloads/dlitem.php?id=76&category=jiblmjournal

Miller, N. (2010b). Modern Geometry II. *Journal of Inquiry-Based Learning in Mathematics, 19*. https://jiblm.org/downloads/dlitem.php?id=79&category=jiblmjournal

Mohr-Schroeder, M., Ronau, R. N., Peters, S., Lee, C. W., & Bush, W. S. (2017). Predicting student achievement using measures of teachers' knowledge for teaching geometry. *Journal for Research in Mathematics Education, 48*(5), 520–566.

Murray, E., & Star, J. R. (2013). What do secondary preservice mathematics teachers need to know? *Notices of the American Mathematical Society, 60*(10), 1297–1299.

National Council of Teachers of Mathematics (NCTM). (2000). *Principles and standards for school mathematics*. National Council of Teachers of Mathematics.

National Governors Association Center for Best Practices [NGA], & Council of Chief State School Officers [CCSSO]. (2010). *Common core state standards for mathematics*. http://www.corestandards.org/Math/

Nirode, W. (2013). Don't sacrifice geometry on the Common Core altar. *Mathematics Teacher, 107*(3), 168–171.

Schoenfeld, A. H. (1988). When good teaching leads to bad results: The disasters of 'well-taught mathematics' courses. *Educational Psychologist, 23*(2), 145–166.

Schoenfeld, A. H. (1994). What do we know about mathematics curricula? *Journal of Mathematical Behavior, 13*(1), 55–80.

Sinclair, N., & Bruce, C. D. (2015). New opportunities in geometry education at the primary school. *ZDM, 47*(3), 319–329.

Sinclair, N., Bussi, M. G. B., de Villiers, M., Jones, K., Kortenkamp, U., Leung, A., & Owens, K. (2016). Recent research on geometry education: An ICME-13 survey team report. *ZDM, 48*(5), 691–719.

Steele, M. D. (2013). Exploring the mathematical knowledge for teaching geometry and measurement through the design and use of rich assessment tasks. *Journal of Mathematics Teacher Education, 16*(4), 245–268.

Szydlik, S. (2001). The hyperbolic toolbox: Non-Euclidean constructions in Geometer's SketchPad. *Journal of Online Mathematics and its Applications, 1*(3). Retrieved May 12, 2022, from https://www.maa.org/press/periodicals/loci/joma/the-hyperbolic-toolbox-abstract

Szydlik, S. (2021). People and Clubs: An Axiomatic System. *GeT: The News!, (2)*2. https://www.gripumich.org/v2-i2-w2021/#people-andclubs:-an-axiomatic-system

Thompson, P. W. (2016). Researching mathematical meanings for teaching. In L. English & D. Kirshner (Eds.), *Handbook of international research in mathematics education* (pp. 435–461). London: Taylor and Francis.

Tucker, A., Burroughs, E., & Hodge, A. (2015). A professional program for preparing future high school mathematics teachers. In P. Zorn (Ed.) *2015 CUPM Curriculum guide to majors in the mathematical sciences*. Mathematical Association of America. Online available at: https://www.maa.org/sites/default/files/HighSchoolMathematicsTeachersPASGReport.pdf

Venema, G., Baker, W., Farris, F., & Greenwald, S. (2015). Geometry course report. In P. Zorn (Ed.), *2015 CUPM curriculum guide to majors in the mathematical sciences*. Mathematical Association of America. Available at: https://www.maa.org/sites/default/files/Geometry.pdf

Wu, H. (2011). The mis-education of mathematics teachers. *Notices of the AMS, 58*(3), 372–384.

Zorn, P. (Ed.). (2015). *CUPM guide to majors in the mathematical sciences*. Mathematical Association of America. https://www.maa.org/sites/default/files/pdf/CUPM/pdf/CUPMguide_print.pdf

CHAPTER 10

MAKING ADVANCED MATHEMATICS WORK IN SECONDARY TEACHER EDUCATION

Yvonne Lai
University of Nebraska-Lincoln

Nicholas Wasserman
Teachers College, Columbia University

Jeremy F. Strayer
Middle Tennessee State University

Stephanie Casey
Eastern Michigan University

Keith Weber
Rutgers University

Timothy Fukawa-Connelly
Temple University

Alyson E. Lischka
Middle Tennessee State University

We argue that it is essential to provide opportunities for prospective secondary mathematics teachers to connect advanced mathematics content to secondary *mathematics teaching practice* (in addition to connections only to secondary *mathematics*), if advanced mathematics courses are to be useful to these teachers. In light of this argument, we review a selection of curricular materials satisfying two criteria: first, they are written for use in advanced mathematics courses that prospective teachers may take; and second, they feature explicit connections to secondary teaching prac-

The AMTE Handbook of Mathematics Teacher Education: Reflection on Past, Present and Future—Paving the Way for the Future of Mathematics Teacher Education, Volume 5
pages 199–218.

tice. We use this review to highlight essential questions for the future of research and practice in advanced mathematics coursework. We suggest there is much unknown about how teachers can successfully integrate their experiences in advanced mathematics and pedagogical methods.

INTRODUCTION

In many university-based secondary mathematics teacher certification programs in the United States (U.S.), teachers complete multiple courses in advanced mathematics (e.g., Ferrini-Mundy & Findell, 2001; Tatto & Bankov, 2018). Our purposes regarding advanced mathematics coursework are twofold. First, we argue that it is essential to provide opportunities for prospective secondary teachers to connect the content of these courses to the *practice of teaching*, if these courses are to be useful to prospective teachers. Second, we suggest there is much unknown about how prospective teachers can successfully integrate course experiences in advanced mathematics and pedagogical methods. This is a critical issue, especially if instruction in both seeks to shape teachers' images of teaching practice. Our arguments implicate AMTE indicators P.2.1, P.2.2, P.3.1, and P.3.4 (Association of Mathematics Teacher Educators, 2017)[1], and how well teachers can connect experiences across these standards.

Note that throughout this chapter, we use "teacher" to refer to "prospective secondary mathematics teacher," unless otherwise stated. We use "instructor" or "faculty" to refer to those teaching university-level courses.

Advanced mathematics content and practice can illuminate secondary mathematics content and practice (e.g., Baldinger, 2018; Lai & Donsig, 2018; Wasserman & Weber, 2017). Yet despite this relationship, exemplifying ties between advanced and secondary mathematics, let alone integrating advanced mathematics and the practice of secondary teaching, has been challenging (e.g., Wasserman & Weber, 2017; Yan et al., 2022). These connections are "not currently well developed in the profession of mathematics teaching" (CBMS, 2012, p. 14). In this chapter, we:

- Demonstrate that, historically, teachers' pedagogy does not usually improve, or even change, as a result of taking advanced mathematics coursework;
- Synthesize the recent movement to connect advanced mathematics coursework to secondary mathematics *teaching practice*, and why these opportunities are important to provide; and
- Pose essential questions for the future of advanced mathematics courses in secondary mathematics teacher education, especially in view of enabling teachers to integrate experiences in mathematics and methods coursework.

---

[1] These indicators state that undergraduate mathematics courses should attend to mathematics relevant to teaching (P.2.1) and build mathematical practices (P.2.2); and methods courses should instill mathematics as deep and meaningful (P.3.1) and offer practice-based experiences (P.3.4).

Although the field has made initial progress, we have much to learn about providing teachers robust opportunities to connect advanced mathematics and secondary teaching practice.

ADVANCED MATHEMATICS COURSES SHOULD DO MORE TO CONNECT TO SECONDARY TEACHING

Advanced mathematics knowledge can potentially strengthen secondary mathematics knowledge (AMTE SPTM Indicator P.2.1, 2017; CBMS, 2012). Historically, a typical approach has been to follow Felix Klein's (1924/1932) prescription of considering "elementary mathematics from an advanced standpoint" (p. 1). The hope here is that with a more sophisticated understanding of secondary concepts, teachers will take more productive actions when teaching.[2] The challenge, though, is that how this knowledge influences teaching is less clear–what can or should secondary teachers conceive or do differently as a result of this strengthened knowledge?

This lack of clarity impacts teachers, who lament their mathematical preparation was irrelevant to secondary teaching (e.g., Goulding et al., 2003; Ticknor, 2012; Wasserman & Galarza, 2018; Zazkis & Leikin, 2010). Even when prospective teachers develop a richer understanding of the secondary mathematical concepts that they will teach, they may still exit advanced mathematics believing that they may have become better mathematicians, but not better mathematics teachers (Wasserman & Ham, 2013). Regardless, while many teachers may develop stronger personal mathematical practices through advanced coursework (Indicator P.2.2), they still may not see why it matters that mathematics can be deep and meaningful (Indicator P.3.1), nor how their mathematical growth might shape their pedagogical growth (Indicator P.3.4).

REFLECTING ON THE PAST: CONNECTIONS THROUGH *MATHEMATICS*

Mathematics teacher educators have historically connected advanced mathematics to secondary teaching along two dimensions: *mathematics content at the secondary level* and *mathematical practice*. To illustrate the first dimension, *mathematical content*, we turn to specifications for advanced mathematics course content for teachers. Multiple policy documents and syllabi argue that advanced mathematics courses should connect to secondary mathematics content (e.g., CUPM, 1961; CBMS, 2001a; Murray & Star, 2013; Tucker et al., 2015).

Consider the concept of a capstone course for teachers, which is offered by many institutions in the U.S. (Cox et al., 2013), and is recommended by the guid-

---

[2] As a case in point, recognizing that the set of invertible functions under composition form a group provides secondary teachers with a deeper, richer, and more connected notion of inverse functions. And, given the application of inverse functions in solving equations in school mathematics, it is reasonable to think that teachers would benefit from this strengthened knowledge base.

ing document *Mathematical Education of Teachers, I* (CBMS, 2001a). The various designs of capstone courses give insight into the kinds of connections that mathematics faculty see between advanced mathematics and secondary mathematics content. In their review of then-present-day capstone coursework for secondary teachers, Murray and Star (2013) found that connections took the form of generalizations or abstractions of secondary mathematical ideas (e.g., geometric transformations, factoring polynomials), or specific uses of advanced mathematics to define concepts that secondary students may encounter (e.g., $2^{\sqrt{5}}$ as an illustration of defining mathematical expressions with limits). These connections can also be viewed as interpretations of Klein's (1924/1932) notion of school mathematics from an advanced standpoint.

As for the second dimension, in the 1990s, scholars such as Cuoco et al. (1996) called attention to the importance of *mathematical practice* in teaching and learning mathematics. Recommendations for mathematics departments followed suit. As the report from CBMS (2001a) reasoned, "Teachers need to learn to ask good mathematical questions, as well as find solutions, and to look at problems from multiple points of view. Most of all, prospective teachers need to learn how to learn mathematics" (p. 8). The report theorized that, consequently, mathematics teachers must experience mathematical practices in their own learning, for then they will be more likely to foster such experiences when teaching secondary mathematics. This argument resonates with various findings from the perspective of teachers; understanding what mathematics is, and reminding them what learning mathematics feels like, were primary ways teachers found such coursework to be relevant (Baldinger, 2018; Even, 2011; Hoffman & Even, 2019).

A secondary teacher's practice benefits from robust mathematical practice and knowledge of secondary content from an advanced standpoint (Baumert et al., 2010; Sword et al., 2018). Nonetheless, teacher educators must consider: Do these two aspects alone constitute sufficient connection from advanced mathematics coursework to teachers' future work? Empirical studies establish that this potential has been unmet. Begle (1979) found that secondary students' performance was associated with neither the number of tertiary mathematics courses taken by teachers nor the average grade received by teachers in these courses. Monk (1994) also studied relationships between student performance and teacher course taking. He found that at the secondary level, there was only negligible effect after the first four mathematics courses–meaning that courses considered "advanced" had little effect. More recently, qualitative results suggest the situation has not changed. Goulding et al.'s (2003) and Zazkis and Leikin's (2010) surveys found that many teachers report their mathematical preparation is disconnected from teaching. In a study of an abstract algebra course, Ticknor (2012) found that even when secondary teachers wanted to do well in the course, and the instructor saw connections between abstract algebra and secondary mathematics, the teachers still perceived the course as irrelevant to teaching. Wasserman et al. (2018) articulated concrete reasons, given by prospective and in-service teachers, for why their knowledge of real analysis did not inform their teaching, even when they understood the material. In the next section, we provide an overview of efforts centered on a third approach to connections—*teaching practice*—which holds promise for addressing the shortcomings just discussed.

AN EMERGING PROGRAM: CONNECTIONS TO SECONDARY MATHEMATICS *TEACHING PRACTICE*

In response to both the promise of advanced mathematics to shape teachers' practice and the observed inefficacy of advanced mathematics courses to do so, secondary mathematics teacher educators have pushed for connecting advanced mathematics to teaching practice (Álvarez et al., 2020a; Artzt et al., 2011; Bremigan et al., 2011; Heid et al., 2015; Lai, 2019; Lischka et al., 2020; Wasserman et al., 2017). In this section, we characterize how various mathematics teacher educators have developed connections from advanced mathematics to *teaching practice*. We ground this characterization in a review of curricular materials and reports of enacting such materials. We show that the approaches used to connect advanced mathematics coursework to secondary teaching practice are consistent with constructs from practice-based teacher education (Forzani, 2014). In particular, the approaches resemble representations of practice, approximations of practice, and decompositions of practice (Grossman et al., 2009). Then, drawing from examples of each type of these pedagogies of practice, we raise issues for the field to consider in moving forward. In particular, we problematize how teachers may be supported–or not–to integrate their experiences in mathematics and methods courses.

MATERIALS SELECTION, AUTHOR POSITIONALITY, AND BOUNDARIES OF THIS REVIEW

We sought to review curricular materials satisfying two criteria: (1) they were intended for use in advanced mathematics courses that secondary teachers may take, and (2) they sought to make explicit connections to secondary mathematics teaching practice. We also restricted this review to materials developed in the U.S.; we reasoned that AMTE is an organization whose members work primarily in the U.S., and this handbook is concerned with implications for the enactment of the AMTE standards (2017).

The authors of this chapter are also developers of such curricular materials, namely, the MODULE(S²) materials (Lai, Strayer, Casey, Lischka) and ULTRA materials (Wasserman, Weber, Fukawa-Connelly). We believe in the approach of connections via teaching practice, and have collectively written in various venues to advocate for this viewpoint. Our investment comes from our own review of the empirical and theoretical literature, as described in the above sections. In this chapter, we have chosen to include our own materials in our review. In doing so, we aimed to revisit our own materials relative to others and constitute part of an emerging movement.

In searching for curricular materials, we looked for both the materials themselves as well as reports of their enactment. We considered textbooks commonly used in capstone courses (as reviewed by Cox et al., 2013), those published by professional organizations of mathematicians, and the literature reporting the use of these materials and textbooks in advanced mathematics courses. As well, we conducted a Fastlane search for materials created with the support of the U.S. National Science Foundation. Table 10.1 displays these materials and reports. Note that in this table, "algebra" refers to the study of number systems, functions, or relations, whereas "abstract algebra" refers to the study of mathematical groups, rings, fields, or related constructs.

We now describe the approaches evinced by our review in terms of pedagogies of practice.

Representations of Practice

The pedagogy of practice known as *representations of practice*, which we henceforth refer to as "representations," allows novices to observe aspects of practice (Grossman et al., 2009). When teachers engage with representations, they can develop ways of noticing and understanding teaching practice. Representations can vary in comprehensiveness and authenticity, and can take a variety of forms, including short narratives of teaching scenarios, videos, animations, and case studies. Ball and Bass used representations, in the form of videos of instruction as well as written descriptions of teaching scenarios, to educate the broader mathematical community on the specialized nature of mathematical knowledge for teaching (e.g., Ball & Bass, 2003; CBMS, 2001b). The influence of their work can be found in a variety of teacher education and professional development materials today, across grades K–16.

The types of representations of practice found in the reviewed materials are shown in Table 10.2. Materials showed these representations of practice and then engaged teachers in evaluating, describing, or reflecting upon features of the representation of practice.

Evaluating student contributions was by far the most prevalent structure for engaging with representations of practice and evaluating teacher contributions was the second most common. A standard format for evaluating student contributions is showing teachers a sample of student work or dialogue, and then asking teachers to identify strengths and weaknesses of the mathematics shown (e.g., Bremigan et al., 2011; Casey et al., 2021a; Hauk et al., 2017, 2018; Mathematical Education of Teachers as an Application of Undergraduate Mathematics [META Math], 2020a; Sultan & Artzt, 2011; Wasserman et al., 2022). Figure 10.1 shows two examples.

The stance of evaluating raises the issue of how teachers frame discourse in terms of personal knowledge. In most materials reviewed, the prospective teachers were positioned to evaluate student and teacher talk with advanced mathematics content learned in the course. But teachers do not only view classroom discourse through a mathematical lens; they also consider implications for students' affect, beliefs, and relative position. Ideally, teachers are able to fluidly transfer lenses and integrate inferences from their observations. In one moment, they may

TABLE 10.1. Reviewed Materials and Reports of Their Enactment

Name of Authors or Project	Materials Reviewed	Topics Addressed
Bremigan, Bremigan, & Lorch	Bremigan et al. (2011)	Algebra
Buchbinder & McCrone	Samples of materials and description of implementation as described in Buchbinder and McCrone (2018, 2020)	Number Theory Geometry Use of conditionals
Capstone Mathematics	Hauk et al. (2017, 2018)	Algebra, Abstract Algebra, Geometry
Enhancing Explorations in Functions for Preservice Secondary Mathematics Teachers Project	Samples of materials and description of implementation in Álvarez et al. (2020b)	Algebra
Mathematical Education of Teachers as an Application of Undergraduate Mathematics (META Math)	MAA META Math (2020a, 2020b, 2020c, 2020d, 2020e, 2020f, 2020g, 2020h, 2020i)	Abstract Algebra Calculus Discrete Mathematics Proof Statistics
Mathematics of Doing, Understanding, Learning, and Educating for Secondary Schools (MODULE(S²))	Lai & Hart (2021), Hart & Lai (2021), Aubrey et al. (2021), Casey et al. (2021a, 2021b, 2021c), Alibegović & Lischka (2021a, 2021b, 2021c), Anhalt et al. (2021a, 2021b, 2021c)	Abstract Algebra, Algebra, Geometry, Mathematical Modeling, Statistics
Mathematical Understanding for Secondary Teaching (MUST)	Heid et al.(2015)	Algebra Abstract Algebra Geometry Statistics Proof by induction
Sultan & Artzt	Sultan & Artzt (2011) Description of implementation in Artzt et al. (2011)	Algebra Geometry Statistics
Upgrading the Learning and Teaching of Real Analysis (ULTRA)	Materials retrieved from http://ultra.gse.rutgers.edu/. Descriptions of implementation in Wasserman and McGuffey (2021), Weber et al. (2020), Fukawa-Connelly et al. (2020), Wasserman et al. (2022), McGuffey et al. (2019), Wasserman et al. (2019).	Real analysis
Usiskin, Peressini, Marchisotto, & Stanley	Usiskin et al. (2003) Description of implementation in Winsor (2009)	Algebra Geometry Abstract algebra

TABLE 10.2. Categories of Representations of Practice

Category	Description
Student contributions	Depictions of student talk, work, or thinking, where teachers are invisible.
Teacher-student interactions	Show a teacher and a student or multiple students
Written curriculum and assessments	Most commonly include actual or hypothetical contents of text-books, and also include lesson plans and standardized assessment tasks
Personal experience with secondary level tasks	Teachers are asked to notice features of their own experiences solving a secondary level task
Educator's teaching of a task	An educator models teaching a task for prospective teachers, and the experience of this task itself becomes an object to notice

consider mathematical implications. The next moment, they may wonder about consequences for students' perception of their agency. Later, they may reflect on these evaluations. We hypothesize that in methods courses, teachers may be asked to use a pedagogical lens to evaluate discourse, and also a mathematical lens that does not include advanced mathematics knowledge. If this hypothesis is true, then there is a disconnect between how teachers learn to evaluate representations of practice in methods courses and in advanced mathematics courses.

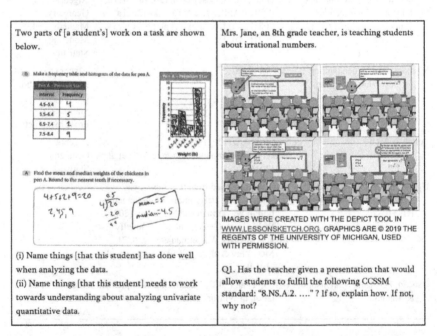

FIGURE 10.1. Examples of representations of practice (Casey et al., 2021a, p. 175; Wasserman et al., n.d.)

Whether teachers are asked to evaluate, describe, or reflect upon representations of practice, there is also a question of authenticity to actual secondary mathematics teaching practice, and how uses of representations influence teachers' developing notions of teaching. Notice that we problematize authenticity–rather than, say, dismissing inauthenticity out of hand. Positive effects are reported both by projects that sought to create representations typifying recurrent classroom interactions (Strayer et al., 2021; Wasserman & McGuffey, 2021) and those where the materials featured representations that may not be as typical (Sultan & Artzt, 2010; Winsor, 2009), such as, "A student wants to know if the expression $(\sqrt{2})^{\curlywedge}(\sqrt{2})^{\curlywedge}(\sqrt{2})$ is rational or irrational. What would you say?" (Sultan & Artzt, 2010, p. 252). There is not empirical evidence at this point to say that one stance is unconditionally better than the other.

In view of the reviewed sources, we hypothesize that implementation is critical to how and whether teachers draw connections to teaching practice. In Artzt et al.'s (2011) and Winsor's (2009) studies, prospective teachers taught their peers, and hence they interacted with their peers-as-learners. In the projects reported by Strayer et al. (2021) and Wasserman and McGuffey (2021), teachers were given time to reflect on representations and produce a reply to typical secondary students' thinking, but there is no actual interaction between teachers and other persons. More examination is needed to identify how teachers' experiences are mediated by implementation.

Approximations of Practice

The pedagogy of *approximations* of practice allow novices to simulate practice so they can attend to particular aspects of practice, rather than all aspects of the complex, relational practice that is teaching (Grossman et al., 2009). Approximations may vary in authenticity and complexity (see Table 10.3).

Across the multiple secondary education projects reviewed, the majority of approximations featured the practices of responding to students' mathematical contributions, evaluating students' work, and explaining content (e.g., Álvarez et al., 2020b; Alibegović & Lischka, 2021b; Aubrey et al., 2021; Bremigan et al., 2011; Buchbinder & McCrone, 2020; Hauk et al., 2017; META Math, 2020b; Sultan & Artzt, 2010; Wasserman et al., 2022). For instance, the Capstone Math materials provided teachers with a set of student work samples and asked teachers to work in groups to "sort the work into categories that represent different ways of thinking and/or difficulties" (Hauk et al., 2017, p. 9). Throughout the MODULE(S²) materials, teachers are asked to consider samples of student work in response to a task, and then to either write a narrative describing how they would "conduct a whole class discussion which will allow you to elicit student thinking, ... with specific use of the [students'] work," or to record a video of themselves "providing a response to [the students]" that highlights the student's thinking as well as engages the student in the intended mathematics (see Figure 10.2).

TABLE 10.3. Categories of Approximations of Practice

Approximation Category	Description
Responding to students' mathematical contributions	Attend to features of students' contributions, interpret them, and then determine how to respond
Evaluating students' work	Determine or quantify mathematical validity of students' work
Explaining content	Produce a written or spoken account of how one might explain a mathematical idea such as a procedure, concept, or connection
Teaching a full lesson	Either teaching a lesson of the advanced mathematics course in which teachers were enrolled, or teaching a lesson to secondary students

These examples constitute approximations of practice because they simulate recurrent work in teaching. In these examples, this work includes diagnosing student conceptions, orchestrating discussion, and posing questions to students. They are also approximations in that they are artificial: the teachers have far more time and latitude to come up with and revise their ideas than they would have in front of actual students. At the same time, these examples raise the issue of diversifying the teaching practices being approximated. Instruction is more than explaining content and responding to student work. It also includes setting up norms and routines for disciplinary work, setting learning goals, and adjusting goals and instruction, among other practices.

One way to weave in more teaching practices is to teach full lessons. We found two types of this pedagogy. First, as reported by Artzt et al. (2011) and Winsor

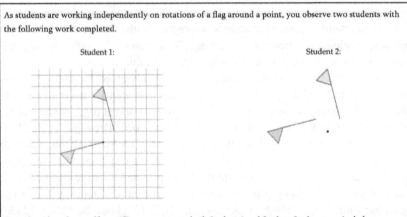

As students are working independently on rotations of a flag around a point, you observe two students with the following work completed.

Student 1:

Student 2:

Record a video of yourself providing a response to both Student 1 and Student 2 where you include a summary of what each student might be thinking and what is worthwhile or reasonable about that student's thinking, and a response to each student that does one or more of the following: helps the student finish their thought, prompts the student to investigate an error, or helps the student move forward in their thinking.

FIGURE 10.2. Example of approximation of practice (Alibegović & Lischka, 2021b)

(2009), teachers might teach a session of the capstone course to their peers. In both cases, the teaching experience imparted a deeper appreciation of the depth of content knowledge entailed in teaching a topic. Further, teachers believed that this lesson learned would apply to their own future secondary teaching. In Buchbinder and McCrone's (2020) case, the teachers planned and taught actual secondary class lessons featuring proving and reasoning. The teachers in Buchbinder and McCrone's study also developed an appreciation of the content knowledge demands of teaching. Further, they replicated models of tasks that they had experienced in their own advanced mathematics course. In all of these cases, teaching full lessons constituted an approximation of practice: the teachers received far more support and planning time than a practicing teacher would.

Within the materials reviewed, the scope of approximations of practice was either relatively small, as exemplified by the approximation shown in Figure 10.2, or relatively large, as in teaching a full lesson. Moreover, the range of teaching practices was narrow. To diversify the teaching practices that are approximated, it may be necessary to design new types of approximations of practice. Such new approximations may at once leverage advanced mathematics and also give teachers the opportunity to approximate practices such as setting up discourse routines or adjusting goals and instruction. Such designs remain an open problem for mathematics teacher education.

Decompositions of Mathematical Practice with Implications for Teaching Practice

The pedagogy of *decompositions* of practice supports educators by providing maps of teaching practice that can be used to attend to prospective teachers' enactment of practice (Grossman et al., 2009). Decompositions of practice in the reviewed materials were rare. However, there were several decompositions of *mathematical* practice rather than teaching practice across the materials. These included decompositions of the ways that secondary students may understand congruence, similarity, or dilation (Hauk et al., 2018) or decomposing the sufficiency or insufficiency of considering examples in proving or disproving certain conditional statements (Buchbinder & McCrone, 2020). These decompositions of mathematical practice have implications for teaching practice and may serve as resources for both advanced mathematics and methods instructors in articulating mathematical practice in teaching.

REFLECTIONS ON THE CURRENT STATE
AND PUSHING THE FIELD FORWARD

We began this chapter by describing historical approaches to connect advanced mathematics and secondary mathematics teaching. Yet, despite the rationales underlying these historical approaches, they largely were ineffective: many teachers viewed advanced mathematics courses as irrelevant to teaching. A practice-based approach to making advanced mathematics courses work for teacher education takes into account this lesson: educators must go further than connecting advanced mathematics to what teachers need to *know*; they must connect advanced mathematics to what teachers actually *do*.

Promising Evidence for a Practice-Based Approach

We see two points of evidence in support of adopting pedagogies of practice as a way to connect advanced mathematics and secondary mathematics teaching practice. First, mathematics faculty have been willing to adopt materials with this approach. Some of the reviewed materials are also among the most common textbooks for capstone courses (Cox et al., 2013); Álvarez et al. (2022) reported that instructors using the META Math materials used the tasks featuring teaching scenarios; additionally, as authors of some of the reviewed materials, we know firsthand of faculty from across the US using the materials, and for multiple years.

Second, in contrast to earlier studies (e.g., Goulding et al., 2003; Zazkis & Leikin, 2010), pilot studies of practice-based approaches to advanced mathematics courses suggest that teachers are able to describe connections between advanced mathematics and their future teaching (cf., Fukawa-Connelly et al., 2020; McGuffey et al., 2019; Wasserman & Galarza, 2018). Artzt et al. (2011) and Winsor (2009) both reported that the teachers in their courses developed an appreciation for how mathematical proofs and underlying reasoning could inform their diagnosis of students' thinking as well as their responses to students' questions. Moreover, when teachers are teaching secondary students, they may incorporate ideas from their advanced mathematics courses. Wasserman and McGuffey (2021) of the ULTRA project observed former students of their project teaching in their own secondary classrooms. They found that teachers attributed some teaching moves to their experiences in real analysis as designed by the ULTRA project. Buchbinder and McCrone (2020) found that prospective teachers used task designs they had encountered in their advanced mathematics courses when they were asked to design and teach a proof-based lesson for secondary students while student teaching. In open-ended responses to MODULE(S^2) surveys, multiple teachers who took advanced mathematics courses using these materials pointed to the usefulness of the approximations of practice.

Pushing the Field Forward: Essential Questions for the Future of Advanced Mathematics Courses for Teachers

Looking forward while holding a view across the pedagogies of practices used, we call attention to five areas to move the field forward in connecting advanced mathematics to secondary mathematics teaching practice in meaningful ways.

(1) *Methodological and theoretical considerations for how knowledge of advanced mathematics—especially mathematical practice—shapes secondary teaching practice.* Across all reviewed projects, we saw connections via both math-

ematical content and mathematical practice to the work of teaching secondary mathematics. The prominent connection across all the materials was via mathematical content. In such instances, it was possible because the advanced mathematical content gave a framework for viewing the secondary content as a special case. That is, studying advanced mathematics helped to situate the secondary content in a mathematically meaningful way. By contrast, when the materials made connections through mathematical practice, the connection was by way of analogy, in drawing on how processes engaged in doing mathematics could be productive for teaching mathematics as well. Notably, these instances appeared primarily in more recent materials, suggesting that the opportunity presented by, and the need to be explicit about, these mathematical practices in advanced coursework– and their relationship to teaching practice–is something that has become more evident over time.

The inclusion of practice connections raises questions about how to best discuss mathematical practices in these courses. Mathematics educators have identified various similarities between advanced and secondary mathematics, and how these similarities may inform teaching. But mathematical practice evolves across K–20 years (Stylianou et al., 2009; Tall, 1991). Mathematics teacher educators thus far appear to leverage similarities, but not differences, in mathematical practice. Yet, an advanced mathematics course must aim to engage teachers in advanced mathematical practice.

We ask: How can differences between advanced mathematics practice and secondary mathematics practice be a resource rather than a limitation to meaningful connections between advanced mathematics and teaching practice?

(2) *Diversifying depictions of practice and issues of inclusion.* As noted earlier, there is a narrow band of teaching practices featured in approximations of practice in advanced mathematics courses. A predominant one was explaining content, which also mirrors the way many situations in teaching were represented. This limitation in the way pedagogies of practice were accomplished in the reviewed materials may reinforce the (undesirable) notion that teaching equates to explaining. Thus, we ask: How can this practice-based approach be diversified to include additional teaching practices? How can the diversity of teaching–beyond just explaining content–be captured and incorporated into advanced mathematics coursework? In exploring such questions, it is also essential to identify ways that do not overload advanced mathematics instructors nor expect too much from teacher candidates who are novices. Only one reviewed project leveraged clinical experiences (Buchbinder & McCrone, 2020). Exploring further whether there might be clinical experiences that meaningfully leverage knowledge of advanced mathematics to, for instance, support students in constructing sound geometric proof, or connect algebra and geometry.

Furthermore, with only two exceptions, the reviewed projects and reports did not appear to intentionally take on issues of social justice and equity in their use of pedagogies of practice. Indeed, the broader literature on teacher education suggests that equity and social justice may be a blind spot in practice-based teacher education more generally (Kavanagh & Danielson, 2014). We ask: Are there ways to take on issues of social justice and equity in advanced mathematics courses, given the complexity of the issue? As a smaller point, but one connected to these broader issues, we note that rarely did materials prompt teachers to identify mathematical strengths of student work without it being paired with identifying weaknesses of student work. Doing so may promote a deficit-based perspective about students as mathematical thinkers, rather than an asset-based perspective. Framing pedagogies of practice may be one realm where issues about equity and inclusion can be taken up.

(3) *Identifying explicit ways to acknowledge and discuss pedagogical aspects of mathematics teaching practice.* Multiple projects featured an explicit way to characterize and discuss pedagogical (and not just mathematical) aspects of teaching. The specific details differed widely across these three projects, but the key point is that in each, the discussion of pedagogy had some explicit operationalization. In META Math, a collection of five different types of connections guided the design of the materials, and prescribed aspects of what teachers should be able to do with their mathematical knowledge in their teaching (Álvarez et al., 2020a). Sultan and Artzt (2010) aimed for teachers to create lessons that aligned with ten dimensions of mathematics teaching competence. In MODULE(S²), responses to teaching situations were considered with respect to Rowland and colleagues' (2013, 2016) Knowledge Quartet and Simon's (2006) notion of mathematical understanding (Lai et al., 2021). When asking teachers to engage in a specific pedagogy of practice, even while primarily discussing mathematics, these projects drew on these conceptions to structure pedagogical conversations and design pedagogies of practice. The MUST project designed situations around the mathematical context of teaching, and specified some of the components of this aspect of teaching (Heid et al., 2015). In ULTRA, a set of eight written teaching principles explicitly guided any conversations about teaching; the principles were described in relation to both mathematical and pedagogical teaching practice (Wasserman et al., 2022).

We ask: What can we learn about how discussions of pedagogy can be incorporated into advanced mathematics courses from projects that did so? How might discussions of pedagogy in mathematics courses differ from discussions of pedagogy in methods courses? What do potential differences mean for integrating teachers' experiences in advanced mathematics and methods courses?

(4) *Differentiating and coordinating how pedagogies of practice are taken up in mathematics content and methods courses.* Initial studies indicate the promise of connecting advanced mathematics to teaching practice via pedagogies of practice. Yet we recognize that pedagogies of practice, as taken up in mathematics courses, may necessarily differ from pedagogies of practice as taken up in methods courses. The pedagogies of practice reviewed here do not always resemble the pedagogies of practice used in methods courses. Where there is overlap, the content and practices are more specific to the intended advanced mathematics

than would likely be possible in a methods course. On the one hand, it is reasonable that the pedagogies of practice have qualitatively different characteristics: methods courses and advanced mathematics courses have different objectives, and different objectives will lead to different task design. On the other hand, these differences raise a question of how prospective teachers, who will encounter these different pedagogies of practice, will integrate rather than silo their teacher preparation experiences. We ask: How can pedagogies of practice developed in mathematics content and methods courses be meaningfully coordinated? What funding and support structures might be needed to foster the collaboration amongst mathematicians, mathematics educators, teachers, and teacher educators that is necessary to do this interdisciplinary work?

In raising these questions, we call for mathematics teacher educators and other stakeholders to learn from current research and practice so as to innovate in critical new directions to improve the enactment of the AMTE Standards for Preparing Teachers of Mathematics (2017) for secondary teacher education.

ACKNOWLEDGEMENT

Most of all, we are grateful for the imagination and dedication of the developers who created materials reviewed in this chapter (especially those who do not include ourselves!). The work of these developers provides a starting point for future research and practice. The writing of this chapter was partially funded by the National Science Foundation Improving Undergraduate STEM Education multi-institutional collaborative grants DUE-1726707, 1726098, 1726252, 1726723, 1726744, and 1726804, and DUE-1524739, 1524681, and 1524619. The views expressed in this chapter are those of its authors, and do not necessarily represent those of the National Science Foundation.

REFERENCES

Alibegović, E., & Lischka, A. E. (2021a). *MODULE(S²): Geometry for secondary mathematics teaching, Module 1: Axiomatic system.* Mathematics Teacher Education-Partnership.

Alibegović, E., & Lischka, A. E. (2021b). *MODULE(S²): Geometry for secondary mathematics teaching, Module 2: Transformational geometry.* Mathematics Teacher Education-Partnership.

Alibegović, E., & Lischka, A. E. (2021c). *MODULE(S²): Geometry for secondary mathematics teaching, Module 3: Similarity.* Mathematics Teacher Education-Partnership.

Álvarez, J. A., Arnold, E. G., Burroughs, E. A., Fulton, E. W., & Kercher, A. (2020a). The design of tasks that address applications to teaching secondary mathematics for use in undergraduate mathematics courses. *The Journal of Mathematical Behavior, 60,* online first. https://doi.org/10.1016/j.jmathb.2020.100814

Álvarez, J. A., Jorgensen, T., & Beach, J. (2020b). *Using multiple scripting tasks to probe preservice secondary mathematics teachers' understanding.* Paper presented at the 14th International Congress on Mathematics Education, Shanghai, China.

Álvarez, J. A., Kercher, A., Turner, K., Arnold, E. G., Burroughs, E. A., & Fulton, E. W. (2022). Including school mathematics teaching applications in an undergraduate abstract algebra course. *PRIMUS, 32*(6), 685–703.

Anhalt, C., Cortez, R., & Kohler, B. (2021a). *MODULE(S²): Mathematical modeling, Module 1: The mathematical modeling process and purpose.* Mathematics Teacher Education-Partnership.

Anhalt, C., Cortez, R., & Kohler, B. (2021b). *MODULE(S²): Mathematical modeling, Module:, Incorporating real data in mathematical modeling.* Mathematics Teacher Education-Partnership.

Anhalt, C., Cortez, R., & Kohler, B. (2021c). *MODULE(S²): Mathematical modeling, Module 3: Diverse perspectives in mathematical modeling.* Mathematics Teacher Education-Partnership.

Artzt, A. F., Sultan, A., Curcio, F. R., & Gurl, T. (2011). A capstone mathematics course for prospective secondary mathematics teachers. *Journal of Mathematics Teacher Education, 15*(3), 251–262.

Association of Mathematics Teacher Educators. (2017). *Standards for preparing teachers of mathematics.* amte.net/standards.

Aubrey, J., Patterson, C., Hart, J., & Tuttle, J. (2021). *MODULE(S²): Algebra for secondary mathematics teaching, Module 3: Fields and number systems.* Mathematics Teacher Education-Partnership.

Baldinger, E. E. (2018). Learning mathematical practices to connect abstract algebra to high school algebra. In N. Wasserman (Ed.), *Connecting abstract algebra to secondary mathematics, for secondary mathematics teachers* (pp. 211–239). Springer.

Ball, D. L., & Bass, H. (2003). Making mathematics reasonable in school. In J. Kilpatrick, W. G. Martin, & D. Schifter (Eds.), *A research companion to principles and standards for school mathematics* (pp. 27–44). National Council of Teachers of Mathematics.

Baumert, J., Kunter, M., Blum, W., Brunner, M., Voss, T., Jordan, A., Klusmann, U., Krauss, S., Neubrand, M., & Tsai, Y. (2010). Teachers' mathematical knowledge, cognitive activation in the classroom, and student progress. *American Educational Research Journal, 47*(1), 133–180.

Begle, E. G. (1979). *Critical variables in mathematics education: Findings from a survey of the empirical literature.* Mathematical Association of America and National Council of Teachers of Mathematics.

Bremigan, E. G., Bremigan, R. J., & Lorch, J. D. (2011). *Mathematics for secondary school teachers.* Mathematical Association of America.

Buchbinder, O. & McCrone, S. (2018). Taking proof into secondary classrooms—Supporting future mathematics teachers. In T. E. Hodges, G. J. Roy, & A. M. Tyminski (Eds.), *Proceedings of the 40th annual meeting of the North American Chapter of the International Group for the Psychology of Mathematics Education* (pp. 711–714). University of South Carolina & Clemson University. https://www.pmena.org/pmenaproceedings/PMENA%2040%202018%20Proceedings.pdf

Buchbinder, O., & McCrone, S. (2020). Preservice teachers learning to teach proof through classroom implementation: Successes and challenges. *The Journal of Mathematical Behavior, 58*, 100779.

Casey, S., Ross, A., Maddox, S., & Wilson, M. (2021a). *MODULE(S²): Statistical knowledge for teaching, Module 1: Study design and exploratory data analysis.* Mathematics Teacher Education-Partnership.

Casey, S., Ross, A., Maddox, S., & Wilson, M. (2021b). *MODULE(S²): Statistical knowledge for teaching, Module 2: Statistical inference.* Mathematics Teacher Education-Partnership.

Casey, S., Ross, A., Maddox, S., & Wilson, M. (2021c). *MODULE(S²): Statistical knowledge for teaching, Module 2: Statistical association.* Mathematics Teacher Education-Partnership.

Conference Board of the Mathematical Sciences. (2001a). *The mathematical education of teachers I.* American Mathematical Society and Mathematical Association of America.

Conference Board of the Mathematical Sciences. (2001b). *National summit on the mathematical education of teachers: Meeting the demand for high quality mathematics education in America.* Summary of the Event. https://www.cbmsweb.org/archive/NationalSummit/summary_article.htm

Conference Board of the Mathematical Sciences. (2012). *The mathematical education of teachers II.* American Mathematical Society and Mathematical Association of America.

Committee on the Undergraduate Program in Mathematics. (1961). *Recommendations for the training of teachers of mathematics.* Mathematical Association of America.

Cox, D. C., Chesler, J., Beisiegel, M., Kenney, R., Newton, J., & Stone, J. (2013). The status of capstone courses for pre-service secondary mathematics teachers. *Issues in the Undergraduate Mathematics Preparation of School Teachers, 4*, 1–10.

Cuoco, A., Goldenberg, E. P., & Mark, J. (1996). Habits of mind: An organizing principle for mathematics curricula. *Journal of Mathematical Behavior, 15*(4), 375–402.

Even, R. (2011). The relevance of advanced mathematics studies to expertise in secondary school mathematics teaching: Practitioners' views. *ZDM Mathematics Education, 43*, 941–950.

Ferrini-Mundy, J. & Findell, B. (2001). *The mathematical education of prospective teachers of secondary mathematics: Old assumptions, new challenges.* In *CUPM Discussion Papers about Mathematics and the Mathematical Sciences in 2010: What should students know?* (pp. 31–41). Mathematical Association of America. https://maa.org/sites/default/files/pdf/CUPM/math-2010.pdf

Forzani, F. M. (2014). Understanding "core practices" and "practice-based" teacher education: Learning from the past. *Journal of Teacher Education, 65*(4), 357–368.

Fukawa-Connelly, T., Mejia-Ramos, J. P., Wasserman, N., & Weber, K. (2020). An evaluation of ULTRA: An experimental real analysis course built on a transformative theoretical model. *International Journal of Research in Undergraduate Mathematics Education, 6*(2), 159–185.

Goulding, M., Hatch, G., & Rodd, M. (2003). Undergraduate mathematics experience: Its significance in secondary mathematics teacher preparation. *Journal of Mathematics Teacher Education, 6*(4), 361–393.

Grossman, P., Compton, C., Igra, D., Ronfeldt, M., Shahan, E., & Williamson, P. W. (2009). Teaching practice: A cross-professional perspective. *Teachers College Record, 111*(9), 2055–2100.

Hart, J., & Lai, Y. (2021). *MODULE(S²): Algebra for secondary mathematics teaching, Module 2: From numbers, to powers, to logarithms.* Mathematics Teacher Education-Partnership.

Hauk, S., Hsu, E., & Speer, N. (2017). *Transformational geometry teaching notes.* Capstone Math Project.

Hauk, S., Hsu, E., & Speer, N. (2018). *Covariation and rate: Flag and bike teaching notes.* Capstone Math Project.

Heid, M. K., Wilson, P. & Blume, G. W. (Eds.). (2015). *Mathematical understanding for secondary teaching: A framework and classroom-based situations.* Information Age Publishing.

Hoffman, A., & Even, R. (2019). The contribution of academic mathematics to teacher learning about the essence of mathematics. In M. Graven, H. Venkat, A. Essien, & P. Vale (Eds.), *Proceedings of the 43rd Conference of the International Group for the Psychology of Mathematics Education* (Vol. 2, pp. 360–367). PME.

Klein, F. (1924/1932). *Elementary mathematics from an advanced standpoint: Arithmetic, algebra, analysis.* Macmillan and Co., Limited.

Lai, Y. (2019). Accounting for mathematicians' priorities in mathematics courses for secondary teachers. *The Journal of Mathematical Behavior, 53*(4), 164–178.

Lai, Y., & Donsig, A. (2018). Using geometric habits of mind to connect geometry from a transformation perspective to graph transformations and abstract algebra. In N. Wasserman (Ed.), *Connecting abstract algebra to secondary mathematic, for secondary mathematics teachers* (pp. 263–290). Springer.

Lai, Y., & Hart, J. (2021). *MODULE(S²): Algebra for secondary mathematics teaching, Module 1: Concepts of relations and functions.* Mathematics Teacher Education-Partnership.

Lai, Y., Lischka, A., & Strayer, J. (2021). Characterizing prospective secondary teachers' foundation and contingency knowledge for definitions of transformations. In D. Olanoff, K. Johnson, & S. Spitzer. (Eds.) *Proceedings of the forty-third annual meeting of the North American Chapter of the International Group for the Psychology of Mathematics Education* (pp. 437–446). Philadelphia, PA.

Lischka, A., Lai, Y., Strayer, J., & Anhalt, C. (2020). MODULE(S²): Developing mathematical knowledge for teaching in content courses. In A. E. Lischka, B. R. Lawler, W. G. Martin, & W. M. Smith (Eds.), *The mathematics teacher education partnership: The power of a networked improvement community to transform secondary mathematics Teacher Preparation* (AMTE Professional Book Series, Vol. 4, pp. 119–141). Information Age Publishing.

Mathematical Education of Teachers as an Application of Undergraduate Mathematics. (2020a). *Groups of transformations: Abstract (Modern) Algebra I.* The MAA META Math Annotated Lesson Plans.

Mathematical Education of Teachers as an Application of Undergraduate Mathematics. (2020b). *Logarithms and isomorphisms: Abstract (Modern) Algebra I.* The MAA META Math Annotated Lesson Plans.

Mathematical Education of Teachers as an Application of Undergraduate Mathematics. (2020c). *Solving equations in Z_n: Abstract (Modern) Algebra I.* The MAA META Math Annotated Lesson Plans.

Mathematical Education of Teachers as an Application of Undergraduate Mathematics. (2020d). *Finding inverse functions and their derivatives: Calculus I.* The MAA META Math Annotated Lesson Plans.

Mathematical Education of Teachers as an Application of Undergraduate Mathematics. (2020e). *Newton's Method: Calculus I.* The MAA META Math Annotated Lesson Plans.

Mathematical Education of Teachers as an Application of Undergraduate Mathematics. (2020f). *Foundations of divisibility: Introduction to proof or discrete mathematics.* The MAA META Math Annotated Lesson Plans.

Mathematical Education of Teachers as an Application of Undergraduate Mathematics. (2020g). *The binomial theorem: Introduction to proof or discrete mathematics.* The MAA META Math Annotated Lesson Plans.

Mathematical Education of Teachers as an Application of Undergraduate Mathematics. (2020h). *Developing a margin of error through simulation: Statistics.* The MAA META Math Annotated Lesson Plans.

Mathematical Education of Teachers as an Application of Undergraduate Mathematics. (2020i). *Variability: Statistics.* The MAA META Math Annotated Lesson Plans.

McGuffey, W., Quea, R., Weber, K., Wasserman, N., Fukawa-Connelly, T., & Mejia-Ramos, J. P. (2019). Pre- and in-service teachers' perceived value of an experimental real analysis course for teachers. *International Journal of Mathematical Education in Science and Technology, 50*(8), 1166–1190.

Monk, D. H. (1994). Subject area preparation of secondary mathematics and science teachers and student achievement. *Economics of Education Review, 13*(2), 125–145.

Murray, E., & Star, J. R. (2013). What do secondary preservice mathematics teachers need to know? *Notices of the AMS, 60*(10), 1297–1299.

Rowland, T. (2013). The Knowledge Quartet: The genesis and application of a framework for analysing mathematics teaching and deepening teachers' mathematics knowledge. *Sisyphus, 1*(3), 15–43.

Rowland, T., Thwaties, A., & Jared, L. (2016). *Analysing secondary mathematics teaching with the Knowledge Quartet.* Paper presented at the 13th International Congress on Mathematics Education. Hamburg, Germany.

Simon, M. A. (2006). Key developmental understandings in mathematics: A direction for investigating and establishing learning goals. *Mathematical Thinking and Learning, 8*(4), 359–371.

Strayer, J. F., Adamoah, K., & Lai, Y. (2021). *Prospective secondary mathematics teachers' expectancy and value for teaching practices: Comparing across content areas.* Proceedings of the 2021 Mathematics Teacher Education Partnership Conference.

Stylianou, D. A., Blanton, M. L., & Knuth, E. J. (2009). *Teaching and learning proof across the grades.* Routledge.

Sultan, A., & Artzt, A. F. (2010). *The mathematics that every secondary school math teacher needs to know.* Routledge.

Sword, S., Matsuura, R., Cuoco, A., Kang, J., & Gates, M. (2018). Leaning on mathematical habits of mind. *Mathematics Teacher, 111*(4), 256–263.

Tall, D. (Ed.). (1991). *Advanced mathematical thinking* (Vol. 11). Springer.

Tatto, M. T., & Bankov, K. (2018). The intended, implemented, and achieved curriculum of mathematics teacher education in the United States. In M. T. Tatto, M. C. Rodriguez, W. M. Smith, M. D. Reckase, & K. Bankov (Eds.), *Exploring the mathematical education of teachers using TEDS-M data* (pp. 69–133). Springer.

Ticknor, C. S. (2012). Situated learning in an abstract algebra classroom. *Educational Studies in Mathematics, 81*(3), 307–323.

Tucker, A., Burroughs, E., & Hodge, A. (2015). *A professional program for preparing future high school mathematics teachers*. Program Area Study Group Report for the Mathematical Association of America. https://www2.kenyon.edu/Depts/Math/schumacherc/public_html/Professional/CUPM/2015Guide/Program%20Reports.html

Usiskin, Z., Peressini, E., Marchisotto, A., & Stanley, D. (2003). *Mathematics for high school teachers: An advanced perspective*. Prentice-Hall.

Wasserman, N., Fukawa-Connelly, T., Villanueva, M., Mejia-Ramos, J. P., & Weber, K. (2017). Making real analysis relevant to secondary teachers: Building up from and stepping down to practice. *PRIMUS, 27*(6), 559–578.

Wasserman, N., Fukawa-Connelly, T., Weber, K., Mejia-Ramos, J. P., & Abbott, S. (2022). *Understanding analysis: Connections for secondary mathematics teachers*. Springer.

Wasserman, N., & Galarza, P. (2018). Exploring an instructional model for designing modules for secondary mathematics teachers in an abstract algebra course. In N. Wasserman (Ed.), *Connecting abstract algebra to secondary mathematics, for secondary mathematics teachers* (pp. 335–361). Springer.

Wasserman, N., & Ham, E. (2013). Beginning teachers' perspectives on attributes for teaching secondary mathematics: Reflections on teacher education. *Mathematics Teacher Education and Development, 15*(2), 70–96.

Wasserman, N., & McGuffey, W. (2021). Opportunities to learn from (advanced) mathematical coursework: A teacher perspective on observed classroom practice. *Journal for Research in Mathematics Education, 52*(4), 370–406.

Wasserman, N., & Weber, K. (2017). Pedagogical applications from real analysis for secondary mathematics teachers. *For the Learning of Mathematics, 37*(3), 14–18.

Wasserman, N., Weber, K., Fukawa-Connelly, T., & McGuffey, W. (2019). Designing advanced mathematics courses to influence secondary teaching: Fostering mathematics teachers' "attention to scope." *Journal of Mathematics Teacher Education, 22*(4), 379–406.

Wasserman, N., Weber, K., Villanueva, M., & Mejia-Ramos, J. P. (2018). Mathematics teachers' views about the limited utility of real analysis: A transport model hypothesis. *Journal of Mathematical Behavior, 50*, 74–89.

Weber, K., Mejia-Ramos, J. P., Fukawa-Connelly, T., & Wasserman, N. (2020). Connecting the learning of advanced mathematics with the teaching of secondary mathematics: Inverse functions, domain restrictions, and the arcsine function. *Journal of Mathematical Behavior, 57(1)*, 100752.

Weber, K., Wasserman, N., Mejia-Ramos, J. P., Cohen, A., & Fukawa-Connelly, T. (n.d.). *ULTRA: Upgrading learning for teachers in real analysis*. https://ultra.gse.rutgers.edu

Winsor, M. S. (2009). One model of a capstone course for preservice high school mathematics teachers. *PRIMUS, 19*(6), 510–518.

Yan, X., Marmur, O., & Zazkis, R. (2022). Advanced mathematics for secondary school teachers: Mathematicians' perspectives. *International Journal of Science and Mathematics Education, 20*(3), 553–573.

Zazkis, R., & Leikin, R. (2010). Advanced mathematical knowledge in teaching practice: Perceptions of secondary mathematics teachers. *Mathematical Thinking and Learning, 12*(4), 263–281.

SECTION III

UTILIZING TECHNOLOGY TO SUPPORT THE LEARNING OF MATHEMATICS

CHAPTER 11

PREPARING SECONDARY PROSPECTIVE TEACHERS TO TEACH MATHEMATICS WITH TECHNOLOGY

Jennifer N. Lovett
Middle Tennessee State University

Allison W. McCulloch
University of North Carolina at Charlotte

Lara K. Dick
Bucknell University

Charity Cayton
East Carolina University

Hollylynne S. Lee
NC State University

Karen F. Hollebrands
NC State University

The Association of Mathematics Teacher Educators' (2017) *Standards for Preparing Teachers of Mathematics* (SPTM) has set aspirational goals for the preparation of well-prepared beginning teachers of mathematics that includes teaching with mathematics action technologies. These standards require that beginning teachers of mathematics not only be proficient users of technologies, but also understand how to use technology in meaningful ways to support students' thinking. In this chapter we describe where we are as a field with respect to meeting the goals of the SPTM

The AMTE Handbook of Mathematics Teacher Education: Reflection on Past, Present and Future—Paving the Way for the Future of Mathematics Teacher Education, Volume 5
pages 221–244.

related to teaching with technology. With problems of practice and theory related to the work of MTEs in preparing secondary mathematics teachers to teach mathematics with technology as a guidepost, we lay out foundational work in this area (the past), followed by the current efforts by three projects to address the practical and theoretical challenges MTEs are facing. Finally, we describe our vision for the work that is needed down the road for us to make headway towards meeting the aspirational goals laid out in the AMTE SPTM.

INTRODUCTION

The Association of Mathematics Teacher Educators' (AMTE, 2017) *Standards for Preparing Teachers of Mathematics* (SPTM) Indicator C.1.6 (*Use Mathematical Tools and Technology*) states that "well-prepared beginning teachers of mathematics are proficient with tools and technology designed to support mathematical reasoning and sense making, both in doing mathematics themselves and in supporting student learning of mathematics" (p. 11). Furthermore, prospective teachers should "make sound instructional decisions about when such tools enhance teaching and learning, recognizing both the insights to be gained and possible limitations of such tools (NCTM, 2012, p. 3)" (AMTE, 2017, p. 12). This requires well-prepared beginning secondary teachers of mathematics (referred to as prospective teachers [PSTs] moving forward) to not only be proficient users of technologies, but also to understand how to use technology in meaningful ways to support students' thinking.

The purpose of this chapter is to describe where the field of mathematics education is in meeting the goals of the SPTM related to teaching with technology and the areas of focus mathematics teacher educators (MTEs) must address to meet these goals. With problems of practice and theory related to the work of MTEs in preparing secondary mathematics teachers to teach mathematics with technology as a guide, we discuss how three National Science Foundation funded teams are currently addressing these critical issues. In addition, we lay out a vision for work that is needed to meet the aspirational goals laid out in the AMTE SPTM.

CONCEPTUAL FRAMEWORK

While learning to use technology tools is important, it is not sufficient for learning to teach with them. Roschelle et al. (2017) observed that "learning how to use technology to advance learning is too often separated from the core of learning to teach mathematics effectively" (p. 859). The AMTE SPTMs (2017) emphasizes this important connection. Indicator C.1.6 in the SPTM articulates what a well-prepared beginning secondary teacher of mathematics should learn about teaching mathematics with technology. Furthermore, PSTs should be proficient in making "sound decisions about when such tools enhance teaching and learning, recognizing both the insight to be gained and possible limitations of such tools" (p. 11). It goes on to specify that PSTs should be "prepared to use 'mathematical action technologies' (cf. NCTM, 2014, p. 79), powerful tools that will be a part of the lives of the students they teach" (p.12). Thus, PSTs need ample experience engaged in both working with mathematical action technologies as learners and in considering how to position these technologies to support student learning.

With the connection to the core of learning to teach with technology in mind, we interpret C.1.6 to have four components as shown in Figure 11.1. The fourth aspect, *use mathematical action technologies as a teacher of mathematics to support mathematical sense-making and reasoning,* suggests that once one decides whether or not using a mathematical action technology is appropriate for a particular learning goal, they must also have the knowledge and skills related to planning and implementing effective and equitable lessons. These skills are articulated in the indicators of Standards C.2 (*Pedagogical knowledge and practices for teaching mathematics*) and C.3 (*Students as learners of mathematics*).

To frame the specific skills both implicitly and explicitly articulated in the SPTM related to PSTs' doing mathematics and teaching mathematics with mathematical action tools, we interpret C.1.6 as assuming the indicators articulated for C.2 and C.3.[1] For example, C.2.2 says that PSTs plan for effective instruction. When technology is used in a lesson, planning includes (but is not limited to) making decisions about what technology tools would best support the learning goals, how to group students (or not) with devices to access the technology, how to gain access to students' mathematical work with technology, and how to facilitate discussions about how students use technology to develop their mathematical understanding. Similarly, PSTs should be able to anticipate and attend to students' thinking about mathematics content (C.3.1) when students are using mathemati-

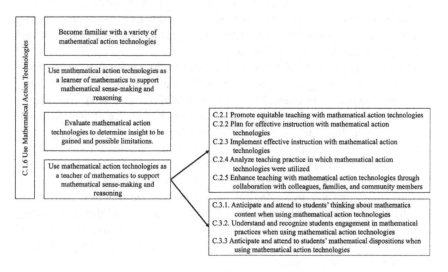

FIGURE 11.1.   Framing the Goals for Preparing Teachers to Teach Mathematics with Technology Adapted from AMTE 2017 (McCulloch, Leatham, et al., 2021)

cal action technologies. Such a skill requires that PSTs coordinate what students articulate in their verbal and written mathematical work with the ways that students engage with technology that is informing their thinking.[2]

PAST, PRESENT, AND FUTURE OF PREPARING PSTS TO TEACH WITH TECHNOLOGY

In their introduction to the theme issue of *Journal of Teacher Education* on innovative uses of technology in teacher education, Borko et al. (2009) noted that teaching teachers to use technology effectively is a "wicked problem" (p. 3). We see this "wicked problem" as an issue that MTEs have faced in the past, are currently facing, and will continue to face in the future due to the ever-evolving state of technology tools available to support secondary students' learning of mathematics. In the following sections we discuss how the field has addressed this in the past, how three teams are currently addressing this problem, and how we believe the field can move forward.

Past

Following the emergence of mathematical action technologies in the 1980s and 1990s, the field began to understand how these technologies could be used to support student learning of mathematics, as well as how to prepare PSTs to teach with these technologies. At this time the prevailing technologies that were believed to be most important to incorporate in PST courses were graphing calculators, spreadsheets, and dynamic geometry environments. Studies suggest that PST preparation during this time period was focused on exploring mathematics, with little to no emphasis placed on making pedagogical decisions or on analyzing students' mathematical thinking in technology-mediated learning environments[3] (Kersaint et al., 2003; Leatham, 2006).

The prevailing theoretical framework related to the work of preparing PSTs to teach with technology at this time was Koehler and Mishra's (2005) Technological Pedagogical and Content Knowledge (TPACK) (Yigit, 2014).[4] This framework describes TPACK, the integration of teachers' knowledge of content, pedagogy, and technology, as a specialized knowledge needed for effectively using technology to teach specific content matter. When considering the development of TPACK in the context of preparing mathematics PSTs, Lee and Hollebrands (2008) noted it is most essential to focus on the intersections of content (mathematical) knowledge with technological and pedagogical knowledge (Figure 11.2). Specifically, this integrated approach is based on a model of the relationships among these four types of knowledge, where mathematical knowledge is the largest and foundational set. They refer to this as Technological Pedagogical Mathematical Knowledge (TPMK).

One of the major concerns of MTEs at this time was the lack of curricular materials available for preparing PSTs for teaching mathematics with technol-

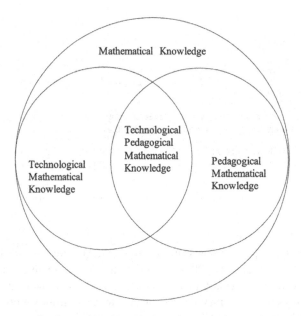

FIGURE 11.2. Technological Pedagogical Mathematical Knowledge (Lee & Hollebrands, n.d.)

ogy. One project that addressed this issue was the Preparing to Teach Mathematics with Technology (PTMT) project. Through a series of grants funded by the National Science Foundation (NSF) the team created three sets of materials: one focused on Teaching Data Analysis and Probability (Lee et al., 2010), one focused on Teaching Geometry (Hollebrands & Lee, 2012), and one focused on Teaching Algebra (McCulloch et al., 2015) that used an integrated approach to preparing PSTs for teaching with technology. This integrated approach was embodied in the materials by first having PSTs examine a content question with a technology tool (labeled *Engaging in Content*), then providing specific steps for how to take particular actions with the technology when needed, and finally answering pedagogical questions (labeled *Considerations for Teaching*). Considerations for Teaching questions were aimed at developing PSTs' understanding of how technology and various representations could support students' mathematical or statistical thinking (see Figure 11.3). These questions sometimes included video clips of students engaged in mathematics or statistics tasks using technology tools.

The PTMT materials have been used by over 300 MTEs across the country, in a variety of course contexts (e.g., content courses, methods courses, math-specific technology courses), impacting over 8000 prospective or practicing teachers and are available on a free online portal (http://go.ncsu.edu/ptmt). Research on the impact of the PTMT curricular materials indicated that PSTs were able to actively use advanced features of technology to solve tasks (e.g., Lee et al., 2012; McCull-

CONSIDERATIONS FOR TEACHING

Q7. In what ways could this sketch and these questions be helpful to eighth-grade students who are learning the triangle inequality theorem? How might this activity confuse students?

Q8. Develop a rationale for why the teacher selected the particular examples that he did. Are there any that you would change? Explain.

Q9. Are there any sets of side lengths that you think students will have difficulty with? What explanations and conjectures do you anticipate students will create?

FIGURE 11.3. Sample Considerations for Teaching Questions from the PTMT Materials

och et al., 2019) and plan lessons using a variety of technology tools (Hollebrands et al., 2016; McCulloch et al., 2018). In addition, PSTs were able to make sense of students' work with technology by examining video cases when given the opportunity to do so, though they needed support to move from simply describing students' work and comparing it to their own toward interpreting students' thinking and restructuring their own, naive, models of student thinking to include what they learn through the experiences (Wilson et al., 2011).

Present

Since work on the PTMT project began, both the technologies that are available and the infrastructure necessary for classroom implementation have changed considerably (Niess & Roschelle, 2018). Mathematical action technologies have become more intuitive and user friendly, more are free, and many are web-based. Recommendations continue to include the incorporation of such technologies in the teaching of secondary mathematics (e.g., Common Core State Standards [National Governors Association Center for Best Practices and the Council of Chief State School Officers, 2010]); Principles to Actions [NCTM, 2014)]). In addition, recommendations for the preparation of secondary mathematics teachers (e.g., Conference Board of Mathematical Sciences, 2012; AMTE, 2017) have become more explicit that PSTs need to have experiences with mathematical action technologies not only as a learner, but also as a teacher.

To understand how U.S. secondary mathematics teacher preparation programs have responded to these recommendations, in 2018 McCulloch, Leatham, et al. (2021) completed a nationwide survey of accredited secondary mathematics teacher preparation programs. They found that by and large programs are providing PSTs opportunities to engage with a variety of mathematical action technologies as learners of mathematics. However, quite a few programs indicated that they provided these opportunities rarely or never, and that the technologies they use are outdated. Additionally, they found that while most programs have PSTs plan for effective mathematics instruction that uses technology, few programs provide opportunities for PSTs to implement their lessons through formal or informal field experiences or for recognizing, understanding, anticipating, and attending to students' mathematical practices with technology. These findings suggest it would be helpful if MTEs had access to high quality instructional materials that provide opportunities to not only engage with mathematical action technologies as learners of mathematics, but also engage in recognizing, understanding, anticipating, and attending to students' mathematical practices and approximations of practice (e.g., Grossman et al., 2009; McDonald et al., 2013) with respect to implementing effective instruction with technology.

In response to the changing landscape, this group of authors has branched out from the original PTMT projects to continue to address areas of critical need in different ways. While there are other MTEs doing similar work as well, in the following sections we describe three other NSF-funded projects that explicitly build from the PTMT projects and the ways in which they address the components of the AMTE SPTMs related to teaching with technology (i.e., C.2-Tech and C.3-Tech).

Forging Connections Through the Geometry of Functions Using Technology

The Forging Mathematical Connections Through the Geometry of Functions (Forging Connections [FC]) project is developing curriculum materials for capstone content courses that focus on strengthening the mathematical background of PSTs through technology-rich investigations. Using a geometric approach to functions, the FC team created units focused on function, geometric transformations, similarity and congruence, and other more advanced topics (e.g., trigonometry, calculus, complex numbers). Emphasizing the components of the SPTMs related to using technology as a learner (C.1.6), the project materials use Web Sketchpad activities (KPC Technologies, 2019) as a means to ground teachers' understanding in a visual, sensorimotor, and embodied approach to the mathematics.

The use of one's body in learning mathematics has drawn more attention as researchers focus on sensorimotor actions and their connection to conceptual understanding. Knowledge and understanding are grounded in goal-oriented sensorimotor schemes; formal knowing is the discursive signification and representational elaboration of these schemes (e.g., de Freitas & Sinclair, 2014; Hutto et al., 2015). For example, as PSTs learn about functions by directly interacting with them by moving a mouse or a finger, their understandings are intertwined with their physical actions and corresponding movements on the screen (Figure 11.4).

With theoretical work in embodied cognition guiding their design, each lesson presents a series of PST learning activities on a single webpage providing access to dynamic mathematics using Web Sketchpad. With Web Sketchpad, teachers and curriculum designers can decide on a task and create a "websketch" that con-

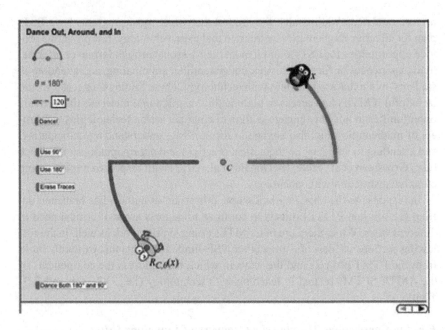

FIGURE 11.4. Example of a Rotate Function Websketch in which PSTs Act on the Penguin

tains only a set of carefully chosen tools specific to the proposed task. As curriculum designers, this enabled the team to design activities with the tools we think are most relevant so that PSTs can immediately start working on the task without the distraction of unnecessary tools. To illustrate, consider different pages from a "construct a rhombus" task the FC team has used in workshops with MTEs shown in Figure 11.5. Each page of the websketch contains a small set of tools that can be used to construct a rhombus, thereby providing PSTs opportunities for mathematical reasoning and sensemaking, as well as opportunities to reflect on how the tools themselves supported their reasoning and sensemaking.

While the FC materials emphasize learning mathematical content through engaging with carefully designed websketches, they also include activities designed

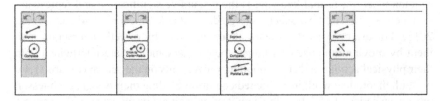

FIGURE 11.5. Example of Specific Tools Made Available for Four Different Rhombus Constructions

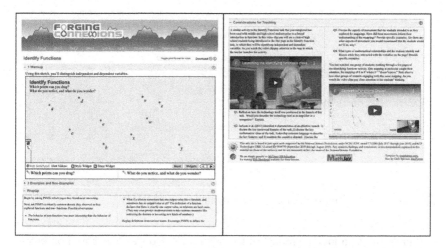

FIGURE 11.6.   PST View of the Identify Functions Lesson

to provide PSTs opportunities to think about pedagogical issues related to the mathematics and technology in each unit. For example, the first lesson of the materials is shown in Figure 11.6. The blue bars are expandable sections that each include a websketch and series of questions to guide the PSTs' mathematical investigation. The yellow bars, the Considerations for Teaching section, has two videos for PSTs to analyze. These videos show high-school students engaged in this very same lesson. Watching and analyzing such videos help PSTs to reflect on the mathematics of the lesson not only through their own experiences as a learner, but also through the perspective of secondary students.

Reflecting on McCulloch, Leatham, et al.'s (2021) framing of the AMTE SPT-Ms (Figure 11.1), the FC materials emphasize the first three components of C.1.6. and in addition, provide opportunities for PSTs to analyze teaching practices (C.2.4-Tech) to anticipate and attend to students' thinking (C.3.1-Tech), and to understand and recognize students' engagement in mathematical practices (C.3.2-Tech) on similar websketch activities. PSTs who have used these materials in capstone courses have shown improved understanding of function and geometric transformations. Of particular importance was that by the end of the course, PSTs' views of function were expanded to include geometric transformations and their approaches to congruence proofs shifted to include instances of transformations (Ozen-Unal et al., 2022). To learn more about the FC project and to access all of the materials see https://geometricfunctions.org/fc/.

Extending Preparation of Teachers with Technology to Statistics

The Enhancing Statistics Teacher Education through E-Modules (ESTEEM) project began by extending the prior PTMT materials focused on data analysis

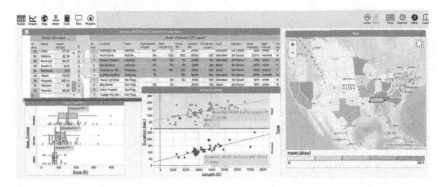

FIGURE 11.7.   Example of Linked Representations in CODAP

and probability to include more modern approaches to statistics based on investigative cycles with large datasets (10–20 variables and 150 or more cases) using the statistical action technology CODAP. Initially, the ESTEEM team expanded the existing CODAP technology, through a collaboration with the CODAP developers at the Concord Consortium, to include common "mathematical actions" needed for teaching secondary statistics within a drag-and-drop interface with linked multiple representations (e.g., data table, graphs, map) (Figure 11.7). Then, ESTEEM enhanced the previous PTMT materials, adding a module titled, "Statistical Investigations" to the PTMT portal (http://go.ncsu.edu/ptmt) to include a stronger attention to learning about the investigative process with multivariate data with new features in CODAP.

Next, the ESTEEM team created online modules for content or methods courses to prepare PSTs for teaching statistics both with and without technology. The ESTEEM materials consist of three interconnected modules and two independent assignments. The modules build critical understanding about statistical concepts, develop pedagogical knowledge about teaching statistics, and provide support for using CODAP as a tool to support students' learning. Details on the design of the ESTEEM materials are discussed in Lee et al. (2021) and all materials are available for free in the ESTEEM portal (http://go.ncsu.edu/esteem).

Foundational to the design of the ESTEEM materials is developing PSTs' TPMK, in this case their Technological Pedagogical Statistical Knowledge (Lee & Hollebrands, 2011) in which knowledge of statistics is foundational. To assist PSTs with understanding how students reason with statistics and develop their understandings the ESTEEM project integrated the Students' Approaches to Statistical Investigations [SASI] framework (Lee & Tran, 2015, see Figure 11.8). At the core of the framework is the cycle of statistical investigation and productive statistical habits of mind are woven throughout. The framework also describes growth in statistical sophistication as described by GAISE (Franklin et al., 2007) and guidance of reasonable expectations for students at each level. The SASI

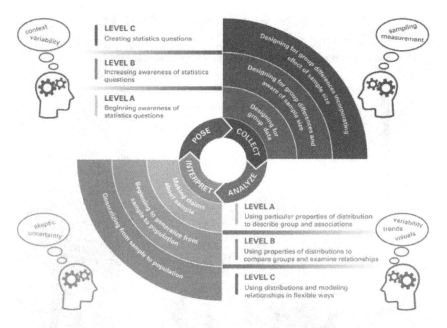

FIGURE 11.8.   The Students' Approaches to Statistical Investigations (SASI) Framework

framework is referred to throughout the ESTEEM materials to explicitly support PSTs in interpreting students' statistical understanding, particularly when using CODAP.

PSTs have multiple opportunities in each module to engage directly with data using CODAP to learn how different actions on and with data, and with various data representations, can provide opportunities for reasoning about statistical ideas. For example, the first module emphasizes the differences between mathematics and statistics as well as how to support students in learning to reason statistically. It includes activities concerning launching meaningful statistical tasks, the statistical investigation cycle, and fostering discussions around students' statistical thinking. PSTs also learn about different features of technology-supported lessons in statistics and develop skills throughout the modules that they apply to create a lesson plan that uses CODAP in a real data context to address students' learning of specific statistics standards.

The ESTEEM project created and curated a collection of videos that illustrate whole class discussions, and student-student and student-teacher interactions when technology is being used. For example, the first module contains a series of videos including: 1) an episode depicting the launch of a task in a middle school classroom using a point-of-view video of a roller coaster ride to contextualize students' understanding of the phenomena being investigated, 2) a video of students

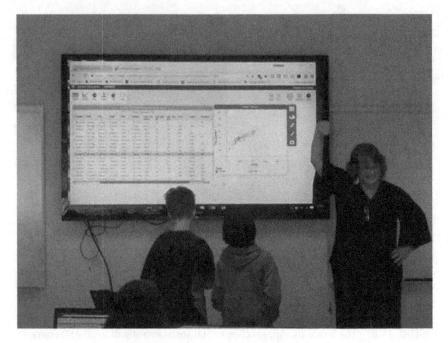

FIGURE 11.9.   Still Picture from the Video of a Whole Class Discussion

exploring data with CODAP using different actions on the data and representations, and 3) a video in which the teacher is facilitating a whole class discussion (Figure 11.9) d using the five practices for orchestrating a productive mathematics discussion (Smith & Stein, 2018).

Reflecting on McCulloch, Leatham et al.'s (2021) framing of the SPTMs, the ESTEEM materials emphasize the first three components of C.1.6. In addition, through the use of whole class video ESTEEM provides opportunities for PSTs to plan for effective teaching (C.2.2-Tech), analyze teaching practices (C.2.4-Tech), and analyze student thinking (C.3.1-Tech, C.3.2-Tech). Analysis of lesson plans created by PSTs who have used these materials have shown the PSTs were developing an ability to design tasks that can give students appropriate experience using large, real data sets and connecting statistical thinking to the context of data. In addition, when describing the task launch in their lesson plans, the majority of PSTs contextualized the data and investigation, and prompted students to make a personal connection with the data's context; however, few of the PSTs' task launches motivated a need for a driving statistical question to engage students in the task. PSTs' lessons illustrated they still struggled with keeping a focus on statistical practices, investigations, and uncertainty in claims, as well as utilizing advanced features of CODAP to support students' work (Casey et al., 2020).

Examining Students' Practices in Technology-Mediated Learning Environments

The Preparing to Teach Mathematics with Technology–Examining Student Practices (PTMT-ESP) project set out to develop curriculum materials that focused on examining students' practices in technology-mediated learning environments by creating video-enhanced resources. The PTMT-ESP team has developed and continues to refine eight modules situated in the context of algebra and functions to complement the PTMT Algebra materials. The materials are designed to increase PSTs' knowledge of students' understanding, thinking, and learning with technology through noticing (Jacobs et al., 2010) and begin to develop the skills needed to orchestrate productive mathematical discussion (Smith & Stein, 2018). To achieve this, each module begins with PSTs engaging with the technology-mediated mathematics task as a learner, then examining middle school and high school students' work to notice their mathematical thinking and consider pedagogical implications (see Lovett et al., 2020 for a description of the module design principles). To facilitate this, every activity includes at least one video clip of a pair of students engaging in a technology-enhanced task (many include a series of video clips).

Similar to the original PTMT materials, the theoretical foundation for the PTMT-ESP materials is the development of PSTs' TPACK. Since the goal is to focus on examining students' mathematical practices with technology in the context of algebra and functions, concentration is placed on the third component of TPACK as described by Niess (2005), "knowledge of students' understanding, thinking, and learning with technology in a particular subject" (p. 511). Such knowledge must be purposely developed via teaching practice. To guide this work, we draw on Jacobs et al.'s (2010) lens of teacher noticing of students' thinking and Wilson et al.'s (2011) categories of PSTs' model building process for making sense of student thinking in technology-mediated learning environments. The prior identifies the component skills that make up and act as a scaffold for the practice of noticing students' thinking, and the latter adds the importance of PSTs reflecting on their TPACK in light of examining student work so they can reconcile their observations and inferences of student's thinking with their own understandings. Specifically, we drew upon these ideas, as well as the theory of semiotic mediation (e.g., Bartolini Bussi & Mariotti, 2008; Mariotti, 2000, 2013), to develop the Noticing Students' Mathematical Thinking in Technology-Mediated Environments framework [NITE] (Bailey et al., 2022; Dick et al., 2022) to guide PSTs as they do this important work.

The NITE framework explicates the complex process of teacher noticing of student thinking (Jacobs et al., 2010) in technology-mediated learning environments. Within the NITE framework (Figure 11.10), attention to and interpretation of students' spoken and written mathematical thinking are separated from attention to and interpretation of the students' engagement with the technology to highlight the importance of coordinating the two. The "decide how to respond" com-

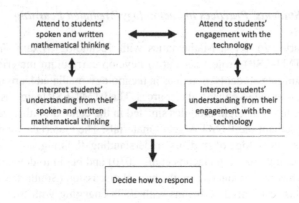

FIGURE 11.10. The Noticing in Technology Environments (NITE) (Dick et al., 2021)

In the video you will see a pair of HS students currently enrolled in an Integrated Math 1 course working on a task in which they are being <u>introduced</u> to parameters of quadratic functions in vertex form (i.e., $f(x) = a(x - h)^2 + k$). Before going any further, take a moment to get familiar with the

 <u>Desmos Task</u>.

Now that you are familiar with the task, you are going to analyze a video of a pair of students, Sara and Julian, working on the same task. As was noted above, Sara and Julian are currently enrolled in an Integrated Math 1 course. They have been working with quadratic functions but have not been formally introduced to vertex form. This task serves as their introduction.

▶ In this video <u>Sara and Julian</u> are working on page 2 and then 8 of the Desmos Quadratic Parameters Activity.

Q1. Attend to (i.e., describe in detail) how the students determined the effect of the h slider on the graph of the quadratic function.

Q2. Interpret the students' current (rough draft) understanding of the h parameter. Provide evidence from the video to support your claims.

FIGURE 11.11. Example Module 1

ponent is separated from the other components in order to balance the importance of focusing on both spoken and written mathematical thinking, and technology-engagement, prior to making instructional decisions; in addition, when making an instructional decision, one must consider how to position the technology (or not) in their response to support the student in moving their mathematical thinking forward.

The PTMT-ESP materials introduce the NITE framework to PSTs as part of a foundational module. Subsequent modules use the NITE framework to scaffold PSTs' practice of noticing student thinking in technology-mediated learning environments. Across the eight modules, the PSTs engage in activities that increase in sophistication from noticing students' spoken and written work and engagement with the technology, to using the foundations of noticing to facilitate mathematical discussions within classrooms. For example, in Module 1 after introducing the NITE framework PSTs have opportunities to engage in the component skills of attending to and interpreting student thinking as is shown in Figure 11.11. In Module 3, PSTs build on their noticing to make informed decisions with respect to selecting and sequencing students' strategies for a whole class discussion (Figure 11.12). Throughout the eight modules there are 46 video clips that represent diverse student populations (e.g., different ethnicities, religious cultures, gender expressions, primary languages) and settings (i.e., rural, suburban, urban).

Reflecting on McCulloch, Leatham et al.'s (2021) framing of the SPTMs, the PTMT-ESP materials provide PSTs the opportunity to engage in all aspects of C.1.6, including many of the components in C.2-Tech and C.3-Tech. Specifically, with the support of the NITE framework PSTs have opportunities to engage in the component skills of attending to and interpreting student thinking (C.3.1-Tech, C.3.2-Tech, C.3.3-Tech) and build upon their noticing to make informed instructional decisions (C.2.3-Tech). To date, studies have shown that 1) examining students' practices on technology-mediated mathematics tasks deepens PSTs mathematical knowledge (Lovett et al., 2019) and 2) explicit introduction to the NITE framework improves PSTs practice of noticing student thinking in technology-mediated learning environments (Bailey et al., 2022). To learn more about the PTMT-ESP project and access all materials visit http://go.ncsu.edu/ptmtportal.

Summary

The FC, ESTEEM, and PTMT-ESP projects have worked to address the need for materials that support MTEs' work in different course models for secondary PSTs (i.e., content courses, methods courses, math-specific technology courses). These projects are developing and highlighting a variety of free, web-based mathematical action technologies alongside authentic videos with an aim toward providing opportunities for PSTs to anticipate, attend to, and interpret student thinking, as well analyze effective teaching practices and engage in approximations of practice. FC and ESTEEM both provide materials that support PSTs learning specific mathematics/statistics content while also considering the ways in which

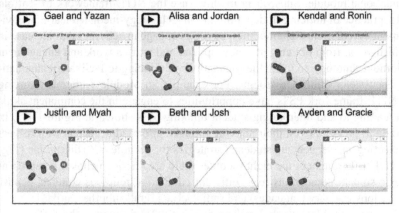

Q3. Click on each pair's graph below to watch video of the students explaining their strategy for creating their first graph (Each video will open in a new window tab). As you watch think closely about the students' reasoning related to the learning goals. As a reminder, the learning goals are:
- Sketch accurate graphs to show how a variable changes over time (i.e., the qualitative features of the relationship).
- Distinguish between a *graph* of a relationship between variables in a scenario and a *picture* of the scenario.

Table 2: Student Video Clips

| Gael and Yazan | Alisa and Jordan | Kendal and Ronin |
| Justin and Myah | Beth and Josh | Ayden and Gracie |

Imagine you have decided to have a whole class discussion about their graphs so far, with the intent of having them go back and refine their graphs after the discussion. (You don't want the discussion to "give away" the big ideas, but to promote thinking about critical features in the simulation and graphs.) For this discussion you have time for at most three pairs to share their thinking.

FIGURE 11.12.   Example Module 3

technology can support student learning of that content. In addition, both of these projects have worked directly with the technology developers to change the tools themselves to address needs in the field. In contrast, PTMT-ESP is situated in a specific content strand but is focused on supporting PSTs in developing their practice with respect to noticing student thinking in technology-mediated learning environments, eliciting student thinking, and facilitating small group and whole class discussions. Whether the content is foregrounded or backgrounded, collectively these projects have provided MTEs with much needed materials that can be used to prepare PSTs with respect to teaching with technology as described in indicators SPTM C.2-Tech and C.3-Tech.

Vision for the Future

To better prepare our beginning teachers to teach mathematics with technology it will take the work of all MTEs—not just those who research or specialize in teaching with technology. We need all MTEs to provide opportunities for PSTs to engage in tasks aligned with C.2-Tech and C.3-Tech, as well as those who specialize in teaching with technology to keep moving the field forward. In the following sections we lay out our vision for MTEs preparing PSTs and for those MTEs engaging in research or curriculum development related to technology.

Implications for ALL MTEs Preparing PSTs

An important aspect of preparing PSTs to teach mathematics using technology as a tool is for MTEs to incorporate commonly used mathematical action technology into their courses. In the past this meant software that required a license to be purchased and downloaded which research has shown is a barrier to teacher use, not only because of the cost, but also because they are not accessible on a variety of platforms, including students' individual devices (McCulloch et al., 2018). Currently, more practicing teachers are incorporating free, web-based software that works on all platforms (e.g., chromebooks, handheld devices), such as GeoGebra, Desmos, CODAP and Web Sketchpad. The features of these software are constantly evolving and new software is always under development. Thus, it is crucial that MTEs continue their own professional development to ensure they incorporate the latest technologies with PSTs. Such professional development can be accessed through webinars, workshops, and seeking information from the technology's websites.

It is necessary for MTEs to consider how to address all aspects of the SPTM within PST coursework. This means providing opportunities for PSTs that reach beyond the initial indicators of C.1.6 to include the important pedagogical practices articulated in C.2-Tech and C.3-Tech. The materials from the three projects discussed in this chapter have been designed specifically for use by MTEs. Thus, a first step is for MTEs to familiarize themselves with curricular materials such as these and others that have been developed and are ready for use in courses designed for prospective teachers (e.g., methods courses, content courses, math-specific technology courses). These materials provide opportunities for PSTs to focus on students' mathematical thinking and for approximations of the teaching practice of teacher noticing in technology-mediated learning environments.

As can be seen in the projects like those described in this chapter, much of these aims can be accomplished in content and methods courses. However, to really get at C.2-Tech PSTs need to move beyond approximations of practice and have opportunities to practice implementation in real classrooms with real students—formal and informal field experiences where students are working in technology-mediated learning environments are necessary. Finding quality field placements is difficult. Finding quality field placements in which students are using mathematical action technologies is even more difficult. Even so, there are examples of models for informal field experiences that provide opportunities for MTEs and clinical educators to collaborate in supporting PSTs learning to teach with technology in real classrooms (see McCulloch et al., 2020; Meagher et al., 2011).

IMPLICATIONS FOR MTES ENGAGING IN TECHNOLOGY RELATED RESEARCH AND DEVELOPMENT

Practically speaking, for MTEs to prepare PSTs and enact C.1.6, C.2-Tech, and C.3-Tech with technology, the field is in need of additional curriculum development, a focus on issues of equity, and additional research. The projects described in this chapter collectively provide a foundation on which to build. However, they all focus on secondary content and are limited in the pedagogical aspects they aim to develop and reinforce. At the secondary level, more curriculum materials are needed examining student-student interactions, student-teacher interactions, teacher moves, and whole class discussions across *all* content areas. In addition, there has not been much work in this realm in the context of K–8 mathematics.

Additionally, an oversight in all of the work discussed in this chapter is the explicit attention to equity. It is imperative that PSTs have the opportunity to witness all students engaging with mathematical action technologies regardless of how students have been tracked, their different ethnicities, religious cultures, gender expressions, and/or primary languages. It is important for PSTs to consider how teachers' position the technology to elicit and build on *all* students' thinking. While including videos that position diverse students as powerful doers of mathematics is a starting point, viewing diverse students is not enough. PSTs need opportunities to interrogate existing inequities and the ways in which using technology in mathematics instruction can perpetuate them or disrupt them. This will require the use of critical theories to guide both curriculum development and research.

Along with the use of critical theories to guide this work, there are additional theoretical issues we should consider as a field. In their analysis of a collection of frameworks MTEs report using in their work, McCulloch , Leatham, et al. (2021) found that MTEs are using theories that frame how students learn mathematics with technology, the design and evaluation of mathematics technology tools and tasks, how teachers use technology to teach mathematics, and how teachers learn to use technology to teach mathematics. Looking at the theoretical underpinnings of the projects described in this chapter, we see a variety of theories that ground material development, as well as frameworks intended to be used explicitly with PSTs to support their learning (e.g., NITE, SASI). Theoretical frameworks that guide the work of MTEs and PSTs have been very important to these projects, and we suggest that more are needed. Developing such theories in mathematics teacher education will help us a) conceive of ways to improve our work, b) describe, interpret, explain, and justify both MTE and PST activity, c) transform problems of our practice into research questions, and d) generate knowledge that moves the field forward (English & Sriraman, 2009).

In order to move the field forward, it is important to consider both what is not being framed in the frameworks we use, and the ways in which we are building theory. For example, McCulloch, Leatham, et al. (2021) noted that the field has yet to explicitly frame issues of equity related to teaching mathematics with technology and classroom mathematics discourse. Since then there have been efforts to frame equity from a perspective of access (e.g., McCulloch, Lovett, et al., 2021) and from the perspective of identity and justice (e.g., Suh et al., 2022). However, this is only the beginning of equity work that is needed in the area of teaching and learning mathematics with technology. We posit that there are additional approaches that can highlight both how students use and learn with mathematical action technologies, alongside how teachers use these technologies to support their students' mathematical understandings. Regardless of what direction the field takes in working to further theoretical development, we believe it is imperative that this work be completed alongside efforts to ensure both prospective and practicing teachers of mathematics be provided opportunities to experience the power of students' learning mathematics with mathematical action technologies.

In 1994, Kaput and Thompson, when reflecting on how research in mathematics education could inform the teaching and learning of mathematics in technology-mediated learning environments, asked if in the future MTEs "would assume responsibility for shaping the roles of new technologies in school mathematics, seeing that uses of new technologies reflect the substance of their curricula and pedagogical ideals, and if researchers will turn issues raised by new technologies into researchable questions" (p. 683). The present work of projects like ESTEEM and FC have begun to assume responsibility for helping technology designers to shape the features that drive students' experiences when engaging the new technologies. MTEs who have expertise in teaching and learning mathematics with technology should seek opportunities for such collaborations. In addition, it is imperative that MTEs with specific technology expertise work to ensure that those technologies are being used in accordance with what the field would regard as sound pedagogical ideals. This might be accomplished through curriculum development or professional development for practicing teachers or MTEs.

Even though a large portion of this chapter has focused on the development and evaluation of curriculum materials to support the work of MTEs, that does not diminish the importance of the role of research on teaching and learning of mathematics in technology-mediated learning environments. Moving forward, it is important that we move away from research focused solely on evaluating the effectiveness of specific curriculum materials and also consider issues raised by the use of these new technologies in instruction. For example, there is a need to focus research around student discourse when students are working together using mathematical action technologies. Questions like, "How do students' discussions as they engage with the technology provide evidence of their mathematical understanding?" and "How can teachers best facilitate student discussions when they must attend to both how students are using technology and their verbal discussions?" should be studied.

CONCLUSION

As we think about the future, we challenge *all* MTEs to continue to incorporate a variety of mathematical action technologies into their courses and to ensure PSTs have opportunities to practice the pedagogical intricacies of teaching with technology as is articulated in the goals of C.2-Tech and C.3-Tech. As mathematical action technologies continue to evolve and become even more ubiquitous in K–12 classrooms, it is imperative that PSTs are prepared to use them in ways that position all students as doers of mathematics.

ACKNOWLEDGMENTS

The PTMT project was supported by the National Science Foundation [NSF] under grants DUE 0442319, 0817253, and 1123001 awarded to NC State University. The PTMT-ESP project is supported by NSF under grant DUE 1820998 awarded to Middle Tennessee State University, grant DUE 1821054 awarded to University of North Carolina at Charlotte, grant DUE 1820967 awarded to East Carolina University, and DUE 1820976 awarded to NC State University. The NSF also supported the ESTEEM project (DUE 1625713) and the Forging Connections project (DUE 1712280) with grants awarded to NC State University. Any opinions, findings, and conclusions or recommendations expressed herein are those of the principal investigators and do not necessarily reflect the views of the National Science Foundation.

ENDNOTES

1. Note that we have not included Standard C.4, *Social contexts of mathematics teaching and learning*, in this framework. This choice was made because this standard is related to the social context of the classroom broadly and teacher knowledge related to this standard is the same (and equally important) regardless of whether or not technology is being used in a particular lesson.
2. Going forward we will refer to these indicators according to their number adding "tech" to the end of those in C.2 and C.3 to indicate our interpretation of them in the context of teaching with technology (i.e., C.2.1-Tech).
3. The term technology-mediated learning environment refers to an environment in which students are engaged with mathematical action technologies to support their learning of the mathematics.
4. For descriptions of other commonly used frameworks including SAMR and IGS see McCulloch et al. (2020).

REFERENCES

Association of Mathematics Teacher Educators. (2017). *Standards for preparing teachers of mathematics*. amte.net/standards

Bailey, N. G., Yalman Ozen, D., Lovett, J. N., McCulloch, A. W., Dick, L., & Cayton, C. (2022). Using a framework to develop preservice teacher noticing of students' mathematical thinking within technology-mediated learning. *Contemporary Issues in Technology and Teacher Education, 22*(3), 511–541.

Bartolini Bussi, M. G., & Mariotti, M. A. (2008). Semiotic mediation in the mathematics classroom: Artifacts and signs after a Vygotskian perspective. In L. D. English, & D. Kirshner (Eds.), *Handbook of international research in mathematics education* (2nd ed., pp. 746–783). Routledge.

Borko, H., Whitcomb, J., & Liston, D. (2009). Wicked problems and other thoughts on issues of technology and teacher learning. *Journal of Teacher Education, 60*(1), 3–7. https://doi.org/10.1177/0022487108328488

Casey, S., Hudson, R., Harrison, T. R., Barker, H., & Draper, J. (2020). Preservice teachers' design of technology-enhanced statistical tasks. *Contemporary Issues in Technology and Teacher Education, 20*(2). https://citejournal.org/volume-20/issue-2-20/mathematics/preservice-teachers-design-of-technology-enhanced-statistical-tasks

Conference Board of the Mathematical Sciences. (2012). *The mathematical education of teachers II*. American Mathematical Society & Mathematical Association of America.

De Freitas, E., & Sinclair, N. (2014). *Mathematics and the body: Material entanglements in the classroom*. Cambridge University Press. http://dx.doi.org/10.1017/cbo9781139600378

Dick, L. K., Lovett, J. N., McCulloch, A. W., Cayton, C., Bailey, N. G., & Yalman Ozen, D. (2022). Preservice teacher noticing of students' mathematical thinking in a technology-mediated learning environment. International *Journal for Technology in Mathematics Education, 29*(3), 129–142. 10.1564/tme_v29.3.02

Dick, L. K., McCulloch, A. W., & Lovett, J. N. (2021). When students use technology tools, what are you noticing? *Mathematics Teacher: Learning and Teaching PreK-12, 114*(4), 272–283. https://doi.org/10.5951/MTLT.2020.0285

English, L., & Sriraman, B. (2009). Theory and its role in mathematics education. In S. L. Swards, D. W. Stinson, & S. Lemons-Smith (Eds.), *Proceedings of the 31st annual meeting of the North American Chapter of the International Group for the Psychology of Mathematics Education* (pp. 1621–1624). Georgia State University.

Franklin, C., Kader, G., Mewborn, D., Moreno, J., Peck, R., Perry, M., & Schaeffer, R. (2007). *Guidelines for assessment and instruction in statistics education (GAISE) Report: A Pre-K–12 curriculum framework*. American Statistical Association. http://www.amstat.org/education/gaise

Grossman, P. L., Competition, C., Igra, D., Ronfeldt, M., Shahan, E., & Williamson, P. W. (2009). Teaching practice: A cross-professional perspective. *Teachers College Record, 111*(9), 2005–2100. https://doi.org/10.1177/016146810911100905

Hollebrands, K. F., & Lee, H. S. (2012). *Preparing to teach mathematics with technology: Geometry*. Kendall Hunt Publishing Company.

Hollebrands, K. F., McCulloch, A. W., & Lee, H. S. (2016). Prospective teachers' incorporation of technology in mathematics lesson plans. In M. Niess, S. Driskell, & K.

Hollebrands (Eds.), *Handbook of research on transforming mathematics teacher education in the digital age* (pp. 272–292). IGI Global.

Hutto, D. D., Kirchhoff, M. D., & Abrahamson, D. (2015). The enactive roots of STEM: Rethinking educational design in mathematics. *Educational Psychology Review, 27*(3), 371–389. http://dx.doi.org/10.1007/s10648-015-9326-2

Jacobs, V. R., Lamb, L. L. C., & Philipp, R. A. (2010). Professional noticing of children's mathematical thinking. *Journal for Research in Mathematics Education, 41*(2), 169–202. https://doi.org/10.5951/jresematheduc.41.2.0169

Kaput, J. J., & Thompson, P. W. (1994). Technology in mathematics education research: The first 25 years in the JRME. *Journal for Research in Mathematics Education, 25*(6), 676–684. https://doi.org/10.5951/jresematheduc.25.6.0676

Kersaint, G., Horton, B., Stohl, H., & Garofalo, J. (2003). Technology beliefs and practices of mathematics education faculty. *Journal of Technology and Teacher Education, 11*(4), 567–595. https://www.learntechlib.org/primary/p/2222/

Koehler, M. J., & Mishra, P. (2005). What happens when teachers design educational technology? The development of technological pedagogical content knowledge. *Journal of Educational Computing Research, 32*(2), 131–152. https://doi.org/10.2190/0EW7-01WB-BKHL-QDYV

KPC Technologies. (2019). *WebSketch tool library.* https://geometricfunctions.org/fc/tools/library/

Leatham, K. R. (2006, January). *A characterization of the preparation of preservice mathematics teachers to teach mathematics with technology* [Paper presentation]. Tenth Annual Conference of the Association of Mathematics Teacher Educators, Tampa, FL, USA.

Lee, H. S., Harper, S., Driskell, S. O., Kersaint, G., & Leatham, K. (2012). Teachers' statistical problem solving with dynamic technology: Research results across multiple institutions. *Contemporary Issues in Technology and Teacher Education, 12*(3). https://www.learntechlib.org/primary/p/39310/

Lee, H. S., & Hollebrands, K. (2008). Preparing to teach mathematics with technology: An integrated approach to developing technological pedagogical content knowledge. *Contemporary Issues in Technology and Teacher Education, 8*(4), 326–341. https://www.learntechlib.org/p/28191/

Lee, H. S., & Hollebrands, K. F. (2011). Characterising and developing teachers' knowledge for teaching statistics with technology. In C. Batanero, G. Burrill, & C. Reading (Eds.), *Teaching statistics in school mathematics-Challenges for teaching and teacher education* (pp. 359–369). Springer.

Lee, H. S. & Hollebrands, K. F. (n.d.). *A framework to guide a research-based approach to teacher education materials.* Retrieved August 20, 2021, from https://www-data.fi.ncsu.edu/wp-content/uploads/2019/12/28150208/ptmt_guiding_framework.pdf

Lee, H. S., Hollebrands, K. F., & Wilson, P. H. (2010). *Preparing to teach mathematics with technology: Data analysis and probability.* Kendall Hunt Publishing Company.

Lee, H. S., Hudson, R., Casey, S., Mojica, G., & Harrison, T. (2021). Online curriculum modules for preparing teachers to teach statistics: Design, implementation, and results. In K. F. Hollebrands, R. Anderson, & K. Oliver (Eds.), *Online learning in mathematics education* (pp. 65–93), Springer.

Lee, H. S., & Tran, D. (2015). Statistical habits of mind. In *Teaching statistics through data investigations MOOC-Ed.* Friday Institute for Educational Innovation: NC

State University, Raleigh, NC. https://fi-courses.s3.amazonaws.com/tsdi/unit_2/Essentials/Habitsofmind.pdf

Lovett, J. N., Dick, L. K., McCulloch, A. W., & Sherman, M. F. (2019). Preservice mathematics teachers' professional noticing of students' mathematical thinking with technology. In L. Liu & D. Gibson (Eds.), *Research highlights in technology and teacher education 2018* (pp. 71–79). AACE—Association for the Advancement of Computing in Education.

Lovett, J. N., McCulloch, A. W., Dick, L. K., & Cayton, C. (2020). Design principles for examining student practices in a technology-mediated environment. *Mathematics Teacher Educator, 8*(3), 120–133. https://doi.org/10.5951/MTE.2020.0007

Mariotti, M. A. (2000). Introduction to proof: The mediation of a dynamic software environment. *Educational Studies in Mathematics, 44*, 25–53. https://doi.org/10.1023/A:1012733122556

Mariotti, M. A. (2013). Introducing students to geometric theorems: How the teacher can exploit the semiotic potential of a DGS. *ZDM Mathematics Education, 45*, 441–452. https://doi.org/10.1007/s11858-013-0495-5

McCulloch, A. W., Bailey, N., Fye, K., & Scott, G. (2020). Creating a third-space for learning to design technology-based math tasks. *Mathematics Teacher Educator, 9*(1), 7–11.

McCulloch, A. W., Hollebrands, K. F., Lee, H.S., Harrison, T., & Mutlu, A. (2018). Factors that influence secondary mathematics teachers' integration of technology in mathematics lessons. *Computers and Education, 123*, 26–40. https://doi.org/10.5951/MTE.2020.0011

McCulloch, A. W., Leatham, K. R., Lovett, J. N., Bailey, N. G., & Reed, S. D. (2021). How we are preparing secondary mathematics teachers to teach with technology: Findings from a nationwide survey. *Journal for Research in Mathematics Education, 52*(1), 94–107. https://doi.org/10.5951/jresemathteduc-2020-0205

McCulloch, A. W., Lovett, J. N., Dick, L. K., & Cayton, C. (2021). Positioning each and every student as a mathematical explorer with technology. *Mathematics Teacher: Learning and Teaching PK–12: Special Issue on Digital Equity and the Digital Divide, 114*(10), 738–749, https://doi.org/10.5951/MTLT.2021.0059

McCulloch, A. W., Lee, H. S., & Hollebrands, K. (2015). *Preparing to teach mathematics with technology: An integrated approach to algebra* (1st ed.). NC State University. http://ptmt.fi.ncsu.edu

McCulloch, A. W., Lovett, J. N., & Edgington, C. (2019). Transforming preservice teachers' understanding of function using a vending machine applet. *Contemporary Issues in Technology and Teacher Education, 19*(1). https://www.citejournal.org/volume-19/issue-1-19/mathematics/designing-to-provoke-disorienting-dilemmas-transforming-preservice-teachers-understanding-of-function-using-a-vending-machine-applet/

McDonald, M. A., Kazemi, E., & Kavanagh, S. S. (2013). Core practices and pedagogies of teacher education: A call for a common language and collective activity. *Journal of Teacher Education, 64*(5), 378–386. https://doi.org/10.1177/0022487113493807

Meagher, M., Edwards, T. E., & Ozgun-Koc, A. (2011). Project CRAFTeD: An adapted lesson study partnering preservice mathematics teachers with a master teacher. In L. R. Wiest & T. Lambert (Eds.), *Proceedings of the 33rd Annual Meeting of the North*

American Chapter of the International Group for the Psychology of Mathematics Education (pp. 496–504), University of Nevada, Reno.

National Council of Teachers of Mathematics. (2012). *NCTM CAEP mathematics content for secondary: Addendum to the NCTM CAEP Standards 2012*. Author.

National Council Teachers of Mathematics. (2014). *Principles to actions: Ensuring mathematical success for all*. Author.

National Governors Association Center for Best Practices and the Council of Chief State School Officers. (2010). *Common core state standards for mathematics*. NGA Center and CCSSO. http://www.corestandards.org.

Niess, M. L. (2005). Preparing teachers to teach science and mathematics with technology: Developing a technology pedagogical content knowledge. *Teaching and Teacher Education, 21*, 509–523. https://doi.org/10.1016/j.tate.2005.03.006

Niess, M. L., & Roschelle, J. (2018). Transforming teachers' knowledge for teaching mathematics with technologies through online knowledge-building communities. In T. E. Hodges, G. J. Roy, & A. M. Tyminski (Eds.), *Proceedings of the 40ᵗʰ Annual Meeting of the North American Chapter of the International Group for the Psychology of Mathematics Education* (pp. 44–62). University of South Carolina & Clemson University.

Ozen-Unal, D., Hollebrands, K., McCulloch, A., Scher, D., & Steketee, S. (2022). Prospective high school mathematics teachers' uses of diagrams and geometric transformations while reasoning about geometric proof tasks. *International Journal of Technology in Mathematics Education, 29*(1), 13–23, 10.1564/tme_v29.1.02.

Roschelle, J., Noss, R., Blikstein, P., & Jackiw, N. (2017). Technology for learning mathematics. In J. Cai (Ed.), *Compendium for research in mathematics education* (pp. 853–876). National Council of Teachers of Mathematics.

Smith, M. S., & Stein, M. (2018). *Five practices for orchestrating productive mathematics discussions*. National Council of Teachers of Mathematics.

Suh, J., Roscoioli, K., Leong, K. M., & Tate, H. (2022). Transformative technology for equity-centered instruction. In *Proceedings of the 33rd International Conference for the Society for Information Technology and Teacher Education* (pp. 1392–1400). Association for the Advancement of Computing in Education (AACE). https://www.learntechlib.org/awards/SITE/2022/.

Wilson, P. H., Lee, H. S., & Hollebrands, K. (2011). Understanding prospective mathematics teachers' processes for making sense of students' work with technology. *Journal for Research in Mathematics Education, 42*(1), 42–67. https://doi.org/10.5951/jresematheduc.42.1.0039

Yigit, M. (2014). A review of the literature: How Pre-service mathematics teachers develop their technological, pedagogical, and content knowledge. *International Journal of Education in Mathematics, Science and Technology, 2*(1), 26–35. https://ijemst.net/index.php/ijemst/article/view/9/9

CHAPTER 12

USING A VIDEO-TAGGING TOOL TO SUPPORT FACILITATORS OF VIDEO-BASED PROFESSIONAL DEVELOPMENT

Margaret Walton
Janet Walkoe
University of Maryland

Research has shown that it is important for teachers to learn how to notice students' mathematical thinking. Video-based interventions, like video clubs, can support teachers' noticing development. We argue that is also important for the facilitators of video clubs to learn to notice student *and* teacher thinking. We examine how a video-tagging tool supported a novice facilitator (the first author) in noticing teacher thinking as she led a video club with middle school math and science pre-service teachers (PSTs). We found that PSTs' video tags acted as a scaffold to help Margaret attend to and interpret PSTs' ideas so that she could respond in ways that supported PSTs' noticing skills. These results have implications for understanding the differences between facilitator noticing and teacher noticing and for helping facilitators in learning to notice.

The AMTE Handbook of Mathematics Teacher Education: Reflection on Past, Present and Future—Paving the Way for the Future of Mathematics Teacher Education, Volume 5
pages 245–264.

INTRODUCTION

Math educators generally agree about the importance of teachers attending to the disciplinary substance of students' mathematical ideas in order to support their learning (e.g., Carpenter et al., 1989; Empson & Jacobs, 2008; Jacobs et al., 2007). Specifically, when teachers *notice*–or attend to, interpret, and decide how to respond to students' thinking, their insights can inform instruction (Empson & Jacobs, 2008; Jacobs et al., 2007; Schoenfeld, 2011). Video clubs have played a prominent role in supporting teachers' learning to notice the substance of students' thinking. During a video club, a group of in-service teachers or pre-service teachers (PSTs) meet to watch video clips of classroom instruction and discuss important moments of student thinking seen in the clips. Video clubs have been found to support teachers in attending to the disciplinary substance of student thinking and in reasoning about that thinking, rather than simply evaluating the correctness of students' ideas (Sherin & van Es, 2005, 2009; Walkoe, 2015).

Video clubs are typically led by a facilitator who sets the course for teacher learning in several ways, including developing teacher goals, selecting video clips, and leading the teacher discussion (Borko, Jacobs et al., 2014; Kang & van Es, 2019; Tekkumru-Kisa & Stein, 2017). Leading the discussion is a particularly important responsibility because, through the discussion, facilitators can encourage teachers to refine their thinking about student ideas, and help teachers gain a range of perspectives on students' mathematical understanding. This work requires facilitators to *notice teachers' thinking* (Amador, 2021; Borko, Koellner et al., 2014; Kazemi et al., 2011; Lesseig et al., 2017; van Es, 2011).

Yet, while teacher noticing has been widely researched, fewer efforts have focused on what it means for *facilitators to notice* and how to support facilitator noticing. For instance, a number of researchers have reported on the process of teacher noticing, the varying characteristics of teachers who are learning to notice, and professional development (PD) that can help teachers strengthen their ability to notice (Jacobs et al., 2010; Sherin & Han, 2004; Sherin & van Es, 2005; van Es, 2011). However, the analogous topics for facilitator noticing are in the nascent stages of investigation (Amador, 2016, 2020, 2021; Borko, Koellner et al., 2014; Lesseig et al., 2017).

Further exploration of these topics is needed because video clubs and other video-based learning opportunities for PSTs and in-service teachers have become increasingly popular (Gaudin & Chaliès, 2015). As this type of PD expands, the demand for high quality facilitators rises. In line with AMTE Standard P.3.5, *Provide Effective Mathematics Methods Instructors* (AMTE, 2017), it is imperative for these facilitators to build their noticing knowledge and skills so that they in turn can provide learning opportunities for teachers to develop their noticing. Understanding how facilitators notice teacher thinking, and what kind of supports might help, is essential for designing tools and PD to support facilitator noticing.

The purpose of this chapter is to begin to explore facilitator noticing and the tools that can help novice facilitators notice teacher thinking in the context of a video club. We examine our experience of facilitating a video club with middle school math and science pre-service teachers. During this video club, PSTs watched and annotated a video clip. The first author, Margaret, who is a novice facilitator, used insights from PSTs' video annotations and comments while leading a discussion that aimed to support PSTs in developing their ideas related to student thinking. We argue that the video annotations scaffolded Margaret's ability to notice PSTs' thinking, which contributed to the rich discussion we saw.

BACKGROUND

Teacher Noticing and Video Clubs

The cognitive process by which teachers identify and make sense of important classroom moments is often referred to as *teacher noticing*. While varying definitions exist, many researchers describe teacher noticing as *attending to, interpreting*, and *deciding how to respond* to students' mathematical thinking (Jacobs et al., 2010). Within the complexity of a classroom, it is important for a teacher to be able to *attend to* or recognize key moments of student mathematical thinking. Once a teacher attends to a moment of student thinking, they can use their knowledge of teaching to *interpret* or make sense of the moment. Finally, a teacher must decide how to *respond* to the student's thinking based on their interpretation of what the student understands.

Research has shown that teachers can learn to notice significant classroom events, and that video clubs can be productive in supporting teacher noticing. Specifically, many teachers who participate in video clubs become more attuned to student mathematical thinking, learn to reason more deeply about students' mathematical understandings, and use evidence to support their reasoning (Jacobs et al., 2010; Sherin & Han, 2004; Sherin & van Es, 2005; Sherin & van Es, 2009; Walkoe, 2015).

Facilitation of Video-Based PD

Just as teachers are instrumental in their students' learning, PD facilitators play an integral role in teacher learning. Indeed, during video-based PD, researchers have argued that facilitator actions can help teachers learn to use video evidence to support their ideas about student thinking, and can help them take a more analytic stance when discussing student thinking (Borko et al., 2008; Castro Superfine et al., 2019; Goldsmith & Seago, 2008). For example, Castro Superfine and colleagues (2019) conducted a study that examined how actions by the facilitator related to the quality of elementary PSTs' noticing in a mathematics content course. The authors found that when the facilitator engaged in actions like pushing PSTs to explain their ideas about student thinking, clarifying PST thinking, or highlighting particular PST comments, a higher percentage of PSTs' resulting responses included specific evidence from the video. Such evidence included moments when students explained their solution strategies or an instance where a student identified patterns. Given these results and others, facilitators can play an important role in developing teachers' noticing skills.

Like Castro Superfine's and colleagues' (2019) study, work on facilitation of video-based PD often focuses on facilitator actions that support teachers to pay attention to student thinking (Borko, Jacobs et al., 2014; Kang & van Es, 2019; Tekkumru-Kisa, & Stein, 2017; van Es et al., 2020, 2014; Zhang et al., 2011). Many of these efforts involve frameworks that detail the work of facilitators, including setting goals for teachers, selecting relevant video clips, and promoting productive teacher discussion about student thinking.

For example, van Es and colleagues (2014) identify four practices that facilitators use to support teachers in analyzing student mathematical thinking more deeply during discussions including: orienting the group to the video analysis task, sustaining an inquiry stance, maintaining a focus on the video and the mathematics, and supporting group collaboration. Each practice has associated talk moves, such as lifting up teacher ideas or distributing participation. They found when these moves were used with a video club of 4th and 5th grade teachers, participants took more of an inquiry stance toward student thinking. For instance, when the facilitator highlighted particular quotes from the students in the video the teachers looked more closely at the mathematical discourse between students. The teachers also grounded their claims about student ideas in evidence from the video. The moves also fostered group collaboration and gave multiple teachers opportunities to share their varying perspectives on student thinking. Other researchers have found that similar moves produce similar teacher outcomes (Borko, Jacobs et al., 2014; Tekkumru-Kisa, & Stein, 2017; Zhang et al., 2011).

The work above is important for understanding facilitators' planning and enactment that leads to productive video-based PD, such as video clubs. However, most of this research is based on *observations* of experienced facilitators. It does not detail *how facilitators learned* these actions, or how they know, especially during a teacher discussion, when to use various talk moves. We argue that the field needs to explore facilitator learning in order to support novice facilitators and inform facilitator PD, a currently underexplored area.

Several studies that examine facilitator learning have found that novice facilitators can have difficulty leading in-service teacher discussions. Specifically, novice facilitators can struggle with pressing teachers to explain their comments, making connections between teachers' ideas, and helping teachers to build on one another's ideas (Borko, Koellner et al., 2014; Elliott et al., 2009; Jackson et al., 2015). Our study focuses on a novice facilitator's work with PSTs, but we suspect that similar difficulties arise. It is important to not only know the actions that facilitators should take to support teachers learning to notice, but also how they can learn to enact those practices. One way to look at that process might be in terms of how facilitators learn to notice, which is discussed in the next section.

Early Work in Facilitator Noticing

Research in facilitator noticing falls more broadly under how teacher educators support teachers' professional learning. Other researchers have examined the work of teacher educators through the interactions among teacher educators, PSTs, and the practices, knowledge, and skills for instructing teachers (Shaughnessy et al., 2016). In this context, teacher educators need to know the practices, knowledge, and skills for K–12 instruction, but beyond that, they need to understand the best ways to present this content to PSTs (Castro Superfine & Li, 2014). We agree and argue that teacher noticing is an essential skill for K–12 teachers. In order for teacher educators, like facilitators of video clubs, to support teachers in learning to notice student thinking, they also need to be able to notice student thinking. However, facilitator noticing should go beyond only noticing student thinking. Yet, few efforts have delved deeply into what facilitator noticing entails (Amador, 2021), how it differs from teacher noticing (Borko, Koellner et al., 2014; Kazemi, 2011; Lesseig et al., 2017), and how a novice facilitator might learn to notice (Amador, 2021; Carlson et al., 2017; Lesseig et al., 2017).

Amador (2021) noted that facilitators need to notice both student *and* teacher thinking. In terms of student thinking, facilitators' focus is similar to teachers; they need to be able to attend, interpret and determine a respose to students' mathematical understandings. Given that video club facilitators lead work with teachers, they also need to be able to notice productive moments of *teacher thinking*. Several researchers have started to highlight distinctions between noticing student thinking and teacher thinking. One major difference seems to be that, while the goal of teacher noticing is to support students' learning of mathematics, facilitator noticing should focus on the development of teachers' specialized content knowledge (SCK) and pedagogical content knowledge (PCK) (Borko, Koellner et al., 2014; Elliott et al., 2009; Kazemi et al., 2011; Lesseig et al., 2017). That is, facilitators should support teacher learning by working to make sense of and respond to teachers' ideas that relate to mathematical knowledge that is specific to teaching mathematics, such as the connections between the varying ways to solve math problems (Ball et al., 2008). In addition, facilitators should pay attention to how teachers reason about student thinking and how it connects to student learning goals (Shulman, 1986).

The studies above begin to detail what is involved in facilitator noticing, and how facilitator and teacher noticing diverge, notably the differences in the learning goals associated with each. Yet, facilitators, particularly novice facilitators, still need to know *how* to address the different goals for teachers. From a noticing perspective, this means becoming adept at attending to and interpreting the types of teacher thinking that help teachers build on their mathematical understandings, their ideas about student thinking, and the connections between the two. Attending to and interpreting these aspects of teacher thinking, we argue, is necessary for facilitators to respond to teachers with actions discussed in the facilitation literature, such as the talk moves identified by van Es et al. (2014). In this chapter, we will focus on these important aspects (attending to and interpreting) of facilitator noticing of teacher thinking.

Tools for Facilitator Noticing

Researchers in the noticing field have often emphasized the affordance of different tools and scaffolds in helping teachers learn to notice. In addition to video, researchers have used targeted questions, frameworks, and animations to help teachers identify important moments of student thinking, relate student thinking to mathematical content, and learn how to ask students questions about their thinking (Amador et al., 2017; Roth McDuffie et al., 2014; Santagata, 2011; Stockero et al., 2017; Walkoe, 2015; Walkoe & Levin, 2018;). Tools such as these and others might help facilitators learn to notice in similar ways.

Several studies have begun to explore tools that support PD facilitator noticing (Amador, 2021; Carlson et al., 2017). Amador (2021) designed a protocol that helped 16 mathematics teacher educators develop their ability to notice student thinking through building a learning model based on clinical interviews. Building the learning model likely contributed to many of the teacher educators' progress in making more sophisticated interpretations of student thinking and in making better direct connections between their claims about student thinking and students' written or verbal evidence.

This body of literature is helpful in demonstrating that tools can be helpful in developing facilitators' noticing skills. Another tool that we think shows promise in this regard is video annotation. In research on teacher noticing, video annotation has been primarily used for data collection and analysis of teacher thinking (Stockero et al., 2017; Walkoe, 2015; Walkoe et al., 2020). One of the reasons video annotation tools have been beneficial in this respect is because teachers' annotations can provide valuable insights into teachers' thinking about the video.

In our work we have been exploring the possibilities video annotation tools might offer practitioners, such as teacher educators. In particular, we explore how video annotation tools can be used to help facilitators lead a video club discussion. In the same way annotation tools help researchers gain insight into what teachers notice in a video (e.g., Walkoe & Levin, 2018; Walkoe et al., 2020), they can also provide teacher educators with this information. If facilitators could see what teachers annotate as they watch a video, then facilitators could plan ahead based on those annotations, rather than waiting for those ideas to come out in the discussion, as typically happens in a video club (Sherin & van Es, 2005, 2009). Having access to teacher thinking ahead of time can help facilitators think about how they might orchestrate the discussion or what moves they might make, based on ideas they see. In this chapter, we demonstrate how an annotation tool gave Margaret access to teacher thinking and supported her experience as a novice facilitator leading a video club discussion.

METHODS

Participants and Context

The video club took place in a middle grades (grades 4–9) math and science methods class. The class session included four undergraduate PSTs and two PSTs seeking their master's degree education. All PSTs were in the midst of their teaching internship, which was the culminating field experience of their program. Due to the COIVD-19 Pandemic, the class was virtual (including the video clubs) and met via Zoom (www.zoom.us).

Margaret was a doctoral student in a mathematics education program at the same university. Margaret facilitated several other video club sessions over that academic year but was still inexperienced in terms of supporting PSTs productive video club conversations.

ANOTEMOS

The video annotation platform we used is called Anotemos (Herbst et al., 2019; www.anotemos.org). Anotemos allows users to watch video and drop "pins" (i.e., tags) at any timestamp. Users can then comment on the moments they pinned. Pins and comments can be made public, allowing different users, including the facilitator, to see everyone's pins.

Video Club Planning and Enactment

The video clip came from a tenth-grade high school Geometry class. In the clip, the students engage in a whole-class discussion during which they try to explain how the area of a square changes when the perimeter increases by a factor of *n*. Most of the video centers around three student explanations. Student A explains the relationship in terms of the dimensions of a square. He appears to argue that, because one has to multiply both the length and the width by a certain number, the resulting area will increase by "that number times itself." Student B demonstrates how the square's area increases by using his hand and calculator to show how the shape changes as the length and width double. Finally, Student C comes to the chalkboard and draws the scenario of a square doubling in length and width, showing the resulting area increasing to four times the original square. Overall, it seems that the first student thinks about the task algebraically and in terms of the formula for the area of a square. In contrast, the two other students appear to be thinking geometrically about how the square grows.

We developed a discussion protocol to guide the video club. The protocol had three goals for PSTs. Many researchers have emphasized the importance of goal setting for teachers as a way for facilitators to guide teacher discussions and help teachers progress in their noticing development (Borko, Jacobs et al., 2014; Jackson et al., 2015; Tekkumru-Kisa & Stein, 2017). The first two goals addressed

PSTs' PCK related to noticing student thinking. For the first goal, we aimed for PSTs to reason about *Moments of Potential Action*. These were instances in the video when students were confused or their understanding was unclear. We wanted PSTs to consider what students might be thinking in these moments, and to also determine how they might respond if they were the teacher.

The second goal was meant to support Multimodal Teacher Noticing (MMTN) (Walkoe et al., 2023). The literature abounds on the importance of multimodal student thinking (gestures, actions, voice prosody, etc.) and how it both reveals aspects of and supports student thinking (e.g., Arzarello, 2006; Gallese & Lakoff, 2005; Goldin-Meadow, 2003; McNeill, 1992; Nathan & Alibali, 2011; Radford, 2009). We aimed for PSTs to attend to the multiple ways that students expressed their reasoning, going beyond written and verbal evidence. We especially wanted PSTs to connect what the students in the video were thinking with what they were doing (e.g., How did their gestures supplement or clarify their thinking?)

Our final goal aimed to enhance PSTs' SCK. We sought for PSTs to see distinctions in students' mathematical thinking. Specifically, we wanted PSTs to make sense of the differences between the students' explanations of the relationship between the perimeter and area of a square and to identify the different algebraic and geometric thinking being used to describe the same situation. For each goal, we developed several questions that the facilitator might ask to guide PSTs.

The video club began with an introduction to the task, which was led by Janet. During this time, the PSTs worked on the task themselves and discussed their thinking about the task with each other. PSTs then watched the video clip individually using Anotemos, which allowed them to simultaneously pin and comment. After watching the video, the participants came back together over Zoom to engage in a discussion about student thinking. The discussion was primarily facilitated by Margaret, though Janet occasionally asked questions. Immediately after the video club, the two of us met via Zoom to debrief the facilitation experience. We discussed Margaret's experience using Anotemos to help her lead the conversation.

Data Analysis

The data we collected included video recordings of the video club and our debrief after the video club, Margaret's handwritten notes during the video club, and the PST video tags. The heart of our analysis centered on the transcripts from the video club recording. Table 12.1 provides an example of our different rounds of coding.

We first transcribed both video clubs and chunked them into idea units (Jacobs & Morita, 2002), based on the topic of conversation. We coded the transcripts to determine which idea units from each video club aligned with the video clubs' overall goals discussed in the previous section. The names of these codes were *Moments of Potential Action, Multiple Modes of Evidence,* and *Distinctions in Students' Thinking.* Idea units were given one of the goal codes if they directly ad-

TABLE 12.1.   Example of Idea Unit from Video Club Discussion Transcript

Idea Unit	Transcript	Goals
8	[00:33:45.06] Margaret: Ok. Got it like between what they're thinking and maybe just like the quick word misusage they might be having. Ok. Great. Um, as you guys were tagging, I was looking through your tags and, Chelsea, you had one really interesting tag at 2:54 where you say that the student, "Sees shapes are two dimensions." Do you remember that? (Lifting up)	Distinctions in students' thinking
	[00:34:13.04] Chelsea: Yes, so like one of the students recognizes that, like, the squares they are talking about are two dimensional. And he doesn't explicitly say it, it's sort of more implied that if the shapes were three dimensional then maybe this wouldn't work or there would be a different pattern. I think that the question the teacher asked was like, "Why does this work?" So he involved, like, the two dimensional aspect in his answer.	
	[00:34:45.02] Margaret: Ok and how did you sort of know, like, do you kind of remember during the video? How do you know he was deciding between two and three dimensions or that he was, you know, really thinking about these two different dimensions? (Pressing)	
	[00:34:59.15] Chelsea: Because then he said that like his process of, like, expanding the image was he said, "You have to multiply each dimension." So he's like, "You have to multiply one dimension and then you have to multiply the other one."	

dressed a goal, or were part of the conversation that built toward directly addressing a goal. For example, the idea unit in Table 12.1 was coded as *Distinctions in Students' Thinking* because, while the PST Chelsea (all names for PSTs are pseudonyms) did not compare different student thinking in this idea unit, her response contributed to a different PST comparing students' thinking in the next idea unit.

Once we identified the idea units that corresponded with the goals, we examined what facilitator moves were made during those ideas units based on van Es et al.'s (2014) framework (most of these moves were made by Margaret, though Janet also occasionally participated). These moves were added in parentheses within the transcript at the place where they occurred (Table 12.1). For example, in Table 12.1 Margaret used the facilitation move *Lifting Up* when she pointed out Chelsea's video tag about Student A's focus on the two-dimensional nature of a square. Margaret then used *Pressing* when she asked Chelsea to discuss her evidence for knowing that Student A was thinking about the two dimensions.

Finally, we used the two recorded debriefs and Margaret's handwritten notes to triangulate what we observed in the transcripts to what we experienced during and directly after the video club. We discuss the results of the video club discussion in the next section.

FINDINGS

Overall, the video club discussion closely aligned to the goals of the video club. Nine of 11 idea units were devoted to topics that related to one of the three video club goals. Throughout the video club, the video tags helped Margaret enact talk moves, like those described by van Es et al. (2014) and others, to engage multiple PSTs in rich discussion about student thinking. Below, we provide some examples of how video tagging supported video club facilitation.

Pressing and Connecting PST Thinking

One of the ways that video tags helped us was by allowing us to see the ways that PSTs thought about particular moments in the video *before* the discussion, as opposed to hearing ideas for the first time as they are expressed in-the-moment during the video club. In Anotemos, tags and comments populate in real time, meaning facilitators can get a sense of PSTs' ideas about the video before the discussion starts. This affordance gave Margaret time to determine how PSTs' thoughts related to the goals of the video club and to each other, and to develop questions that pressed and connected PSTs' thinking.

For example, while PSTs watched the video and tagged it, Margaret noticed that several PSTs began to attend to the different algebraic and geometric thinking in the video. When tagging a moment about Student A, Chelsea wrote, "Student uses formulas for a square and triangle to explain his thinking." Another PST, Carey, wrote a tag about Student C that said, "Thinking about the actual shape of the shape." While PSTs continued to watch and tag, Margaret wrote questions related to these tags that asked PSTs to build on their thinking about student ideas and addressed the video club goal, *Distinctions in Student Thinking.* One of those questions, directed at Chelsea was, "Is [Student A's use of formulas] to explain his thinking different from the other two students?" A similar question, for Carey was, "Is [Student C's visual explanation] different than [Student B's]?"

The questions that Margaret developed in response to the tags came in handy during the video club discussion. Early in the discussion Carey brought up her own tag:

Carey: At like, from like four minutes on they were all trying to... I put, "Thinking of the actual shape of the shape..." So, like, they were all trying to visualize and then they didn't realize you could actually draw it until someone drew it and then they were like, "Oh you can draw it."

Margaret: Ok. Carey, that's really interesting. What do you mean by, "They were all trying to visualize?" Like, what were they trying to do?

Carey: Um, so starting with [Student A] I think it was... They were talking about this five by five figure and it's like adding five here, adding five there, so when you multiply 5 by 5, it's 25 and that's where that's coming from, instead of it being, like, 10 because you're not adding five plus five, you're multiplying those two *n*'s. And I could tell that [Student A] was trying to create this visual out loud, and then it went to [Student B], who was also trying to explain it verbally, but then couldn't. And it ended with [Student C] going up to the board and drawing, I think it was a two by two or a three by three that he ended up drawing. But, like they brought, this like, what they were trying to verbalize into a visual, so it was almost more tangible.

Several other students replied to Carey's comment, including one student who noted that Student A talked about the square in terms of its dimensions. At this point, Margaret was able to ask a question similar to the one she planned based on Carey's tag.

Margaret: So is this, do you all think, like doing this sort of dimension thing, is this different than what the [Student C] did on the board or is it the same? Like, are they thinking in the same way? Um, what do you think?

Zach: I would say it's the same.

Margaret: Ok. Why do you think that, Zach?

Zach: Just cause even in the way he drew it, it was just very much, like, that same extension method.

Margaret: Ok, so the idea of extending. So, on the board, we sort of saw that extending [when Student C] actually drew it out, right? How did you see that Zach, that idea of extension, with either of the other two students?

Zach: [Student B] with the calculator he, like, took the one and then physically separated and like then made it two, to extend.

Here, after Margaret pressed Zach to expand on his thinking, Zach noted that Student B's thinking was similar Student C's; Student B just used physical gesture as opposed to a drawing. Zach's comment built on Carey's original, more general, comment and addressed the *Distinctions in Student Thinking* goal of the video club.

Margaret did not ask the exact questions that she wrote down in her notes but using tags to identify PSTs' varying thinking ahead of time allowed Margaret to create a plan, which she could loosely follow as the discussion unfolded. In this case, the plan helped Margaret press Carey to expand her explanation about students visualizing the square and connect Carey's and others' ideas by asking them to compare the students' thinking, which was one of the goals of the video club.

FIGURE 12.1. Screenshot of Anotemos Pins
Note. Anotemos' setup allows the facilitator identify clusters of pins.

Distributing Participation

Video tagging also supported Margaret in distributing participation among PSTs. First, Margaret was able to use Anotemos to identify clusters of pins that indicated moments in the video that multiple PSTs tagged (Figure 12.1). These clusters were a signal that a number of PSTs attended to those video moments, and it could be a place for PSTs to offer multiple perspectives on student thinking, or a place for PSTs to build on each other's ideas.

Margaret pointed out one of these clusters to PSTs during the video club. The cluster also occurred at a moment in the video that was related to the goal *Reason about Moments of Potential Action.* During this moment, a student in the video showed some confusion over the difference between 3^3, 3^2, and 3×3 and said that three raised to itself is equal to nine.

> **Margaret:** A lot of you pinned around, like 26, 30 seconds, all around there. Um, that was when the discussion was sort of first getting going. Can anyone sort of tell me what was sort of going on there for you? Why did you find that interesting?
>
> **Chelsea:** So, something I pinned around that time was, the one girl that was talking, she was confusing three times three with three squared.
>
> **Margaret:** Ok.

Chelsea: So, like, she was saying that it was supposed to be three to the power of three. What she was really thinking was three times three.

Margaret: Ok. How do you know? Do you have evidence for that? How do you know she was thinking that?

Chelsea: Because she later corrects herself.

Margaret: Got it. So, she later corrects herself. Did anyone else notice anything about that?

Melanie: Yeah, it relates to the language they were using. So, she says "Three to itself," that's what she says. And then [students in the class] have to be like, "That's three times three." And what [the class has] been saying is "Something to the power of," like two to the power of two or something times itself and that's where, like, she has a combination. Like, she takes the two words and she does three to the power of itself instead of three times itself or three squared, which would be to the power of two. And so, I just saw that, like, as an intersection of that language. Like, she kind of knew what she was talking about but the language wasn't clear.

Here, Margaret pointed out the cluster of tags and invited PSTs to share their thoughts. Chelsea began and provided some general ideas about the student's confusion. Melanie then built on Chelsea's response by offering a more sophisticated interpretation of the same instance, that the student in the video conflated the language associated with multiplication and exponents and offered evidence to support her claim.

Margaret continued by asking a question that addressed the *Moments of Potential Action* goal:

Margaret: What do you think, um and this is for anyone... How might you respond to that as a teacher? If you saw that happening with a student?

Carey: I think I might ask them to explain out loud or on the board, whichever is preferred or going on, um, what three to the power of three is. Like folding it all out almost so then she would probably realize halfway through a sentence that it's three times three times three and then she would realize, "Oh that's that extra three."

Margaret: Nice, that's great Carey. Any other ideas about that? How might you respond to this student?

Melanie: I mean, I really like Carey's idea... A thought I had was to be like, "Ok what happens when it's a factor of four? Would we do four to the power of four then? Like does that sound right? Or just seeing if it's, like a misunderstanding of, like to the power of, or if it's just a misuse of the language in the moment.

As the conversation progressed, Margaret guided PSTs to move from reasoning about the instance in the video to deciding how they might respond if they were the teacher. At this point, multiple PSTs offered their perspectives on how they could support the student in clarifying her ideas about multiplication and exponents. A number of PSTs contributed to the conversation, and the discussion ultimately engaged multiple participants. The annotation tool allowed Margaret to identify a cluster of tags and invite PSTs into the conversation.

DISCUSSION

The purpose of this chapter is to highlight how tools, such as video annotation tools, can support novice facilitators in leading more substantive video club discussions with PSTs. We found that when PSTs tagged videos before discussing them in a video club Margaret, as a novice facilitator, could use the tags to plan questions that pressed and connected PSTs' thinking in service of addressing the goals of the video club. The tagging tool also allowed Margaret to identify video moments tagged by multiple PSTs. These clusters were helpful in drawing multiple participants into the conversation.

The talk moves that Margaret used are actions that others have argued support PSTs to learn to notice and respond to student thinking in instruction (Borko, Jacobs et al., 2014; Tekkumru-Kisa, & Stein, 2017; van Es et al., 2014; Zhang et al., 2011). It is notable that Margaret executed these moves as a *novice*, rather than an experienced facilitator. We argue that the video tags acted as a scaffold for Margaret to help her better attend to and interpret PST thinking so that she could respond to PSTs contributions using the talk moves that enriched the conversation and invited more PSTs to participate. We explain more below.

Windows into Teacher Thinking

Much of the teacher noticing literature explores how to give teachers windows into student thinking, through artifacts like video and student work, that they can use to inform instruction (e.g., Jacobs et al., 2010; Sherin, 2011; Sherin & Han, 2004). This method has helped teachers focus on student thinking and learn to reason about that thinking based on their knowledge of content and students (Sherin & van Es, 2009). It is plausible that artifacts, like video tags and others, could provide similar support to video club facilitators as they learn to notice teacher thinking.

Specifically, video tags gave Margaret a window into teacher thinking that allowed her to notice teacher perspectives in real time as PSTs watched the video *before* the discussion. The extra few minutes are important in terms of facilitator noticing because, as a novice, Margaret might not be able to attend to and interpret PST thinking as quickly as an expert (Borko, Jacobs et al., 2014; Lesseig et al., 2017). As Margaret read the video tags at the beginning of the video club, she could filter them for PST thinking that aligned to the video club goals. In terms of facilitator noticing, this filtering seemed to have aspects of both attending and interpreting. The list of PST comments on the Anotemos platform helped Margaret focus on PST thinking and quickly find comments related to the video club goals. Additionally, Margaret made sense of PSTs' ideas in light of the goals. Margaret's use of the video club goals as a guide in noticing PST thinking is critical because the goals were created to support PSTs' SCK and PCK development related to teacher noticing, which others have highlighted as a potential purpose of facilitator noticing (Borko, Koellner et al., 2014; Elliott et al., 2009; Kazemi et al., 2011; Lesseig et al., 2017). Due to the video tags, Margaret already had an idea of which PST ideas she should pursue that aligned to the video club goals and could support PSTs' noticing learning prior to the discussion.

Once Margaret knew which ideas to pursue, she could also use video tags to plan her responses to PSTs and execute facilitation moves recognized as supporting PSTs' noticing learning (van Es at al., 2014; Borko, Jacobs et al., 2014; Tekkumru-Kisa, Stein, 2017). For instance, Margaret had questions ready to press Carey when she brought up her own comment. Margaret also saw the opportunity to invite more PSTs to talk about *Moments of Potential Action* based on a cluster of tags. In other words, video tags helped Margaret determine how to respond to PST thinking in ways that can support them in learning to notice.

Implications for Video Club Facilitators and Facilitator Learning

We posit that tools that give windows into teacher thinking have the potential to support both experienced and novice video club facilitators. For video club facilitators in general, artifacts, such as video tags, can be analogous to student artifacts that teachers use to guide student discussions (Stein et al., 2008). While we used video tags to monitor PST thinking, there are a variety of instructional methods, like small group discussions, or brain dumps on chart paper or white boards, or teachers' own mathematical work (Borko & Koellner et al., 2014; Lesseig et al., 2017), that could also give facilitators an early look into how teachers begin to make sense of student thinking in video clips. Such previews of teacher thinking can enhance facilitator noticing, giving them more time to attend, interpret, and plan responses to teacher thinking that could lead to richer teacher discussions and wider teacher participation.

We also argue that tools, like video tags, could be valuable in supporting novice video club facilitators to learn to notice. Using teacher video tags as a scaffold could help novices understand the differences between teacher noticing and facilitator noticing, and shift their focus to pursuing goals that develop teachers' PCK and SCK related to teacher noticing of student thinking. We are currently developing a facilitator PD that tests this hypothesis and incorporates Anotemos as a way for novice facilitators to practice analyzing teacher tags and use them in planning potential video club discussions.

CONCLUSION

The study here focuses on facilitator noticing, but our work contributes to the broader call for research on teacher educator learning (Borko et al., 2011; Elliot et al., 2009; Sherin & van Es, 2017). This study is one example that suggests that teacher educators need knowledge, skills, and dispositions beyond those needed for K–12 instruction and that more research is necessary to understand what teacher educators need to effectively support teachers (Borko, Koellner et al., 2014; Castro Superfine & Li, 2014; Shaughnessy et al., 2016). We shed light on how tools, like video tagging, might work to scaffold an essential practice for teaching teachers: noticing student *and* teacher thinking. While this work is a good start, future efforts can help us better understand how teacher educators facilitate video-PD with teachers, including understanding each facilitator noticing component in greater detail, the role that noticing student thinking plays in facilitator noticing, and how facilitators learn to notice. These insights are essential for designing learning opportunities that prepare teacher educators to lead effective PD experiences for both pre-service teachers and in-service teachers (AMTE, 2017). Such advances and research on these ideas can help to move the field of mathematics teacher education forward.

ACKNOWLEDGEMENTS

This material is based upon work supported by the National Science Foundation under Grant No. DRL 17-537. Any opinions, findings, and conclusions or recommendations expressed in this material are those of the author(s) and do not necessarily reflect the views of the National Science Foundation.

REFERENCES

Amador, J. (2016). Professional noticing practices of novice mathematics teacher educators. *International Journal of Science and Mathematics Education, 14*(1), 217–241. https://doi.org/10.1007/s10763-014-9570-9

Amador, J. M. (2020). Teacher leaders' mathematical noticing: Eliciting and analyzing. *International Journal of Science and Mathematics Education, 18*(2), 295–313. https://doi.org/10.1007/s10763-019-09956-5

Amador, J. M. (2021). Mathematics teacher educator noticing: examining interpretations and evidence of students' thinking. *Journal of Mathematics Teacher Education, 25,* 163–189. https://doi.org/10.1007/s10857-020-09483-z

Amador, J. M., Estapa, A., Araujo, Z. de, Kosko, K. W., & Weston, T. L. (2017). Eliciting and analyzing preservice teachers' mathematical noticing. *Mathematics Teacher Educator, 5*(2), 158–177. https://doi.org/10.5951/mathteaceduc.5.2.0158

Arzarello, F. (2006). Semiosis as a multimodal process. *Revista Latinoamericana de Investigación en Matemática Educativa RELIME, 9*(Extraordinario 1), 267–299.

Association of Mathematics Teacher Educators. (2017). *Standards for preparing teachers of mathematics.* https://amte.net/standards

Ball, D. L., Thames, M. H., & Phelps, G. (2008). Content knowledge for teaching: What makes it special? *Journal of Teacher Education, 59*(5), 389–407. https://doi.org/10.1177/0022487108324554

Borko, H., Jacobs, J., Eiteljorg, E., & Pittman, M. E. (2008). Video as a tool for fostering productive discussions in mathematics professional development. *Teaching and Teacher Education, 24*(2), 417–436. https://doi.org/10.1016/j.tate.2006.11.012

Borko, H., Koellner, K., & Jacobs, J. (2011). Meeting the challenges of scale: The importance of preparing professional development leaders. *Teachers College Record,* 11–14.

Borko, H., Jacobs, J., Seago, N., & Mangram, C. (2014). Facilitating video-based professional development: Planning and orchestrating productive discussions. In Y. Li, E. A. Silver, & S. Li (Eds.), *Transforming mathematics instruction multiple approaches and practices* (pp. 259–281). https://doi.org/10.1007/978-3-319-04993-9_16

Borko, H., Koellner, K., & Jacobs, J. (2014). Examining novice teacher leaders' facilitation of mathematics professional development. *Journal of Mathematical Behavior, 33*(8), 149–167. https://doi.org/10.1016/j.jmathb.2013.11.003

Carlson, M. A., Heaton, R., & Williams, M. (2017). Translating professional development for teachers intro professional development for instructional leaders. *Mathematics Teacher Educator, 6*(1), 27–39.

Carpenter, T. P., Fennema, E., Peterson, P. L., Chiang, C.-P., & Loef, M. (1989). Using knowledge of children's mathematics thinking in classroom teaching: An experimental study. *American Educational Research Journal, 26*(4), 499–531. https://doi.org/10.3102/00028312026004499

Castro Superfine, A., Amador, J., & Bragelman, J. (2019). Facilitating video-based discussions to support prospective teacher noticing. *Journal of Mathematical Behavior, 54,* 1–18. https://doi.org/10.1016/j.jmathb.2018.11.002

Castro Superfine, A., & Li, W. (2014). Exploring the mathematical knowledge needed for teaching teachers. *Journal of Teacher Education, 65*(4), 303–314. https://doi.org/10.1177/0022487114534265

Elliott, R., Kazemi, E., Lesseig, K., Mumme, J., Carroll, C., & Kelley-Petersen, M. (2009). Conceptualizing the work of leading mathematical tasks in professional development. *Journal of Teacher Education, 60*(4), 364–379. https://doi.org/10.1177/0022487109341150

Empson, S. B., & Jacobs, V. R. (2008). Learning to listen to children's mathematics. In D. Tirosh & T. Wood (Eds.), *International handbook of mathematics teacher education: Volume 2* (pp. 257–281). Sense Publishers.

Gallese, V., & Lakoff, G. (2005). The brain's concepts: The role of the sensory-motor system in conceptual knowledge. *Cognitive Neuropsychology, 22*(3–4), 455–479. https://doi.org/10.1080/02643290442000310

Gaudin, C., & Chaliès, S. (2015). Video viewing in teacher education and professional development: A literature review. *Educational Research Review, 16,* 41–67. https://doi.org/10.1016/j.edurev.2015.06.001

Goldin-Meadow, S. (2003). *Hearing gesture: How our hands help us think.* Harvard University Press.

Goldsmith, L. T., & Seago, N. (2008). Using video cases to unpack the mathematics in students' thinking. In M. S. Smith & S. N. Friel (Eds.), *Cases in mathematics teach-*

er education: Tools for developing knowledge needed for teaching. Association of Mathematics Teacher Educators.

Herbst, P., Chazan, D., & Lavu, S. (2019). *Anotemos.* Web-based collaborative software tool for the annotation of video. Disclosed to the Office of Technology Transfer, University of Michigan.

Jackson, K., Cobb, P., Wilson, J., Webster, M., Dunlap, C., & Appelgate, M. (2015). Investigating the development of mathematics leaders' capacity to support teachers' learning on a large scale. *ZDM Mathematics Education, 47*(1), 93–104. https://doi.org/10.1007/s11858-014-0652-5

Jacobs, J. K., & Morita, E. (2002). Japanese and American teachers' evaluations of videotaped mathematics lessons. *Journal for Research in Mathematics Education, 33*(3), 154–175. https://doi.org/10.2307/749723

Jacobs, V. R., Franke, M. L., Carpenter, T. P., Levi, L., & Battey, D. (2007). Professional development focused on children's algebraic reasoning in elementary school. *Journal for Research in Mathematics Education, 38*(3), 258–288. https://doi.org/10.2307/30034868

Jacobs, V. R., Lamb, L. L. C., & Philipp, R. A. (2010). Professional noticing of children's mathematical thinking. *Journal for Research in Mathematics Education, 41*(2), 169–202. https://doi.org/10.5951/jresematheduc.41.2.0169

Kang, H., & van Es, E. A. (2019). Articulating design principles for productive use of video in preservice education. *Journal of Teacher Education, 70*(3), 237–250. https://doi.org/10.1177/0022487118778549

Kazemi, E., Elliott, R., Mumme, J., Carroll C. Lesseig, K., & Kelley-Peterson, K. (2011). Noticing leaders' thinking about videocases of teachers engaged in mathematics tasks in professional development. In M. Sherin, V. Jacobs, & R. Phillip (Eds.), *Mathematics teacher noticing: Seeing through teachers' eyes* (pp. 188–203). Routledge.

Lesseig, K., Elliott, R., Kazemi, E., Kelley-Petersen, M., Campbell, M., Mumme, J., & Carroll, C. (2017). Leader noticing of facilitation in videocases of mathematics professional development. *Journal of Mathematics Teacher Education, 20*(6), 591–619. https://doi.org/10.1007/s10857-016-9346-y

McNeill, D. (1992). *Hand and mind: What gestures reveal about thought.* University of Chicago press.

Nathan, M. J., & Alibali, M. W. (2011). How gesture use enables intersubjectivity in the classroom. In G. Stam & M. Ishino (Eds.), *Integrating gestures: The interdisciplinary nature of gesture* (pp. 257–266). John Benjamins.

Radford, L. (2009). Why do gestures matter? Sensuous cognition and the palpability of mathematical meanings. *Educational Studies in Mathematics, 70*(2), 111–126. https://doi.org/10.1007/s10649-008-9127-3

Roth McDuffie, A., Foote, M. Q., Bolson, C., Turner, E. E., Aguirre, J. M., Bartell, T. G., Drake, C., & Land, T. (2014). Using video analysis to support prospective K–8 teachers' noticing of students' multiple mathematical knowledge bases. *Journal of Mathematics Teacher Education, 17*(3), 245–270. https://doi.org/10.1007/s10857-013-9257-0

Santagata, R. (2011). From teacher noticing to a framework for analyzing and improving classroom lessons. In M. Sherin, V. Jacobs, & R. Philipp (Eds.), *Mathematics teacher noticing: Seeing through teachers' eyes* (pp. 152–168). Routledge.

Schoenfeld, A. (2011). Noticing matters. A lot. Now what? In M. Sherin, V. Jacobs, & R. Philipp (Eds.), *Mathematics teacher noticing: Seeing through teachers' eyes* (pp. 2–24). Routledge.

Shaughnessy, M., Garcia, N., Selling, S. K., & Ball, D. L. (2016). What knowledge and skill do mathematics teacher educators need and (how) can we support its development? In M. B. Wood, E. E. Turner, M. Civil, & J. A. Eli (Eds.), *Proceedings of the 38th annual meeting of the North American Chapter of the International Group for the Psychology of Mathematics Education* (pp. 813–820). The University of Arizona.

Sherin, M. G. (2011). The develop of teachers' professional vision in video clubs. In R. Goldman, R. Pea, B. Barron, & S. J. Derry (Eds.), *Video research in the learning sciences* (pp. 383–395). Lawrence Earlbaum Associates.

Sherin, M. G., & Han, S. Y. (2004). Teacher learning in the context of a video club. *Teaching and Teacher Education, 20*(2), 163–183. https://doi.org/10.1016/j.tate.2003.08.001

Sherin, M. G., & van Es, E. A. (2005). Using video to support teachers' ability to notice classroom interactions. *Journal of Technology and Teacher Education, 13*(3), 475–491.

Sherin, M. G., & van Es, E. A. (2009). Effects of video club participation on teachers' professional vision. *Journal of Teacher Education, 60*(1), 20–37. https://doi.org/10.1177/0022487108328155

Shulman, L. S. (1986). Those who understand: Knowledge growth in teaching. *Educational Researcher, 15*(2), 4–14. https://doi.org/10.3102/0013189X015002004

Stein, M. K., Engle, R. A., Smith, M. S., & Hughes, E. K. (2008). Orchestrating productive mathematical discussions: Five practices for helping teachers move beyond show and tell. *Mathematical Thinking and Learning, 10*(4), 313–340. https://doi.org/https://doi.org/10.1080/10986060802229675

Stockero, S. L., Rupnow, R. L., & Pascoe, A. E. (2017). Learning to notice important student mathematical thinking in complex classroom interactions. *Teaching and Teacher Education, 63*, 384–395. https://doi.org/10.1016/j.tate.2017.01.006

Tekkumru-Kisa, M., & Stein, M. K. (2017). A framework for planning and facilitating video-based professional development. *International Journal of STEM Education, 4*(1). https://doi.org/10.1186/s40594-017-0086-z

van Es, E. A. (2011). A framework for learning to notice student thinking. In M. Sherin, V. Jacobs, & R. Phillip (Eds.), *Mathematics teacher noticing seeing through teachers' eyes* (pp. 134–151). Routledge. https://doi.org/10.4324/9780203832714

van Es, E. A., Tekkumru-Kisa, M., & Seago, N. (2020). Leveraging the power of video for teacher learning. In S. Llinares & O. Chapman (Eds.), *International handbook of mathematics teacher education: Volume 2* (pp. 23–54). Leiden, The Netherlands: Koninklijke Brill NV. https://doi.org/10.1163/9789004418967_002

van Es, E. A., Tunney, J., Goldsmith, L. T., & Seago, N. (2014). A framework for the facilitation of teachers' analysis of video. *Journal of Teacher Education, 65*(4), 340–356. https://doi.org/10.1177/0022487114534266

Walkoe, J. (2015). Exploring teacher noticing of student algebraic thinking in a video club. *Journal of Mathematics Teacher Education, 18*(6), 523–550. https://doi.org/10.1007/s10857-014-9289-0

Walkoe, J., & Levin, D. (2018). Using technology in representing practice to support pre-service teachers' quality questioning: The roles of noticing in improving practice. *Journal of Technology and Teacher Education, 26*(1), 127–147.

Walkoe, J., Sherin, M., & Elby, A. (2020). Video tagging as a window into teacher noticing. *Journal of Mathematics Teacher Education, 23*(4), 385–405. https://doi.org/10.1007/s10857-019-09429-0

Walkoe, J., Williams-Pierce, C. C., Flood, V. J., & Walton, M. (2023). Toward professional development for multimodal teacher noticing. *Journal for Research in Mathematics Education, 54*(4), 279–285.

Zhang, M., Lundeberg, M., & Eberhardt, J. (2011). Strategic facilitation of problem-based discussion for teacher professional development. *Journal of the Learning Sciences, 20*(3), 342–394. https://doi.org/10.1080/10508406.2011.553258

CHAPTER 13

USING MANIPULATIVES IN FACE-TO-FACE, HYBRID, AND VIRTUAL EARLY CHILDHOOD AND ELEMENTARY MATHEMATICS METHODS COURSES

Aidong Zhang
Louisiana State University Shreveport

Carrie Cutler
University of Houston

This chapter examines the implementation of manipulatives in face-to-face, hybrid, and virtual instruction. We utilized our Methods Course Instructional Re-Design (MCIRD) model to re-envision our teaching practices amidst the COVID-19 pandemic. Our MCIRD model drew from the Technological Pedagogical Content Knowledge (TPACK) Framework (Mishra & Koehler, 2006), Universal Design for Learning (UDL) (CAST, 2018) and the Standards for Preparing Teachers of Math-

The AMTE Handbook of Mathematics Teacher Education: Reflection on Past, Present and Future—Paving the Way for the Future of Mathematics Teacher Education, Volume 5
pages 265–289.

ematics (AMTE, 2017). Within a context of preparing teacher candidates (TCs) to create and implement developmentally appropriate mathematical learning experiences for children, we share two case studies embedded within the broader discussion of why and how to effectively implement concrete and virtual manipulatives. Case Study 1 details an elementary teacher educator's processes implementing virtual manipulatives in a synchronous online course. Case Study 2 details an early childhood teacher educator's incorporation of both concrete and virtual manipulatives in face-to-face, hybrid, and virtual courses. Based on our course design, we characterize face-to-face courses as meeting in person, hybrid courses as including both face-to-face and virtual components, and virtual as being held fully online. By exploring the integration of manipulatives in multiple modes of instruction, we offer practical suggestions for teacher educators engaged in this work. Finally, we identify what we do not yet understand about using manipulatives in different modes of instruction and how we can advance the field.

INTRODUCTION

Mathematics methods courses prepare teacher candidates (TCs) to create and implement developmentally appropriate mathematical learning experiences for children. As the COVID-19 pandemic disrupted educational institutions at every level, early childhood and elementary mathematics educators realized the need for approximations of the rich hands-on experiences that typified in-person courses. We faced considerable uncertainty in the transformation of our course delivery methods, assignments, and TCs' interactions with one another and with hands-on materials such as mathematics manipulatives. Specifically, we grappled with how to maintain the quality of our courses as exemplars of best practices for teaching children mathematics as well as a laboratory-style rehearsal space for novice early childhood teachers. The changes the COVID-19 pandemic made to teacher preparation programs have persisted for some time (Ellis et al., 2020), and in some contexts continues to persist. A new reality has included students and instructors becoming ill or entering quarantine at unpredictable times, necessitating time-sensitive adjustments to courses. While there has been a return to traditional face-to-face courses, the substantial investment in resources to support remote learning drives conversations at universities regarding expansions of hybrid and online courses, as well as fully online programs.

This study of using manipulatives in multiple modes of instruction has practical applications for mathematics teacher education programs operating in a fluid educational environment. Two immediate concerns surfaced as we began to revise our courses. First, since early childhood educators have thrived in traditional face-to-face classroom settings with developmentally appropriate practices, we understandably resisted moving to online spaces (Martin et al., 2019; Mills et al., 2009). We understood how our pedagogy performed in face-to-face settings but struggled to translate practices and engage teacher candidates in a virtual environment (Cutri et al., 2019; Livers & Piccolo, 2020). Second, we realized that considerations for our practice extended beyond our own teaching; they also encompassed preparing TCs to manage online educational formats about which we ourselves were still learning (Downing & Dyment, 2013). As such, we sought to explore ways to maintain the philosophical frames for our face-to-face courses during the pivot to virtual and hybrid instruction. We considered mathematics methods classes as workspaces for TCs to experience best practices related to hands-on learning of early childhood mathematics concepts, developmentally appropriate practice, instructional expectations, and promotion of just and equitable mathematics education for all learners. As such, we joined apprehensive educators across the globe who stepped out of their comfort zones to set aside the teddy bear counters in favor of virtual blocks.

This chapter has three foci. First, we review the rich history and extensive research support for manipulatives in early childhood classrooms and mathematics method courses. Second, we report the results of two reflective self-studies that examined approaches, challenges, and concerns related to using manipulatives in three different settings (i.e., face-to-face, hybrid, virtual) particularly considering the disruption caused by the COVID-19 pandemic. Finally, we describe takeaways for teacher educators implementing manipulatives in various modes of instruction.

THEORETICAL FRAMEWORK AND LITERATURE REVIEW

Children are inherently mathematical, using content and process skills informally but adeptly in everyday activities and to solve problems (NCTM, 2000). Ginsburg (2006) noted that young children's mathematics resembles mathematicians' work as they ask questions, invent solutions, and play with mathematics, often through hands-on activities with manipulatives. The kinesthetic experience offered by manipulatives develops understanding of mathematics concepts through multiple representations and positions learning around the constructive processes of problem-solving, communication, and reasoning. In this section, we examine manipulatives from a historical perspective, explore justifications for implementing them with young children and preservice teachers, and provide support for using manipulatives in multiple instructional modes for early childhood and elementary methods courses.

Historical Foundations of Manipulatives in Mathematics Education

Humans have applied simple tools to solve mathematics problems since ancient times. For example, the Chinese abacus, also known as *suan pan* in Chinese, was used as a counting and calculation aid as early as the 2nd century BC and included in the Chinese elementary mathematics curriculum as late as 2001. In the early 19th century, Friedrich Fröebel, considered the father of kindergarten, developed mathematical manipulatives or "gifts" for young children meant to stimulate understanding of the world through exploration of spatial relationships or properties by constructing with blocks (Reinhold et al., 2017). In the early 20th century, Italian educator Maria Montessori (1870–1952) designed hands-on learning materials to help young children learn basic mathematical concepts through progressively sequenced activities leading to abstract reasoning. Georges Cuisenaire (1891–1975), a Belgian educator, developed Cuisenaire Rods in the 1950s to provide concrete representations of number relationships. In the mid-20th century, manipulatives appeared in traditional elementary mathematics instruction (van Engen, 1949). Piaget (1952) believed that children developed logico-mathematical knowledge on a concrete level before advancing to symbolic conceptualizations. While manipulatives have traditionally been deemed concrete objects, technological advancements and the expansion of online learning, especially during the COVID-19 pandemic, have prompted greater use of virtual manipulatives as tools for supporting mathematics learning (Aliyyah et al., 2020)

Types of Manipulatives

Mathematics manipulatives are physical objects designed to concretely represent abstract mathematical ideas (Moyer et al., 2005). They may facilitate the exploration, acquisition, or investigation of mathematical concepts or processes by "drawing on perceptual (visual, tactile, or, more generally, sensory) evidence" (Bartolini & Martignone, 2014, p. 365). For the purposes of this study, we identified manipulatives as either concrete or virtual tools that engage children in exploration, visualization, and representation of mathematical ideas.

Concrete Manipulatives

Concrete manipulatives (CMs), also called concrete materials, concrete objects, physical manipulatives, or physical materials, assist teachers and students in modeling or simulating mathematical phenomena by (re)constructing concepts via a physical representation of the problem (D'Angelo & Iliev, 2012; Loong, 2014). Essential to a comprehensive and developmentally appropriate mathematics education program for young children, concrete materials support fine motor skills alongside mathematics learning by activating multiple learning modalities (NAEYC, 2020). Children experience counting, sorting, comparing, spatial reasoning, and problem solving among other early mathematics objectives as they manipulate objects like cubes, counters of various forms, base-ten blocks, pattern blocks, attribute blocks, geoboards, two- and three-dimensional shapes and more.

Virtual Manipulatives

Virtual manipulatives (VMs) are interactive web-based visual representations of concrete manipulatives or tools. VMs fall under the larger umbrella of representations used to assist in modeling, representing, or simulating mathematical phenomena (Moyer et al., 2005). While not offering a fully concrete experience, VMs have some advantages over CMs. They are readily available, provide immediate visual feedback, and may be more motivating for some students. They require no storage space or sanitization, cannot become broken or lost, and are generally free through online portals. VMs can be used by individual students, small groups, or for teacher demonstration and are accessible to any user with internet access and a compatible device. Certain VMs, such as tangrams and geoboards, closely resemble their CM counterparts and have been shown to promote as much engagement as CMs (e.g., Weng et al., 2020).

Why Use Manipulatives in Mathematics Methods Courses

Manipulatives are widely regarded as essential to developing children's understanding in mathematics (Willingham, 2017). These tools stimulate interest in mathematics (Sutton & Krueger, 2002), support mathematical reasoning (Stein & Bovalino, 2001), deepen understanding (Witzel & Allsopp, 2007), and connect abstract ideas to concrete representations (Uribe-Flórez & Wilkins, 2017). The implementation of manipulatives in teacher preparation courses builds on these strengths to include three intertwined benefits which align with the AMTE's Standards for Preparing Teachers of Mathematics (SPTM) (2017). First, using manipulatives in methods courses promotes TCs' understanding of mathematical concepts (Elaboration of Indicator C.1.1). Second, TCs see firsthand how to use manipulatives with children (Indicators C.1.6 and P.3.5). Third, using manipulatives gives TCs insight into how children learn mathematics (Indicator EC.7).

Deepen TCs' Mathematical Content Knowledge

The SPTM (AMTE, 2017) state that "well-prepared beginning teachers of mathematics at the early childhood level have a deep understanding of the mathematical concepts and processes important in early learning as well as knowledge beyond what they will teach" (Elaboration of C.1.1). Manipulatives are appropriate tools for all ages and grade levels and support the goal of strong content knowledge among early childhood educators (NCTM, 2000).

In mathematics methods courses, manipulatives have been shown to improve TCs' conceptual and procedural understanding as well as overall achievement by connecting real world situations to mathematical representations, thereby encouraging cooperative problem-solving efforts (Carbonneau et al., 2013; National Research Council, 2001). They facilitate discussion by offering visual representations of student thinking and demonstrate that mathematics problems can be symbolized in varied ways (Riccomini et al., 2015). Hunt et al. (2011) found advantages to incorporating both concrete and virtual manipulatives. When conceptual understanding was developed first with CMs, the subsequent use of VMs facilitated transition to abstract representations. The use of VMs allowed instructors to excel in providing immediate feedback, sustained engagement, and school-home connections through their accessibility from numerous devices. CMs appeared to be more effective for building TCs' conceptual understanding with VMs used to reinforce those concepts. Hunt et al. also noted that incorporating both types of manipulatives not only helped TCs build conceptual understanding but also provided them with sound pedagogical strategies for future use.

Model How TCs May Use Manipulatives with Children

Well-prepared mathematics teachers must be able to use mathematical tools and technology (AMTE, 2017, Indicator C.1.6). TCs become competent in selecting and implementing manipulatives that fit the developmental levels of children (Smith, 2009; Van de Walle et al., 2018) in part by using the tools themselves. NCTM's (2014) Effective Mathematics Teaching Practices promote equitable access to resources and materials that maximize students' learning potential and urge teachers to engage students in making connections utilizing representations that deepen their understanding of mathematics. When TCs are introduced to these ideas, manipulatives may be used to demonstrate their classroom implementation. For example, a methods instructor may give TCs the task of constructing parallelograms using tangrams but offer inequitable resources to groups of TCs. While one group receives a full set of tangrams, another is given an incomplete set that makes the task impossible, providing a simplified demonstration of how access to resources affects learning. The methods instructor may also discuss how to arrange the physical environment to promote access to manipulatives, management tips to ensure safety (e.g., keeping manipulatives away from mouths), and developmentally appropriate expectations (e.g., allowing time for free exploration before expecting children to use manipulatives in a prescribed manner).

Mathematics educators should instruct TCs on "research and practice regarding pedagogical and equity issues associated with effectively teaching each and every student" (AMTE, 2017, Indicator P.3.5; NAEYC, 2019). Manipulative-integrated methods courses help TCs explore how students from all backgrounds, including multilanguage learners and children with special needs, can access fundamental concepts through hands-on activities that allow for cooperative work, mathematical communication, and differentiated instruction. Giving TCs opportunities to adapt or adjust manipulatives to meet students' individual needs provides valuable applications for best practices in differentiated learning. For example, TCs may design a mathematics workstation with easily modified materials that can be adjusted based on a child's individual needs.

Help TCs Understand How Children Learn Mathematics

Teacher candidates at the early childhood and elementary school level should be given opportunities to see mathematical situations through children's eyes (AMTE, 2017, Indicator EC.7). Therefore, when designing lessons for mathematics methods courses, teacher educators should orchestrate experiences where TCs can feel some of the wonder encountered by young mathematicians. For example, teacher educators can help TCs compare the engagement they feel while completing a worksheet where they trace outlines of shapes to their interest during a hands-on activity when they create 2D shapes by using drinking straws as the sides and balls of clay as the vertices of shapes. TCs' experiences with hands-on learning during methods courses may also contribute to their personal beliefs about developmentally appropriate practice (NAEYC, 2020). When TCs experience mathematics instruction that incorporates manipulatives, they see how hands-on learning supports all the domains of a child's development. For example, stringing ten beads on a bracelet and sliding the beads to discover the addends for ten improves fine motor skills while also supporting the visualization of the parts compared to the whole. Speaking while acting out a join or separate scenario with manipulatives supports linguistic development. Ongoing, rich mathematical experiences where children reason, problem solve, connect, represent, and communicate their mathematical ideas (NCTM, 2000) enhance cognitive development.

How to Use Manipulatives in Multiple Modes of Instruction

With rapidly changing social conditions and expanding modes of instruction, mathematics educators need flexible approaches for using concrete and virtual manipulatives to support TCs' mathematics learning in face-to-face, hybrid, and virtual instructional settings.

Manipulatives in Face-to-Face Instruction

Face-to-face courses aim to provide TCs with a laboratory-style setting for enacting best practices. The Concrete-Representational-Abstract (CRA) framework exemplifies one such practice. CRA begins with students manipulating CMs such as linking cubes then moving to the representational level by drawing images like tallies or dots before advancing to exclusive use of abstract symbols. CRA instruction allows students to associate one stage of the process with the next (Jones & Tiller, 2017). Through repeated experiences with this pedagogical cycle, TCs build personal conceptual understanding but also recognize how children may internalize increasingly abstract concepts and processes. TCs should realize that using CMs is only a step toward mastering mathematical concepts and that children's guidance through the CRA process can be complex and nuanced. Although manipulatives play a significant role in learning, their representational natures do not carry the meaning of the mathematical idea (Sarama & Clements, 2016). As such, TCs may benefit from CMs to build meaning initially and to understand children's learning, but they must reflect on and discuss their actions with manipulatives to make connections among the tools, the concepts, and the learning processes.

Manipulatives in Virtual Instruction

While Concrete Manipulatives (CMs) hold a significant role in face-to-face mathematics classrooms, access to CMs limits their implementation in virtual instruction. Thus, educators rely upon websites with VMs, software applications, and other virtual tools to design interactive and engaging activities (Bouck et al., 2021; Young, 2017). The mathematical, cognitive, and pedagogical fidelity of the tool should be considered when selecting a VM (Zbiek et al., 2007). The teacher educator may ask: To what degree does the VM faithfully represent the underlying mathematical properties of that object? How well does the VM reflect the user's cognitive actions and choices? To what extent does the VM allow students to behave in accordance with the nature of mathematical learning and pedagogical best practices? Integrating VMs helps TCs develop a critical eye for VMs and their pedagogical considerations (Moyer et al., 2005).

Manipulatives in Hybrid Instruction

Hybrid models of instruction combine the accessibility of CMs afforded by face-to-face classes with the flexibility of virtual instruction incorporating VMs. TCs in hybrid courses may attend some classes in person and others asynchronously or synchronously online. Methods class instructors who are engaged in hybrid teaching can plan face-to-face lessons rich with hands-on experiences supplemented by virtual learning using VMs or introduce VMs during face-to-face classes. When TCs recognize how quickly technologies emerge and develop facilities for analyzing VMs' potential and limitations for students' learning, they may confidently apply VMs in their future classrooms (AMTE, 2017).

METHODOLOGICAL APPROACH

This section details our self-reflective case study research as mathematics teacher educators were tasked with adapting our face-to-face mathematics methods courses to virtual and hybrid settings where manipulatives could still play an essential role in TCs' learning. We describe the components of our instructional redesign, instructional strategies, and approaches using CMs and VMs in our courses.

Research Design

A self-reflective case study was utilized as a strategy of inquiry to explore the experiences of two teacher educators who taught elementary and early childhood mathematics courses during the switch to virtual learning. Case study research begins with the desire to derive an in-depth, personal understanding of a case set in its real-world context and assumes that examining the context and other complex conditions related to the case are integral to producing new learning about behavior and its meaning (Yin, 2012). Our experiences revolved around our use of manipulatives with the underlying goal of examining how "our practice as teacher educators could be improved" (Bullough & Pinnegar, 2001, p. 14).

Contexts and Participants

We conducted our studies in two four-year university-based teacher education programs in the Southwestern part of the United States where mathematics methods courses fulfill state teaching certificate requirements. Lee (pseudonym) teaches courses to preservice early childhood educators while Stewart (pseudonym) works with teacher candidates seeking certification in early childhood through grade six. Both have substantial teaching experience in face-to-face settings where they use manipulatives to provide a laboratory-like setting for TCs to practice techniques for instructing young children. These techniques include inquiry-based learning, socially mediated learning through workstations, and peer feedback on a TC's practice teaching.

Methods Course Instructional Re-Design Model

The instructional design approach for re-envisioning methods courses for an online format drew from the Technological Pedagogical Content Knowledge (TPACK) model (Mishra & Koehler, 2006) and Universal Design for Learning (UDL) (CAST, 2018). The TPACK and Methods Course Instructional Re-Design (MCIRD) model are shown in Figure 13.1.

TPACK differentiates between technological knowledge (TK), pedagogical knowledge (PK), and content knowledge (CK). Personal knowledge of the subject matter and best practices for conveying this information to learners comprise content knowledge. Skills revolving around managing classroom norms, planning instruction, and assessing learning produce pedagogical knowledge. The intersection between these two, pedagogical and content knowledge, provides guidance to improve teaching practices by creating stronger connections between the content and the pedagogy used to communicate it. The intersection of the three forms of

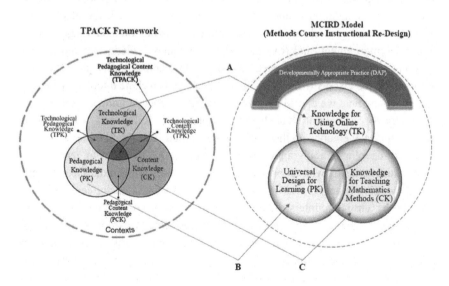

FIGURE 13.1.   TPACK Framework (Image ©2012 by tpack.org) and Methods Course Instructional Re-Design Model

knowledge is TPACK. Blending and balancing technology, content, and pedagogy are key to implementing TPACK.

The MCIRD model was derived from the TPACK framework but specifically targeted our synchronous online teaching of mathematics methods. The MCIRD model includes three forms of essential knowledge: Knowledge for Using Online Technology, Universal Design for Learning, and Knowledge for Teaching Mathematics Methods. First, as shown in Figure 13.1, Arrow A highlights the correspondence between TPACK's technological knowledge domain and knowledge for using online technology, specifically in a synchronous setting. Second, UDL represented the pedagogical knowledge domain of MCIRD with Arrow B identifying our commitment to promoting multiple means of engagement, representation, and expression when teaching and learning mathematics. Arrow C shows how knowledge for teaching mathematics methods content fulfilled TPACK's content knowledge domain. Importantly, as illustrated by the overarching umbrella labeled Developmentally Appropriate Practice (DAP) as context within the MCIRD model, we deemed expectations for young children's online and mathematics learning as protectively superimposing all other components of the MCIRD model (Bredekamp & Copple, 2009; NAEYC, 2020).

We utilized our MCIRD model to re-envision our teaching practices in mathematics methods. By purposefully blending our TK (online technology), PK (Universal Design for Learning), and CK (mathematics methods courses) together, we sought approaches to support (a) content knowledge of and best practices for teaching early childhood mathematics, (b) pedagogical knowledge of effective methods for managing classroom norms, effective instruction, and assessment in an online environment, and (c) pedagogical content knowledge for integrating technology in the mathematics methods course. In planning and leading the synchronous online classes where VMs would be used as technical tools to promote TCs understanding, we considered how pedagogical content knowledge could be effectively accessed, improved, shared, and observed. To this end, we identified areas of the mathematics methods course curriculum that could be leveraged to exemplify developmentally appropriate practices (DAP Context) in early childhood mathematics as well as principles of UDL. In a simple example, TCs explored counting concepts with homemade CMs (like beans or dry cereal pieces) followed by virtual counters, ticking off a box next to knowledge for teaching mathematics methods by promoting engagement through exploration. Using a variety of tools attended to UDL by providing multiple means of representation as well as to the early childhood practice of using tactile manipulations to aid understanding of mathematics. Moving between the CM and VM representations also modeled pedagogical knowledge for managing classroom routines in a synchronous online environment.

Case Studies

In this study, we acted as researchers as well as participants by bringing our individual perspectives to the foreground. Our data consisted of observations, journal entries, and videotaped virtual class sessions held during Spring 2020, Spring 2021, and Fall 2021 semesters. We collected polling results and Google Jamboard pages on which TCs collaborated in breakout rooms to examine content knowledge of and best practices for teaching children mathematics. With respect to the use of words as data to support data interpretation, we consider the influence of the collaborative and iterative teacher educator experience to be highly informative. Nonetheless, our interpretations may or may not be generalizable across similar teaching situations and are highly dependent on the ways teacher educators' particular experiences, past and present, affect their use of manipulatives and perceptions related to those experiences.

Case Study 1: Stewart's Experiences

Early March 2020: posted instructions for a daily warm-up task greeted TCs as they spilled into the university classroom.

> Work with a group of 3–4 peers. Choose a set of manipulatives. Count out 15 manipulatives and place them on your table. Show all the ways the 15 objects can be put into four piles so that each pile has a different number of objects in it. How can you be sure you have found all possible solutions? (Burns, 2015)

TCs selected tubs of counters—cubes, tiles, teddy bears, dinosaurs—and set to work. Later, TCs modeled strategies and shared their solutions with the whole class. Manipulatives provided TCs an entry into mathematics, allowed for collaborative work, facilitated mathematical communication, and suggested the need for a systematic approach to finding all possibilities. More importantly, TCs learned how they may use manipulatives with young children. The TCs' work exemplified UDL pedagogical knowledge (PK) for teaching mathematics methods through engaging and explorative tasks.

Packing up the materials after class, I was unaware this would be the final in-person meeting of the semester. The pivot to online learning would curtail my pedagogical go-to of hands-on learning. Recognizing this untested moment in higher education, I initiated a self-study to record, examine, and learn from my missteps and successes.

Using Manipulatives in Virtual Classes. Buoyed with confidence in my knowledge of teaching mathematics methods, I ambitiously redesigned my course for a synchronous online format. The Blackboard Learn learning management system and selected supportive technologies facilitated synchronous experiences with virtual manipulatives. Zoom, Google Jamboard, and breakout rooms enabled collaborative work as students shared their screens and displayed VMs. However, I was new to these tools and had not integrated VMs regularly during face-to-face methods courses. Could I maintain pedagogical fidelity to best practices considering my novice level of knowledge for implementing online technology (TK)? I geared my focus toward selecting and becoming skilled with appropriate tools that captured CMs' key features. These aims aligned with DAP and the 2017 AMTE SPTM Indicators C.1.6, C.2.2, C.2.3, and EC.7 as detailed in Table 13.1. Practical application of the MCIRD model centered on a) attending to knowledge for using online technology through the effective use of tools such as VMs,

TABLE 13.1. Examples of Enactments of AMTE Standards for Preparing Teachers of Mathematics (2017) Through Virtual Manipulatives

AMTE Standards for Preparing Teachers of Mathematics Indicators Aligned to Stewart's Integration of VMs	Specific Examples of How the Course Promoted Each Standard
Indicator C.1.6. Use Mathematical Tools and Technology. Well-prepared beginning teachers of mathematics are proficient with tools and technology designed to support mathematical reasoning and sense making, both in doing mathematics themselves and in supporting student learning of mathematics.	TCs viewed and discussed one another's representations using VMs through screen sharing. TCs used discussion boards to share reasoning, reflect, and receive feedback from their peers and instructor online. TCs debriefed technology experiences via online polling tools and Google Jamboards.
Indicator C.2.2. Plan for Effective Instruction. Well-prepared beginning teachers of mathematics attend to a multitude of factors to design mathematical learning opportunities for students, including content, students' learning needs, students' strengths, task selection, and the results of formative and summative assessments.	TCs manipulated virtual base ten pieces to compose and decompose 2- and 3-digit numbers through rapid iterative practice with regrouping, a skill with which some TCs struggled.
Indicator C.2.3. Implement Effective Instruction. Well-prepared beginning teachers of mathematics use a core set of pedagogical practices that are effective for developing students' meaningful learning of mathematics.	Using VMs that closely resembled their CM counterparts, TCs used virtual rekenreks to explore the compositions for the number 10, acted out join and separate problems using virtual counters, and used virtual pattern blocks to find shapes that combine to compose a hexagon. TCs engaged with virtual coins to practice adding coin values and explicitly trade up to the next coin. TCs used virtual linking cubes to play Nim games and identify patterns that led to a winning strategy.
Indicator EC.7. Seeing Mathematics Through Children's Eyes. Well-prepared beginning teachers of mathematics at the early childhood level are conversant in the developmental progressions that are the core components of learning trajectories and strive to see mathematical situations through children's eyes.	TCs used virtual attribute blocks to describe 2D shapes using informal language. TCs then built 1-, 2-, and 3- difference trains, telling how each block's attributes differed from the block before it.

b) supporting TCs' sharing of mathematical thinking in a visual format thereby attending to the UDL section of the MCIRD, and c) facilitating a meaningful debrief of the technology experiences and mathematics content thereby leveraging my existing knowledge for teaching the mathematics methods course.

Engaging via Tools and Technology. Tapping the Supplemental Materials for AMTE's Standards (2018), specifically Thomas and Edson's (2019) Framework for Evaluating Digital Instructional Materials, helped me see how to integrate the Effective Mathematics Teaching Practices (NCTM, 2014) and the appropriate level of technology integration. According to Thomas and Edson, instructional materials might be utilized at three levels of integration—replace, amplify, or transform (Hughes et al., 2006). In the synchronous class, just as they would have done with CMs in a face-to-face class, TCs explained their thinking by sharing their screen to show their work with VMs. In this way, digital technology *replaced* CMs but maintained established instructional practices, student learning processes, and content goals (Hughes et al., 2006). At the *amplification* level, technology increases efficiency or productivity of students' learning. TCs experienced this using virtual base ten pieces for rapid composing or decomposing of sets, maximizing their iterative experiences with place value understanding. At the *transformation* level, digital tools or platforms transform the instructional method and learning process by facilitating collaborative efforts. TCs shared access to VMs or had individual access with the capability for screen sharing while explaining their thinking in breakout rooms. Groups then reported their results and strategies to the class.

The *replace, amplify, or transform* levels helped me see ways to attend to UDL while integrating VMs into mathematics tasks. For example, virtual pattern blocks, counters, geoboards, and base ten blocks replaced but closely resembled their concrete counterparts, allowing TCs to use the VM in ways similar to how they might use CM (Burris, 2013) but also increased flexibility and precision (Sarama & Clements, 2009). Table 13.2 shows how VMs modified a selection of activities using the MCIRD model as a guide.

Sharing Mathematical Thinking via VMs. Making mathematical thinking visual presents unique challenges to teachers and teacher educators, particularly in a virtual environment. In the synchronous online format, I attended to the MCIRD element of knowledge for using online technology by integrating technology (TK) as a tool to promote TCs' reflection, reasoning, collaboration, and problem solving. The primary tools for presenting thinking visually included screen sharing and collaborative work with shared VMs. For example, individual TCs used virtual tangrams to compose triangles, rectangles, and trapezoids. TCs then compared their designs with classmates in a breakout room. A drawback to this approach emerged when TCs used tablets or mobile phones rather than computers to log on to synchronous classes, making it difficult for them to access VMs or demonstrate their work. Through trial-and-error, I found digital tools compatible with a wide range of devices, further supporting the MCIRD model's knowledge for using online technology (TK).

TABLE 13.2. Adjustments to Methods Course Tasks to Integrate Virtual Manipulatives (VMs)

Mathematics Task	Description of Task and Prompt	CM Used	VM Used
Explore Number Composition and Decomposition	TCs directly model using rekenreks to show ways to compose numbers 1–10. How many ways can you make the number ten?	TCs use the top and bottom rows of beads to show number compositions	TCs use personal virtual rekenreks to compose sets then share screens. https://apps.mathlearningcenter.org/number-rack/
Create Pattern Block Fraction Books	TCs separate a whole into equal parts by using the hexagon pattern block piece to represent a whole. TCs explore ways smaller pattern blocks compose the whole. For example, two trapezoids compose one hexagon. What blocks can you use to make a whole?	TCs trace the hexagon pattern block on pages of a paper booklet. TCs cover the hexagon on each page with a different combination of blocks to show fractional portions of the whole then label the fractions.	TCs use virtual pattern blocks then label the fractional pieces with the text tool. https://apps.mathlearningcenter.org/pattern-shapes/
Build Numbers Using Base Ten Blocks	TCs use base ten blocks to compose two-digit and three-digit numbers. If you have 10 blocks, what numbers can you make?	TCs use concrete base ten blocks to find outcomes possible when using 10 blocks of any quantity.	TCs use virtual base ten blocks to rapidly compose and decompose numbers while finding possible combinations for using 10 blocks. https://apps.mathlearningcenter.org/number-pieces/
Model Cognitively Guided Instruction Problems	TCs explore how children at the direct modeling level use manipulatives to concretely represent numerical quantities within a problem. How might young children use counters to act out a join or separate problem?	TCs use linking cubes to model problems and write number sentences with paper and pencil.	TCs use virtual linking cubes to represent problems and the text tool to write number sentences. https://www.didax.com/apps/unifix/

Strategize with Nim Logic Game	TCs play with a partner. They place 7 counters on the table. Each player can pick up one or two counters. Play continues until all the counters are picked up. The goal is to not pick up the last counter. Play several times to discover a winning strategy by identifying patterns in the play. Who will always be the winner—the person who plays first or the person who plays second?	TCs use cubes to explore strategies to avoid picking up the final cube.	Virtual Nim games can be played online versus the computer or against a classmate. Short, repeated rounds of play help TCs identify patterns that lead to a winning strategy. https://education.jlab.org/nim/s_gamepage.html
Build and Describe Attribute Block Trains	TCs identify all the ways attribute blocks can be different from one another (color, shape, thickness, and size). TCs make a 2-difference train by making each successive block different from the previous block in exactly two ways. Each time TCs lay down a block, they tell how it is different from the block before it. TCs then make a 3-difference train and a 1-difference train. Which train was the hardest to build?	TCs use attribute blocks to describe attributes of two-dimensional shapes using informal language.	TCs use virtual attribute blocks to build trains. The virtual attribute blocks differ in the same ways concrete blocks do including thickness which is indicated by shading on the edge of the shape. https://oervm.s3-us-west-2.amazonaws.com/AttributeBlocks/index.html
Learn Coin Values with the Coin Trading Game	TCs draw three columns and label them penny, nickel, and dime. Partners take turns rolling a die and placing coins on their boards to match the value shown on the die. Players must show, in any column, the least number of coins possible. For example, as soon as you have enough to "trade up" 5 pennies for a nickel—do it. The first player with 5 dimes wins (Burns, 2015). How can you identify individual coins by name and value and describe relationships among them?	TCs use real or plastic pennies, nickels, and dimes to practice adding coin values and explicitly trading up to the next coin.	TCs use virtual coins, offering an optional grid representation for coin values which may serve as a scaffold for recognizing and combining coin values. https://apps.mathlearningcenter.org/money-pieces/

Debriefing VM Experiences and Mathematics Content. In face-to-face classes, I was able to scan the room while TCs worked in small groups, purposefully selecting who would debrief tasks in the class discussion. Determined to replicate this best practice, centered in the MCIRD element for knowledge for teaching the mathematics methods course, I relied on Google Jamboard, Zoom's polling and chat features, and online graphic organizers to orchestrate debriefing. These tools allowed individual TCs to give feedback about technology and mathematics experiences. For example, after working in breakout rooms, TCs responded to prompts such as, "Contribute 3 terms to the word cloud describing a child at the rational counter level" or "Rate your opinion of the virtual pattern blocks used in this activity on a scale of 1–10." In another lesson, I placed small groups of students in breakout rooms and had them contribute to a Jamboard like the one found in Figure 13.2. These digital tools provided a rapid, formative assessment and prompted TCs to think critically about technology and mathematics. These quick looks allowed the instructor to see all students' contributions at the same time and contextualized follow up discussions debriefing the tool's efficacy.

Case Study 2: Lee's Experiences

Prior to the pandemic, I had been exclusively using concrete manipulatives in traditional face-to-face mathematics methods courses. The Spring 2021semester began with my course being delivered fully online through the Moodle course

Group Five- Productive Struggle and Evidence of Student Thinking in The Band Concert
How might we check in on student thinking and struggles and adjust our instruction?

We can ask students different levels of questions and based on those responses we can see what students really understand.	We can ask the students to explain the method of solving the strategy in their own words, or also ask them questions as to how different strategies can be used to reach the same answer.	We can check on a student by asking them to make connections to other strategies/ and discuss with others in order to get them to explain their thinking.	Formative assessments/q uestioning

On what lines of the lesson do you find evidence of this?

Lines 59-81 whole class discussion where the students have to explain the reasons the strategies connect to each other.	Productive struggle 26-31	Lines 35-36	Lines 71-76 it asks students to make connections to other strategies.	52-53 Mr. Harris asked the presenting students to explain what they had done and why 53 and to answer questions posed by their peers.

FIGURE 13.2. Google Jamboard Image of Small Group Debrief of Sample Mathematics Lesson, The Band Concert

management system, a free open- source package designed for online teaching. Halfway through the semester, the course transitioned to hybrid teaching. It allowed TCs to attend in person or join classes virtually if they had to quarantine due to exposure to COVID-19. . In Fall 2021, I transitioned back to face-to-face instruction with mandatory social distancing and mask requirements. The use of manipulatives in each teaching format presented unique challenges.

Using Manipulatives in Face-to-Face Classes. In my traditional face-to-face classes where the aim was to prepare TCs to understand the content knowledge (CK) and pedagogical knowledge (PK), I used CMs to help TCs understand how children learn foundational mathematical concepts and develop mathematical thinking skills. For example, I relied heavily on modeling concepts using manipulatives, first explaining and demonstrating a mathematical concept using CMs then giving TCs time to practice as a group or individually. Finally, I expected each TC to be able to solve problems independently using the manipulatives to explain concepts. Lessons ended with TCs reflecting on their understanding of content knowledge (CK) or pedagogical knowledge (PK) from the day's lesson.

Before the pandemic, I stored tubs of manipulatives in a classroom closet accessible to TCs. When my course was required to observe social distancing, I prepared an individual manipulative kit for each TC. I sanitized the manipulatives after each class and stored them in the classroom closet. These kits promoted TCs' feelings of ownership which were amplified when TCs used wooden blocks to make dice for their differentiated "Roll and Add" addition mathematics station. Since TCs could not share the manipulatives due to social distance requirements, they had to make their own set of dice for different learning levels or age groups. Each of them was given five blank wooden blocks. They made five dice: one dice with numbers 0 to 2, one with numbers 1 to 3, one with 1 to 6, one with 5 to 10, and one with 10 to 15. At the end of the lesson, TCs reflected on their understanding of content knowledge (CK) or pedagogical knowledge (PK) from making their own set of dice for differentiated addition mathematics stations and discussed when and why differentiated manipulatives might be appropriate.

Using Manipulatives in Virtual Classes. Like educators across the globe, I had limited time to transition to an online platform due to the rapid emergence of COVID-19 during the middle of the spring 2020 semester. The Moodle course management system and Zoom conferences facilitated interaction with and between TCs via virtual office hours and discussion boards. In their weekly reflections, TCs shared that they enjoyed their participation in a community of virtual learners where discussion boards allowed them to reflect, interact, and receive feedback from their peers and instructor online (TK). Feeling more comfortable with CMs than VMs, I planned to provide each TC an individual manipulative kit they could take home and use during synchronous classes online (TK). However, COVID protocols prohibited this practice. Instead TCs gathered everyday household objects such as food items and small toys for manipulatives. We used these items to model simple operations, classifying, patterning, and so on. When TCs reflected on their experiences, they acknowledged the unconventional manipulatives helped them understand mathematics content knowledge (CK) while also bridging real-world everyday mathematical situations.

I introduced VMs to TCs when courses transitioned to a synchronous online format; however, I felt challenged by my own technological and pedagogical shortcomings. With limited lead time prior to the change and lack of technology resources, I used VMs only supplementally. When planning to use manipulatives in virtual classes, I was confident with utilizing UDL. I was comfortable demonstrating multiple ways to represent mathematical concepts and making learning experiences engaged to diverse learners. My greatest challenge stemmed from an incomplete understanding of how to integrate technology into teaching mathematics methods courses, the knowledge for using online technology (TK) element of our MCIRD Model. Even though I was able to deliver the knowledge of teaching mathematics methods through the online course management system Moodle, I was not proficient in blending the content knowledge (CK) and technological knowledge (TK). When reflecting on this limitation, I recognized that teachers' success in creating DAP mathematical learning experiences may be hindered if they cannot purposefully blend the knowledge of using online technology, UDL, and teaching mathematics methods together.

Using Manipulatives in Hybrid Classes. Courses taught during the spring 2021 semester followed a hybrid format that allowed TCs to attend in person or join classes virtually if they had to quarantine due to exposure to COVID-19. I found this format helpful in showing TCs that they could use CMs in a virtual setting. Hybrid classes allowed me to blend my pedagogical knowledge of UDL and knowledge of teaching mathematics methods through creative course assignments. For example, TCs designed mathematics learning stations with multiple forms of CMs then presented to peers. On the day TCs presented their stations to the class, all students attended the class via Zoom to allow everyone, even those under quarantine, to participate on a level playing field. TCs created their CMs using common household items and recycled materials. Some examples included cereal, candy, recycled water bottles, children's toys, books, and so on. This experience demonstrated how CMs can be used during virtual learning, a skill TCs may need in their future teaching. In the hybrid setting, VMs that closely resembled their CMs such as counters and geoboards supplemented when CMs were unavailable. To address this, TCs learned to search for, evaluate, and use online tools as a platform to enrich and expand ways of engaging children in their learning while supporting equity and access. Blending knowledge for using online technology tools (TK) and knowledge of teaching mathematics methods exposed TCs to DAP with concrete and virtual manipulatives.

FINDINGS

Cross-study analysis of artifacts, journals, and reflections yielded several themes around our experiences implementing concrete and virtual manipulatives in face-to-face, hybrid, and virtual settings. First, our personal knowledge of using technology influenced implementation. For example, Stewart's initial fears about using VMs were alleviated by integrating digital tools in specific ways—to replace, amplify, or transform. Lee felt her lack of experience with VMs prevented her from implementing them in hybrid or virtual settings. Instead, she relied primarily on CMs kits or unconventional manipulatives such as food items or small toys. Importantly, while multiple frameworks give structure and direction when transitioning from CMs to VMs (CAST, 2018; Mishra & Koehler, 2006; Thomas & Edson, 2019), they cannot compensate for lack of technological knowledge.

Second, both CMs and VMs seemed to enhance TCs' content knowledge and ability to express mathematical thinking as evidenced in the TCs' explanations while sharing visual representations regardless of the setting. We found that in face-to-face courses, TCs were able to use both concrete and virtual manipulatives to create visual representations, demonstrate understanding, and explain mathematical thinking though VMs were used less often than CMs. When no manipulatives were used, we noted TCs' explanations lacked specificity and clarity. Similarly, in hybrid or virtual settings, TCs were able to translate their own experiences with VMs (content knowledge) to their role as teachers (pedagogical content knowledge) when they demonstrated for their peers. For example, Figure 13.3 shows the image of a TC's virtual coin board used for The Coin Trading Game (described fully in Table 13.2). In this partner game, TCs spun a virtual die and worked individually with their own VMs before justifying their moves to a partner. This transfer between TCs' content knowledge and pedagogical content knowledge supported our intent to model best practices for early childhood and

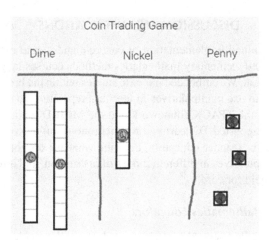

FIGURE 13.3.   Image of Virtual Coins Used by TCs to Play Coin Trading Game.
Note. Applet found at https://apps.mathlearningcenter.org/money-pieces/

elementary teacher education and exemplified the MCIRD model's knowledge for teaching mathematics methods.

Third, manipulatives appeared to promote engagement and collaboration in all settings. For example, TCs who were given personal manipulative kits or gathered household items to use as manipulatives demonstrated engagement similar to what we have seen in face-to-face classes with commercial manipulatives shared among peers. We found that a face-to-face setting allowed TCs to work collaboratively while sharing materials, but screen sharing and collaborative platforms such as Moodle or Google Jamboard also facilitated collaboration and discussion. A choice of CMs or VMs, as is possible in a hybrid setting, maximized flexibility, accessibility, and motivation. TCs who had difficulty during online sessions received reinforcement during face-to-face sessions. Regardless of the setting, we found improved collaboration when TCs used shared VMs or viewed images of CMs through screen sharing. After these collaborations, they seemed more willing to participate in class discussions, producing a richer debrief of the technology experiences as well as the mathematical content.

Finally, while VMs eliminated some constraints of CMs such as distribution or sanitization issues, VMs presented their own shortcomings. Instructors needed time to find and familiarize themselves with VMs and to adapt mathematics tasks to the capacities of the VMs. TCs required reliable internet and compatible devices to use VMs as well as capability for screen sharing or camera access. Jamboards and Moodles had to be created to support collaboration and discussion. The availability of technology did not guarantee effortless implementation of VMs in methods courses. Nevertheless, integrating VMs gave TCs varied experiences that supported understanding of how and why manipulatives are used in teaching and learning mathematics.

DISCUSSION AND IMPLICATIONS

This chapter examined implementation of concrete and virtual manipulatives in early childhood and elementary mathematics methods courses taught in multiple modes of instruction. We embedded our case studies within the broader discussion of why and how to use manipulatives in teacher preparation courses. We shared our adaptation of the TPACK framework and the MCIRD model and described how our model supported TC learning in synchronous online contexts. Here we offer takeaways for teacher educators, describe what we do not yet understand about using manipulatives in different modes of instruction, and offer suggestions for moving the field forward.

Takeaways for Mathematics Educators

Our results distilled several takeaways we hope will provide stepping-stones for mathematics teacher educators endeavoring to prepare TCs to become competent in using manipulatives in different instructional settings. We found that in face-to-face courses, teacher educators can leverage CMs and VMs to increase TCs' engagement with one another by promoting discussion and placing a lens on diverse ways of representing problems. Similarly, in hybrid settings, CMs created by TCs using everyday objects can promote ownership over mathematical thinking and promote expression of mathematical thinking. In addition, VMs modeled during in-person hybrid classes can then be elaborated upon during online sessions. In virtual environments, CMs are more difficult to employ, but VMs that closely resemble their CM counterparts offer TCs with tools to express mathematical thinking and represent problems visually, thus enhancing engagement in synchronous online learning.

We further suggest that teacher educators incorporate or adapt frameworks such as the TPACK framework (Mishra & Koehler, 2006) and UDL (CAST, 2018) into their course design. We used our MCIRD model to maintain fidelity to UDL, advance our knowledge for using online technology, and apply our knowledge for teaching the mathematics methods course to the synchronous online format. Mathematics teacher educators may likewise create strong connections between technological knowledge (TK), pedagogical knowledge (PK), and content knowledge (CK) as they integrate technology to plan effective learning experiences for TCs. To provide TCs with a model of pedagogical flexibility, teacher educators must understand how to replace, amplify, or transform mathematical and pedagogical content in virtual and hybrid mathematics courses (Hughes et al., 2006). Teacher educators should strive to incorporate VMs that facilitate TCs' content knowledge of mathematics, engagement, collaboration, and discussions (Hattie et al., 2016). When teacher educators provide TCs with opportunities to collaborate through shared CMs and digital platforms, TCs become familiar with a variety of manipulatives utilized in multiple modes of instruction. Even if mathematics teacher educators lack expertise with online technology tools, they can extend discussions of mathematics learning to also include debriefs of TCs' technological experiences. Integrating VMs may help TCs understand how and why manipulatives are used in teaching mathematics. Finally, teacher educators' own technological knowledge influences the degree to which VMs are integrated into their mathematics methods courses; therefore, substantial effort may be required to research and develop proficiency with VMs, and the techniques and technologies associated with them.

What We Do Not Yet Know and Moving Forward

These self-reflective case studies were prompted by a sudden and largely improvised conversion to online learning. However, building upon our experiences, researchers may further clarify several related concerns for using concrete and virtual manipulatives in face-to-face, hybrid, and virtual settings. Because of the multi-faceted nature of using manipulatives and their impact on TCs learning outcomes, future studies should examine TCs' perceptions, attitudes, and experiences surrounding manipulatives in different instructional settings. Even if TCs have successful, diverse experiences with CMs and VMs in methods courses, we are unsure whether this transfers to full implementation of either type of manipulative in TCs' field-based classrooms (Maschietto & Trouche, 2010). The lasting impact of TCs' experiences in methods courses could be investigated after they become credentialed teachers.

We suggest that longitudinal investigations involving comparisons of TCs' applications of concrete and virtual manipulatives be conducted to clarify the effectiveness of pedagogical approaches. We also wonder about ways to empower TCs in their own pedagogical content knowledge, by giving them greater responsibility for researching, evaluating, and selecting CMs and VMs to use for specific mathematical objectives or activities. Finally, what barriers must TCs overcome for successful implementation of concrete and virtual manipulatives in elementary classrooms and how can mathematics methods courses provide anticipatory support for overcoming those barriers?

REFERENCES

Aliyyah, R. R., Rachmadtullah, R., Samsudin, A., Syaodih, E., Nurtanto, M., & Tambunan, A. R. S. (2020). The perceptions of primary school teachers of online learning during the COVID-19 pandemic period: A case study in Indonesia. *Journal of Ethnic and Cultural Studies, 7*(2), 90–109.

Association of Mathematics Teacher Educators. (2017). *Standards for preparing teachers of mathematics.* https://amte.net/standards

Association of Mathematics Teacher Educators. (2018). *Supplementary materials for AMTE standards.* https://www.amte.net/sptm/supp

Bartolini, M. G., & Martignone, F. (2014). Manipulatives in mathematics education. In Lerman, S. (Ed.), *Encyclopedia of mathematics education* (pp. 365–372). Springer. https://doi.org/10.1007/978-94-007-4978-8_93

Bouck, E. C., Anderson, R. D., Long, H., & Sprick, J. (2021). Manipulative-based instructional sequences in mathematics for students with disabilities. *Teaching Exceptional Children, 54*(3), 178–190. https://doi.org/10.1177/0040059921994599

Bredekamp, S., & Copple, C. (Eds.). (2009). *Developmentally appropriate practice in early childhood programs serving children from birth through age 8.* National Association for the Education of Young Children.

Bullough, R. V. Jr., & Pinnegar, S. (2001). Guidelines for quality in autobiographical forms of self-study research. *Educational Researcher, 30*(3), 13–21.

Burns, M. (2015). *About teaching mathematics: A K–8 resource (4th Ed.).* Math Solutions.

Burris, J. T. (2013). Virtual place value. *Teaching Children Mathematics, 20*(4), 228–236.

Carbonneau, K. J., Marley, S. C., & Selig, J. P. (2013). A meta-analysis of the efficacy of teaching mathematics with concrete manipulatives. *Journal of Educational Psychology, 105*(2), 380–400. https://doi.org/10.1037/a0031084

CAST. (2018). *Universal design for learning guidelines version 2.2* [graphic organizer]. Author.

Cutri, R. M., Whiting, E. F., & Bybee, E. R. (2019). Knowledge production and power in an online critical multicultural teacher education course. *Educational Studies, 56*(1), 54–65.

D'Angelo, F., & Iliev, N. (2012). *Teaching mathematics to young children through the use of concrete and virtual manipulatives*. https://files.eric.ed.gov/fulltext/ED534228.pdf.

Downing, J. J., & Dyment, J. E. (2013). Teacher educators' readiness, preparation, and perceptions of preparing preservice teachers in a fully online environment: An exploratory study. *The Teacher Educator, 48*(2), 96–109, DOI: https://doi.org/10.1080/08878730.2012.760023

Ellis, V., Steadman, S., & Mao, Q. (2020). Come to a screeching halt: Can change in teacher education during the COVID-19 pandemic be seen as innovation? *European Journal of Teacher Education, 43*(4), 559–572.

Ginsburg, H. P. (2006). Mathematical play and playful mathematics: A guide for early education. In D. G. Singer, R. M. Golinkoff, & K. Hirsh-Pasek (Eds.), *Play = Learning: How play motivates and enhances children's cognitive and social-emotional growth* (pp. 145–165). Oxford University Press. https://doi.org/10.1093/acprof:oso/9780195304381.003.0008

Hattie, J., Fisher, D., Frey, N., Gojak, L. M., Moore, S. D., & Mellman, W. (2016). *Visible learning for mathematics, grades K–12: What works best to optimize student learning*. Corwin Press.

Hughes, J., Thomas, R., & Scharber, C. (2006). Assessing technology integration: The RAT-replacement, amplification, and transformation-framework. In C. Crawford, R. Carlsen, K. McFerrin, J. Price, R. Weber, & D. Willis (Eds.), *Proceedings of SITE 2006–Society for Information Technology & Teacher Education International Conference* (pp. 1616–1620). Association for the Advancement of Computing in Education.

Hunt, A. W., Nipper, K. L., & Nash, L. E. (2011). Virtual vs. concrete manipulatives in Mathematics teacher education: Is one type more effective than the other? *Current Issues in Middle Level Education, 16*(2), 1–6. https://files.eric.ed.gov/fulltext/EJ1092638.pdf

Jones, J. P., & Tiller, M. (2017). Using concrete manipulatives in mathematical instruction. *Dimensions of Early Childhood, 45*(1), 18–23.

Livers, S. D., & Piccolo, D. L. (2020). Using a critical perspective to transition an elementary mathematics methods course to a virtual learning experience. In R. E. Ferdig, E. Baumgartner, R. Hartshorne, R. Kaplan-Rakowski, & C. Mouza, C. (Eds.), *Teaching, technology, and teacher education during the COVID-19 pandemic: Stories from the field* (pp. 393–399). Association for the Advancement of Computing in Education (AACE).

Loong, E. Y. K. (2014). Fostering mathematical understanding through physical and virtual manipulatives. *The Australian Mathematics Teacher, 70*(4), 3–10.

Martin, F., Budhrani, K., & Wang, C. (2019). Examining faculty perception of their readiness to teach online. *Online Learning Journal, 23*(3), 97–119. doi:10.24059/olj.v23i3.1555

Maschietto, M., & Trouche, L. (2010). Mathematics learning and tools from theoretical, historical and practical points of view: The productive notion of mathematics laboratories, *ZDM-The International Journal on Mathematics Education, 42*(1), 33–47.

Mills, S. J., Yanes, M. J., & Casebeer, C. M. (2009). Perceptions of distance learning among faculty of a college of education. *Journal of Online Learning and Teaching, 5*(1). https://jolt.merlot.org/vol5no1/mills_0309.htm

Mishra, P., & Koehler, M. J. (2006). Technological pedagogical content knowledge: A new framework for teacher knowledge. *Teachers College Record, 108*(6), 1017–1054.

Moyer, P. S., Niezgoda, D., & Stanley, J. (2005). Young children's use of virtual manipulatives and other forms of mathematical representations. In W. J. Masalski & P. C. Elliott (Eds.), *Technology-supported mathematics learning environments: Sixty-seventh yearbook* (pp. 17–34). NCTM.

National Association for the Education of Young Children. (2019). *Advancing equity in early childhood education.* NAEYC. https://naeyc.org/resources/position-statements/equity.

National Association for the Education of Young Children. (2020). *Developmentally appropriate practice. Position statement.* NAEYC National Governing Board. https://www.naeyc.org/sites/default/files/globally-shared/downloads/PDFs/resources/position-statements/dap-statement_0.pdf

National Council of Teachers of Mathematics. (2000). *Principles and standards for school mathematics.* NCTM.

National Council of Teachers of Mathematics. (2014). *Principles to actions: Ensuring mathematical success for all.* NCTM.

National Research Council. (2001). *Adding it up: Helping children learn mathematics.* National Academy Press. https://doi.org/10.17226/9822

Piaget, J. (1952). *The child's concept of number.* Humanities Press.

Reinhold, S., Downton, A., & Livy, S. (2017). Revisiting Friedrich Froebel and his gifts for Kindergarten: What are the benefits for primary mathematics education? In A. Downton, S. Livy, & J. Hall (Eds.), *We are still learning! Proceedings of the 40th Annual Conference of the Mathematics Education Research Group of Australasia* (pp. 434–441). MERGA.

Riccomini, P. J., Smith, G. W., Hughes, E. M., & Fries, K. M. (2015). The language of mathematics: The importance of teaching and learning mathematical vocabulary. *Reading & Writing Quarterly: Overcoming Learning Difficulties, 31*(3), 235–252. https://doi.org/10.1080/10573569.2015.1030995

Sarama, J., & Clements, D. H. (2009). "Concrete" computer manipulatives in mathematics education. *Child Development Perspectives, 3*(3), 145–150.

Sarama, J., & Clements, D. H. (2016). Physical and virtual manipulatives: What is "concrete"? In P. S. Moyer-Packenham (Ed.), *International perspectives on teaching and learning mathematics with virtual manipulatives* (pp. 71–93). Mathematics Education in the Digital Era 7. DOI 10.1007/978-3-319-32718-1_4

Smith, S. S. (2009). *Early childhood mathematics* (4th ed.). Pearson Education

Stein, M. K., & Bovalino, J. W. (2001). Manipulatives: One piece of the puzzle. *Mathematics Teaching in the Middle School, 6*(6), 356–359.

Sutton, J., & Krueger, A. (2002). *EDThoughts: What we know about mathematics teaching and learning.* Mid-continent Research for Education and Learning.

Thomas, A., & Edson, A. J. (2019). A framework for mathematics teachers' evaluation of digital instructional materials: Integrating mathematics teaching practices with technology use in K–8 Classrooms. *Contemporary Issues in Technology and Teacher Education, 19*(3), 351–372.

Uribe-Flórez, L., & Wilkins, J. (2017). Manipulative use and elementary school students' mathematics learning. *International Journal of Science and Mathematics Education, 15*(8), 1541–1557.

Van de Walle, J. A., Karp, K. S., & Bay Williams, J. M. (2018). *Elementary and middle school mathematics: Teaching developmentally* (10th ed.). Pearson.

van Engen, H. (1949). An analysis of meaning in arithmetic. II. *The Elementary School Journal, 49*(7), 395–400. http://www.jstor.org/stable/998900

Weng, C., Otanga, S., Weng, A., & Tran, K. N. P. (2020). Effects of tangrams on learning engagement and achievement: Case of preschool learners. *Journal of Computer Assisted Learning, 36*(4), 458–467. https://doi.org/10.1111/jcal.12411

Willingham, D. T. (2017). Ask the cognitive scientist: Do manipulatives help students learn? *American Educator, 41*(3), 25–30.

Witzel, B. S., & Allsopp, D. (2007). Dynamic concrete instruction in an inclusive classroom. *Mathematics Teaching in the Middle School, 13*(4), 244–248.

Yin, R. K. (2012). *Applications of case study research* (3rd ed.). SAGE Publications, Inc.

Young, J. (2017). Technology-enhanced mathematics instruction: A second order meta-analysis of 30 years of research. *Educational Research Review, 22,* 19–32. http://dx.doi.org/10.1016/j.edurev.2017.07.001

Zbiek, R. M., Heid, M. K., Blume, G. W., & Dick, T. P. (2007). Research on technology in mathematics education: The perspective of constructs. In F. Lester (Ed.), *Handbook of research on mathematics teaching and learning* (Vol. 2, pp. 1169–1207). Information Age.

SECTION IV

RECONCEPTUALIZING TEACHER PREPARATION PROGRAMS

SECTION IV

RECOGNITION, TRAINING, TEACHER
PREPARATION FOR DREAMS

CHAPTER 14

INCORPORATING LEARNING-TO-TEACH TRAJECTORIES AND IDENTITY DEVELOPMENT IN A MODEL UNDERGRADUATE TEACHER PREPARATION PROGRAM

Alice F. Artzt
Frances R. Curcio
Alan Sultan
Queens College of the City University of New York

This chapter describes how the learning trajectories and the resulting identity transformations of preservice teachers are supported and developed through engagement in an innovative, comprehensive, carefully sequenced four-year undergraduate secondary mathematics teacher preparation program. In this diverse, urban, public university, candidates' learning trajectories are tracked from before they even begin the program to the end when they reflect on their learning and the shift in their identities as college students to beginning teachers. The value of the cohort model

The AMTE Handbook of Mathematics Teacher Education: Reflection on Past, Present and Future—Paving the Way for the Future of Mathematics Teacher Education, Volume 5
pages 293–312.

is emphasized as a main feature for enabling multiple components of the program that facilitate candidates' learning and development into highly effective secondary mathematics teachers. It is hoped that the specific comprehensive secondary mathematics teacher preparation program described in this chapter will influence the development of future programs and contribute to improving and advancing the field.

INTRODUCTION

The purpose of this chapter is to provide a vision for a highly effective four-year undergraduate secondary mathematics teacher preparation program (student earns bachelor's degree and certification in 4 years). The goal of the program is to prepare teacher leaders who have a passion for and deep understanding of mathematical concepts, are able to engage students in active learning using problem solving and real-life examples, and technology. Given the complexity of developing a mathematics teacher, beyond focusing on just the components of a program, we describe how a program integrates and attends to the learning trajectories[1] and resulting identity[2] transformations of young college freshmen as they become teachers responsible for students who are likely not much younger than themselves. The vision we propose is one supported by research as well as a successful model that has served our local, diverse, urban community for over 25 years. "A well-prepared beginning teacher of mathematics" is one who undergoes a teacher development continuum that begins with pre-preservice, preservice, and then in-service experiences (AMTE, 2017, p. 4). We examine the features of a high-level secondary mathematics teacher preparation program by discussing the recruitment, preparation, retention, and supportive reflection activities of teacher candidates. Before describing the program features, theoretical underpinnings that support its design are discussed.

THEORETICAL UNDERPINNINGS

In order to prepare young college freshmen to become teachers in four short years, they must experience a steep learning trajectory and undergo major shifts in their identities. Research on learning trajectories and identity transformations has taken place from various perspectives. Peressini et al. (2004) took a situative perspective to trace the learning trajectories of teachers during the early stages of their careers. Pascarella and Terenzini (1991) examined student change in terms of personal identity and determined that in college "students successfully resolve identity-related issues, become more positive about their academic and social competencies..." (p. 202). Although researchers have examined trajectories of teacher identity in their last year of college or first years of teaching (Avraamidou, 2014; Friesen & Besley, 2013; Losano et al., 2018; Richmond et al., 2011; Walkington, 2005), research related to the development of teacher identity from early stages of teacher preparation in college is sparse.

A common theme in this literature suggests that many actors play a role in developing one's learning trajectory and identity as a teacher. Teacher candidates learn about teaching from their experiences as school and college students, from their peers, professors, mentors, supervisors, and during their fieldwork and student teaching (Richmond et al., 2011). At all levels, school students, pre- and in-service teachers learn about mathematics, mathematics for teaching, and teaching from different communities of practice (Jimenez-Silva & Olson, 2012; Wenger, 1998). Recently, Goos (2020) reviewed research that takes a social perspective of learning communities that contributes towards one's identity as a mathematics teacher. Voskoglou (2019) took the perspective of examining "the importance of the communities of practice for teaching and learning mathematics" (p. 386). Working in academic cohorts has also promoted retention and increased the success rates of students (Lei et al., 2011). Moreover, reflecting on their thoughts and practice in teaching have been effective means through which to learn and examine one's developing identity (Richmond et al., 2011; Walkington, 2005). Goos and Bennison (2008) reported the value of a community of practice that extends beyond graduation into teachers' first few years of teaching.

Supported by this research, we describe an example of a program that has produced more than 325 exemplary secondary mathematics teachers with a retention rate of over 92% in 5 years of teaching. Through innovative recruitment, preparation and retention, and reflective strategies that are aligned with the research literature we can support candidates' learning trajectories and identity formations. For an overview of the many components of this program see Table 14.1.

PRE-PRESERVICE CANDIDATES AND RECRUITMENT

Recruiting graduating high-school students into a teacher preparation program allows them to become part of the pre-service communities that will support their learning and formation of their teaching identities throughout their four+ years of college. The importance of early indoctrination into teaching as a means for determining their suitability for the profession and staying interested in teaching has been documented (Lucas & Robinson, 2002). It is for this reason that we recruit students while they are in their senior year of high school and work with them as soon as they begin college. To maximize the chances that the students who are recruited will complete the program, we require that they must show good aptitude in mathematics as well as a disposition for helping others (AMTE, 2017; Artzt et al., 2007). Two types of recruitment, indirect and direct, play a role in students' inclinations to become mathematics teachers.

Indirect Recruitment

We recognize that the learning-to-teach trajectory occurs well before a student enters a teacher preparation program. According to Lortie's (1975) theory of the "apprenticeship of observation," teacher candidates start learning how to teach when they are school students who spend years of observing their teachers and developing their perceptions of teaching as well as their views about mathematics.

TABLE 14.1. Overview of Key Components of the Program and Beyond

Recruitment	Preparation/Retention	Extracurricular Activities	Supportive Reflection Activities	Graduates' Involvement in Program
Indirect	Cohort model	Attendance at monthly seminars	Journal entries	Share professional experiences at an annual seminar
Populate local schools with graduates	Learning community	Presentations at selected seminars	Carefully structured observation reports	Presentations at host conference
Opportunities offered to tutor during high school	Faculty meetings	Attending and assisting at professional conferences	Portfolios	Serve as cooperating teachers
Role models reveal their love of mathematics	Small group conferences with students	Tutoring Club		Serve as student-teaching supervisors
Role models reveal their love of teaching	Mathematics and Education Departments collaborate	Annual math excursions		
	Carefully sequenced course-work	Field Day		
Direct	Innovative field work			
Develop brochures				
Personal letters to high schoolers				
Undergrads return to their high schools to meet with interested students				
Host conference for high schoolers				

According to the type of teaching they experience, this can work for or against them as future teachers. Therefore, an indirect method of recruiting students is to populate the local schools with our excellent graduates who make it known that they love mathematics and teaching and recommend teaching as a gratifying and fulfilling profession. For example, 14 of our graduates are employed in one of our local large high schools. Over the years we have had 31 students join our program from that same school. Of those 31 students, 4 are now teaching there. The chairperson has claimed that the key to the success of his math department is that he now only hires our graduates and they all work together as a family. In his words, "I just wanted to pay tribute to what you have done. Our high school math department would not have been this strong had it not been for your program" (J. Chou, personal communication, 15 Nov. 2018). In support of his observation, research indicates that practicing teachers learn from their colleagues and perform better the more they collaborate (AMTE, 2017; Kanold et al., 2008).

Moreover, the activities in which students participate while in high school or lower grades is another indirect recruitment method that also influences students' chosen career path and contributes towards their identity development. For example, in our program, part of the application process is to write an essay in which they are asked to describe their interests and commitment to the teaching of mathematics. Out of 220 applicants, 83% mentioned that their experiences in tutoring their peers or others in mathematics contributed to their desire to teach. Giving school students the chance to experience the satisfaction of teaching is a first step in the shift of their identity from student to teacher at an early age, and is in fact, a seemingly worthwhile method of recruitment used by all of the graduates of our program.

Direct Recruitment

Other more direct methods of recruiting high school students into a program take careful planning and innovative tactics. Sending brochures to guidance counselors, administrators, graduates, and teachers has always been an effective advertising method. A personal letter to students who have high grades in mathematics and have already been accepted to our college has been one of our most effective recruitment methods. Moreover, we require our undergraduates to go in pairs to make in-person visits to the high schools from which they graduated to speak to classes of talented mathematics students. Our undergraduates are comfortable doing this since they are members of the same cohort and know each other well. Such methods have been effective in implanting the idea in high school students' minds of considering building on their identity from being a student to possibly also being a teacher.

Once a year we hold a conference inviting mathematically talented high school students and their mathematics teachers to our college to meet our current undergraduates and participate in interactive innovative lessons taught by our graduates or other exemplary mathematics teachers (Artzt et al., 2007). As part of their workshops, the presenters speak about why they love to teach mathematics. The conference is not only a recruitment device, but it serves as a professional teacher preparation opportunity for the undergraduates. Additionally, undergraduates participate in the planning and implementation of the conference, giving them experience in working together to organize events and supervising the adolescents in attendance. These experiences contribute to their evolving identity as a person with authority, responsible for students and high-level tasks. In his journal following the conference, one of the students wrote:

> This was a spectacular experience. I did feel like one of the leaders. I've never had so much on my shoulders for one event and the fact that everything worked out the way it should have worked out, I am beyond happy. I was working on this conference for so long and to see it come together was one of the best feelings. Thank you for this experience. (November, 2016)

Pre-preservice experiences and exposures are all part of a young person's learning trajectory whether or not they decide to become a mathematics teacher. But, if they do take that path, one can say with legitimacy that the identity transformation to become a mathematics teacher occurred well before entrance into a teacher preparation program and is often the result of both direct and indirect methods of recruitment.

PREPARATION AND RETENTION

As an example of the research cited previously and good practice, we have found that the cohort model, carefully sequenced coursework, meaningful fieldwork, and communities of practice beyond classwork are essential components for a successful program.

The Cohort Model

The needs and interests of developing teachers must be addressed at many levels if they are to complete a program, become certified, and spend their careers in education. Accordingly, a main strength of our program, as reported by our undergraduates, is that the students are placed in cohorts and take their mathematics and education courses together. The value of a cohort model for undergraduate teacher preparation, most especially in mathematics cannot be overestimated. A cohort can be described as a group of about 10–25 students who begin the program together, take their classes together, and end the program at the same time (Barnett et al., 2000; Barnett & Caffarella, 1992; Maher, 2005). Its benefits for retaining students and improving success in school and graduation rates have been documented in the literature (Richmond et al., 2011). Once cohorts are established in a program it allows for many other important program components to occur that contribute to quickening and enhancing both the learning trajectory and identity transformation of the teacher candidates. As an example, we describe the many features of our program that are only enabled through the use of the cohort model.

Supportive Learning Community for Preservice Candidates and Professors

The need for a sense of "belongingness" to a group has been documented as an essential part of learning for all students (Osterman, 2000). To ensure that entering students feel part of our community, upon acceptance into the program they are assigned mentors who are undergraduates in the program (Yomtov et al., 2017). The pairing is based on being alumni of the same high schools, residents of close neighborhoods, or interests in similar hobbies. Having a mentor from the very start arouses a sense of belonging needed to begin the program with some level of security and optimism. Entering as a cohesive group of students allows for an immediate support system (Unzueta et al., 2008) that they rely on throughout their college life. This is especially important for students who take difficult mathematics courses and need to, and do, form study groups to succeed. Throughout the program, their cohort accompanies them through their different learning trajectories and in fact, contributes to their identity as a mathematics teacher as they work together to accomplish the best results. In fact, the cohorts continue beyond graduation. The students become so close that regardless of the school in which they teach, they feel comfortable calling on one another for ideas and for support. They have chat groups and use other types of social media not only to keep in touch with their own cohort but with other cohorts as well. According to what many of our graduates have told us informally, it is this connection with other members of their "family" that contributes to their high rate of retention in teaching years after they graduate.

To maintain close contact with the undergraduates, once per semester, the program director hosts small-group conferences with the students in which they can feel free to share any concerns they have about the program or other aspects of their college experiences. In a recent survey, 75% of the students claimed that it was the close learning community of students and their professors and staff that they most liked about the program. For example, one student wrote in her end-of-year portfolio, "It would be impossible to imagine myself taking classes without my friends and exceptional professors. Together through the challenging times, we were able to persevere."

Moreover, once per month we hold a meeting with all of the professors who are teaching the different cohorts of students in the program. The professors share and compare their observations discussing ways that they might better meet the needs of students and ways that their courses might be changed or even reordered. Oftentimes the professors gain different perspectives of their students from these meetings, which ultimately influences their approach in dealing with them. The more professors understand their students the better they can help shape their identities in positive ways. As a bonus, much like the preservice teachers, the professors enjoy the feeling of belonging and claim they most enjoy teaching the courses in our program. They claim that the students are usually more motivated than others, and they feel they are doing more than just teaching one isolated course where they never see the students again. Simply put, they like to feel part of the program and its mission.

Coursework

The collaboration of the mathematics and education departments and faculty are at the heart of the design and implementation of the mathematics and education courses that comprise the program.

Collaboration of the Mathematics and Education Departments

The original idea of the cohort model was to block students into their classes so that they could feel comfortable with the people around them, form study groups and ultimately succeed, especially in their mathematics courses with hand-picked professors who would be most suitable for them. However, such collaboration can only occur with the cooperation of the mathematics department and the secondary education department. Support for the collaboration of mathematics and education professors has been long in coming. For too many years mathematicians were disconnected from math education at the middle and high school levels since their focus was on isolated college-level mathematics. Mathematicians are now seeing their role changing and understand how taking a part in what is taught in college can make a difference in the competence of prospective middle and high school teachers. The report of the Conference Board of the Mathematical Sciences (Feuer et al., 2013), indicates that:

> At institutions that prepare teachers or offer professional development, teacher education must be recognized as an important part of a mathematics department's mission and should be undertaken in collaboration with mathematics education faculty. More mathematics faculty need to become deeply involved in PreK–12 mathematics education by participating in preparation and professional development for teachers and becoming involved with local schools or districts. (p. 19)

In a cohort model the collaboration within and between both the mathematics and education departments is essential. For example, in one cohort the mathematics and education courses cannot be offered at overlapping times. The courses must be scheduled so that they do not interfere with other activities. Most importantly, professors must be assigned who can model some level of student-centered instruction and who have an investment in helping the students learn the material so that they can be better teachers. Moreover, professors have to be willing to attend meetings and share student grades with the program director so if problems arise, they can be investigated and dealt with quickly. Such collaborative efforts can only occur if the chairs of the education and mathematics departments are on board with the program and are committed to its mission and strategies. As a result of collaborative efforts and activities of both departments, a coherent and logically sequenced program of study that suits the learning trajectory of prospective mathematics teachers can be developed. For an example of our program, see Table 14.2.

TABLE 14.2. Sequence of Mathematics and Education Courses

Freshman Year		Sophomore Year	
Semester 1	**Semester 2**	**Semester 3**	**Semester 4**
Calculus/Differentiation Psychology of Mathematics Learning	Calculus/Integration Discrete Math	Calculus/Infinite Series Probability & Statistics	Multivariable Calculus Methods of Mathematical Statistics Mathematical Models Foundations of Ed
Junior Year		**Senior Year**	
Semester 5	**Semester 6**	**Semester 7**	**Semester 8**
Differential Equations Linear Algebra Problem Solving	Algebraic Structures H. S. Math from an Advanced Standpoint Language, Diversity, & Development	College Geometry Cognition, Teaching, & Technology Methods of Teaching Math Student Teaching 1 (HS/ JHS)	Foundations of Geometry Curriculum & Assessment Student Teaching 2 (JHS/HS)

Mathematics Courses

The sequencing of the mathematics courses took many years to develop. For example, when the program first began, we required students to take calculus as well as discrete mathematics which emphasizes proof in their first semester. We soon realized that although proof is the foundation of mathematics, entering freshmen are too overwhelmed by college-level calculus to take it at the same time. We also found that delaying abstract algebra and linear algebra until students' junior year produced better results than having them take it earlier, since students gained more mathematical sophistication by then. In line with the mission of the program, wherever possible the mathematics courses include applications to real life and incorporate the use of technology. All of the mathematics courses in the program meet the Council for the Accreditation of Educator Preparation (CAEP, 2013) and the National Council of Teachers of Mathematics (NCTM, 2000, 2020) requirements for mathematics teachers. The courses are open to all students in the college but the mathematics department saves seats for the students in our program. An innovative capstone mathematics course that requires the joint efforts of a mathematician and mathematics educator is highlighted (Artzt et al., 2012).

This capstone course entitled, *Secondary School Mathematics from an Advanced Standpoint*, is unique to our program and is offered to students in their upper junior year, before they begin student teaching (done in senior year). There is a dual mission to this capstone course. The first is that students see the connections between their college courses and the content taught in middle and high school mathematics, as well as the connections between and among the different mathematical content areas (Sultan & Artzt, 2018). The second mission is for students to get a first try at teaching a class by designing and implementing lessons and designing and assessing homework assignments and quizzes. In a sense this course is analogous to a teaching lab, when candidates are well on their way to shifting their identity from students to beginning teachers. Through the many experiences they have had in their courses, in their fieldwork, and in their professional activities, the candidates have the knowledge base and dispositions to begin to try to create and implement hands-on, high-tech, student-centered, meaningful mathematics lessons in a protected environment. With close guidance, support, and supervision of the professor, students work with their peers and are scaffolded in the process of getting a deep understanding of mathematical concepts in the process of learning how to teach. This course thus serves as a bridge to the mathematics methods course.

Education Courses

The mathematics and education professors work together to coordinate courses so that students are not overwhelmed and that the sequence of education courses makes sense and eases the transition from student to teacher. Students are enrolled in an educational psychology course, *The Psychology of Learning Mathematics*, as soon as they enter the program. We highlight this course because it is most unique to the program and only the students in the program are permitted to take it. There are several reasons the course is offered early in the students' college study. First, it is taught by the program director, giving her a chance to get to know all of the students and through extensive cooperative-learning activities, giving the students the opportunity to get to know one another well. Although the substance of the course material is rooted in the traditional beginning concepts of educational psychology, the context directs students to focus on how they have learned and are learning mathematics. The expectation is that through this examination of the ways of learning, they will improve their own learning strategies of the demanding college mathematics courses that will follow. They also focus on how adolescents learn mathematics and through the special fieldwork described below, they can decide early in their college career whether or not they enjoy working with adolescents.

The education department allots special sections of education courses for our students. For example, when they take the course, *Foundations of Education,* they work on developing a philosophy of teaching mathematics related to social issues and diversity. When they later take the course, *Language, Diversity and Development,* they focus on the role of language in the learning of mathematics. In their senior year, when they take the course, *Cognition, Teaching, and Technology,* they focus on the specific technologies used for learning mathematics (e.g., Desmos, Geogebra).

FIELDWORK

Requiring fieldwork in schools has been an integral part of preservice teacher preparation, but depending on its quality, its effectiveness varies. Simply expecting candidates to accumulate hours in classrooms where there are many happenings competing for an observer's attention does not contribute to building critical observation skills. Novice observers need guidelines to help them focus on critical aspects of the lesson supportive of the theme(s) in course assignments. Fieldwork that is carefully designed to support program philosophy and course goals and objectives while at the same time focusing on the thinking and sense-making of the observer, has the potential to influence and shape candidates' beliefs about teaching, teaching practices, and identity (Clift & Brady, 2006; Goodman, 1985). Darling-Hammond describes such a program with coursework and fieldwork as being coherent and integrated (2006). Selecting field sites is a critical and difficult process, and an example below shows how when done well, its impact is powerful.

Selecting Field Sites

One of the challenges for designing and implementing meaningful field experiences that are sensitive to diverse student populations is finding schools that are willing to accept observers and student teachers. Moreover, exemplary teachers willing to open their doors and contribute to the development of future teachers are critical in this endeavor (AMTE, 2017). Partnerships between colleges/universities and schools are ongoing collaborative relationships that benefit all constituents involved (AACTE, 2017; Trubowitz et al., 1984; Trubowitz & Longo, 1997) and provide a collegial environment for ongoing preservice and in-service teacher development (AACTE, 2017; AMTE, 2017; NCTM, 2000, 2020).

An Example. One such college-school partnership is situated in the most ethnically diverse community in America (Gamio, 4 July 2019). Early and frequent fieldwork in a variety of settings (AMTE, 2017; Clift & Brady, 2006) is one characteristic of our undergraduate secondary mathematics teacher preparation program. During the course of their first semester, a cohort of freshmen enrolled in *The Psychology of Learning Mathematics*, spend ten three-hour sessions with a college faculty member and an exemplary eighth-grade teacher who is a graduate of the program (Curcio et al., 2005). Each session corresponds with the psychological principles and theories discussed in the course. The entire class of freshmen spend the first hour in a pre-lesson discussion, reviewing the psychological constructs they expect to observe. The teacher shares her pre-lesson thoughts, highlighting typical misconceptions, and different directions in which she may take the lesson, depending on her ongoing assessment of the students. During the second hour, the freshmen arrange themselves around the perimeter of the room to observe the students during instruction. As the semester progresses and the students become familiar with the observers, the freshmen join the eighth graders in their small groups to listen, to question, and to assist as necessary. During the third hour, the eighth graders move on to their next class and the freshmen remain with the teacher for a debriefing of the lesson. The college instructor serves as a moderator to assist in maintaining the focus of the observation.

Appropriate related fieldwork at a variety of levels in multiple classrooms is required during the junior and senior years (CAEP, 2013). Student teaching occurs in the last two semesters; first in pairs, and then solo. All fieldwork includes attention to appropriate mathematical content (AMTE, 2017). Meaningful fieldwork in a natural setting bridges theory and practice (Curcio, Artzt, with Porter, 2005). Furthermore, candidates gradually progress from observers in the back or the side of the classroom, to working with individual students or small groups of learners, to student teaching as they plan for instruction and manage entire classes.

Impact. The developmental nature of its design must support the learning trajectory of teacher candidates. Candidates typically find field experiences to be a valuable and influential part of their professional preparation (Ronfeldt, 2012). Memorable experiences as early as freshman year often inform program graduates with ways of dealing with classroom dilemmas (DeSousa, 2020). Meaningful, coordinated field experience is an essential component of teacher preparation in developing reflective practitioners (Artzt et al., 2015; Clift & Brady, 2006).

Communities of Practice Beyond Classwork

The complexity of teaching is well matched by the complexity of preparing a teacher. In light of the literature on communities of practice in education (Van Zoest & Bohl, 2005; Voskoglou, 2019) it is clear that having candidates take a heavy load of mathematics and education courses, even with the innovative fieldwork described above, is not sufficient for the full preparation of a teacher. Teacher candidates need to have exposure to other professional communities that usually only beginning teachers have. To enrich their learning and expand their identities as teachers they need experiences outside of what the typical university program provides. For example, in our program attendance at monthly seminars and professional conferences, and tutoring experiences are a few of the extracurricular experiences required of the candidates.

Monthly Seminars

The cohort setting allows for the program to control the students' schedules so that all of the members of the program can congregate at specific times. For example, once per month the four undergraduate cohorts meet in topical monthly seminars conducted by participants, faculty, and off-campus guests who highlight such topics as the latest technology for teaching, interesting teaching approaches, and how mathematics is related to different areas (e.g., dance, art, magic). Exposing developing teachers to experts in other areas of the professional community is similar to engaging them in various communities of practice which is a widely supported practice for teacher development (e.g., Goos, 2020; Van Zoest & Bohl, 2005). The freshmen in the program receive advice about how to succeed in the program from upper classmen and witness accomplishments of the seniors. Throughout the program, every student is given the chance to give presentations in front of the 80 or more students present at each seminar.

Most special, once per year, graduates from all different cohorts speak at a seminar to share their varied experiences in teaching. According to the journal comments of our students, experiences such as these have had a great impact on them. They have seen how passionate people in the field are about their work, even after many years in the profession. They have heard the honest challenges of beginning teachers and learned how these challenges have been overcome by experienced teachers who were once standing in their very own shoes. Moreover, they learn how the graduates who have been teaching many years have taken on leadership roles in the profession (e.g., becoming chair people, presenting at conferences, acting as cooperating teachers, writing articles, forming and leading math teams, etc.). In essence, undergraduates are witnesses of the learning trajectories and evolving identities of their former peers who are working professionals. In their journals, the undergraduates who attend these seminars have said such things as, "I found the seminar to be extremely beneficial to hear the stories and advice from people who are experiencing teaching, because that is going to be us in just a few years." "I always enjoy this seminar the most. Hearing the alumni speak, gives me hope for the future. I look forward to hopefully being on the panel next year."

Attending Professional Conferences

Another way of involving our teacher candidates in the professional mathematics community is having them attend two mathematics education conferences per year. One is held at a local State University of New York college and the other is hosted by our program and held at our college. After attending a professional off-site conference, one student wrote in his journal, "From the moment I got off the bus and entered the conference, I felt like a teacher. I couldn't believe how all of these teachers got together just to find out new ways of teaching mathematics." His comment in the journal made it clear how attending these other communities impacted the students' identities and accelerated their learning trajectories. Although this particular student had other career opportunities, after attending the conference, he was fully committed to teaching and to the program. Now, 22 years later, he is the principal of a public high school in New York City.

Attending and facilitating the conference at our college, "Celebrating Mathematics Teaching," has similar effects on the candidates. Escorting high school students to the different sessions and speaking at the panel discussion at the end of the conference allows candidates to participate in a different type of professional community and take on different roles, and thereby identities, than they would normally have in their college work. For example, one student wrote about how dramatically her identity and learning trajectory changed after serving as the master of ceremonies in front of over 400 high school students.

> As I stood at that podium in the fall of my senior year, ready to address the audience, I saw one of my professor's gesture to me from the front row. The college president arrived and wanted to address the conference attendees before we began the forum. I took the liberty to introduce him myself. ...So here I was, finally starting to think of myself as a teacher. (Kimyagarov, 2020, p. 34)

According to Peressini et al., (2004), taking on such a professional identity affects what the novice teacher learns in the teacher preparation program.

Tutoring, an Identity-Enhancing Activity

As a service to the school communities, candidates are in charge of a club that offers low-cost tutoring to local middle and high school students. Running a club gives the candidates a chance to participate in leadership roles that contribute to their learning trajectories as beginning teachers who take on adult roles in dealing with students and their parents. Essentially, this is another learning community that is facilitated by the cohort model that contributes to candidates' vision of themselves as professionals and as teachers while in the program.

SUPPORTIVE REFLECTION ACTIVITIES

Since the work by Schön (1983, 1987), reflection has taken on a critical role in teaching and in teacher education. Moreover, Freese (2006), Urzúa and Vasquez (2008), and Warin et al. (2006) all point out how reflection is an essential part of the development of one's identity. Thus, reflection is a major component of our program, both in structured ways of reflecting on the lessons candidates teach, and in the end-of year portfolio that they are required to create (Artzt et al., 2015). The program is based on a framework that suggests that a teacher's knowledge and beliefs about students, mathematics and teaching mathematics are what drive instruction, as well as their goals for what they want and how they want their students to learn (Artzt & Armour-Thomas, 1999). Throughout the program the students are required to reflect on their learning trajectories and changing identities through the lens of this framework. That is, as one's identity evolves, so do their knowledge, beliefs, and goals with regard to mathematics, students, and pedagogy. For example, in their year-end portfolio, freshmen, sophomores, and juniors are specifically asked how their knowledge, beliefs, and goals have changed from the previous years[3]. In general, their responses are indicative of how they transitioned from concentrating on themselves as learners to how they envision themselves as teachers. For example, 12 of the current 17 freshmen wrote about how they loved mathematics and enjoyed working with others. Thirteen of the current 14 sophomores wrote about what they wanted to teach students and how they wanted to incorporate real-life examples in their instruction. All of the current 21 juniors wrote about how they wanted their students to understand mathematical concepts and also make an emotional connection with them. Nevertheless, as the assignment required, all of the 52 respondents wrote about their changes with respect to the learning and teaching of mathematics. For selected statements of such changes, see Appendix A. Reporting one's changing ideas over time is a way of solidifying the shifting identity of our students as they quickly grow from adolescents to young adults with more mature views required of teachers who enter classrooms fraught with difficult issues.

MOVING FORWARD: A CRITICAL EXAMINATION

When planning, implementing, and evaluating a coherent and cohesive mathematics teacher education program, collegial relationships among school-college personnel are essential, as are the logistics involved (AACTE, 2017; AMTE,

2017). As pointed out in this chapter, collaboration is at the heart of the success of a mathematics teacher preparation program, and yet it is also most difficult to accomplish. Suggestions for moving forward and overcoming these obstacles are offered.

Personnel

It is critical that a respectful, collegial working relationship exists among mathematics and education faculty. Mathematicians and mathematics educators may have different perspectives and goals (Sultan & Artzt, 2005). Collaborative and collegial discussions need to occur to make sure that all professors involved in the teacher education program are committed to the common mission of preparing teachers who have a passion for and deep understanding of mathematical concepts and are able to engage their students in active learning. Coordinating course offerings and schedules, attending to candidates' progress and resolving their difficulties, and collaborating to review and evaluate the attainment of program objectives requires ongoing interactions.

To enable fieldwork that is carefully structured with faculty involvement as described above, education faculty may be given course-related credit for accompanying candidates in the field to supervise, monitor, and moderate discussions. If faculty view this role as a research-related opportunity or a potential to develop a manuscript, it may be viewed as a productive assignment. Another way to bring college faculty into the schools would be to offer the field-related course at the school if the school facilities are able to support such an arrangement. The fieldwork could be offered during the afternoon and the course could be offered after the grades 6–12 students are dismissed. This would allow the faculty member to be onsite to participate in the observations, to moderate the pre- and post-lesson discussions, and teach the course.

School teachers have many administrative-imposed, non-instructional responsibilities. Giving up two prep periods for pre- and post-lesson discussions with observers may be too demanding for some teachers. However, others may view the opportunity as a step toward becoming an adjunct instructor at the college. If institutional support is available and if allowed by the host school, the host teacher may be allotted a stipend, the equivalent of a class coverage, for each prep period dedicated to pre- and post-lesson discussions.

Logistics

Scheduling college courses for a teacher preparation program involves planning semesters in advance. Such advanced planning should ensure that the mathematics courses and education courses are offered at convenient times for the students in different cohorts, and taught by professors who are committed to the program.

College schedules and school schedules are often thought of as being in different "time zones." Schedules need to be coordinated, especially if the host teacher is to conduct pre- and post-discussions. At the end of the academic year, when teachers' and students' programs are created for the following school year, college faculty, the host teacher, and the school principal (or a designee) need to discuss the logistics.

Furthermore, there are many school activities that may seem to develop "overnight" and flexibility in adjusting to schedule changes may be needed. Maintaining open lines of communication between college faculty and the host teacher is essential.

By adding a cohort of undergraduate observers in a classroom with middle or high school students, room capacity may present an obstacle. Safety must always be a top priority. If the classroom cannot accommodate the total number of students and observers, an alternative venue (e.g., a larger classroom) will be needed.

Although the current program has experienced much success as indicated by the many school leaders who seek to employ our graduates, there are many challenges and obstacles faced by trying to realize the vision of an exemplary teacher preparation program. College faculty, college and school administrators, and school teachers, who are committed to the partnership (which goes beyond preservice fieldwork and beyond the scope of this chapter), communicate, plan together, and explore creative ways to overcome obstacles. Work needs to be done to resolve the difficulties of coordinating the efforts of education faculty with mathematics faculty, partnering with local schools, and finally obtaining institutional support to offer teacher candidates important extra-curricular professional activities.

The complexity of mathematics teacher preparation is well acknowledged, and as documented in this chapter, many constituents are involved in making it a successful endeavor. It is our responsibility to provide programs that support the learning trajectories of undergraduates and facilitate their identity transition from college students to classroom teachers. The secondary teacher preparation program described in this chapter is one example of such a program that can serve as a model to help to support program development and move the field forward.

ENDNOTES

1. For the purpose of this chapter, learning trajectories are defined as "empirically-grounded patterns of student thought and behavior that are observed as students move from naïve to sophisticated reasoning" as they become secondary mathematics teachers (Confrey et al., 2020, p. 2).
2. "Identity development...can be best characterized as an ongoing process, a process of interpreting oneself as a certain kind of person and being recognized as such in a given context" (Oruc, 2013, p. 207).
3. Seniors are excluded since they produce e-portfolios for job interview purposes.

APPENDIX A: SELECTED STATEMENTS RELATED TO CHANGES IN KNOWLEDGE, BELIEFS, AND GOALS FROM CANDIDATES' ANNUAL PORTFOLIOS

Knowledge	Beliefs	Goals
Reading my first portfolio made me emotional, as I can see how insecure I was in my academic life and now I can see that I am accepting that I may not always understand everything the first time, but with perseverance and dedication, I know the end knowledge is attainable. (Arly, junior)	In my last portfolio, I believed that I was a determined mathematics student who studied to excel in my courses. I do believe that I am a determined student, but I also believe that I want to drop the importance I place on excelling… This semester instead of focusing on excelling in my course, I focused more on learning the material to use in the future. Students should be motivated to learn. I hope that I can be more motivated to learn than motivated to get A's. (Paidi, sophomore)	In previous years, my goals for my future students were to have them love math and get 100% as their grades. This year, my goals changed to having the students realize that math is actually important. I understand that grades aren't a reflection of how well a student understands math. I want my students to seek improvement rather than perfection. (Paul, junior)
Last year I learned that students really need reinforcement over time to help remember mathematical concepts. This still holds true, but I also now know that it is important to show the real-world value of a concept to truly remember and appreciate the lesson. (Elizabeth, sophomore)	I used to view mathematics as an easy subject that almost everyone could understand with the proper guidance, but I no longer see it that way. Mathematics can only be taught if the students themselves want to learn. We as educators cannot have students learn mathematics through lectures but we must find ways that engage them and make them see math as an exciting topic that has endless possibilities. (Malesa, sophomore)	The type of teacher I want to be has grown in comparison to previous years with respect to the impact I want to have on my students. I want them to look at me as a role model and to understand that you do not need to be perfect in life. Based on this past year, I have grown to want to be more enthusiastic and inspiring in my daily classroom life, motivating students to want to learn and share ideas with each other. It is a truly beautiful sight when you are able to see students work together to learn. (Rasheed, junior)
My knowledge of how students learn mathematics has changed this year because I now know how to create lessons that are more inclusive. My courses this year have taught me how to teach students who may have disabilities, ELL speakers, or may have specific learning needs. I now realize how important it is to be able to meet the needs of all students. (Emily, junior)	My beliefs have changed in that I can now clearly see that I can make a difference in not only students' academic but also their personal lives… It is extremely important to the academic, emotional, and mental well being of each student to treat each student delicately. Teachers must be able to see through the façade that teenagers put up and see into the heart of the student. (Dina, sophomore)	This year everything seemed to culminate in a broader view of what I want my students to learn. As a math teacher, I want my students to learn to love and truly understand mathematics. More importantly, I want my students to learn about themselves. With all of the racism and hate that is present in the world today, I aim to create an inclusive and respectful classroom for all. So, my goals about what I want my students to learn have not necessarily changed, but rather evolved into a more complete existence based on what has changed over the past years. (Daniel, junior)

ACKNOWLEDGEMENT

Gratitude and recognition go to Tara Wachter for reading through all of the student portfolios and extracting the key patterns and representative student comments that support the work reported in this chapter.

REFERENCES

American Association of Colleges of Teacher Education. (2017). *A pivot toward clinical practice, its lexicon, and renewing the profession of teaching: Executive summary.* Author.

Artzt, A. F., & Armour-Thomas, E. (1999). A cognitive model for examining teachers' instructional practice in mathematics.: A guide for facilitating teacher reflection. *Educational Studies in Mathematics, 40*(3), 211–235.

Artzt, A. F., Armour-Thomas, E., Curcio, F. R., & Gurl, T. J. (2015). *Becoming a reflective mathematics teacher* (3rd ed.). Routledge.

Artzt, A. F., & Curcio, F. R., with Weinman, N. (2007, Spring). Teachers need to sell mathematics teaching: Reaching out to excellent high school students. *NCSM Journal of Mathematics Education Leadership, 10*(1), 4–7.

Artzt, A. F., Sultan, A., Curcio, F. R., & Gurl, T. (2012). A capstone mathematics course for prospective secondary mathematics teachers. *Journal of Mathematics Teacher Education, 15*(3), 251–262.

Association of Mathematics Teacher Educators (AMTE). (2017). *Standards for preparing teachers of mathematics.* https://amte.net/standards

Avraamidou, L. (2014). Tracing a beginning elementary teacher's development of identity for science teaching. *Journal of Teacher Education, 65*(3), 223–240.

Barnett, B. G., Basom, M. R., Yerkes, D. M., & Norris, C. J. (2000). Cohorts in educational leadership programs: Benefits, difficulties, and the potential for developing school leaders. *Educational Administration Quarterly, 36*(2), 255–282.

Barnett, B. G., & Caffarella, R. S. (1992, October). *The use of cohorts: A powerful way for addressing issues of diversity in preparation programs* [Paper presentation]. Annual meeting of the University Council for Educational Administration, Minneapolis, MN. ERIC Document Reproduction Service No. ED 354627

Clift, R., & Brady, P. (2006). Research on methods courses and field experiences. In M. Cochran-Smith & K. Zeichner (Eds.), *Studying teacher education: The report on the AERA panel on Research and Teacher Education* (pp. 309–424). Erlbaum.

Confrey, J., Toutkoushian, E., & Shah, M. (2020). Working at scale to initiate ongoing validation of learning trajectory-based classroom assessments for middle grade mathematics. *Journal of Mathematical Behavior, 60*(2), 1–18.

Council for the Accreditation of Educator Preparation (CAEP). (2013). *CAEP accreditation standards.* http://caepnet.org/~/media/Files/caep/standards/caep-2013- accreditation-standards.pdf

Curcio, F. R., & Artzt, A. F., with Porter, M. (2005). Providing meaningful fieldwork for preservice mathematics teachers: A college-school collaboration. *Mathematics Teacher, 98*(9), 604–609.

Darling-Hammond, L. (2006). Constructing 21st century teacher education. *Journal of Teacher Education, 57*(6), 300–314.

DeSousa, D. (2020). My unexpected happiness. In A. F. Artzt & F. R. Curcio (Eds.), *The inspirational untold stories of secondary mathematics teachers* (pp. 25–30). Information Age Publishing, Inc.

Feuer, M. J., Floden, R. E., Chudowsky, N., & Ahn, J. (2013). *The evaluation of teacher programs: Purposes, methods, and policy options* (ED 565694). ERIC. https://files. eric.ed.gov/fulltext/ED565694.

Freese, A. (2006). Reframing one's teaching: Discovering our teacher selves through reflection and inquiry. *Teaching and Teacher Education, 22*(1), 100–119.

Friesen, D., & Besley, C. (2013). Teacher identity development in the first year of teacher education: A developmental and social psychological perspective. *Teaching and Teacher Education, 36*, 23–32.

Gamio, L. (4 July 2019). Where America's diversity is increasing fastest. Retrieved 10 July 2021 from *Axios*, https://www.axios.com/where-americas-diversity-is-increasing-the-fastest-ae06eea7-e031-46a2-bb64-c74de85eca77.html

Goodman, J. (1985). What students learn from early field experiences: A case study and critical analysis. *Journal of Teacher Education, 36*(6), 42–48.

Goos, M. (2020). Communities of practice in mathematics teacher education. In S. Lerman (Ed.), *Encyclopedia of mathematics education* (pp. 107–110). Springer.

Goos, M., & Bennison, A. (2008). Developing a communal identity as beginning teachers of mathematics: Emergence of an online community of practice. *Journal of Mathematics Teacher Education, 11*(1), 41–60.

Jimenez-Silva, M., & Olson, K. (2012). A community of practice in teacher education: Insights and perceptions. *International Journal of Teaching and Learning in Higher Education, 24*(3), 335–348.

Kanold, T., Tonchef, M., & Douglas, C. (2008, Summer). Two high school districts recite the ABCs of professional learning communities. *Journal of Staff Development, 19*(3), 22–27.

Kimyagarov, I. (2020). A journey in defining my inner teacher. In A. F. Artzt & F. R. Curcio (Eds.), *The inspirational untold stories of secondary mathematics teachers* (pp. 31–36). Information Age Publishing, Inc.

Lei, S., Gorelick, D., Short, K., Smallwood, L., & Wright-Porter, K. (2011). Academic cohorts: Benefits and drawbacks of being a member of a community of learners. *Education, 131*, 497–504.

Lortie, D. J. (1975). *Schoolteacher: A sociological study.* University of Chicago Press.

Losano, L., Fiorentini, D., & Villareal, M. (2018). The development of a mathematics teacher's professional identity during her first year teaching. *Journal of Mathematics Teacher Education, 21*(3), 287–315.

Lucas, T., & Robinson, J. (2002, October/November). Promoting the retention of prospective teachers through a cohort for college freshman. *High School Journal, 86*(1), 3–14.

Maher, M. (2005). The evolving meaning and influence of cohort membership. *Innovative Higher Education, 30*(3), 195–211.

National Council of Teachers of Mathematics. (2000). *Principles and standards for school mathematics.* Author.

National Council of Teachers of Mathematics. (May, 2020). *Standards for the preparation of secondary mathematics teachers.* https://www.nctm.org/Standards-and-Positions/CAEP- Standards/

Oruc, N. (2013). Early teacher identity development. *Procedia—Social and Behavioral Sciences, 70*, 207–212.

Osterman, K. F. (2000). Students' need for belonging in the school community. *Review of Educational Research, 70*(3), 323–367.

Pascarella, E. T., & Terenzini, P. T. (1991). *How college affects students: Finding and insights from twenty years of research.* Jossey-Bass.

Peressini, A., Borko, H., Romangnano, L., Knuth, E., & Willis, C. (2004). A conceptual framework for learning to teach secondary mathematics: A situative perspective. *Educational Studies in Mathematics, 56,* 67–96.

Richmond, G., Juzwik, M. M., & Steele, M. D. (2011). Trajectories of teacher identity development across institutional contexts: Constructing a narrative approach. *Teachers College Record, 113*(9), 1863–1905.

Ronfeldt, M. (2012). Where should student teachers learn to teach?: Effects of field placement school characteristics on teacher retention and effectiveness. *Educational Evaluation and Policy Analysis, 34*(1), 3–26.

Schön, D. A. (1983). *The reflective practitioner.* Basic Books.

Schön, D. A. (1987). *Educating the reflective practitioner: Toward a new design for teaching and learning in the professions.* Jossey-Bass.

Sultan, A., & Artzt, A. F. (2005). Mathematicians are from Mars, math educators are from Venus: The story of a successful collaboration. *Notices of the AMS, 52*(1), 48–53.

Sultan, A., & Artzt, A. F. (2018). *The mathematics that every secondary school math teacher needs to know* (2nd ed.). Routledge.

Trubowitz, S., Duncan, J., Fibkins, W., Longo, P., & Sarason, S. (1984). *When a college works* with *a public school: A case study of school-college collaboration.* Institute for Responsive Education.

Trubowitz, S., & Longo, P. (1997). *How it works—Inside a school-college collaboration.* Teachers College Press.

Unzueta, C., Moores-Abdool, W., & Donet, D. (2008, March). *A different slant on cohorts: Perceptions of professors and special education doctoral students* [Paper presentation]. Annual meeting of the American Educational Research Association, Miami, FL. ERIC Document Reproduction Service Number ED 500897

Urzúa, A., & Vasquez, C. (2008). Reflection and professional identity in teachers' future-oriented discourse. *Teaching and Teacher Education, 24*(7), 1935–1946.

Van Zoest, L. R., & Bohl, J. V. (2005). Mathematics teacher identity: A framework for understanding secondary school mathematics teachers' learning through practice. *Teacher Development, 9*(3), 315–345.

Voskoglou, M. G. (2019). Communities of practice for teaching and learning mathematics. *American Journal of Educational Research, 7*(6), 386–391.

Walkington, J. (2005). Becoming a teacher: Encouraging development of teacher identity through reflective practice. *Asia-Pacific Journal of Teacher Education, 33*(1), 53–64.

Warin, J., Maddock, M., Pell, A., & Hargreaves, L. (2006). Resolving identity dissonance through reflective and reflexive practice in teaching. *Reflective Practice, 7*(2), 233–245.

Wenger, E. (1998). *Communities of practice: Learning, meaning and identity.* Cambridge: Cambridge University Press.

Yomtov, D., Plunkett, S. W., Efrat, R., & Marin, A. G. (2017). Can peer mentors improve first-year experiences of university students? *Journal of College Student Retention: Research, Theory and Practice, 19*(1), 25–44.

CHAPTER 15

WORKING WITH MATHEMATICS TEACHER CANDIDATES TO DISMANTLE TYPICAL PATTERNS OF POWER, PRIVILEGE, AND OPPRESSION IN MATHEMATICS CLASSROOMS

Teresa K. Dunleavy
Seattle, WA

Elizabeth A. Self
Nashville, TN

We need to learn about how teacher candidates (TCs) experience, understand, and enact justice-oriented practices, as well as how such practices work together across a teacher education program (TEP). That is to say, we need to understand how, if at all, TCs develop a justice-oriented lens through which they see the practices they learn about in their TEP. This chapter examines four teacher education practices that, as a composite, aim to build on one another toward a deeper understanding

The AMTE Handbook of Mathematics Teacher Education: Reflection on Past, Present and Future—Paving the Way for the Future of Mathematics Teacher Education, Volume 5
pages 313–330.

of justice-oriented teaching and learning. We explore: (a) community agreements (ComA), (b) instructor-TC letter writing, (c) live-actor simulated encounters, and (d) blogging. While each one of these practices exists on its own in various formats across teacher education, we are particularly interested in how practices are used across a TEP, under a justice-orientated framing, as a means to dismantle typical patterns of power, privilege, and oppression in mathematics classrooms. We suggest that the work is unending, as we discuss our goal for TCs to learn to dismantle the systemic pattern of inequities and oppression in our current schools and in society.

INTRODUCTION

Students need mathematics teachers who *simultaneously* believe in each student's capacity for brilliance *and* who can attend to the role that power, privilege, and oppression play in mathematics learning (Shah & Coles, 2020). The AMTE Standards for the Preparation of Teachers communicate the need for teachers to learn about the social and historical contexts for learning mathematics (Standard C.4, 2017). Because the social and historical contexts for today's mathematics classrooms involve the perpetuation of systemic oppressions that are centuries old, novice teachers need opportunities to confront and attend to their ideas about power, privilege, and oppression. This chapter contributes an understanding of one teacher education program's [(TEP)'s] attention to systemic inequities in mathematics through the use of justice-oriented practices. We define *justice-oriented practices* as classroom practices that attend to the social and political structures that shape the classroom environment and address them in ways that interrupt systemic oppressions.

Teacher candidates need to uncover whether and how biases are playing a role in how they interpret what mathematics is and how students learn mathematics. And, even if we agree that justice-oriented practices are needed in teacher education, there remains a need to understand *how* these kinds of practices are developed across a TEP. That is, there is a need to understand how teacher candidates (TCs) develop the lens through which they see the practices they are learning about in their TEP. Given that a TEP offers a justice-oriented framing for teaching and learning, we need to understand how TCs: (1) experience practices as justice-oriented when they are modeled, (2) understand where the need for justice-oriented practices is coming from, and (3) work to enact justice-oriented practices in their own settings. This chapter examines four practices that, individually or as a composite, might not be executed as justice-oriented. And yet, when their framing is embedded in a justice-oriented program, such practices build on one another toward a deeper understanding of what justice-oriented teaching and learning means. In this chapter, we share *how* secondary mathematics TCs engage in a justice-oriented framing around: (a) community agreements (ComA), (b) instructor-teacher candidate letter writing, (c) live-actor simulated encounters, and (d) blogging. While we acknowledge that each one of these practices exists on its own in various formats across teacher education, this chapter unpacks using them in concert, under a justice-oriented framing, as a means to dismantle typical patterns of power, privilege, and oppression in mathematics classrooms. We suggest that the work is unending as TCs learn to dismantle systemic pattern of inequities and oppression in schools and in society.

ENGAGING COMMUNITY AGREEMENTS TO RESIST WHITE SUPREMACY CULTURE

The practices in the secondary TEP discussed in this chapter are grounded in a justice-oriented framing. One of the first ways that students experience the grounding for that framing is by engaging in a set of ComA that seek to dismantle typical patterns of power, privilege, and oppression in mathematics. Over several years leading up to 2019, faculty in the program noticed that, even while seeking to create a more just learning environment, instructors and TCs often reinforced forms of oppression in the process of establishing and enforcing classroom norms. For example, during a typical norm-setting activity at the start of a course, TCs at this predominantly-white institution often shared norms they had learned in prior schooling, including, "Be respectful." Absent a critical perspective that attends to the histories of what "respect" has meant in schools for hundreds of years, this kind of norm can be taken up in ways that reinforce white supremacy culture (Leonardo & Porter, 2010). Classroom norms in these cases are also often established by a "majority rules" approach, directly undercutting a justice orientation. Critical analysis of this programmatic practice revealed that establishing norms by a majority-rules approach offered TCs with dominant identities ways to protect themselves and their often-oppressive ideas. Faculty further found that student-produced norms could reify schooling as it has always happened (Peterson, 2014), could limit TCs' own learning, and as such, could reify typical patterns of mathematical power and whiteness. Faculty concluded that co-construction of classroom norms failed to provide a vision of classroom interaction that resisted typical patterns of white supremacy culture (Leonardo & Porter, 2010).

The What, Why, and How of Community Agreements

The analysis and critique of our typical norm-setting patterns led to the development of ComA. The ComA were created by faculty who regularly led conversations with TCs about systemic oppression and were informed by critiques of common guidelines for social justice education (Applebaum, 2014; Ayers, 2014; Fujiyoshi, 2015; Sensoy & DiAngelo, 2014), as well as by justice-oriented organizations and facilitators (e.g., AORTA, 2017; Hurst, 2019). Additionally, as a way of both pushing against oppressive perspectives that faculty are better knowledge-holders than students, and as a way of being responsive to the needs of TCs, ComA undergo constant revisions based on feedback from TCs. This set of agreements serve as a vision of what is necessary in order to engage both *in* and *for* justice-oriented education.

During the 2020–21 school year, the four Community Agreements (ComA) were:

1. *Move up*—into a role of speaking or listening more, if you notice yourself doing less of one and more of the other. Move up what you notice about patterns of speaking and listening to the attention of others so it can be addressed as needed.
2. *Press on*—vague and unclear comments, as well as oppressive terminology and language. Press on your own feelings of anger, sadness, and discomfort as you seek to understand others and their ideas.
3. *Fall forward*—stay open to learning in moments when you have negatively impacted another or the group. Own the harm and seek to minimize it. Do not rely on those harmed to provide the new learning.
4. *Leave space*—for quiet, for those not being heard from, for uncertainty and lack of clarity, for play, curiosity, and creative thinking, for growth, in yourself, the group, and others.

ComA allowed the program to change the process for establishing what eventually became the classroom norms. After introducing the ComA, students are prompted to discuss how the agreements are different from norms. Students are provided sample situations and groups are invited to role play situations that would refer the group back to the agreements. Envisioning situations that engage the ComA are part of the work of disrupting a rigid epistemic stance of what mathematics is, who engages in it, and how engagement happens. TCs who enter mathematics focused on answers that are right and quick may otherwise not see why such agreements are important (Dunleavy, 2018). Taking time to explore why the ComA matter helps orient TCs to the TEP's goal of embracing students' humanity, refocusing our efforts not just on what is learned, but on the process of learning as well.

How ComA Function in Mathematics Teacher Education

In the mathematics secondary TEP, TCs revisit the ComA often, including when the instructor wants to draw TCs' attention toward students' mathematical engagement. The pointed attention to the ComA invites conversations about what sustains and what works to dismantle typical patterns of power, privilege, and oppression in mathematics. In Spring 2021 *Introduction to Mathematical Literacies* course, the instructor shared, *"Norms aren't developed by what we say we want to d–norms are developed by what we actually do, in practice."* This class serves as an example of how the ComA led to norm development. About halfway through the course, the students participated in the series of activities that surrounded a live-actor simulated encounter. During the class session in which the encounter was debriefed (described in more detail in section c), the instructor began by referencing the ComA and reminding TCs, *"We have been able to develop a great classroom community, but that doesn't mean we're 'there.' Just because we liked each other and we have good banter—that doesn't mean we've pushed ourselves and one another to grow in particular ways."* The instructor then invited the TCs to keep the ComA in mind during the debrief, naming each one at a time, while asking the TCs to think about how they would engage in conversations that would push their learning that day. While most of the TCs responded by asking for honesty, openness, and vulnerability, several specifically named their positioning as well, in one case writing into the chat, *"I recognize that I did not respond well to the situation, and that my identity as a white[1] person is, in part, responsible for that. I want to learn from my mistakes instead of dwelling on them."* Consistent with this student's reply, TCs in that course worked to embrace the community agreement to *fall forward* in their learning. Also visible during the debrief were moments of *moving up,* where TCs vocally passed an opportunity to talk by inviting and/or waiting on a classmate to speak up. Over the course of the debrief of that simulated encounter, some TCs showed a willingness to engage a range of "what if's" alongside what happened during their encounter, *leaving space* for peers to consider a range of outcomes as well.

CENTERING MATHEMATICS TEACHER CANDIDATES' HUMANITY THROUGH INSTRUCTOR: TC LETTER WRITING

Using writing to engage TCs in learning about mathematics is not new; some scholars have engaged TCs in writing to pen pals (Phillips & Crespo, 1996), others have invited TCs to write *to* mathematics (Cohen, 2016), and still others have engaged in the common Mathematics Autobiography assignment (e.g., Drake, 2006; Kalinec-Craig et al., 2019; Marshall & Chao, 2018). And while all of these examples offer opportunities for TCs to move beyond solving algorithms, the field of mathematics is still striving to find ways to use TCs' writing to intentionally disrupt typical classroom patterns of interaction.

In this TEP, the practice of instructor-student letter writing was used under the justice-oriented framing as a way to lay the foundation for building relationships, for instructors and TCs to see one another's humanity, and to break down barriers for what typically counts as mathematics. In particular, because candidates were asked to explore their own identities, the instructors saw it as critical for them to share about who they were as people. Another feature of this practice was the two-way street of TCs writing letters to instructors and instructors writing letters back to our TCs. The letters, which could be written, typed, or video recorded, were exchanged several times across a TC's TEP experience. The hope is to build on a TC's identity exploration, particularly as it relates to unpacking *how* they will attend to the systems of oppression that surround the teaching of mathematics. Over time, ongoing letter writing set a foundation for: (1) intentionally inviting TCs to encounter and revisit the multi-faceted, complex identities they bring to the work of learning to teach mathematics (Aguirre et al., 2013; Dunleavy et al., 2020), (2) honoring TCs' individuality and humanity (Gutiérrez, 2017), (3) building TC-instructor relationships, and (4) fostering opportunities for deep, meaningful learning that breaks the boundaries of the typical classroom experience.

The What, Why, and How of Instructor-Teacher Candidate Letter Writing

Student-Instructor letter writing is framed as a tool for building a community of learners who take academic risks, who seek a sense of belonging, who make sense of their own identity, and who work to push themselves and their colleagues to grow (Curd et al., 2019). When instructors and TCs utilize letter writing several times throughout the TEP, the letters are framed as check-in letters. Introduced as a practice that TCs will also enact with their students, the goal of letter writing involves acknowledging students' humanity and aiming to create spaces that build community and that contribute to developing teachers who place students as individuals at the center of practice.

How Letter Writing Functions in Mathematics Teacher Education

We have found instructor-TC letter writing to be totally different from introductory surveys, exit slips, and mathematics autobiographies. In this approach, the instructor's directions might begin, *"I want to get to know who you are as a person, as a mathematician, and as a learner. Tell me anything you think I should know in order to be a good instructor for you..."* By framing the letter as open-ended and off-script, TCs often share more about themselves than we would ever guess. Additionally, because TCs are introduced to the ComA first, they may approach letter writing with that framing, sharing moments of *falling forward,* and embracing the uncertainty of what they do not yet know. The open format also offers students the freedom and space to express themselves in the ways they choose; over the years, students have shared anecdotes, poetry, creative videos, music, stories, and more. Across all formats, students have shared statements of worry, excitement, and gratitude.

The group of TCs in the 2021 *Introduction to Mathematical Literacies* course used their letters to engage in one-on-one conversations with their instructor on: (1) mathematics in the world, (2) themselves as humans, and (3) their thoughts on teaching as they worked with students. One white TC[2] wrote about how they were making sense of mathematics in the world as a place of systemic inequities (1): *"I am thinking I want to start out in an urban setting, since issues of inequity and discrimination in education have become really important to me."* Another white TC shared, *"I can't wait to learn more about how we think about math now, how we should reframe it, and how we can make math teaching as fun as possible, and also in a way that helps disrupt power structures."* One of the TCs of color wrote:

As far as I can tell one of the main focuses of this class seems to be exploring the intersectionality between mathematics and equity, which is an area I would want to possibly explore further in my future career, especially in terms of the intersectionality between mathematics and race.

The instructor wrote back to each student asking how they would bring their identities into teaching mathematics. The TCs' letters were also used to frame how the community understood the systems of power embedded in mathematics learning contexts.

A second way that TCs in this TEP engaged in letter writing involved talking about themselves as humans (2)—how they are doing and what are feeling or experiencing personally, sometimes with respect to their overlapping, intersectional identities. In one letter, a student shared:

> I am currently on day 5 of being symptomatic with COVID-19…As a young, white, generally healthy woman, sure, there was a 'potential' for me to become gravely ill with this virus, but receiving a positive COVID-19 test definitely did not make me fear for my life as it may others.

In another case, a student shared about themselves as a learner, "*I would like you to know that I have recently been struggling with anxiety, which has been exacerbated by the pandemic.*" While these students talked about the significant role that the pandemic was playing in their lives, they also revealed thoughts on their racial identities and mental health in the context of the pandemic. This information offered opportunities for the instructor to be sensitive to how the class would interact with and support one another through the course.

A third way of engaging letter writing involved TCs sharing about how they were experiencing the work of learning to teach (3). In one case, a TC shared a 10-minute video check-in letter—talking in depth about how she was feeling and what she was experiencing as a TC working with high school mathematics students. The video format allowed her to not only share in depth about one of her students, but to also wonder about how she might extend this connection to more students. She shared:

> I am amazed by one of my students in particular, and I just want to shout her out, just because, she is just so brave, and unafraid and unphased by, like, coming off mic and asking her questions, like, 'okay I have a question' and [she is] really just so brave about that, which is amazing. And so, I guess right now, I'm kind of wondering how to extend beyond just one student.

This TC revisited their thinking on this student in their blog (discussed in the next section). Open-ended letter writing affords a range of opportunities for instructors to connect one-on one with TCs. One of the central benefits of letter writing in the context of a justice-oriented program was that deep, insightful comments were often embedded in unexpected places. Inviting students to submit letters in multiple formats (e.g., written, audio, video) shares more of the control with TCs on the kinds of connections they may want to make with instructors. These choices support the aim learn about and support the communities to grow while learning about justice-oriented practices.

ENGAGING TEACHER CANDIDATES IN RECOGNIZING OPPRESSIONS IN THE MATHEMATICS CLASSROOM THROUGH LIVE-ACTOR SIMULATED ENCOUNTERS

Many scholars have featured teaching scenarios as a way to practice learning to teach. For example, Shaughnessy and Boerst (2018) placed TCs' attention to students' mathematical thinking at the center of their scenarios, while Gutiérrez et al. (2017) acknowledge teaching as inherently political and rehearse with their TCs for the politics of teaching mathematics. In *live-actor simulated encounters,* teacher candidates engage with live actors around a teaching scenario for which they have received a bit of information. While recognizing the multiple ways of rehearsing teaching, live-actor simulated encounters offer TCs the chance to engage with live actors in moments of mathematics learning that are inherently political, complex, and for which they will not initially bring a lot of experience (Self & Stengel, 2020). Live-actor simulated encounters also intend to build on the classroom community established by the ComA and the instructor-TC relationships deepened through letter writing. And while previous scholars have engaged TCs in role play in order to prepare for the work of teaching, the hope with this TEP is for TCs to experience the components of a live-actor simulated encounter as a way to be pressed, in the moment, on the choices they make, and in so doing, to *fall forward* as they learn from these choices. The encounter used in the mathematics TEP associated with the *Mathematical Literacies* course intended to emphasize how mathematics and systems of oppression are interconnected, and how all teachers and students are either complicit in replicating or dismantling these systems of oppression on a moment-to-moment basis.

The What, Why, and How of Live-Actor Simulated Encounters

TCs in this TEP often engage in the mathematics live-actor simulated encounter after having completed encounters in other courses. In the *Introduction to Mathematical Literacies* course, TCs play the role of a 10th-grade teacher who holds a meeting for two students who have had a disagreement while working together on a groupwork task. Building on the work of Gutiérrez et al. (2017), the encounter seeks to engage secondary mathematics TCs in a typical, yet complex teaching situation that keeps both the content of mathematics and that of systemic oppression fully present. The two students featured in the encounter, played by actors, are called Matthew and Luciana. The encounter takes place over five stages (described below). Across these stages, TCs are invited to discover whether and how they embrace(d) each student's identities, including their histories, genders, and languages. In the iteration, the role of Matthew is identified as a white, monolingual, cis-gendered English-speaking 10th grade student who was seen, and saw himself, as someone who was good at mathematics, and who successfully advanced through mathematics courses. The role of Luciana is identified as a 10th grade Latina student who is multilingual and cis-gendered, and who was new to taking an advanced mathematics course. By asking the TC to play the role of the teacher, they engage, in real time, in a meeting between the two students, making choices about the tensions that arise regarding the students' identities, their approaches to mathematics, and their group's learning.

The stages of the encounter include: (1) TCs reading an initial scenario describing Matthew and Luciana and responding to a set of questions about how they are making sense of what the scenario's core issues are; (2) interacting (on subsequent days) with live actors playing the roles of Matthew and Luciana in a real-time setting, which is videorecorded; (3) reacting (immediately following their encounter) with one or two peers who engaged separately with the scenario, comparing their experiences with the actors (audio-recorded); (4) reviewing videos individually (a day or two after the encounter takes place) and responding to questions that encourage a reframing of the core issues in the encounter; and (5) engaging in classroom discussion (the next class session) around salient moments of power, privilege, and oppression (Self & Stengel, 2020). Across the stages of the encounter, the goal is to leverage TCs' opportunities to re-watch their scenarios, in order to slow down this moment in time as a point for learning, deep reflection, and personal growth. A second aim involves TCs comparing their experiences with others, in ways that point to their own identities and positioning. The specifics of this approach force TCs to respond to a situation for which there is (video) evidence of their actions. Further, because TCs complete their encounters individually, rather than publicly, we find that their defenses are lower going into the group debrief (stage 5 of the cycle), knowing they can share what they choose to share while being accountable to what happened.

Nodding to the ways that the four practices discussed in this chapter are intertwined, live-actor simulated encounters are built into coursework throughout the TEP. The Matthew and Luciana simulation, for example, follows readings and discussion in the *Mathematical Literacies* course about positioning, complex instruction (Cohen & Lotan, 1997), grouping practices, and productive classroom discourse. Following the encounter, TCs read scholars such as Jackson and Delaney (2018) and Sengupta-Irving (2014), who highlight the nature of equity in learning communities and offer new perspectives as TCs watch their videos back. The encounters also live on in the TEP as a shared text, as TCs refer back to them in subsequent courses and field experiences. The goal, then, is not for TCs to rehearse aspects of an anticipated encounter, but rather to complicate, question, and unlearn their immediate (live) responses in ways that expand their understanding of what issues of oppression are at play, alongside what responses are possible. This approach allows for addressing systemic forms of oppression that are evident in and across mathematics education. This opportunity, often prior to significant field experiences, can expand TCs' vision of what counts as justice-oriented mathematics.

What TCs Learn from Live-Actor Simulated Encounters Work in Mathematics Teacher Education

When the 2021 *Introduction to Mathematical Literacies* students debriefed the Matthew and Luciana simulation, they talked in detail about how their own identities were present in how they approached the interaction, both in terms of the biases and the alliances they formed with the students, as well as the resources their identities brought to the encounter. Early in the debrief, for example, one student stated that he could relate to Luciana, but that he found it more difficult to connect with Matthew, whom he described as demeaning towards Luciana. Another TC talked specifically about how, as a Black woman teacher candidate who is *"especially vilified in every form,"* there is complexity in managing the tensions between Matthew and Luciana. That TC talked about how she felt she could support Luciana and push Matthew to keep learning. A Latina classmate echoed this viewpoint, adding that she might turn to white colleagues in that moment as well.

The teacher educators' role in the debrief involves allowing TCs to explore their experiences, while finding places to nudge them into *falling forward, moving up, leaving space, and pressing on*. One thing that was noted in the 2021 debrief was that none of the white-identifying TCs talked specifically about how their race shaped their interactions with Matthew and Luciana—until a Black woman TC in the class brought it up, saying, *"Those who perpetuate white supremacy are more likely to listen to a white person."* At that point, some of the white TCs started talking about how they felt about Matthew and Luciana. This debrief revealed the potential for TCs of color to articulate how their racial and ethnic identities impacted them as the teacher in this encounter, while simultaneously reinforcing the need for white TCs to grapple with how they centered whiteness while trying to resolve a student conflict. While the debrief is often a place for students to lead the conversation as a way to reveal their deep learning, as instructors in this TEP, we addressed our own racial identities early and often. Many of the instructors in the TEP during the time of this writing identified as white; naming whiteness in the TEP leadership allowed us to acknowledge the systemic complexity of discussing power and race in mathematics learning and allowed us to frame how we learn with and from students of color as we aim to disrupt systemic inequities. As such, one goal for instructors during this debrief is to attend to and mediate conversations in which some TCs center the role of race and racial identity while other TCs demonstrate colorblind and color-evasive tendencies.

We believe the experience of the live-actor simulated encounter is not meant to contribute to "answers" or "perfect" resolutions, which is why we do not aim for the TCs to feel like they have resolved the conflict between Matthew and Luciana by the time the stages of the encounter are complete. Rather, we hope for the experience to allow for ongoing conversations about teacher responsibilities, centering whiteness (or not), and understanding other acts of oppression, complicity, or resistance to carry forward into the rest of the course and into the candidate's progress across the TEP.

Some TCs recognized that the practical strategies they might use in a situation like this one are dependent on the goals of the encounter. Early on in the debrief, one TC invited the class to focus on strategies that would resolve the conflict. The strategies then discussed included: deciding who should talk first, figuring out how to get students talking, and discussing whether to validate each student's experience. Later in the debrief, TCs reflected that conflict resolution may not be the most important. One TC shared that early on in the process, they thought, "*Well, I have to resolve something. So let me focus on the mathematical resolution,*" but as they continued to debrief the encounter, they shifted to thinking that it was not realistic to resolve everything in ten minutes. For many students, the intended goals of the interaction shifted during the debrief when the instructor shared a video from the actors who played Matthew and Luciana. The actors shared about their experiences acting as students, noting in particular the things TCs did and did not take up. These reflections shifted the debrief such that TCs thereafter discussed "missed opportunities." For example, Matthew had an issue with how Luciana used Spanish to describe a mathematics growing pattern, describing her method as "less-efficient." When the actors reflected that some TCs had attended to this issue and some had not, TCs pointed to comments Matthew made that now stood out as important in understanding the role of white supremacy in the interaction. Ultimately, TCs were left with the opportunity to reflect on how they used the time in the encounter as directly related to what they saw as the goals. By confronting and reflecting on their choices, the TCs had opportunities to recognize and confront the forms of oppression present.

Across the five stages of the simulated encounter cycle, the TCs faced opportunities to reflect on how systemic oppression, including white supremacy, are not always obvious at first glance. One of the several examples of microaggressions (Kohli & Solórzano, 2012) from the encounter occurred when Matthew repeatedly called Luciana "Lucy." The TCs' initial description of the students included a reference to Luciana that stated, "*whom some students call Lucy.*" Many TCs did not initially recognize this moment of whitewashing Luciana's Latina name as an example of white supremacy. During the live encounter, when either a TC or Matthew call her Lucy, Luciana might have responded, "*My name is Luciana,*" offering yet another moment for TCs to reflect on what name Luciana preferred. Similarly, while several TCs talked about Matthew's complaints about Luciana using Spanish as problematic, they did not also initially name his request for English as an example perpetuating white supremacy. Once TCs were able to make these connections, they were ready for more direct conversations about how to interrupt and dismantle the white supremacy present in the interaction. Interestingly, once TCs recognized how white supremacy was at the root of the conflict, they realized how much harder that made it to reach any kind of resolution by the end of the encounter, saying in one case, "*I felt like it wasn't resolved. And I don't know how I could really fix that.*" Comments like these point to the complicated nature of what TCs perceived to be immediate concerns related to teaching mathematics and the larger, complex issues of oppression that exist in classrooms.

DISMANTLING NOTIONS OF MATHEMATICAL SMARTNESS THROUGH BLOGGING

When TCs enter student teaching, teacher educators become interested in the ways that they engage, reflect on, and implement what they have learned (Aguirre et al., 2013). Tied to one of their methods courses and alongside student teaching, TCs in this TEP engaged in weekly blogging. While inquiry that explicitly invites TCs to attend to their identity is called for (Marshall & Chao, 2018), TCs still need significant opportunities to engage with the ways their identity impacts their teaching practice (Dunleavy et al., 2020). Blogging was framed as yet another way to engage in a justice-oriented framing toward teaching, drawing on the community norms built through the ComA, the TC-instructor relationship started or continued by the letter writing, and the critical perspectives highlighted through the experiences involved in the simulated encounters. The hope was that, as an activity that takes place across the culminating courses of their program, blogging offers opportunities to synthesize across and between the justice-oriented practices that TCs learned about throughout their program.

The What, Why, and How of Blogging

TCs start weekly blogging when they are near the end of their program, as they begin student teaching, and in alignment with mathematics methods. TCs are offered open-ended options for blog posts, with the idea that TCs are working to synthesize what they have learned about justice-oriented practices. They can do so through many formats, including but not limited to: course weekly questions (e.g., "*How do students' identities matter in our classrooms?*" and/or "*What teacher and student moves shift the mathematical power from teacher to student?*"), something that happened in student teaching, such as what they learned when a student shared a mathematical idea, how the course weekly question relates to their student teaching placement, and/or something else related to their learning to teach. As is the case with letter writing, blogging involves the opportunity for students to choose what to blog about—and the hope is that by writing off-script they will engage in deep, explorative reflections in how they are experiencing learning to teach mathematics. TCs in this TEP utilized blogging to engage in an open-ended exploration of their own identities, their experiences of learning to teach, and how they are working through and toward understanding the systems of power and privilege that are at play in mathematics classrooms. Different from journal writing, TCs describe that blogging, even in "private mode," affords a commitment to take a definitive stance or wonder on what they are learning about teaching. Blogging also offers an opportunity to build a reflective bridge between coursework and fieldwork. TCs who use blogging to bridge the gap between what they are learning and what they are teaching experience blogging as a connective tissue throughout student teaching.

What TCs Learn from Blogging in Mathematics Teacher Education

On their own, blogs allow TCs to reflect on their emerging ideas through the lens of classroom practices. As a culminating activity across the TEP, blogs can afford a place for TCs to make sense of the justice-oriented practices that they have learned about. One student posed a blog question to themselves inspired by a weekly question from a methods course, *"How do we develop students' disciplinary agency?"* The student reflected, *"It starts with knowing. Know the strengths of your students. Know the ways they like to play and work. Then, put it all together."* The student then returned to this post when processing a teaching moment in which a student revised a conjecture while articulating a mathematical idea. Another student blogged about how videos, chats, and Nearpod assignments offered them opportunities to get to know their students and they said, *"the work, emotions, and explanations my students had about their own identity, what it meant to be smart, ... and differences correlated to real world events mathematically and socio-emotionally."* Another student used their blog to superimpose metaphors from *Game of Thrones* onto what they were learning about teaching challenging mathematics, bringing in a storyline about riding hundreds of miles as a metaphor for what it was like to create challenging mathematics tasks for students. And another student reflected that their understanding of students allowed them to connect to the power and beauty of mathematics, saying:

> I want my students to recognize their power as mathematicians and participate in math that is personally and socially meaningful to them. It is because of this that I must also know where my students come from, their backgrounds, and how they are positioned in society.

Students cite their own and one another's blogs throughout their time in student teaching, making sense of and connecting their individual, personal experiences about their journey in learning to teach toward justice-oriented learning.

TAKEAWAYS FOR MATHEMATICS TEACHER EDUCATORS

In this chapter, we shared four practices that were framed by a justice-oriented approach to learning to teach mathematics: community agreements (ComA), instructor-TC letter writing, live-actor simulated encounters, and blogging. And while we have acknowledged that each of these practices exists in teacher education spaces in various formats, our hope has been to learn about how our secondary mathematics TCs learn to intertwine and engage these practices across a TEP under a justice-orientated framing and as a means to dismantle typical patterns of power, privilege, and oppression.

Takeaway: Experiencing Justice-Oriented Practices when they are Modeled

As teacher educators, we seek to engage the ComA and letter writing to ultimately alter not only what TCs learn, but also their experience while learning. Student feedback indicates that the ComA and letter writing have influenced our students' orientation towards how they experiencing justice and what they hope to do with their own students. In particular, Black, Latino/a/x, LGBTQ+, and dis/abled TCs have shared when they have felt, seen and affirmed by their peers and instructors. And when unproductive conflict arises, some TCs have spoken up to address these issues. White and male TCs have also engaged in moments of recognizing their power and privilege. And while there are many ways a TEP might enact justice-oriented ComA and letter writing, we offer three potential takeaways:

1. Typical group-produced community norms tend to focus on what individual TCs need to feel safe, while justice-oriented ComA and instructor-TC letters aim to orient TCs toward both what the community needs and who individuals are, while resisting the replication of forms of oppression.
2. ComA and letters only offer guidance for how to engage and only yield new forms of interaction when they are explicitly talked about and regularly revisited.
3. When TCs recognize the benefits of ComA and letter writing, some begin to ask for them in other contexts, including other courses, fieldwork, and in their personal relationships.

Ultimately, enacting ComA and letter writing cannot dismantle power, privilege, and oppression alone, but their consistent use across a justice-oriented TEP can set the groundwork in ways that reshape both what is learned and the experience TCs have while learning.

Takeaway: Understanding Where Justice-Oriented Practices are Coming From

The live-actor simulated encounter is just one example of an enactment that pushes TCs to confront their own assumptions about teaching mathematics embedded in a complex system of oppression. Evidence of learning can be seen in how TCs talk about the encounter throughout their TEP. In addition to referencing salient identities, both their own and those of the students, TCs move from trying to solve the conflict in the encounter to better understanding and envisioning responses that address the larger issues at play. The hope is to support TCs to persist in their efforts at dismantling these systems without positioning themselves as rarified saviors. While there are many forms of enactment that one might use, we have been exploring the following:

1.   Simulated encounters are particularly powerful when the full humanity of both teacher and student are present, without flattening them in ways that make salient identities invisible. TCs have demonstrated an understanding of how power, privilege, and oppression play out when they are able to grapple with the identities involved.

2.   The kind of embodied learning found in live-actor simulated encounters, particularly when coupled with video recordings that TCs refer back to, offer the opportunity for deep reflection on the evidence of what their choices were. Engaging in this way affords TCs the opportunity to confront what may be deficit, limiting, or unproductive.

3.   Mathematics as a content area is heavily tied up in the replication of systemic oppressions related to race, gender, language, dis/ability, among other identities. TCs need to understand how these oppressions are specific to mathematics. Selecting mathematics topics that call into questions the epistemologies of math, who is perceived as "*good at math*," and that show the value of expansive thinking around a mathematics problem offer a potential for exploring the ways that one can work to disrupt some of the typical classroom system oppressions.

Takeaway: Working to Enact Justice-Oriented Practices in TCs' Own Settings

While TCs have used blogging in multiple formats, including mapping blog posts onto ideas from pop culture, social media, and intermixing video and writing, our aim is for TCs to utilize bloggings as a reflective space for synthesizing their understanding about justice-oriented practices. The hope is that TCs will start to see themselves in teaching and will use the blogging as a way to find their individual voice. As such, three potential takeaways include:

1.   The format of blogging allows for a TC to be creative in how they synthesize what it means to learn justice-oriented teaching practices.

2.   The act of publishing an individual blog post, even in private mode, facilitates opportunities for TCs to make public their ideas about teaching toward justice. By committing to a post, sharing them with members of cohort, mentor teachers, and instructors, TCs find themselves careful to blog in ways that others can access, rather than capturing a moment of learning only for themselves.

3.   Blogging weekly across the period of student teaching affords the opportunity for students to return to ideas that they may have started thinking

about early on in the TEP, affording rich ways to synthesize learning across courses and time throughout the TEP.

CONCLUSION

Mathematics education remains a white supremacist space, in which there are still far too few PK–12 students who find themselves successful. We have used this chapter to explore four practices that one secondary mathematics TEP engages using a justice-oriented framing. In particular, we have shared how TCs: interact with one another (via the ComA), practice learning to teach (via the live-actor simulated encounters), and write about their identities they reflect on teaching and learning (via instructor-TC letter writing and blogging). TCs move through these practices as ways of reflecting on who they are, how they are learning to practice the work of teaching, and themselves as teachers. While we are not claiming that any one of these practices is entirely new, we are interested in the nuances of *how* these four practices have been enacted across the TEP as a means of working toward justice-oriented teaching. And in so exploring, how, if at all, the practices, enacted together, might offer opportunities for TCs to learn to dismantle typical patterns of power and oppression in mathematics classrooms. Our goal is that TCs will not enact "business as usual" in mathematics education spaces, but that they will challenge their teaching environment as a political space (Gutiérrez, 2017), in which they have the ability to disrupt typical enactments of power, privilege, and oppression in mathematics spaces—in order to envision a role in working toward productive, systemic change. In this way, we hope this TEP example and our reflections of its impact will help to inspire others and will therefore move the field forward.

ENDNOTES

1. In this chapter, we intentionally capitalize racial and ethnic descriptors such as Black, but not white. Rather than affirming whiteness as a standard, our choice to leave white in the lowercase seeks to center the histories and authority of racially marginalized scholars and students.
2. While sharing a TC's racial identity provides an incomplete picture of an individual's identity, we share this as a small part of the context in which TCs made sense of themselves as teachers.

ACKNOWLEDGEMENTS

We would like to acknowledge other TEP faculty who were a part of the discussions around some of these practices, including Anita Wager, Heather J. Johnson, Andy Hostetler, Melanie Hundley, and Rebecca Peterson.

REFERENCES

Aguirre, J., Mayfield-Ingram, K., & Martin, D. (2013). *The impact of identity in K–8 mathematics: Rethinking equity-based practices.* The National Council of Teachers of Mathematics.

AORTA. (2017, June). *Anti-oppressive facilitation for democratic process: Making meetings awesome for everyone.* https://arts-campout-2015.sites.olt.ubc.ca/files/2019/02/AORTA_Facilitation-Resource-Sheet-JUNE2017.pdf

Applebaum, B. (2014). Hold that thought! A response to "Respect differences? Challenging the common guidelines in social justice education." *Democracy and Education, 22*(2), 1–4.

Ayers, R. (2014). Critical discomfort and deep engagement needed for transformation. A response to "Respect differences? Challenging the common guidelines in social justice education." *Democracy and Education, 22*(2), 6.

Cohen, E. G., & Lotan, R. A. (1997). *Working for equity in heterogeneous classrooms: Sociological theory in practice.* Teachers College Press.

Cohen, M. D. (2016). "Dear Math: I hate you." *For the Learning of Mathematics, 36*(2), 18–19.

Curd, V., Yosef, M., Collins, C., & Dunleavy, T. K. (2019). *Toward emancipatory teaching: Using letter writing in math class* [Workshop]. Annual Southeast Regional meeting of the National Council of Teachers of Mathematics, Nashville, TN.

Drake, C. (2006). Turning points: Using teachers' mathematics life stories to understand the implementation of mathematics education reform. *Journal of Mathematics Teacher Education, 9*(6), 579–608.

Dunleavy, T. K. (2018). High school algebra students busting the myth about mathematical smartness: Counterstories to the dominant narrative "Get it quick and get it right." *Education Sciences, 8*(2), 1–13.

Dunleavy, T. K., Marzocchi, A. S., & Gholson, M. L. (2020). Teacher candidates' silhouettes: Supporting mathematics teacher identity development in secondary mathematics methods courses. *Investigations in Mathematics Learning, 13*(2), 1–16.

Fujiyoshi, K. F. (2015). Becoming a social justice educator: Emerging from the pits of whiteness into the light of love. A response to "Respect differences? Challenging the common guidelines in social justice education." *Democracy and Education, 23*(1), 1–6.

Gutiérrez, R. (2017). Political Conocimiento for teacher mathematics: Why teachers need it and how to develop it. In S. E. Kastberg, A. M Tyminski, A. E. Lischka, & W. B. Sanchez (Eds.), *Building support for scholarly practices in mathematics methods* (AMTE Professional Book Series, Vol. 3, pp. 11–37). Information Age Publishing.

Gutiérrez, R., Gerardo, J. M., Vargas, G., & Irving, S. E. (2017). Rehearsing for the politics of teaching mathematics. In S. E. Kastberg, A. M Tyminski, A. E. Lischka, & W. B. Sanchez (Eds.), *Building support for scholarly practices in mathematics methods* (AMTE Professional Book Series, Vol. 3, pp. 149–164). Information Age Publishing.

Hurst, K. W. [@mochamomma]. (2019, April 6). *Have you ever fallen down, and because you're anticipating the pain do all manner of the things to mitigate the fall?....* [Tweet], Twitter.

Jackson, C., & Delaney, A. (2018). Mindsets and practices: Shifting to an equity centered paradigm. In A. Fernandes, S. Crespo, & M. Civil (Eds.), *Access and equity promoting high-quality mathematics in grades 6–8* (pp. 143–155). NCTM.

Kalinec-Craig, C., Chao, T., Maldonado, L. A., & Celedón-Pattichis, S. (2019). Reflecting back to move forward: Using a mathematics autobiography to open humanizing learning spaces for pre-service mathematics teachers. In T. Bartell, C. Drake, A. McDuffie, J. Aguirre, E. Turner, & M. Foote (Eds.), *Transforming mathematics teacher education.* https://doi.org/10.1007/978-3-030-21017-5_10

Kohli, R., & Solórzano, D. G. (2012). Teachers, please learn our names!: Racial microaggressions and the K–12 classroom. *Race Ethnicity and Education, 15*(4), 441–462.

Leonardo, Z., & Porter, R. K. (2010). Pedagogy of fear: Toward a Fanonian theory of 'safety' in race dialogue. *Race Ethnicity and Education, 13*(2), 139–157.

Marshall, A. M., & Chao, T. (2018). Using math autobiography stories to support emerging elementary teachers' sociopolitical conscious and identity. In S. E. Kastberg, A. M Tyminski, A. E. Lischka, & W. B. Sanchez (Eds.), *Building support for scholarly practices in mathematics methods* (AMTE Professional Book Series, Vol 3, pp. 279–293). Information Age Publishing.

Peterson, B. (2014). Scrutiny instead of silence. *Democracy & Education, 22*(2), Article 7.

Phillips, E., & Crespo, S. (1996). Developing written communication in mathematics through math penpal letters. *For the Learning of Mathematics, 16*(1), 15–22.

Self, E. A., & Stengel, B. S. (2020). *Toward anti-oppressive teaching: Designing and using simulated encounters.* Harvard Education Press.

Sengupta-Irving, T. (2014). Affinity through mathematical activity: Cultivating democratic learning communities. *Journal of Urban Mathematics Education, 7*(2), 31–54.

Sensoy, Ö., & DiAngelo, R. (2014). Respect differences? Challenging the common guidelines in social justice education. *Democracy & Education, 22*(2), 1–10.

Shah, N., & Coles, J. A. (2020). Preparing teachers to notice race in classrooms: Contextualizing the competencies of preservice teachers with antiracist inclinations." *Journal of Teacher Education, 71*(5), 584–599.

Shaughnessy, M., & Boerst, T. A. (2018). Uncovering the skills that preservice teachers bring to teacher education: The practice of eliciting a student's thinking. *Journal of Teacher Education, 69*(1), 40–55.

CHAPTER 16

RETHINKING "THE BASICS"

Toward Mathematics Teacher Preparation for Anti-Racist Classrooms

Evra Marie Baldinger

San Francisco State University

Maria del Rosario Zavala

San Francisco State University

This chapter challenges traditional notions of "the basics" for beginning mathematics teachers, arguing that the commonly recognized narratives therein contribute to the maintenance of white supremacy in mathematics classrooms, and that a re-envisioning of mathematics teacher preparation is necessary for real progress toward greater humanity and justice. We problematize commonplace notions of "the basics" and then review various alternative ways of conceptualizing mathematics teacher preparation. We offer some examples of our own experimentation with these issues in our mathematics methods courses. This chapter contributes to conversations among mathematics teacher educators that interrogate our responsibilities, approaches, and strategies for preparing teachers toward anti-racist math classrooms.

The AMTE Handbook of Mathematics Teacher Education: Reflection on Past, Present and Future—Paving the Way for the Future of Mathematics Teacher Education, Volume 5
pages 331–347.

INTRODUCTION

A few years ago, I (Maria) was in a large conference room with district and teacher education stakeholders discussing the future of teacher education and partnerships between the university and various organizations. I recall the following interaction:

> The meeting began with an opening circle, in which each person spoke briefly about their role in teacher education, why they were at the meeting, and what was important to them in teacher education. I commented that I thought it important for future elementary teachers to be well-versed in issues of social justice that impact local communities and how to make connections between the math they are teaching and these issues. A school board member disrupted the circle to respond, "This social justice stuff is good, but teachers need to be able to make it in the district. They need the basics. I've seen first-hand what happens when they don't have what it takes to just be a good teacher."

As I relayed this story to my co-author, we wondered:

- What are the "basics" to which she referred? What is she assuming (and what do we generally assume) about being a "good new teacher" that leaves little room for issues of social justice? How do these assumptions uphold white supremacy in schools and classrooms?
- Why must we be limited in our vision for new teachers? How might we aim for a more expansive and liberatory notion of "basics," a vision which would prepare teachers to work toward disrupting unjust systems?

Questions about what is fundamental for mathematics teacher preparation are central to the work of mathematics teacher educators (MTEs). As we (Evra and Maria) reflected together, we acknowledged that the ways in which we grapple with these questions must contend with and respond to the inequities endemic to our educational system, and the realities of race and white supremacy that matter for the experiences and learning of children in math classrooms. In the board member's claim that "the basics" that teachers need is distinct from "this social justice stuff" we recognized echoes of long-standing narratives about preparing mathematics teachers which connect to racism in schools and white supremacy culture.

Martin (2008) calls attention to the extent to which policy-level discussions of math education have avoided considerations of race, despite overwhelming evidence of racial disparities. We join him and other scholars (Nasir et al., 2008; Shah & Coles, 2020) in the position that we must name and examine racism in math education to have any hope of dismantling it. Alongside the recognition of racism is the need to take on whiteness, to recognize that white supremacy subjugates not only children from historically marginalized communities but all children, including those who are labeled as white (Battey & Leyva, 2016). This chapter takes up questions of "the basics" of mathematics teacher preparation and puts them in conversation with the work of scholars who illuminate ways in which white supremacy and other forms of injustice show up in our schools. We hope to contribute to conversations about preparing mathematics teachers to engage in liberatory and anti-racist teaching.

To consider how white supremacy culture plays out in mathematics teacher preparation, we draw from the work of Okun and Jones (Okun, 2020; Okun & Jones, 2000), who articulate characteristics of white supremacy culture and offer antidotes for them. For our consideration in this chapter of narratives around "the basics" of teaching and teacher preparation, we have found it useful to focus particularly on the characteristics of white supremacy culture that Okun and Jones name as *power-hoarding, one right way* (the belief that there is one right way to do things), *paternalism* (those holding power control decision-making, assume that they are qualified and entitled to do so, and see no need to inquire into the experiences or viewpoints of those whom their decisions impact), and *either/or thinking* (positioning issues or options as either/or, good/bad, or right/wrong, with little attention to nuance or complexity). We want to acknowledge the risk of engaging ourselves in the "one right way" aspect of white supremacy culture and express our wish that this chapter, rather than claiming to offer "right" answers or simplistic solutions, contributes to ongoing dialogue and collective investigation of the issues we raise.

To contribute to the work of dismantling white supremacy, we suggest that MTEs and those who develop and enact math teacher preparation policies move away from conceptions of teacher preparation which do little to prepare teachers in credential programs to be agents of change. Explicit attention to white supremacy, we suggest, can prepare us and our students to work against the perpetuation of an unjust system for all of our children. We see a need to centralize issues of racism and justice in the design of teacher preparation programs. To that end, we invite other math teacher educators to join us in conversations working to collectively redefine "the basics" of preparing math teachers.

THE BASICS AS A LIMITING NARRATIVE
OF TEACHER LEARNING

Narratives about the "fundamentals" of beginning teaching which we have repeatedly encountered in our work have generally included the assumptions that teachers need to first be well versed in classroom management, content knowledge, and some set of teaching practices which are presumed to be "best." Here we interrogate the narratives surrounding each of these constructs through a lens examining white supremacy culture, as articulated by Okun and Jones.

"Classroom Management" as a Proxy for Control, Power-hoarding, and Cultural Erasure

Creating environments in which large numbers of children can engage productively in learning is one of the great challenges brought on by the development of schools as we know them. In schools and districts, we have generally heard this challenge discussed in terms of the need for teachers to have strong "classroom management." While some interpretations of classroom management start from relational approaches to caring and cultural dimensions of belonging (Noddings, 2012; Valenzuela, 1999), we see and hear classroom management as it gets used in district conversations implying systems and practices meant to exert control over children, with the assumption that this control is necessary for teaching and learning. To accomplish this control, teachers are encouraged to hoard power over children's bodies and behaviors and over their learning. As is consistent with Okun and Jones's (2000) articulation of white supremacy culture, they often do not perceive this power hoarding to exist or, when they do, rationalize it as necessary to pursue the best interests of students.

This hoarding of power contributes to classroom cultures that leave little room for students to bring their own, authentic, cultural ways of being into a classroom (Hand, 2009). It discourages the collective meaning-making that is at the heart of learning environments. It sets up situations in which teachers have few options other than to enact harsh discipline in response to students' ways of being that fall outside the teacher's expectations (Ferguson, 2000).

In school environments situated within communities of color, these narratives of control are particularly reminiscent of a long and sinister history of white supremacy, in which systems of control are justified by the need to "civilize" indigenous communities or communities of color. Further, we find the presumption inherent in these systems that children must be coerced into the "right" behaviors insulting to both children and teachers. It presumes that children are not predisposed to learn, but must be coerced, and that the learning experiences teachers design for children are not enough to inspire learning. It relies on a banking model of education (Freire, 2017), in which the knowledge and cultural ways of being that students bring into the classroom are erased in favor of the more valued knowledge that the teachers hold, which they are responsible for delivering to the children.

Discourse around classroom management is full of the need for situation-proof and teacher-proof systems of warnings, censure, points, threats, and rewards, which, these narratives claim, should be applied "consistently and fairly." We see in this framing the presence of paternalism, in that people with power to create these systems do so without seeking out the viewpoints or experiences of the students or school communities whom their systems impact. In the name of avoiding potential biases, these systems imply that teachers' responses to students should disregard each student's particular needs, experiences, strengths, and challenges. As with other 'color-evasive' classroom systems (Annamma et al., 2017), this disregard for students effectively removes their cultures, communities, particularities, and humanity from the classroom community and renders racially oppressive conditions invisible and unchallengeable. Further, we see that in aiming to be teacher- and situation- proof, these systems require a monolithic approach that devalues the particular ways of being (and thus the humanity) of teachers.

The Privileging of Mathematical Content Knowledge

There is general agreement in math education that teachers should strive for rich and flexible knowledge of the mathematical content that they teach (Ma, 1999), coupled with specialized knowledge for teaching it (Ball et al., 2008; Hauk et al., 2014). Frameworks for pedagogical content knowledge for teaching math ultimately privilege math knowledge as a basis to build teaching knowledge. District and state teacher certification policy generally prioritizes new teachers demonstrating particular kinds of mathematical knowledge above other forms of knowledge, such as knowledge of the histories or cultural conditions of the communities in which they intend to teach.

This math knowledge prioritization serves as a barrier to teacher preparation programs' ability to both recruit and prepare teachers to receive credentials and support novice teachers in developing practice that promote justice and humanity in classrooms. Math content exams for teachers (such as Praxis Exams in Mathematics from the Educational Testing Service) focus on content that comes from a European canon, is narrowly conceived, and prizes abstraction and generalization above context and application. The processes by which the canon is enshrined in educational policy are marked by paternalism (Okun & Jones, 2000) in that policy makers decide what "counts" as important math knowledge with little regard for the voices of communities who are impacted by these decisions. Solely "knowing" the content in this canon, or being able to pass the gatekeeper exams, does little to support teachers with the flexibility and richness of knowledge that prepares them to interpret and respond well to children's ways of talking and thinking about math in classrooms. Rather, we argue, a hyper focus on this knowledge may equip them to resort to explaining to children how math works or ignoring children's bids to be recognized as mathematically powerful in non-mainstream ways, further contributing to classroom culture that devalues the mathematical creativity and ingenuity of children.

Presumptions of Best Practices

Recently, math education has articulated "high leverage" teaching practices, or "core practices," which tend to be taken up in teacher education as "best practices" that teachers should presumably learn to implement in their own classrooms. We see these approaches as part and parcel of Okun's articulation of the either/or aspect of white supremacy culture in that they suggest that there is "best" and "not best" teaching and discount the complexity and context-specificity of math teaching. We join other scholars (Philip et al., 2019) to challenge these approaches. We take issue not with the particular teaching practices that are named as "high leverage" or "core" in these approaches, but with the notion that it is possible to articulate practices that are presumed to "work" without attention to the myriad of questions that might arise around the specifics of their implementation.

For instance, the instructional routine known as a "number talk" has made its way into many math classrooms. It is seen as an instructional practice that teachers can learn to implement relatively easily. It appears in many elementary school curricula, framed as a practice that "works." Indeed, this practice has been shown to have the potential to support students to make sense of their own and other students' mathematical ideas and it provides a relatively easy-to-implement, more student-centered alternative to "drill and kill" approaches to arithmetic. However, assuming that number talks are a "best practice," without care for nuance or particulars, can lead to surface-level implementation that does little to support more liberatory math teaching and learning (Murata et al., 2017). Consequently, when number talks do not have the outcomes teachers expect, or are hard for children to engage in, the notion of best practice implies two possible reasons: either the teacher is doing it wrong, or the children are doing it wrong. Considering this (or any) teaching routine a "best practice" hides the nuances of implementation that are part of responsiveness to the particular needs, strengths, and development of a classroom community, all of which matter for powerful math teaching.

Popular narratives about classroom management, content knowledge, and mastery of best practices as "the basics" of teaching and teacher preparation function to uphold and reproduce white supremacy culture. They falsely position novice teachers as not yet professionals, incapable of grappling with complexities in teaching. A preponderance of evidence has made it clear that these arrangements cause harm to children and perpetuate racialized hierarchies of opportunity, achievement, and recognition, with Black, Latinx, Indigenous, and other nondominant groups of students at the bottom (de Saxe et al., 2020; Milner, 2013; Oakes, 2008). We suggest that building teacher preparation from more ambitious visions of math teaching and of learning to teach has the potential to inspire and sustain teachers through many of the challenges of new teaching, giving their work meaning and purpose, and giving them connection to communities of educators dedicated to similar work.

TOWARD A NEW FOUNDATION FOR TEACHING MATHEMATICS

Re-envisioning mathematics teacher education to respond to the critiques we have raised requires us as MTEs to think carefully about what we privilege in our classes and how we create opportunities for teachers to learn to refigure their own classrooms to challenge white supremacy. This goal relates to AMTE's (2017) Standards for Preparing Teachers of Mathematics, indicator C.4.4, which states, "Well-prepared beginning teachers of mathematics understand the roles of power, privilege, and oppression in the history of mathematics education and are equipped to question existing educational systems that produce inequitable learning experiences and outcomes for students" (p. 23). Here we highlight the work of a few scholars who offer alternatives that inspire us in our own work.

Culturally responsive pedagogies and rehumanizing approaches have emerged in mathematics teacher preparation at the forefront of approaches to math teaching that challenge white supremacy and work to reconfigure arrangements of power, privilege, and oppression. Aguirre et al. (2012) draw from ideas of *culturally responsive pedagogy* (Ladson-Billings, 1995) and *culturally responsive teaching* (Gay, 2010) to argue that mathematical knowledge should be understood as culturally situated, embodied, and humanized, rather than disembodied and abstract. Aguirre and Zavala (2013) propose *culturally responsive mathematics teaching*, which privileges community and cultural knowledge, suggesting that in-school math be structured around children's own funds of knowledge (Moll et al., 1992), disrupting white supremacist patterns of the knowledge that is valued in math classrooms. Gutiérrez (2012) conceptualizes two axes of equity in mathematics education: the dominant axis, consisting of access and achievement, and the critical axis, consisting of identity and power. As she articulates a "sociopolitical turn" in math education (Gutiérrez, 2013), she points out that mathematics educators generally attend to the dominant axis and neglect the critical. In our work to take up the critical axis and challenge white supremacy, we consider calls to rehumanize mathematics, in particular for Latinx, Indigenous, and Black children (Goffney & Gutiérrez, 2018). This involves, "recognizing hierarchies in the classroom and society and shifting the role of authority from teacher/text to other students," as well as, "recognizing mathematics is a living practice, where students see mathematics as full of not just culture and history, but power dynamics, debates, divergent answers, and rule breaking" (Gutiérrez, 2018a, p. 5).

Philip (2019) offers a reimagining of teacher preparation around the conception of teaching as principled improvisation. While he developed this work outside of mathematics-specific teacher preparation context (his work was not in discipline-focused teacher preparation courses) we propose that it has much to offer to our considerations of the critical axis of equity in our mathematics methods contexts. His conception of teaching as principled improvisation prioritizes the relational work of teaching, framing teaching as centrally concerned with issues of identity, power, and "taken-for-granted categories such as race" (p. 4) (and, we add, ability level, English language level, special education designation, etc.). According to Philip, "the relational work of teachers necessitates that teachers consider how identity is negotiated and constructed in interaction and how the contributions, experiences, and knowledges of different students are valued or devalued in and through classroom discourse" (p. 4). Conceiving of teaching as principled improvisation positions teaching practices as tools for doing the work of teaching, which is fundamentally relational and therefore dependent on the particularities of the teacher, the students, and the communities involved. (Phillip contrasts this approach with that of practice-focused scholars, who consider practices first and then consider how practices might be modified for different groups of students.

This approach, he points out, assumes a "norm" from which "diverse" groups of students deviate, locating diversity in "otherness" and inviting stereotyping and essentializing groups of students.)

Philip's notion of teaching as principled improvisation aligns well with perspectives and approaches in mathematics teaching and mathematics teacher education that acknowledge and respond to issues of power, oppression, and white supremacy. It offers an alternative approach to teacher learning that centers the complex, intellectual work of teaching, while challenging binaries and one-size-fits all approaches that are at the root of the current limited view of "the basics." When we conceive of teaching as principled improvisation, the need for teacher-centered tools for "classroom management" recedes as power is redistributed to prioritize the development of a community of learners in the classroom. Prioritizing the relational work of mathematics teaching supports us to reconsider what knowledge matters for that work. From centering decontextualized "content knowledge" such as the knowledge demonstrated through "subject matter competency" examination, this conception supports us to consider what teachers might need to know, understand, see, or become aware of in order to both make sense of their own position and role in creating liberatory mathematics classrooms in the particular communities in which they work and to recognize and value the mathematical (and other) experiences, thoughts, and contributions of each of their students and their communities.

The lens of principled improvisation offers useful perspectives about the currently published AMTE standards for teacher preparation (2017). It invites MTEs to consider ways in which the standards take up the complexities and fundamentally improvisational nature of justice-focused teaching. In our read, the primary focus of the language in the first three professional learning standards relates to mathematics content, knowledge about how children learn mathematics, with considerations of equity throughout. We see the issues of social hierarchy, power, and identity construction stated most explicitly in Indicator 4, which includes C.4.4, quoted above. Philip's ideas suggest that this articulation be expanded to more explicitly attend to the relational work of teaching, which includes examination of "taken-for-granted categories such as race," of teachers' own positioning within social hierarchies and of whose experiences, perspectives, and knowledge are valued in our math classrooms. For instance, we might consider ways in which standards 1 to 3 could benefit from more explicit consideration of power, privilege and oppression as they relate to curriculum, pedagogy, and enacting equitable instruction. We understand that such considerations would need to involve multiple stakeholders and are beyond the scope of this chapter or this book, but we suggest that moving into the future of mathematics teacher preparation should include such work.

Philip (2019) invites us to see our own teaching in teacher education contexts also as principled improvisation. This suggests that we consider our own work supporting the development of mathematics teachers to be relational first. It invites us to consider how the "contributions, experiences, and knowledges" of our own students (novice teachers) are "valued or devalued through" our own "classroom discourse." Philip (2019) offers the design principle of supporting teachers in the preservice classroom space to "narrate, re-narrate, and re-vision" (p. 11) their experiences with children in classrooms, demonstrating through examples from his own work ways in which novice teachers' understandings of themselves, of their students, and of teaching deepen through this process.

How might we design experiences that position our students as sense makers (with each other and with us) about teaching? How might we build community, routines, and assignments in our math methods classes that support this type of collective sense-making about math teaching and learning? How might we invite our own students to narrate their experiences with students' mathematical experiences, to re-narrate them in community with other teachers, and to re-vision what is possible for their current and future practice? How might we create opportunities for our students to center the relational work of their own developing mathematics teaching, as they come to understand mathematics learning as expansive, multifaceted, and part of every child's identity and humanity?

ILLUSTRATIONS FROM OUR METHODS COURSES: TWO VIGNETTES

In this section we share vignettes from our math methods courses illustrating ways that we have worked toward reconfiguring our preparation of mathematics teachers to support sense-making of practices that center students' knowledges and identities and support the narration and re-narration of students' communities as mathematically powerful. We teach in a large, public university, in a state where teacher credentialing is a post-baccalaureate degree, in credentialing programs (secondary education and elementary education) that are organized over one or two years, depending on the credential. Our college has a mission statement that includes specific commitments to social justice. (We acknowledge that we thus enjoy relative safety in raising the issues here in our courses.) By sharing our developing practice, we hope to contribute to conversations about reorganizing mathematics methods experiences toward anti-racist teaching.

Learning to Prioritize Community Knowledge: An Example from an Elementary Math Methods Course (by Maria)

Teachers in my elementary methods courses are working on a variety of teaching credentials from multiple-subject credentials for teaching in general education K–8 classrooms to credentials for teaching students with visual impairments. Each teacher in these programs is required to take a one-semester elementary mathematics methods course. Most take this course from me: an American-born Peruvian woman and a mother of two mixed-race, white-presenting children. I base this course on the principles of culturally responsive math teaching, antiracism, and teaching mathematics for liberation. At the beginning of our semester, we make sense of these ideas together, and we explore them more deeply throughout the semester. I also introduce teachers to a variety of classroom routines that can support mathematics teaching.

One concept central to our class is that children bring knowledge (mathematical, linguistic, cultural, etc.) from their home communities that they can draw on to make sense of math in school. I have found that the students in my classes often need opportunities to see that they, as math teachers, also have numerous resources that are useful for their own mathematical sense-making. Many teachers in my courses have been traumatized by the math instruction they experienced as children and young adults. Mathematizing, a process we engage in together, helps them to begin to see their own mathematical competencies–or ways in which they are mathematical people. These new ways of understanding themselves support teachers to in turn recognize the mathematical competencies and identities in children and families. I have developed a community math exploration assignment to give teachers opportunities to both notice such competencies and community connections to classroom mathematics, and to learn to center the lives of their own students in their math teaching.

I learned about math explorations through Turner et al.'s (2012) work on children's funds of knowledge in mathematics. In math explorations, teachers walk through neighborhoods in which their students live, take notes and pictures, and talk with business owners and other community members about anything mathematical, working to uncover the not-so-obvious ways that mathematical reasoning is embedded in their lives. Teachers explore the neighborhoods with the idea that they might aim to feel some of what the students might feel, see what the students might see, take in the scents, etc. Prior to the physical exploration, novice teachers interview a student from the neighborhood about the community and use the student's guidance in their exploration. Next, teachers generate mathematical questions that students could ask and answer related to the community, and then develop a math task around one or more of these questions. In this assignment, teachers create math curriculum that is contextualized in their own students' lives, offer explorations of genuinely problematic questions, and support multiple solution pathways. At this phase in teachers' task design, we talk about introducing choice and open-ended problems, the kinds of problems in which students make and justify decisions. In this way, we talk about mathematical modeling, holding this form of traditionally valued mathematical knowledge (i.e., Common Core State Standards for Mathematical Practice [National Governors Association Center for Best Practices, 2010] Number 4 "Model with Mathematics"), alongside children's community-based knowledge.

I have seen this project support teachers in the relational work of teaching in that teachers reconsider "how the contributions, experiences, and knowledges of different students are valued or devalued" (Philip, 2019; p. 4) in the activity in their math classrooms. I have seen teachers develop their mathematical and pedagogical creativity and make use of innovative and meaningful connections between students' communities and their math lessons. I have seen this project encourage an "opening up" of mathematics curriculum and tasks as teachers learn to listen to and support children in pursuing emergent math ideas that come from the tasks they have created (Drake et al., 2015). In my class, teachers have expressed joy in finding ways to bring more interesting, relatable math problems into their classrooms. They talk about wanting to create something different and more meaningful than the boring and disconnected experiences that they often had as young students in math classrooms.

I have found that in this assignment, teachers are in different places on their journeys of making math tasks responsive to their students. I have worked to take a relational approach to understanding what supports I might provide to nudge teachers along in their journeys. For instance, some tasks come out with overtones of paternalism. To paraphrase a recent example, a teacher wrote in her reflection, narrating her current understanding of the relationship between her task and her students' lives, "this task is about how sugar is bad for us, since I see kids pack a lot of junk in their lunches." She and I had a conversation in which I invited her to re-narrate this relationship, pointing out ways in which her work attended to consequential mathematizing for her students, but also carried assumptions and judgement about students' lives, rather than a sincere desire to learn more about the circumstances surrounding her observations about the "junk" in students' lunches. Her first narration carried the assumption that the situation she observed was the result of ignorance and presumed that by delivering the "right" knowledge, she could get children to eat healthier food. My feedback to the teacher highlighted ways in which her task draft demonstrated her own growing practice of designing math tasks that are relevant to students and also came across as paternalistic, and I pointed to possible directions for revision.

While I have found the math exploration assignment to be powerful in my methods courses, I also struggle with how to support novice teachers to envision incorporating these practices into their ongoing teaching. How and when should they take time to mathematize with their students? How can they build routines around doing this? How and when does it make sense to invest time in writing math tasks that connect to students' lives? But as I grapple with these issues, I am encouraged that I can grapple alongside my students as we make sense of it all together, sorting through a toolbox of principled improvisational practices as we all learn to learn about each other, and to be responsive to the humans whose learning we facilitate.

Learning to Attend to the Mathematical Ingenuity of Each Student: An Example from Secondary Math Methods (Evra)

In my secondary math methods course, I aim to foster ways of being and learning together in our community and with children in classrooms that create opportunities for greater humanity, connection, and dignity (Scott & Philip, 2023).

Toward that end, I include experiences and assignments that invite novice math teachers to "narrate, re-narrate, and re-vision" (Philip, 2019, p. 11) their experiences in math classrooms, as students and as developing teachers. We work together to construct a community of learners in our methods classroom that allows us to make sense together of both theory related to race and white supremacy in math classrooms, and of the experiences novice teachers have with children and adults in math classrooms, working collectively to get better at uncovering the mathematical experiences, ideas, and ingenuity of children.

Toward this end, I have worked with respected colleagues and friends to engage in ongoing interrogation of my own identity (I identify as a white woman and a mother of black daughters), assumptions, and practices, and to design experiences to support novice teachers to engage in their own similar interrogations. Together, we have read and listened to arguments from critical scholars and educators and engaged in discussions and sharing of personal stories, and we have engaged in assignments that create space for the narration, re-narration, and re-visioning that Philip proposes as important for learning together about justice-focused teaching. I describe one such assignment here.

In an effort to prepare teachers to build classroom instruction from an assumption of the brilliance of each of their own students (Baldinger et al., 2021), I have experimented with an assignment that I call a listening interview that I learned about in 2019 from Nicole Louie (personal communication) and adapted in collaboration with Mallika Scott (personal communications). I join other MTEs (Chao et al., 2017) in the hope that we can craft interview experiences that support attention to students' own rich sense-making. The listening interview shares some features of a clinical interview (Hunting, 1997), but with a focus on discovering the ingenuity and nuance in students' thinking, rather than on diagnosing students with respect to any predetermined criteria for that thinking. In uncovering the ingenuity of each student, it counters narrow and deficit-focused narratives about who can be "smart" at math, and in this way supports teachers to work toward building mathematical communities in which each student is authentically valued as a mathematical thinker.

In the methods class, we work together to watch, craft, and practice mathematical interactions with students that come from a place of generosity and curiosity: giving children our admiration, our time, and our authentic desire to learn. We practice asking questions that invite children to share their thinking without judgement or evaluation and that investigate without pushing. We practice listening generously, inviting children to share their ideas in whatever ways of talking, drawing, writing, or gesturing make most sense for them. Novice teachers then conduct these interviews with their own students and reflect on the experience. They bring video of these interviews into class, and engage with their colleagues on further reflection, re-narrating their understanding of the interaction, as we together get better at listening generously to students and identifying evidence of their mathematical ingenuity. As a class, we engage in discussion to share new ideas and perspectives and to re-vision how mathematical learning experiences might be designed to highlight and build from the ingenuity of students.

My students have reported in reflective writing about their experiences with this assignment. One novice teacher said that it supported her to "understand what it means to be smart in math in new ways." Another reported that, in part as a result of this assignment, "I went from a mindset of 'They need to know this material in the most effective way' to 'Students have so much to offer and I should be open to learning from [them].'" Another teacher explained that "the listening interview assignment gave me time to really sit down with tangible evidence of my student's thinking and repeatedly practice listening for understanding and for sense-making."

In discussions, novice teachers reported that it is most difficult for them to learn to be quiet and listen and to let go of the desire to either diagnose or fix students' thinking. In my engagement with their assignments, I have noticed quite a bit of variation in the extent to which they did that successfully, as well as the depth with which they engaged with children's mathematical ingenuity.

As I move forward in my own practice, I am reflecting on these questions: How might I work to connect this assignment more strategically to other aspects of our course which aim to support teachers to make sense together of math teaching which challenges systems and patterns of racism and white supremacy? How might my course design support more opportunities for novice teachers to re-narrate together their understanding of students' mathematical thinking? How might I develop our math methods community to more effectively engage in the collective re-narration and re-visioning that Philip proposes to move toward more liberatory teaching? How might I continue to develop my own attention to the relational work of my teaching, working to further examine my own position in and impact on the community of learners in our methods classroom?

CONCLUSIONS

All of us in mathematics teacher preparation are both complicit in white supremacy and are positioned to work against it. Here we have employed articulations of white supremacy culture provided by Okun and Jones (2000) to help us to see and name its presence in traditional notions of "the basics" of math teacher preparation and have argued that rejecting these notions is an important part of working against white supremacy. Further, we have shared approaches from scholars positioned both within and outside of math education from which we draw inspiration in our ever-evolving work to design mathematics methods courses in ways that challenge racism and white supremacy culture.

We suggest that this work is more powerful when taken on collectively and we call for the field of mathematics teacher education to join with current political and social mo(ve)ments (TODOS, 2020), and to say "¡Ya basta! Enough already!" We also acknowledge that engaging in this kind of sociopolitical work may leave MTEs uncomfortable and unreconciled, and we acknowledge the importance of discomfort in making change. We also acknowledge that MTEs are positioned differently across our institutional settings to take particular kinds of risks. This is also why it is important that we do not attempt to do transformative work alone, but rather with the backing of colleagues and professional organizations such as AMTE.

We suggest also that there is joy to be found in engaging in these important struggles in community. It is possible for us as MTEs and for mathematics teachers to love this engagement, and to feel empowered to try things in teaching that push against norms, bolstered by the knowledge that we are not alone. Rehumanizing mathematics classrooms and working against white supremacy culture is about the experiences of all of the human beings involved: the children who are students in math classrooms and the adults who gain access to greater humanity and connection through their engagement in this work.

REFLECTION QUESTIONS FOR MTES

To support continued exploration of ways to re-figure mathematics teacher preparation to challenge systems of white supremacy and to work toward antiracism, we offer questions for reflection. These are by no means exhaustive but include questions we have asked and will continue to ask of ourselves.

- How can I organize my courses to best support our development as a community of sense-makers? In what ways might I need to attend to my own tendencies toward power-hoarding and positioning folks as either expert or novice with respect to teaching?
- How does my course design (including routines, norms, activities, and assignments) make space for each member of the classroom community to be welcome as their vulnerable, human, particularly ingenious self and to experience being seen and valued in their full humanity?
- How can I design experiences that support teachers in my courses to expect, recognize, and build instruction from the particular mathematical practices and ingenuity of their own students and of their communities?
- How can I design experiences that support teachers in my courses to develop understandings of the educational systems in which math teaching takes place, including acknowledging and challenging white supremacy culture in those systems?
- How can I engage with my own professional community to continue to make sense of anti-racist teaching and working toward the preparation of anti-racist math teachers?

REFERENCES

Aguirre, J. M., Zavala, M., & Katanyoutanant, T. (2012). Developing robust forms of pre-service teachers' pedagogical content knowledge through culturally responsive mathematics teaching analysis. *Mathematics Teacher Education and Development, 14*(2), 113–136

Aguirre, J. M., & Zavala, M. (2013). Making culturally responsive mathematics teaching explicit: A lesson analysis tool. *Pedagogies: An International Journal, 8*(2), 163–190.

Annamma, S. A., Jackson D. D., & Morrison, D. (2017). Conceptualizing color-evasiveness: Using dis/ability critical race theory to expand a color-blind racial ideology in education and society. *Race Ethnicity and Education, 20*(2), 147–162.

Association of Mathematics Teacher Educators (AMTE). (2017). *Standards for preparing teachers of mathematics.* amte.net/standards.

Baldinger, E. M., Johnstone, A., & Palmer, E. (2021). Disrupting inequity by building classroom instruction from the assumption of each student's brilliance. *Social Education, 85*(5), 280–286.

Ball, D. L., Thames, M. H., & Phelps, G. (2008). Content knowledge for teaching: What makes it special? *Journal of Teacher Education, 59*(5), 389–407.

Battey, D., & Leyva, L. A. (2016). A framework for understanding Whiteness in mathematics education. *Journal of Urban Mathematics Education, 9*(2), 49–80.

Chao, T., Hale, J., & Cross, S. B. (2017). Experiences using clinical interviews in mathematics methods courses to empower prospective teachers. In S. E. Kastberg, A. M. Tyminski, A. E. Lischka, & W. B. Sanchez (Eds.), *Building support for scholarly practices in mathematics methods* (pp. 117–131). Information Age Publishing.

de Saxe, J. G., Bucknovitz, S., & Mahoney-Mosedale, F. (2020). The deprofessionalization of educators: An intersectional analysis of neoliberalism and education "reform." *Education and Urban Society, 52*(1), 51–69.

Drake, C., Land, T. J., Bartell, T. G., Aguirre, J. M., Foote, M. Q., McDuffie, A. R., & Turner, E. E. (2015). Three strategies for opening curriculum spaces. *Teaching Children Mathematics, 21*(6), 346–353.

Ferguson, A. (2000). *Bad boys: Public schools in the making of Black masculinity.* The University of Michigan Press.

Freire, P. (2017). *Pedagogy of the oppressed.* Penguin Classics.

Gay, G. (2010). *Culturally responsive teaching: Theory, research and practice* (2nd ed.). Teachers College Press.

Goffney, I., & Gutiérrez, R. (2018). *Annual perspectives in mathematics education: Rehumanizing mathematics for Black, Indigenous, and Latinx students.* NCTM.

Gutiérrez, R. (2012). Context matters: How should we conceptualize equity in mathematics Education? In B. Herbel-Eisenmann, J. Choppin, D. Wagner, & D. Pimm (Eds.), *Equity in discourse for mathematics education: Theories, practices, and policies* (pp. 17–34). Springer.

Gutiérrez, R. (2013). The sociopolitical turn in mathematics education. *Journal for Research in Mathematics Education, 44*(1), 37–68.

Gutiérrez, R. (2018a). The need to rehumanize mathematics. In I. Goffney & R. Gutiérrez (Eds.), *Rehumanizing mathematics for Black, Indigenous, and Latinx students* (pp. 1–10). NCTM.

Hand, V. M. (2009). The co-construction of opposition in a low-track mathematics classroom. *American Educational Research Journal, 47*(1), 97–132.

Hauk, S., Toney, A., Jackson, B., Nair, R., & Tsay, J-J. (2014). Developing a model of pedagogical content knowledge for secondary and post-secondary mathematics instruction. *Dialogic Pedagogy: An International Online Journal, 2*, A16–A40.

Hunting, R. P. (1997). Clinical interview methods in mathematics education research and practice. *The Journal of Mathematical Behavior, 16*(2), 145–165.

Ladson-Billings, G. (1995). Toward a theory of culturally relevant pedagogy. *American Educational Research Journal, 32*(3), 465–491.

Ma, L. (1999). *Knowing and teaching elementary mathematics: Teachers' understanding of fundamental mathematics in China and the United States* (1st ed.). Routledge.

Martin, D. B. (2008). E (race) ing race from a national conversation on mathematics teaching and learning: The national mathematics advisory panel as white institutional space. *The Montana Mathematics Enthusiast, 5*(2–3), 387–398.

Milner, H. R. (2013). *Policy reforms and de-professionalization of teaching*. National Education Policy Center. http://nepc.colorado.edu/publication/policy-reforms-deprofessionalization.

Moll, L. C., Amanti, C., Neff, D., & Gonzalez, N. (1992). Funds of knowledge for teaching: Using a qualitative approach to connect homes and classrooms. *Theory Into Practice, 31*(2). 132–141.

Murata, A., Siker, J., Kang, B., Baldinger, E. M., Kim, H.-J., Scott, M., & Lanouette, K. (2017). Math talk and student strategy trajectories: The case of two first grade classrooms. *Cognition and Instruction, 35*(4), 290–316.

Nasir, N. S., Hand, V., & Taylor, E. V. (2008). Culture and mathematics in school: Boundaries between "cultural" and "domain" knowledge in the mathematics classroom and beyond. *Review of Research in Education, 32*(1), 187–240.

National Governors Association Center for Best Practices, Council of Chief State School Officers. (2010). *Common core state standards for mathematics*. Authors. http://www.corestandards.org/Math/

Noddings, N. (2012) The caring relation in teaching. *Oxford Review of Education, 38*(6), 771–781.

Oakes, J. (2008). Keeping track: Structuring equality and inequality in an era of accountability. *Teachers College Record, 110*(3), 700–712.

Okun, T. (2010). *The emperor has no clothes: Teaching about race and racism to people who don't want to know.* Information Age Publishing.

Okun, T., & Jones, K. (2000). *Dismantling racism: A workbook for social change groups.* dRworks. https://www.whitesupremacyculture.info/

Philip, T. M. (2019). Principled improvisation to support novice teacher learning. *Teachers College Record, 121*(6), 1–32.

Philip, T. M., Souto-Manning, M., Anderson, L., Horn, I., Carter Andrews, D. J., Stillman, J., & Varghese, M. (2019). Making justice peripheral by constructing practice as "core": How the increasing prominence of core practices challenges teacher education. *Journal of Teacher Education, 70*(3), 251–264.

Scott, M., & Philip, T. M. (2023). "We ask so much of these tiny humans": Supporting beginning teachers to honor the dignity of young people as mathematical learners. *Cognition and Instruction. 41*(3), 291–315.

Shah, N., & Coles, J. A. (2020). Preparing teachers to notice race in classrooms: Contextualizing the competencies of preservice teachers with antiracist inclinations. *Journal of Teacher Education, 71*(5), 584–599.

Turner, E., Varley Gutiérrez, M., & Gutiérrez, R. (2012). "This project opened my eyes": Pre-service elementary teachers learning to connect school, community and mathematics. In L. J. Jacobsen, J. Mistele, & B. Sriraman (Eds.), *Mathematics education in the public interest: Equity and social justice* (pp. 183–212). Information Age Publishing.

Valenzuela, A. (1999). *Subtractive schooling: U.S-Mexican youth and the politics of caring.* SUNY Press.

Zavala, M. d. R., Andres-Salgarino, M. B., de Araujo, Z., Candela, A. G., Krause, G., & Sylves, E. (2020). *The mo(ve)ment to prioritize antiracist mathematics: Planning for this and every school year TODOS: Mathematics for all.* https://scholarworks.wm.edu/educationpubs/179

CHAPTER 17

DIVERSIFYING TEACHER PREPARATION PATHWAYS

Kristin E. Harbour
University of South Carolina

George J. Roy
University of South Carolina

Thomas E. Hodges
University of South Carolina

Carolina Teaching Collaborative[1]
University of South Carolina

With stagnant and in many cases declining enrollment in mathematics teacher preparation, diversifying university pathways and creating innovative ways of engaging in professional learning is necessary. The dearth of teacher candidates in STEM fields and the persistent lack of racial diversity in teacher preparation require innovative ways of engaging aspiring educators. With these issues compounded in rural areas across the United States, and specifically within South Carolina, we sought to leverage our district partnerships, expertise of local teachers and university faculty, collaborations with non-profit organizations, and best practices for mathematics teacher preparation to create an innovative, alternative approach to support teacher

The AMTE Handbook of Mathematics Teacher Education: Reflection on Past, Present and Future—Paving the Way for the Future of Mathematics Teacher Education, Volume 5
pages 349–365.

recruitment, preparation, and retention efforts. In this chapter, we outline our innovative alternative approach, Carolina Collaborative for Alternative Preparation (CarolinaCAP), provide an overview of a residency adaption of this plan for degree-seeking students [Carolina Transition to Teaching (CarolinaTtT)], and present early findings across these programmatic undertakings. Further, we present our programmatic description and tentative results to serve as a model for one way in which the mathematics education community can move towards diversifying teacher preparation pathways.

INTRODUCTION

The recruitment, preparation, and retention of teachers of mathematics, and P–12 teachers more generally, is in need of reform. With states' investments in higher education declining (Mitchell et al., 2019), the subsequent financial reforms within institutions of higher education have left vulnerable more conventional colleges of education whose central focus is on the initial and advanced preparation of P–12 school personnel. That is, declining enrollment in initial teacher preparation programs (American Association of Colleges for Teacher Education [AACTE], 2022) has resulted in decreased revenue for colleges of education, making vulnerable the very programs that are needed to construct and elevate the teacher workforce.

This "fight for survival" is juxtaposed against the need for resources to drive innovations that address two central issues in teacher workforce development: (1) the dearth of candidates in STEM fields (Cowan et al., 2016); and (2) the persistent lack of racial diversity in teacher preparation programs (King et al., 2016). These issues are particularly acute in rural regions, with complications arising in: (a) teacher compensation; (b) elimination of incentives aimed at keeping teachers in the classroom; (c) current qualification and demographic makeup up of teachers; and (d) declining number of individuals from traditional teacher preparation pathways (Monk, 2007). Additionally, within mathematics teacher preparation, persistent efforts and focus on the development of teachers' mathematics content and pedagogical knowledge are required. Research has established significant relationships among teachers' knowledge, teachers' instructional practices, and student learning outcomes (e.g., Baumert et al., 2010; Franke et al., 2001; Hill et al., 2005); however, this creates potential concerns within mathematics education as research indicates some teachers have fragmented mathematical knowledge that may negatively influence their mathematics teaching practices (e.g., Ball, 1988, 1990; Ma, 2010). Taken collectively, the lack of mathematics teacher candidates, particularly candidates who have been historically marginalized, the need for developed mathematical content and pedagogical knowledge, and the expectation for colleges to diversify revenue streams creates a situation ripe for innovative recruitment, preparation, and retention reform. In this chapter, we outline the University of South Carolina's innovative, alternative approach to the initial preparation of mathematics teachers that seeks to simultaneously address the need for increased numbers of mathematics teachers alongside strategies to address the racial diversity of the teacher workforce. We present our program as one example that can serve as a model in working towards diversifying teacher preparation pathways within the mathematics education community.

RATIONALE FOR INNOVATIVE
APPROACH TO TEACHER DEVELOPMENT

As the flagship institution in our state, the University of South Carolina's College of Education was well positioned to provide a wide range of expertise when collaborating with school district partners in increasing the number of mathematics teachers and diversifying the teacher workforce. Prior to establishing an alternative pathway, we first sought to understand the landscape of teacher vacancies across the state of South Carolina. Teacher shortage data is often aggregated; however, teacher shortage impacts areas in different ways, such as higher rates in special education, science, and mathematics (Sutcher et al., 2016), as well as in rural communities (Latterman & Steffes, 2017). Although aggregated data can provide an overarching understanding, determining the needs within specific states and local school districts is of utmost importance. Therefore, a grant funded consortium, the South Carolina Teacher Education Advanced Consortium through Higher Education Research (SC-TEACHER), analyzed data related to the number of teacher vacancies and the number of full-time equivalent teaching positions across South Carolina school districts and by subject area (Dickenson et al., 2021).

At the start of the 2020–2021 school year, South Carolina had the greatest teacher vacancies in special education, early childhood/elementary education, and mathematics education, with vacancy rates of 17.6%, 13.6%, and 11.1%, respectively (Dickenson et al., 2021). Additionally, findings indicated that districts in rural locations, districts with high concentration of pupils in poverty (PIP), and districts of smaller size experienced more variation in teacher vacancy. Importantly, "factors often comingled with smaller districts and districts with the most PIP also located in rural of the state" (Dickenson et al., 2021, p. 6). Often, the solution for challenges around the recruiting, preparing, and retaining mathematics teachers in rural communities within South Carolina is to rely on individuals to fill these positions with short-term solutions (e.g., international visiting teacher programs). These temporary "fixes" lead to an ongoing need to address the same teaching positions they are attempting to fill, thus perpetuating the cycle of teacher vacancies in rural communities and for students who are marginalized.

With respect to student achievement on state standardized assessments and end of course assessments in South Carolina, findings revealed statistically significant relationships between teacher vacancies and student achievement, indicating districts with higher teacher vacancy rates were significantly associated with lower student achievement results across all levels (i.e., high, middle, and elementary; Dickenson et al., 2021). These findings are similar to those specific to our state wherein teacher retention in rural communities in South Carolina stem from issues related to ongoing teacher recruitment coupled with challenges in the retention of teachers, especially the early departure of teachers (Fan et al., 2020). Therefore, a strong focus on the induction years (i.e., years 1–3 in the classroom setting) is warranted as we work to create a strong and stable workforce (Dickenson et al., 2021).

Once we determined the needs for South Carolina and local partner districts, we then aimed to establish an alternative pathway that leverages effective practices within mathematics teacher preparation. Next, we provide an overview of these best practices we relied on and integrated within our alternative pathway.

LEVERAGING THE "BEST OF WHAT WE KNOW" IN MATHEMATICS TEACHER PREPARATION

Within teacher preparation research, a common message indicated is the disconnect between the theoretical learning associated with coursework and the authentic experiences associated with P–12 classrooms (Cook et al., 2002; Grossman et al., 2009; Zeichner, 2010). To address this disconnect, numerous policy and reform documents provide recommendations that foreground clinical practice and partnerships in teacher preparation (e.g., AACTE, 2018; Association of Mathematics Teacher Educators [AMTE], 2017; National Council for Accreditation of Teacher Education, 2010). For instance, AMTE's *Standards for Preparing Teachers of Mathematics* (2017) presents five standards for effective preparation programs of mathematics teachers, focusing on the recruitment and retention of high-quality teacher candidates (Standard P.5) by providing high-quality mathematics learning and teaching opportunities (Standards P.2, P.3), with collaborative partnerships (Standard P.1) and clinical settings (Standard P. 4) serving as vehicles for these opportunities to occur.

One practice that collectively addresses mathematics knowledge, partnerships, and clinical settings is site-based methods courses. For more than 20 years, the University of South Carolina has studied the implementation of site-based methods courses as a central design element in teacher preparation (e.g., Hodges & Mills, 2014; Hodges et al., 2017; Thompson & Emmer, 2019). Site-based methods courses are intentionally planned and designed to be offered within P–12 schools, where a university instructor and classroom teacher take part in co-teaching both the methods class and P–12 students. The design of such courses is organized around Oonk et al.'s (2015) concept of *theory-enriched practical knowledge* where teacher candidates routinely engage in carefully crafted learning experiences that provide opportunities for making meaning of theories that originate from classroom practice.

Generally, the methods class begins by meeting with the university instructor separately within the P–12 school to reflect on readings, solve mathematics tasks, and experiment with instructional materials. This is followed by a period of "setting the stage" for entering the P–12 classroom where teacher candidates explore learning progression, learn about student invented strategies, and any relevant content or pedagogical content knowledge needed during the lesson. Teacher candidates then enter the classroom, making observations of the classroom teacher and/or methods instructor. The teacher candidates also partner with one or more students during classroom interactions where they are *looking closely, listening carefully* to the things P–12 students say and do as they engage with mathematical tasks that promote reasoning and problem solving through mathematical and statistical reasoning (Bargagliotti et al. 2020; Consortium for Mathematics and Its Applications [COMAP] & Society for Industrial and Applied Mathematics [SIAM], 2016); NCTM, 2014). These tasks are selected by the methods instructor and classroom teachers so that the P–12 students use mathematics and statistics to explore meaningful questions to them and where they are developmentally along learning progressions. Moreover, the tasks allow for teacher candidates, their methods instructor, and the classroom teacher to notice and dissect both the P–12 students' and teacher actions such as perseverance and supporting P–12 students' reasoning without taking it over for instance (NCTM, 2014). Finally, teacher candidates leave the P–12 classroom and return to their separate space where they engage in a debrief focused on making meaning of children's work and reflect upon their observations.

Coupled with the knowledge of teacher vacancies across the state and districts and knowledge around effective teacher recruitment, preparation, and retention, we sought to develop an innovative approach to teacher development programs aimed at diversifying teacher preparation pathways. We intentionally sought ways to collaborate with school districts more systematically in South Carolina, especially in our rural communities, with their attempts to recruit and retain teachers that were identified as areas of need including special education, elementary education, and mathematics. Below we outline our efforts in developing an innovative approach to teacher preparation, with a keen focus on diversifying the teaching workforce within rural communities and leveraging effective teacher preparation practices. We then present an adaptation of this model which focuses on a residency model, followed by tentative results from these innovative programmatic efforts.

INNOVATIVE APPROACH TO TEACHER DEVELOPMENT

In collaboration with district partners and an understanding of state and district needs, it was apparent that we would need to consider individuals who were not identified by or did not complete traditional initial certification routes (Sanders & West, 2020). As such, we did not set out to address the teaching vacancies in our state by focusing our efforts on just increasing numbers of individuals entering traditional initial certification pathways (i.e., undergraduate education degree), rather we sought a more comprehensive approach to develop pathways in which the recruitment, preparation, and retention could co-exist alongside those traditional teacher preparation programs in our state. This meant that we would have to re-envision our paradigms regarding who could teach mathematics to include individuals who might not have pursued a teaching career in the past. This reframing of our approach to initial certification along with the integration of research about best practices across traditional and alternative pathways into teaching led to the development of an alternative pathway, the Carolina Collaborative for Alternative Preparation (CarolinaCAP).

We designed CarolinaCAP as a program that had, at its core, strategies to break down known barriers (e.g., TNTP, 2020) and embodied a *Grow Your Own* model for alternative pathways. A *Grow Your Own* approach situates teacher preparation as a collaborative endeavor with local communities, showing promise related to the recruitment, preparation, and retention of Teachers of Color (Gist et al., 2019). With CarolinaCAP, we sought to work with diverse communities, particularly rural communities, to allow for "on-the-job" preparation built off our existing program designs. Partner districts, in an effort to recruit more diverse teachers, designed their own recruitment strategies around self-identified areas of need (cf. Men of Charleston Teach program, see https://www.ccsdschools.com/site/default.aspx?PageType=3&DomainID=4&ModuleInstanceID=488&ViewID=6446EE88-D30C-497E-9316-3F8874B3E108&RenderLoc=0&FlexDataID=41630&PageID=1).

Specifically, CarolinaCAP is a non-degree program leading to full licensure for candidates, representing a collaboration among local education agencies and the college of education with the goal to create a high-quality, alternative pathway into the teaching profession. This approach marries the expertise of local teachers, schools, and districts with a non-profit organization, and institution of higher education to design and deliver teacher preparation that is responsive to the unique needs of candidates and districts. CarolinaCAP also aligns with AMTE's vision regarding the significant responsibility, input, investment, and participation of relevant stakeholders in the success of the pathway (AMTE, 2017). Additionally, CarolinaCAP was designed with the tenets of effective teacher development and learning at the forefront. In the following description, one can see the explicit alignment with research-based practices related to teacher development and learning infused in each aspect of this alternative preparation pathway, including, but not limited to: (a) focus on specific subject matter content; (b) emphasis on student understanding; (c) sustained duration; (d) active learning; (e) norms that propagate supportive and challenging collaboration; (f) coherence; (g) modeling; (g) feedback and follow up; (h) coaching and support; and (i) trust (e.g., Borko, 2004; Darling-Hammond et al., 2017; Desimone, 2009).

Within CarolinaCAP, an individual who enrolls with the goal of becoming a teacher is bound by the following requirements prior to being admitted. Early childhood and upper elementary education teacher candidates must possess a bachelor's degree from a regionally accredited institution with a minimum of a 2.75 undergraduate GPA. They must also pass the state required grade band content-based licensure assessment prior to being admitted to the CarolinaCAP. Similarly, middle and secondary mathematics teacher candidates are required to have successfully completed an undergraduate degree; however, the degree must be in mathematics, engineering, statistics, or closely related fields with a minimum of 18 hours of mathematics coursework including a calculus sequence from a regionally accredited university during which they earned at least a 2.5 grade point average. These mathematics teacher candidates also must pass the state required content-based licensure assessment, in this case, Praxis Subject Area Assessment for middle level or secondary mathematics respectively.

Once the initial admission requirements are met, an individual becomes a CarolinaCAP candidate while also being a teacher of record in a partner district. Prior to the start of the school year (i.e., fall semester), CarolinaCAP candidates participate in a summer launch experience, followed by ongoing face-to-face and virtual professional development experiences throughout the school year, and into their induction. The CarolinaCAP candidates receive individualized support from a CarolinaCAP coach who understands the local context of the school and district and, most importantly, provides the necessary support each CarolinaCAP candidate requires to be successful in the classroom. Furthermore, throughout the school year, CarolinaCAP candidates engage in the process for their state certification by completing a series of micro-credentials (MCs) that are individualized to their content area and grade levels. The emphasis on job embedded MCs is a shift from experiences in traditional initial preparation programs (Fong et al., 2016); MCs are tailored to each individual to emphasize the core pedagogical competencies (e.g., intentional questioning, productive mathematical discourse) and content area specific teacher knowledge (e.g., mathematical representational fluency, early fraction understandings) to collectively foster effective teaching and learning of mathematics. Each micro-credential was developed to introduce individuals new to the teaching profession research informed, effective teaching practices (e.g., AMTE, 2017; NCTM, 2014) that emphasize mathematical and statistical proficiency (Bargagliotti et al., 2020; COMAP & SIAM, 2016; National Research Council, 2001).

The inclusion of MCs was designed to recognize skills and abilities in lieu of defined "seat time" as evidenced in a more traditional teacher preparation environment (e.g., Fong et al., 2016; Gish-Lieberman et al., 2021). Job embedded micro-credentials provide real-time learning experiences within authentic classrooms with practicing teachers and students, affording CarolinaCAP candidates the opportunity to witness, take part in, and theorize from actual classroom practice (Oonk et al., 2015). Additionally, MCs carry several advantages over more traditional ways of engaging in professional learning, including:

- Flexibility—a module, flexible style of learning that recognizes candidates' professional learning needs regardless of where, when, or how learning happens (e.g., Hunt et al., 2020)

- Personalization—candidates and mentors can select MCs based upon the candidate's needs, their students' needs, as well as school and district priorities (e.g., Hunt et al., 2020)
- Cost effectiveness—targeted competency-based learning that allows educators to build a career progression and manage costs along the way to further credentials and degrees (e.g., McGreal & Olcott, 2022)
- Results-driven—since MCs rely on evidence-based indicators of meaning shifts in practice, results are observed in real time within the candidate's classroom setting (e.g., Hunt et al., 2020).

Collectively, the MCs align to address what it means to be a well-prepared beginning teacher of mathematics (AMTE, 2017). Together, they provide flexibility and personalization, while emphasizing evidence-based results within a CarolinaCAP candidate's authentic classroom setting by underscoring the importance of mathematical understanding throughout all four standards in this domain (Standards C.1-C.4, AMTE, 2017).

Building upon the micro-credentialing work focused on pedagogical and content preparation, our CarolinaCAP candidates participate in systematic layers of support stemming from a number of stakeholders in supporting them as teacher of record (Standard P. 1, AMTE, 2017). Initially, this happens through personalized support of CarolinaCAP coaches. The coaches spend between 3 to 5 hours a week engaged in systematic support by co-planning, co-teaching, and co-assessing student learning through observations in the CarolinaCAP candidate's classroom. There exists extensive support for the use of coaching as a component of effective professional learning experiences (e.g., Darling-Hammond et al., 2017; Knight 2007) and, specifically, coaching to support the teaching and learning of mathematics (e.g., Fennell, 2017; McGatha et al., 2015). Coaching has been shown to support teacher and student learning (e.g., Gibbons et al., 2017; Harbour et al., 2018; Harbour & Saclarides, 2020; Munson, 2017). As such, the use of individual coaching, as well as group coaching (described below), continues to show the evidence-based approaches integrated within the design and implementation of CarolinaCAP.

Furthermore, candidates have additional time set aside each month to collaborate with CarolinaCAP coaches, college of education faculty, and selected educational experts on a variety of topics aligned to instruction and leadership. CarolinaCAP candidates also participate as members of virtual learning communities. The intent of the virtual communities is to extend the learning opportunities for CarolinaCAP candidates beyond their engagement in MCs, workshops, and professional development experiences. Together, communities of learners build upon learning and experiences of the required coursework and experience a collaborative network within and among the cohorts of candidates. Specifically, CarolinaCAP candidates are invited to participate in two monthly professional learning sessions—one with a focus on micro-credential support and the other with a focus on content and pedagogical specific areas identified by the CarolinaCAP candidates and coaches. Over time, the goal is to involve "veteran" CarolinaCAP candidates who will leverage their CarolinaCAP experiences to serve as facilitators and mentors for new CarolinaCAP candidates.

In addition to the supports provided to CarolinaCAP candidates, the CarolinaCAP pathway also includes the systematic attention needed to address retention, especially during the Induction Period (i.e., first three years in the classroom) of an early career teacher. The Carolina Teacher Induction Program (CarolinaTIP) supports early career teachers with ongoing professional development workshops alongside individualized mentoring and classroom support (Skeen et al., 2020). With the guidance of experienced university coaches, early career teachers learn how to implement effective mathematics pedagogical practices, thus providing high-quality mathematics teaching and learning opportunities within a collaborative and authentic setting (AMTE, 2017). Ultimately, all teachers receive support in their own classroom contexts, where the goal is to address the challenges that contribute to the high rates of early career departure, and South Carolina's teacher shortage.

Adapting the CarolinaCAP Model

Through funding from the U.S. Department of Education's Teacher Quality Partnership program (TQP), we have layered the CarolinaCAP approach with a unique adaptation that includes a year-long residency whereby individuals do not serve as teachers of record rather they serve alongside a mentor teacher in their classroom; this adaptation is known as the Carolina Transition to Teaching (CarolinaTtT; see Figure 17.1) residency. The absolute and competitive priorities addressed in the CarolinaTtT residency were outlined in the Teacher Quality Partnership Grant Program, funded by the U.S. Department of Education, which included the establishment of effective teacher residencies that increase the number of STEM educators in identified opportunity zones as identified by districts where the poverty rate for children aged 5–17 was greater than 20%, and teacher turnover was greater than 15%. The priorities also informed our goals to include improving access to learning opportunities that promote effective mathematics instruction by increasing the recruitment, preparation, and retention of educators who reflect the diversity of students they serve in these rural communities. In the end, we identified South Carolina school districts who met these requirements, and then, subsequently, collaborated with four of the identified districts to develop a residency program with the goal of diversifying the mathematics teaching workforce in these rural areas through a *Grow Your Own* approach.

In the CarolinaTtT residency, teacher candidates work full time alongside the teacher of record to experience the life of a teacher through extensive, school-based experiences within rural schools and communities while also pursuing a graduate degree. The CarolinaTtT candidates, referred to as residents henceforth, are coached and supported by both district and college of education personnel

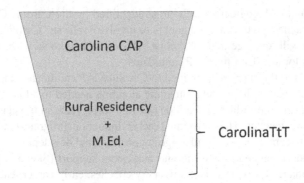

FIGURE 17.1. Relationship Between CarolinaCAP and CarolinaTtT Alternative Pathways. Note: CarolinaCAP is the overarching alternative certification pathway for non-degree seeking students, with CarolinaTtT being an adaptation that marries CarolinaCAP with a residency model for degree-seeking students.

throughout their residency to be responsive educators within localized rural contexts.

Although the initial entrance requirements are the same for CarolinaTtT residency as in CarolinaCAP, including recruiting individuals with degrees outside education, GPA requirements, and summer launch experiences, there is one major difference; once vetted by our partner school district's representatives–the individuals who are accepted into CarolinaTtT do not begin their school year experience as the teacher of record as they do within CarolinaCAP. Rather, the CarolinaTtT candidates participate in a year-long practicum as a resident in a coaching teacher's classroom (i.e., teacher of record's classroom who provides coaching support to residents). A central focus of the designed residency is to experience the life of a teacher through extensive, school-based experiences within schools in rural communities while simultaneously pursuing a graduate degree. Similar to rounds within a teaching hospital, CarolinaTtT residents engage in teaching and learning situated in actual upper elementary classroom settings, supporting the development of *theory-enriched practical knowledge* (Oonk et al., 2015). Residents are provided with multiple layers of coaching support by their coaching teacher, a university supervisor, and CarolinaTtT program faculty. Additionally, residents receive a living wage stipend that can be used to pay for graduate school. Individuals in CarolinaTtT residency also complete a series of MCs that are tailored to their content area and grade band while completing coursework in a hybrid master's program leading to an M.Ed. in Teaching.

Whereas MCs are central to the licensure, the main underpinning of CarolinaTtT residency is embedded in a resident's coursework experience. These clinical experiences align to the AMTE's Standards (2017). More specifically, CarolinaTtT residents have the opportunity to explore how to teach mathematics in an upper elementary classroom (Standard P. 4) while simultaneously learning how to do so in a virtual site-based methods course (Standard P. 3). The authentic classroom experiences in site-based methods courses allow residents to engage with mathematical tasks that promote reasoning and problem solving through mathematical and statistical insight into real-world phenomena (Bargagliotti et al. 2020; COMAP & SIAM, 2016), learn about student-invented strategies, and develop their pedagogical content knowledge through carefully crafted experiences with students (AMTE, 2017; Hodges & Mills, 2014; Hodges et al., 2017; Thompson & Emmer, 2019). Following the M.Ed. in Teaching, CarolinaCAP supports our former residents, now early career teachers, during their induction years (i.e., years 1–3) through ongoing professional development workshops and individualized coaching support as described previously.

EARLY INDICATORS OF SUCCESS

At present, there have been promising results of our *Grow Your Own* alternative pathway. During the first two years of implementation, CarolinaCAP, including the CarolinaTtT residency adaptation, has supported districts in the rural communities in South Carolina in three significant ways: (1) building the collaborative infrastructure needed to address the recruitment efforts identified in the state data regarding teacher shortages; (2) increasing the number of individuals who became teachers in our rural communities; and (3) diversifying the teacher workforce. Specifically, in the initial two-years of implementation, CarolinaCAP has partnered with 17 school district (i.e., 10 of which are rural communities) that have yielded more than 263 applicants in the pipeline, thus working to decrease the identified teacher shortage impacting rural communities (e.g., Dickenson et al., 2021; Latterman & Steffes, 2017). This collaboration with the university and local districts allowed relatively rapid progress despite a challenging recruitment period due to the COVID-19 pandemic. Additionally, these applicants meet the needs for targeted recruitment for teachers (e.g., elementary educators) as identified in the analyses of state teacher shortages (Dickenson et al., 2021). As CarolinaCAP and CarolinaTtT continue to grow, a concerted effort to expand this work into middle level mathematics teachers is underway, aligning with the needs of our partner districts.

CarolinaCAP is also preparing diverse teachers who will meet the needs of diverse districts. Specifically, when considering recruitment efforts, 81% of participating CarolinaCAP and CarolinaTtT residents self-reported as Black, thereby showing positive signs of success in reducing the persistent lack of racial diversity in teacher preparation programs (King et al., 2016) and aligning with research on the effectiveness of *Grow Your Own* approaches in recruiting Teachers of Color (Gist et al., 2019). Furthermore, the ages of candidates in both the CarolinaCAP and CarolinaTtT range from 20 to 60 years old, and many of the individuals held paraprofessional positions in local schools prior to starting the alternative pathway. Importantly, in meeting the needs of our diverse districts, CarolinaCAP is serving districts wherein 71% to 92% of students are eligible for Free and Re-duced Lunch. As one of the main goals of CarolinaCAP and CarolinaTtT residency is to diversify the teaching workforce within rural districts across South Carolina, we see these as promising initial results.

When considering preparation, university faculty's expertise combined with an innovative approach to alternative certification educator pathways has created a unique and successful experience for candidates. For instance, more than 90% of all CarolinaCAP candidates "agreed" or "strongly agreed" that they were capable and prepared as educators, and 100% of candidates indicated they were "moderately prepared" or "very prepared" to teach classes equitably, develop relationships, and be culturally responsive. Additionally, CarolinaCAP support and coaching supports were received very positively across district liaisons, school administrators, and CarolinaCAP candidates. Students and administration within our partner districts also indicated positive perceptions around various topics, including well-being, engagement, and feedback. It behooves us to also note that administrators were extremely positive about CarolinaCAP candidates teaching in adverse conditions, referencing the global COVID-19 pandemic.

Additionally, findings related to the CarolinaTtT adaptation model (i.e., residency model) also indicate positive, tentative results. Through an analysis of observation notes, interviews with district representatives, and a focus group with the grant leadership team regarding the implementation of the CarolinaTtT program, six themes emerged around positive aspects of the development and implementation of the grant-funded project. These themes included: (1a) organized and engaged core leadership team, (b) navigating university and district level policies and systems, (c) understanding residency population and supporting their transition to teaching, (d) embracing conditions and adapting to situations through perseverance and determination, (e) promoting a common vision and philosophy of teaching, and (f) developing a sense of community among stakeholders (D'Amico et al., 2022). Of note on retention efforts, CarolinaCAP and CarolinaTtT have each recently "graduated" cohorts of completers and as a result, there are no data regarding the retention of these candidates and residents respectively. However, during its first four years as this program was established prior to the alternative pathways, CarolinaTIP has had a 98 percent retention rate of the new teachers supported; therefore, we believe successful retention is attainable for CarolinaCAP candidates and CarolinaTtT residents.

While much of these data are promising, challenges do exist. The biggest challenge faced by CarolinaCAP candidates was passing Praxis II, which is required for state certification. Praxis II persists as a barrier, thus continuing to complicate the ways in which we work to recruit, prepare, and retain diverse teachers in many of our rural school districts. Our data parallel previous research, wherein state certification exams create persistent barriers to certification, particularly for Teachers of Color (e.g., Sharp et al., 2019; Shuls, 2018). As such, CarolinaCAP will continue to support candidates who have not yet passed Praxis II in an effort to keep high-quality teachers in classrooms. In addition to the Praxis barrier, the following challenges were identified during the initial implementation of the CarolinaTtT program: (a) shifting to virtual recruitment, learning, and teaching due to the COVID-19 pandemic; (b) establishing a new alternative route to certification pathway at the university; (c) supporting residents with unique needs and financial concerns; (d) navigating changes at the district-level that impacted previous financial agreements; and finally (e) streamlining graduate-level coursework, certification routes, and induction support with linked, yet distinct university-based programs. These data from both CarolinaCAP and CarolinaTtT are critical as we continue in these innovative alternative programs. Through a cyclical process of adaptation, we collect, analyze, and reflect upon our data to continuously refine our efforts (e.g., Bryk et al., 2015); therefore, these data provide us with important reflection points as we refine and adjust CarolinaCAP and CarolinaTtT programmatic efforts to better serve candidates and partner districts.

CONCLUDING THOUGHTS

Universities of higher education, especially colleges of education, are well positioned to provide a wide range of expertise when collaborating with school districts in identifying and supporting individuals through innovative pathways into the teaching profession. We must be clear, we are not suggesting these alternative, innovative pathways replace more traditional ones; rather, we suggest them as complementary pathways to expand the opportunity for individuals to enter the profession. With stagnant, and in many cases, declining enrollment in mathematics teacher preparation coupled with the persistent lack of racial diversity in teacher preparation (e.g., Cowan et al., 2016; King et al., 2016), exacerbated in rural districts (e.g., Monk, 2007) diversifying university pathways and creating new ways of engaging in professional learning provides colleges opportunities to look at innovative ways of engaging educators.

By first analyzing the specific needs of our state and local districts (e.g., Dickenson et al., 2021), we developed both an innovative, alternative pathway for certification for non-degree seeking students and an adaptation of this model through grant funding for degree-seeking students. Tentative results show promise in the areas of recruitment and preparation of high-quality, diverse educators situated in diverse, rural districts. Although this novel approach presents many positive findings, challenges exist (e.g., the state licensure barrier). Data spur continued efforts to refine and adjust our efforts to better serve our candidates and districts. We believe that our approach to alternative teacher development can provide an avenue for other institutions as they continue to develop innovative ways to diversify the teaching workforce through effective and targeted recruitment, preparation, and retention efforts.

ACKNOWLEDGEMENT

The Carolina Transition to Teaching program was supported by a Teacher Quality Partnership Program Grant supported by the U.S. Department of Education, Office of Elementary and Secondary Education, through grant number U336S190031. The content is solely the responsibility of the authors and does not necessarily represent the official views of the funding agency.

ENDNOTE

1. The Carolina Teaching Collaborative is a team of scholars who collaborate to design, implement, refine, and disseminate various teacher preparation and professional development efforts at the University of South Carolina. Members include the authors of this paper and Amber Adgerson, Melissa A. Baker, Catherine Compton-Lilly, Jennifer Crooks-Monastra, Rachelle Curcio, Leigh D'Amico, Jessie D. Guest, Stephen L. Thompson, Carli Toliver, & Hall S. West.

REFERENCES

American Association of Colleges for Teacher Education. (2018, October). *A pivot toward clinical practice, its lexicon, and the renewal of educator preparation: A report of the AACTE Clinical Practice Commission.* https://aacte.org/resources/clinical-practice-commission

American Association of Colleges for Teacher Education (2022, March). *Colleges of education: A National portrait, Second Edition.* https://aacte.org/wp-content/uploads/2022/03/Colleges-of-Education-A-National-Portrait-Executive-Summary.pdf.

Association of Mathematics Teacher Educators (AMTE). (2017). *Standards for preparing teachers of mathematics.* http://www.amte.net/standards

Ball, D. L. (1988) *Research on teaching mathematics: Making subject matter knowledge part of the equation.* https://files.eric.ed.gov/fulltext/ED301467.pdf

Ball, D. L. (1990). The mathematical understandings that prospective teachers bring to teacher education. *Elementary School Journal, 90*(4), 449–466.

Bargagliotti, A., Franklin, C., Arnold, P., Gould, R., Johnson, S., Perez, L., & Spangler, D. (2020). *Pre-K–12 guidelines for assessment and instruction in statistics education (GAISE) report II.* American Statistical Association and National Council of Teachers of Mathematics.

Baumert, J., Kunter, M., Blum, W., Brunner, M., Voss, T., Jordan, A., Klusmann, U., Krauss, S., Neubrand, M., & Tsai, Y. M. (2010). Teachers' mathematical knowledge, cognitive activation in the classroom, and student progress. *American Educational Research Journal, 47*(1), 133–180.

Borko, H. (2004). PD and teacher learning: Mapping the terrain. *Educational Researcher, 33*(8), 3–15. http://www.jstor.org/stable/3699979

Bryk, A. S., Gomez, L. M., Grunow, A., & LeMahieu, P. G. (2015). *Learning to improve: How America's schools can get better at getting better.* Harvard Education Press.

Consortium for Mathematics and Its Applications (COMAP), & Society for Industrial and Applied Mathematics (SIAM). (2016). *GAIMME: Guidelines for assessment & instruction in mathematical modeling education.* Author.

Cook, L. S., Smagorinsky, P., Fry, P. G., Konopak, B., & Moore, C. (2002). Problems in developing a constructivist approach to teaching: One teacher's transition from teacher preparation to teaching. *The Elementary School Journal, 102*(5), 389–413.

Cowan, J., Goldhaber, D., Hayes, K., & Theobald, R. (2016). Missing elements in the discussion of teacher shortages. *Educational Researcher, 45*(8), 460–462.

D'Amico, L. K., West, H. S., Baker, M. A., Roy, G. J., Curcio, R., Harbour, K. E., Thompson, S. L., Guest, J., Compton-Lilly, C., & Adgerson, A. (2022). Using improvement science to develop and implement a teacher residency program in rural school districts. *Theory & Practice in Rural Education, 12*(1), 83–104.

Darling-Hammond, L., Hyler, M. E., & Gardner, M. (2017). *Effective teacher professional development.* Learning Policy Institute.

Desimone, L. M. (2009). Improving impact studies of teachers' professional development: Toward better conceptualizations and measures. *Educational Researcher, 38*(3), 181–199.

Dickenson, T. S., Hodges, T. E., Kunz, G. M., & Garrett, J. J. (2021). Exploring the uniformity of South Carolina teacher vacancies: Policy and practice implications for addressing the teacher shortage in South Carolina. *Working paper series II: What we know about the South Carolina teacher workforce.* https://sc-teacher.org/wp-content/uploads/2021/04/SC-T-Vacancy-paper_FINAL.pdf.

Fan, X., Pan, F., Dickenson, T. S., Kunz, G. M., & Hodges, T. E. (2020). School-level factors associated with teacher retention in South Carolina. *Working paper series II: What we know about the South Carolina teacher workforce.* https://sc-teacher.org/wp-content/uploads/2020/10/WP-2-Retention_FINAL.pdf.

Fennell, F. (2017). We need mathematics specialists now: A historical perspective and next steps. In M. B. McGatha & N. R. Rigelman (Eds.), *Elementary mathematics specialists: Developing, refining, and examining programs that support mathematics teaching and learning* (AMTE Professional Book Series, Vol. 2, pp. 3–18). Information Age Publishing Inc.

Fong, J., Janzow, P., & Peck, D. K. (2016). *Demographic shifts in educational demand and the rise of alternative credentials.* UPCEA. https://upcea.edu/wp-content/uploads/2017/05/Demographic-Shifts-in-Educational-Demand-and-the-Rise-of-Alternative-Credentials.pdf.

Franke, M. L., Carpenter, T. P., Levi, L., & Fennema, E. (2001). Capturing teachers' generative change: A follow-up study of professional development in mathematics. *American Educational Research Journal, 38*(3), 653–689.

Gibbons, L. K., Kazemi, E., & Lewis, R. M. (2017). Developing collective capacity to improve mathematics instruction: Coaching as a lever for school-wide improvement. *The Journal of Mathematical Behavior, 46*, 231–250. https://doi.org/10.1016/j.jmathb.2016.12.002

Gish-Lieberman, J. J., Tawfik, A., & Gatewood, J. (2021). Micro-credentials and badges in education: A historical overview. *TechTrends, 65*(1), 5–7.

Gist, C. D., Bianco, M., & Lynn, M. (2019). Examining grow your own programs across the teacher development continuum: Mining research on teachers of color and nontraditional educator pipelines. *Journal of Teacher Education, 70*(1), 13–25.

Grossman, P., Compton, C., Igra, D., Ronfeldt, M., Shahan, E., & Williamson, P. W. (2009). Teaching practice: A cross professional perspective. *Teachers College Record, 111,* 2055–2100.

Harbour, K. E., Adelson, J. L., Karp, K. S., & Pittard, C. M. (2018). Examining the relationships among mathematics coaches and specialists, student achievement, and disability status: A multi-level analysis using National Assessment of Education Progress data. *The Elementary School Journal, 118*(4), 654–679. https://doi.org/10.1086/697529

Harbour, K. E., & Saclarides, E. S. (2020). Math coaches, specialists, and student achievement: Learning from the data. *Phi Delta Kappan, 102*(3), 42–45. https://doi.org/10.1177/0031721720970701

Hill, H. C., Rowan, B., & Ball, D. L. (2005). Effects of teachers' mathematical knowledge for teaching on student achievement. *American Educational Research Journal, 42*(2), 371–406.

Hodges, T. E., & Mills, H. (2014). Embedded field experiences as professional apprenticeships In K. Karp & A. R. McDuffie (Eds.), *Annual perspectives in mathematics education 2014: Using research to improve instruction* (pp. 249–260). National Council of Teachers of Mathematics.

Hodges, T. E., Mills, H. A., Blackwell, B., Scott, J., & Somerall, S. (2017). Learning to theorize from practice: The power of embedded field experiences. In D. Polly & C. Martin (Eds.), *Handbook of research on teacher education and professional development* (pp. 34–47). IGI Global.

Hunt, T., Carter, R., Zhang, L., & Yang, S. (2020). Micro-credentials: The potential of personalized professional development. *Development and Learning in Organizations, 34*(2), 33–35. https://doi.org/10.1108/DLO-09-2019-0215

King, J. B., McIntosh, A., & Bell-Ellwanger, J. (2016). *The state of racial diversity in the educator workforce.* US Department of Education.

Knight, J. (2007). *Instructional coaching: A partnership approach to improving instruction.* Corwin Press.

Latterman, K., & Steffes, S. (2017, October). Tackling teacher and principal shortages in rural areas. *National Council of State Legislatures, 25*(40). https://www.ncsl.org/research/education/tackling-teacher-and-principal-shortages-in-rural-areas.aspx.

Ma, L. (2010). *Knowing and teaching elementary mathematics: Teachers' understanding of fundamental mathematics in China and the United States* (2nd ed.). Routledge.

McGatha, M. B., Davis, R., & Stokes, A. (2015). *The impact of mathematics coaching on teachers and students* (Brief). National Council of Teachers of Mathematics.

McGreal, R., & Olcott, D. (2022). A strategic reset: micro-credentials for higher education leaders. *Smart Learning Environments, 9*(1), 1–23.

Mitchell, M., Leachman, M., & Saenz, M. (2019). *State higher education funding cuts have pushed costs to students, worsened inequality.* Center on Budget and Policy Priorities.

Monk, D. (2007). *Recruiting and retaining high-quality teachers in rural areas. The Future of Children/Center for the Future of Children, the David and Lucile Packard Foundation, 17,* 155–74. 10.1353/foc.2007.0009

Munson, J. (2017). Examining the efficacy of side-by-side coaching for growing responsive teacher-student interactions in elementary classrooms. In E. Galindo & J. Newton

(Eds.), *Proceedings of the 29th Annual Meeting of the North American Chapter of the International Group for the Psychology of Mathematics Education.* (pp. 471–474).

National Council for Accreditation of Teacher Education. (2010). *Transforming teacher education through clinical practice: A national strategy to prepare effective teachers.* Author.

National Council of Teachers of Mathematics (NCTM). (2014). *Principles to actions: Ensuring mathematical success for all.* Author.

National Research Council. (2001). *Adding it up: Helping children learn mathematics.* National Academy Press.

Oonk, W., Verloop, N., & Gravemeijer, K. P. (2015). Enriching practical knowledge: Exploring student teachers' competence in integrating theory and practice of mathematics teaching. *Journal for Research in Mathematics Education, 46*(5), 559–598.

Sanders, M., & West, T. (2020). Alternative certification in higher education: New initiatives in South Carolina. *Working paper series I: Setting the baseline for South Carolina.* https://sc-teacher.org/wp-content/uploads/2020/01/ALTCERT_Working-Paper_2020.pdf

Sharp, L., Carruba-Rogel, Z., & Diego-Medrano, E. (2019). Strengths and shortcomings of a teacher preparation program: Learning from racially diverse preservice teachers. *Journal of Teacher Education and Educators, 8*(3), 281–301.

Shuls, J. V. (2018). Raising the bar on teacher quality: Assessing the impact of increasing licensure exam cut-scores. *Educational Policy, 32*(7), 969–992.

Skeen, N., Lewis, A. A., Van Buren, C., & Hodges, T. E. (2020). Helping hands for new teachers. *The Learning Professional, 41*(6), 28–32.

Sutcher, L., Darling-Hammond, L., & Carver-Thomas, D. (2016). *A coming crisis in teaching? Teacher supply, demand, and shortages in the U.S.* Learning Policy Institute.

Thompson, S., & Emmer, E., (2019). Closing the experience gap: The influence of an immersed methods course in science. *Journal of Science Teacher Education, 30*(3), 300–319.

TNTP. (2020). *A broken pipeline: Teacher preparation's diversity problem.* Author. https://tntp.org/assets/documents/TNTP_BrokenPipeline_FINAL.pdf

Zeichner, K. (2010). Rethinking the connections between campus courses and field experiences in college and university-based teacher education. *Journal of Teacher Education, 89*(11), 88–89.

CHAPTER 18

OPPORTUNITIES FOR CO-LEARNING EQUITY-ORIENTED MATHEMATICS INSTRUCTION IN THE FIELD EXPERIENCE

Torrey Kulow
Portland State University

Imani Goffney
University of Maryland

Taylor Stafford
University of Washington

Mary Alice Carlson
Montana State University

Ruth Heaton
Teachers Development Group

Kara Jackson
University of Washington

Melinda Knapp
Oregon State University—Cascades

Heather Fink
Portland State University

This chapter offers provisional routines and a tool aimed at supporting teacher candidates and mentor teachers to learn together in becoming equity-oriented mathematics educators. These routines and tool aim to bring to life AMTE's *Standards for Preparing Teachers of Mathematics* (2017) vision that field experiences should be mutually beneficial, "a system of simultaneous growth and renewal for the teacher candidate-mentor teacher-university supervisor team" such that "all participants

The AMTE Handbook of Mathematics Teacher Education: Reflection on Past, Present and Future—Paving the Way for the Future of Mathematics Teacher Education, Volume 5
pages 367–389.

learn and lead while they work on behalf of students" (p. 37). We offer an example of how interactions in the field experience, guided by the routines and tool we are creating, can serve as an authentic, mutually beneficial partnership between teacher education programs and K–12 schools enabling teachers at different points in their careers to work together to develop and deepen their vision, understandings, and practices of equity-oriented mathematics instruction. The routines and the tool intend to support simultaneous learning and growth for both teacher candidates and their mentor teachers in ways that are flexible enough to be implemented in a variety of university and field placement settings. Our exploration focuses on the following questions in the context of a design research study:

1. How can the field experience be a productive site for co-learning equity-oriented mathematics instruction?
2. What are the design features of teacher education tools that support teacher candidates and mentor teachers in co-learning equity-oriented mathematics instruction?

INTRODUCTION

Preparing and supporting teachers in continually becoming equity-oriented mathematics educators is an enduring and critical challenge. For decades, mathematics educators and researchers have advocated for equitable and inclusive education for all students (Association of Mathematics Teacher Educators [AMTE], 2017; National Council of Teachers of Mathematics [NCTM], 1989, 2000; National Governors Association, 2010, 2014). The field has also acknowledged inequitable and dehumanizing experiences and outcomes for historically excluded students, especially for those of Black, Latinx, and Indigenous descent (e.g., Gutiérrez, 2018; National Council of Supervisors of Mathematics & TODOS Mathematics for All, 2016). In spite of more than thirty years of calls for equitable and inclusive educational experiences for all students, opportunity gaps and detrimental experiences in math classrooms for students of color persist (e.g., Berry III, et al., 2014; Martin, 2019).

The challenges around developing equity-oriented mathematics teachers are well documented. Although the teaching force is currently dominated by white female teachers (National Center for Educational Statistics, 2021), many of whom have little or no experience questioning their own teaching practices and a lack of awareness "that some practices benefit some students but inhibit others" (AMTE, 2017, p. 38), all teachers need additional support because equity-oriented teaching practices are unevenly addressed and taught in teacher preparation programs across the country. Although teachers who do not identify as "white female teachers" likely bring a wealth of cultural and linguistic resources from their lived experiences, they still also have much to learn about engaging in equity-oriented teaching because they likely had few and/or poor-quality learning opportunities to connect their cultural and linguistic resources to the work of teaching. This work is dynamic and explicit opportunities for learning are needed to support these teachers with learning to be continually responsive to the ever-shifting contexts in which they work, to the varied needs of the diverse students they work with across their teaching career, and to changes within their own self and identify as they learn from and through the process of becoming equity-oriented.

Further, many teachers have a colonized view of mathematics teaching and learning that is based, at least in part, on their own K–12 experiences (Paris, 2012). Such teachers may have limited understanding of what Black, Latinx and Indigenous students consider to be positive, joyful learning experiences and narrow vision for the range of ways students might intuitively express mathematics competencies (Gay, 2000; Ladson-Billings, 1997, 2014; Leonard et al., 2010; Leonard & Martin, 2013). Presently, most teacher education programs offer a "weak antidote to the powerful socialization into teaching that occurs in teachers' own prior experience as students" (Ball & Cohen, 1999, p. 5). Even explicit attempts to attend to equity in both teacher education programs and professional development programs tend to "tack on" multicultural or culturally responsive practices and offer minimal support for teachers to deeply and meaningfully integrate these important perspectives into their foundational understandings about their work as teachers (Sleeter, 2008; Warren, 2018; Zeichner, 2016). Teachers often have limited opportunities to practice and get support in leading equity-oriented instruction in their own classrooms, in particular teacher candidates working with mentor teachers who vary in their ability to implement and support their teacher candidate in leading equity-oriented instruction (Thompson et al., 2015). These limited opportunities to practice and get support in leading equity-oriented instruction are often due to a variety of contextual constraints including limited professional development, lack of support from colleagues and school structures to engage in this work together, and mandated school curricula and policies that exacerbate, rather than address, inequitable learning opportunities (Darling-Hammond, 2006; Henfield & Washington, 2012). As a result, preparation for instruction that is equity-focused across the professional teacher learning continuum is uneven at best (Mintos et al., 2019; Turner & Drake, 2015), suggesting that both preservice and inservice teachers need additional support if they are to engage in "career-long learning" to teaching mathematics effectively (AMTE, 2017, p. 1).

Our design research project (Cobb et al., 2017) is rooted in the assertion that a teacher's development as an equity-oriented educator is an adaptive, ongoing, deliberate, and collaborative process rather than a process that concludes by reaching an endpoint of mastery with regard to equity. We are particularly interested in investigating how the field experience can be a productive site for teacher candidates *and* mentor teachers to continually work toward becoming equity-oriented educators. Therefore, we are engaged in the iterative design of a tool to support teacher candidates' and mentor teachers' collaborative learning (i.e., co-learning) of equity-oriented mathematics instruction. Our goal is for the tool to support routines which center equity-oriented instruction while bringing to life the vision of the field experience described in AMTE's Standards for Preparing Teachers of Mathematics (2017).

In a mutually beneficial partnership, clinical experiences are designed to support more than just the candidate or to provide extra classroom support for a teacher. The experience can become a system of simultaneous growth and renewal for the teacher candidate-mentor teacher-university supervisor team when they collaborate; all participants learn and lead while they work on behalf of students. (p. 37)

Our exploration focuses on the following questions:

1. How can the field experience be a productive site for co-learning equity-oriented mathematics instruction?
2. What are the design features of teacher education tools that support teacher candidates and mentor teachers in co-learning equity-oriented mathematics instruction?

WHAT IS AN EQUITY-ORIENTED EDUCATOR AND EQUITY-ORIENTED INSTRUCTION?

While the terms *equity-oriented mathematics educator* and *equity-oriented mathematics instruction* are not universally defined and understood, they each imply commitments and practices that should be universally held and enacted by teachers in their local context and community. Our project team currently characterizes an equity-oriented mathematics educator in terms of a teacher's "*vision* for their practice; a set of *understandings* about teaching, learning, and children; *dispositions* about how to use this knowledge; *practices* that allow them to act on their intentions and beliefs; and *tools* that support their efforts" (Hammerness et al., 2005, p. 385).

For our project team, an equity-oriented mathematics educator has a *vision* for mathematics teaching and learning where:

All students, in light of their humanity–personal experiences, backgrounds, histories, languages, and physical and emotional well-being–must have the opportunity and support to learn rich mathematics that fosters meaning making, empowers decision making, and critiques, challenges, and transforms inequities and injustices. Equity does not mean that every student should receive identical instruction. Instead, equity demands that responsive accommodations be made as needed to promote equitable access, attainment, and advancement in mathematics education for each student. (Aguirre et al., 2013, p. 9)

In addition to the content they teach, equity-oriented teachers continually seek to *understand* their students' identities, interests, and assets (e.g., Civil, 1994, 2007; Featherstone et al., 2011; Moll et al., 1992; Paris, 2012; Yosso, 2005), as well as their own implicit biases and deficit perspectives (İnan-Kaya & Rubie-Davies, 2021; Paris, 2016). They enact "ambitious teaching practices" (Kazemi et al., 2009) focused on *practices* that celebrate and draw on students' identities and assets (Aguirre et al., 2013; Aguirre, Turner et al., 2013; Bartell et al., 2017; Celedón-Pattichis et al., 2018), elicit and validate non-dominant forms of mathematics competence (Ukpokodu, 2011), establish equitable participation norms that position all students as capable (Bartell et al., 2017), and critique and dismantle classroom structures of power and privilege that suppress their students' success (Gutiérrez, 2013; Ukpokodu, 2011).

Equity-oriented mathematics educators cultivate *dispositions* that make them receptive to viewing their own classrooms as sites for continued and collaborative professional learning and make regular use of *tools* supportive of their efforts to becoming equity-oriented mathematics teachers in ways that are sustained over time. Framing equity-oriented instruction in terms of a teacher's vision, understandings, practices, dispositions, and tools of teaching invites teachers to be learners across the continuum of professional learning (Feiman-Nemser, 2012). In other words, becoming an equity-oriented mathematics educator is an ongoing, deliberate career-long endeavor, one we characterize as "continually becoming."

Conceptualizing equity-orientation as a career-long endeavor challenges many of the norms typically associated with learning, namely the idea that learning should result in some kind of "mastery" or "arrival." Instead, we aspire to develop tools that support teachers in learning to use an adaptive and ongoing process as they work on a daily basis to be mindful of how they are attending to equity-related aspects of students' experiences in their own classrooms, particularly from their students' perspectives, and continually identify and systematically address problematic aspects of their classroom with the support of others. This stance toward continuous improvement demonstrates responsiveness in that it acknowledges the ever-shifting contexts in which teachers work, the varied needs of the diverse students who come and go across a teacher's career, and changes within teachers themselves as they learn from and through the process of becoming equity-oriented.

CO-LEARNING EQUITY-ORIENTED
INSTRUCTION IN THE FIELD EXPERIENCE

Crucially, we see the work of becoming equity-oriented as necessarily collaborative. The idea that learning is the work of independently acquiring knowledge for oneself is itself an educational norm which stands in the way of equity. Engaging with people with diverse perspectives and experiences (notably colleagues, students, parents, community members) helps a teacher notice and analyze parts of their practice that may be "invisible" to them and supports them in making their intentions explicit and the equitable or inequitable impact of their actions open to collective inquiry. This aligns with what we have long understood about the benefits of collaborative professional learning opportunities within communities of practice (Wenger, 1999).

Given our experiences as teacher educators, we believe that the field experience is a ripe and underutilized space for teachers to continually become equity-oriented math educators because teacher candidates and mentor teachers are typically encouraged and expected to engage in adaptive, ongoing, deliberate, and collaborative work during this component of teacher education programs. Past research has investigated and shared examples of equity-focused mathematics teacher education in initial teacher education (e.g., Aguirre, Turner et al., 2013; Drake et al., 2015; Stoltz et al., 2021) and professional development (e.g., Bartell, 2013; Civil, 1994; Suh et al., 2021; Wager, 2012; Wager & Foote, 2013). In contrast, few studies in the field of mathematics education have examined and described how the field experience in teacher education programs can be a productive site for teacher candidates *and* mentor teachers to continually become equity-oriented educators. Therefore, we provide provisional routines and a tool that supports reimagining the field experience as an opportunity for teacher candidates and mentor teachers to collaboratively learn (i.e., co-learn) equity-oriented instruction.

Conceptualizing the field experience as an opportunity for teacher co-learning is a significant departure from past framings. Historically, the field experience has shown to be an influential and rich site for teacher candidate learning (Borko & Mayfield, 1995; Guyton & McIntyre, 1990; Zeichner, 1996) while its potential as a site for mentor teacher development often goes unrealized (e.g., Bullough & Draper, 2004; Valencia et al., 2009). Moreover, the foci of the teachers' collaborative work varies greatly, based in part on mentor teachers' or institutions' mentoring approach (AMTE, 2017; Borko & Mayfield, 1995; Brooks & Sikes, 1997; Collins et al., 1991; Fernandez & Erbilgin, 2009; Furlong & Maynard, 1995; NCTM, 2014, 2018; Thompson et al., 2015; Valencia et al., 2009) as well as mentor teachers' ability to implement and support their teacher candidate in leading equity-oriented instruction (Thompson et al., 2015).

An emphasis on teacher candidate learning and lack of attention to mentor teacher learning in the field experience may be due, in part, to the enduring prevalence of the apprenticeship mentoring model guiding field experiences (Borko & Mayfield, 1995; Brooks & Sikes, 1997; Dewey, 1965/1904; Valencia et al., 2009). In the apprenticeship mentoring model, the mentor teacher demonstrates instruction for the teacher candidate with the expectation that the teacher candidate will learn to mimic those practices (Furlong & Maynard, 1995). In essence, the apprenticeship model frames the teacher candidate as a learner and the mentor teacher as an exemplar who demonstrates correct ways of thinking about and enacting instruction (Borko & Mayfield, 1995; Valencia et al., 2009). Although this approach is helpful when teaching people to perform tasks with externally visible processes (Collins et al., 1991), it is less useful for helping teacher candidates learn professional acts of teaching with both visible and invisible practices (e.g., interpreting student work, selecting and sequencing student solutions to share during a class discussion, modifying a mathematics task to be more rigorous) (Feiman-Nemser & Beasley, 2007). The pervasiveness of the apprenticeship mentoring model may also impede goals related to developing equity-oriented teachers since teacher candidates may learn to replicate existing harmful practices, thereby perpetuating the existing systems of privilege and oppression in classrooms and educational contexts equity-oriented teacher education is aiming to disrupt.

Our alternative mentoring model positions teacher candidates and mentor teachers as "critical friends," "co-enquirers," and "partners" (Furlong & Maynard, 1995; Males et al., 2010). This mentoring approach is aligned with teacher collaboration in professional development experiences where teachers learn in, from, and for practice (Lampert, 2010) and make visible and analyze both internal and external aspects of their instruction. It also aligns with the notions that good teachers and mentors are good inquirers of practice (Cochran-Smith & Lytle, 2009; Feiman-Nemser, 2012). In this mentoring model, both of the teachers are learners, with the teacher candidate positioned to work in partnership with their mentor teacher. In essence, the teachers assist and sustain one another's learning. In addition, this approach aligns with AMTE's (2017) assertion that "clinical experiences are guided, hands-on, practical applications and demonstrations of professional knowledge of theory to practice, skills, and dispositions through collaborative and facilitated learning" (p. 37).

TOOLS TO SUPPORT TEACHER CO-LEARNING EQUITY-ORIENTED INSTRUCTION

Our project team is conducting a design study to create the conditions in which teacher candidates and mentor teachers are supported to co-learn together in service of equity-oriented mathematics instruction. As Cobb et al. (2017) write, design studies involve work on both theory and practice.

> Pragmatically, they involve investigating and improving a design for supporting learning. Theoretically, they involve developing, testing, and revising conjectures about both learning processes and the means of supporting that learning. (p. 208)

In our case, we are designing social routines and a tool to guide and support teacher candidates' and mentor teachers' co-learning equity-oriented instruction. The routines and tool aim to support mentor teachers and their teacher candidates in productively working on equity-oriented instruction. Further, the routines and tool aim to do so in ways that minimize the power dynamics between teachers, draw on assets of both teachers, and support teachers in developing the humble and vulnerable stance required of lifelong teacher learners (Feiman-Nemser, 2012).

By acting as "objects of mediation" (Thompson et al., 2015, p. 365), the routines and tool prompt teacher candidates and mentor teachers to collaboratively develop and refine their vision, understandings, and practices of teaching as they try "to name and solve problems and tensions related to these practices in their classrooms" (p. 364). Through iterative cycles of testing and refining the tool and surrounding routines, we aim to generate knowledge regarding the conditions that are supportive of teacher candidates and mentor teachers continuing to develop, together, an equity-oriented *vision* for their practice; a corresponding set of *understandings* about teaching, learning, and children; *dispositions* about how to use this knowledge; and *practices* that allow them to act on their intentions and beliefs and examine the impact of their actions on students.

Our tools, referred to as "collaborative learning structures" (CLS) and based on Cameron's (2005) and Thompson et al.'s (2015) descriptions of a "collaborative structure," are designed to enable teachers to work together on the problems inherent in planning and enacting equity-oriented instruction. For Thompson et al., supporting co-learning between preservice and inservice teachers involved balancing asymmetric power dynamics, orienting work toward a finite number of practices with well-designed social resources, tools, and routines, and building communities whose collective work is aimed toward improving teaching for all students.

We are bringing Thompson et al.'s (2015) work into contact with current research-based understandings of equity-oriented mathematics instruction and our own experiences as teacher educators supporting teacher learning. For us, the term "collaborative" signifies "teachers … working together to find solutions" (Cameron, 2005, p. 312) to problems of inquiry or dilemmas arising from instruction (Baker & Knapp, 2019; Lampert, 1985) and the term "structure" signifies the social organization of how the teachers work together and interact. We include the term "learning" to foreground that teachers' collaborative endeavors should promote their professional learning. We outlined a set of design principles (Kirshner & Polman, 2013) to guide our project team's efforts to develop a tool and routines. For our project, the tool and associated routines should:

1. Make visible and investigate "enduring questions" (Cochran-Smith et al., 2008) and practical dilemmas (Lampert, 1985) of equity-oriented instruction (e.g., teachers' vision and values, equitable participation patterns and opportunities for rich learning; leveraging students' mathematical and cultural assets; and facilitating meaningful conversation between students).

2. Center students' experiences and perspectives, especially students' perspectives on their own access, achievement, power and identity (Gutiérrez, 2009) in the classroom.

3. Reflect a humanizing and assets-based perspective (Gutiérrez, 2018) toward all learners by eliciting, drawing on, valuing, and validating teacher candidates' and mentor teachers' perspectives, strengths, knowledge and skills.

4. Enable teachers to interrogate problematic aspects of their own understanding and enactment of equity-oriented instruction and support them in formulating solutions to those problems with guidance and input from others, including students, colleagues, parents, and community members (Zeichner et al., 2015).

5. Support teachers in being humble and vulnerable as they make their own instruction open to collective inquiry (Gibbons et al., 2017; Little, 2002;

Little & Horn, 2007), and consider how they are perpetuating and can potentially disrupt inequities in their classrooms.

In the next section, we describe our initial design of a collaborative learning structure based on these design principles and share tensions that we are considering and encountering as we prepare to support teachers in using the routines and tool.

DESIGNING COLLABORATIVE LEARNING STRUCTURES THAT SUPPORT CO-LEARNING EQUITY-ORIENTED INSTRUCTION IN THE FIELD EXPERIENCE

At present, we are collecting data in three teacher education programs on how teacher candidate-mentor teacher dyads use our first CLS. The dyads are working together in a field placement setting with supplemental support and content from the teacher candidate's mathematics methods course. This CLS intends to support dyads in co-learning how to enact equity-oriented mathematics teaching with a focus on leveraging opportunities for justification and generalization (Bieda & Staples, 2020). We specifically focus on supporting students' mathematical generalization and justification because prior research about equity has shown that mathematics that focuses on procedures continues to dominate secondary math classrooms (Sinicrope et al., 2015), especially in economically disadvantaged districts and schools that primarily serve Black, Latinx, and Indigenous students (Darling-Hammond, 1999; National Center for Education Statistics, 2001). If the acquisition of the knowledge and skills of justification and generalization (Blanton & Kaput, 2005; Stephans et al., 2018; Stylianides, 2008) are goals of math learning in grades 6–12, then secondary math teachers at any career stage must learn, or relearn, to teach math differently and expand what count as justifications and generalizations. They must deepen their own understanding of the math they teach (Ball et al., 2008) to better understand their students' mathematical perspectives and positions on a math learning trajectory (Boston et al., 2018; Smith et al., 2018), and leverage their cultural awareness and sensitivities (Aguirre, 2009; Ladson-Billings, 1995; Leonard et al., 2010) in ways that strengthen their capacities to be intellectually responsive to diverse learners.

Two of the most critical and impactful design decisions we made were (1) to have the CLS guide the dyad's collaborative work to plan, lead, and reflect on lessons together over time, and (2) to focus the dyad's work on co-noticing aspects of their instruction and students' experiences in the classroom in order to collaboratively make instructional decisions. With respect to our first design decision, we are designing the CLS to support the dyad's collaborative lesson planning, enactment, and reflection since these are teaching activities that dyads commonly do together in the field experience and are times when teachers can have focused conversations about aspects of their instruction and classroom. Given that continually becoming an equity-oriented educator is an ongoing process, we frame these teaching activities (lesson planning, enactment, reflection) as interconnected so that they provide opportunities for the teachers to continue making sense of equity-related aspects of their instruction or classroom during all phases of lesson development and enactment over the course of a school day. We intend to have the dyads use the CLS repeatedly across a semester or term to deepen their co-learning across time, so we are developing a structure that provides dyads with *routine* ways of learning together. As demonstrated in Figure 18.1, dyads use the same CLS multiple times to support their investigations into a range of goals. In this way, the dyad's collaborative, equity-focused work is not treated as an isolated event but rather sustained through their ongoing classroom co-inquiry.

With respect to our second design decision, we are designing the CLS to support dyads in *co-noticing* aspects of their instruction and students' experiences in the classroom in order to collaboratively make instructional decisions. Broadly, our intention is to engage dyads in noticing salient instructional moments which advance or constrain equity in order to collaboratively generate and experiment with more equitable practices. A rich body of research in math education suggests that teacher noticing is a central dimension of teaching expertise and has important implications for how teachers attend to students' mathematical reasoning and sensemaking (e.g., van Es & Sherin, 2008) and equity in their classrooms (e.g., van Es et al., 2017). Our conjecture is that supporting dyads in *co*-noticing relevant aspects of instruction easily applies to a range of enduring questions of equity-oriented instruction and, in turn, might support *co*-learning. Co-noticing patterns of student participation, for example, might support teachers in brainstorming ways to elicit ideas from a broader range of students (in particular Black, Latinx, and Indigenous students who often have their brilliance and resources underutilized in mathematics classrooms), or to be more mindful of how assumptions about ability inform which students are asked the richest questions. We also believe that noticing is a necessary precursor to disrupting aspects of

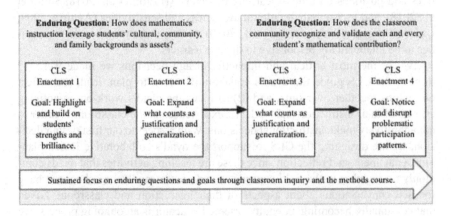

FIGURE 18.1.   Collaborative Learning Structure Use Across Time

teachers' instructional practice which may perpetuate inequities unbeknownst to the teacher. We hope that developing noticing through the lens of equity-oriented instruction will point teachers to elements of their practice that might otherwise go *unnoticed* and inspire them to collaboratively respond to these issues in new ways. Further, when co-noticing, teacher candidates and mentor teachers have the opportunity to share their individual perspective about salient moments of instruction and draw on each teacher's assets when collectively interpreting those moments and determining next steps. Thus, in contrast to collaborative work that elicits and privileges one perspective or interpretation of what is happening in the classroom, co-noticing invites multiple perspectives and interpretations and presses each teacher to understand in new ways how opportunities for equity are advanced and constrained through a lesson.

The CLS currently includes the following three parts: (1) a protocol for planning to co-notice during the lesson, (2) a protocol for co-noticing during the lesson, and (3) a protocol for co-debriefing the lesson. Part 1 is a protocol for planning to co-notice during the lesson. While using this planning to co-notice protocol, the teacher leading instruction first summarizes the upcoming lesson. Next, the dyad selects co-learning goals to focus on during their lesson planning-enactment-debriefing cycle and highlights places in the lesson they feel curious or uncertain about related to their co-learning goal. The goals we have for co-learning, indicated by our design principles, are critical components of equity-oriented math instruction, and are accessible and relevant to teachers in a variety of contexts and stages of their teaching career. The dyad chooses an enduring question related to recognizing, understanding, and disrupting inequitable participation patterns (e.g., *"Who is talking/expressing their ideas and what are they talking about/expressing?," "Who is not talking or expressing ideas? Why might that be?," "How might we create more equitable opportunities for engaging specific students deeply in the content?"*), honoring and making sense of students' diverse ideas (e.g., *"What ideas and partial understandings are students offering? In what ways are these ideas mathematically rich and relevant?," "What might 'buds' or beginnings of justifications and generalizations sound like given the content we are working on? How can we leverage these 'buds'?"*), and creating opportunities for learning collectively (e.g., *"How are students oriented to the brilliance and ideas of their peers?," "How are students learning from one another?," "Who is making sense of other people's ideas and how do I know?"*). Finally, the dyad decides what each teacher will notice/pay attention to during the lesson related to the curiosity/uncertainty shared and when they will pause and "meet up" during the lesson to make sense of and quietly discuss what they are each noticing and collaboratively make instructional decisions. By prompting the dyads to articulate their curiosities and uncertainties and identify moments to confer with each other during the lesson related to their co-learning goal, the CLS is designed to have both teachers' contributions be grounded in practice and, based on shared classroom experiences, interrogate problematic aspects of their understanding and enactment of equity-oriented instruction, and to be humble and vulnerable as they each take turns making their own instruction open to collective inquiry.

Part 2 is a protocol for co-noticing during the lesson. While using this co-noticing protocol, the teachers each record information during the lesson about what they are noticing related to their co-learning goal and then "meet up" with their colleague at the planned moments during the lesson to share what they have noticed as well as plan how to respond and proceed in the lesson based on their goals and what they have noticed. During each "meet up," each teacher shares aspects of the lesson they found interesting or confusing, an idea or question they have about the lesson or students' contributions, and what they think they should do next in the lesson (e.g., how to select, sequence and connect student solutions (Stein et al., 2008), how selected students will share their work, how they will start or lead a whole class discussion about the selected student work, how they can prompt reluctant students to participate or engage more in the lesson). By prompting both teachers to share what they notice, the CLS is designed to have the teachers adopt a humanizing and assets-based perspective by eliciting and valuing what each teacher has noticed during the lesson related to their goal. By prompting the teachers to strategize together what to do next, the CLS is designed to have the teachers formulate solutions to their curiosity or uncertainty with guidance and input from their colleague.

Part 3 is a protocol for co-debriefing the lesson. While using this co-debriefing protocol, the teachers reflect on what they noticed with their colleague (*"What did you/your partner notice that surprised you?," "Is there anything else you noticed that you'd like to share?"*), discuss how a decision they made together went and relates to their co-learning goal (*"Think together about a time when you and your partner made a decision about what to do next with students? How did it go? What might we try differently next time?," "How do you think the instructional decisions you made together related to your focus?"*), and summarize their learning and plan for future co-learning opportunities (*"What new ideas, understandings or questions do you have related to your focus today?," "What stands out to you as most important about today?," "What questions about teaching do you have now?," "How did it feel to try this protocol with your colleague?," "When would it be helpful to try this again? What might we want to focus on or try out?"*). By prompting the dyad to reflect on the what they noticed with their colleague and discuss how a decision they made went and relates to their co-learning goal, the CLS is designed to have both teachers adopt a humanizing and assets-based perspective by drawing on and valuing what each teacher noticed during the lesson, and be humble and vulnerable as they consider how they are perpetuating and can potentially interrupt inequities in their classroom. By prompting the teachers to summarize their learning and plan for future co-learning opportunities, the CLS is designed to have both teachers identify problematic aspects of their own understandings and enactments of equity-oriented instruction that they can interrogate together in the future. By prompting the teachers to summarize their learning, the CLS is designed to have the teachers adopt a humanizing and assets-based perspective by having each teacher reflect on and articulate how they valued and validated what their colleague shared during their experience collaborating. By prompting the teachers to plan for future co-learning opportunities, the CLS is designed to have the teachers identify how they can collaboratively interrogate and formulate possible next steps to face and act upon the ongoing challenges of enacting equity-oriented instruction.

During the 2022–23 school year, we are piloting the CLS in three programs that license mathematics teachers—two programs license secondary teachers and one licenses elementary teachers. During the summer of 2023, we will revise the CLS and continue to develop our theory of teacher co-learning based on our analysis of the data collected during the 2022–23 school year. During the 2023–24 school year, we will continue to study how the CLS is used in these three programs as well as how it is used in six additional teacher education programs. All of the teacher education programs vary in terms of teacher population, program size, and location, as well as length and structure of field placement. We intend to iteratively revise the CLS throughout the duration of the project in light of what the users do and what and how they learn.

AN EXAMPLE OF THE COLLABORATIVE LEARNING STRUCTURE IN USE

Since we are still in the midst of collecting and analyzing data, we provide a brief example of how we envision teachers using the co-debriefing protocol (part 3 of the CLS). The co-debriefing protocol, shown in Figure 18.2, includes three sections as well as prompts for each section. While we see great potential in this protocol with respect to supporting co-learning equity-oriented instruction, we are actively grappling with a number of tensions as we consider our design principles and the current context of pre-service teacher education.

In order to support the teachers in using a humanizing and assets-based perspective, we designed the "co-debriefing" protocol in ways that challenge the "observation debrief" dynamic typical in field experiences which tends to privilege the knowledge and perspective of the mentor teacher. In this protocol, we try to orient teachers to this shift in a number of ways. First, while we understand that a debrief nearly always involves more than one party, we refer to it as a *co*-debrief to signal the meaningful intellectual participation of both teachers. As such, the prompts in the co-debrief protocol elicit observations, ideas and questions from both teachers. Additionally, in an effort to further challenge standard power dynamics we refer to teachers as "partners" and "colleagues" throughout the protocol. As the teachers move through the protocol and reflect on the lesson, we offer non-evaluative language in the prompts and refrain from using tools which orient either teacher to instructional "proficiency." Instead, we ask teachers what "surprises them" or "stands out" about their experiences co-noticing, hoping to elicit ideas from both teachers grounded in genuine curiosity.

Part 3: Co-Debriefing
After the lesson, share some reflections about your noticings with your colleague.
• *What did you/your partner notice that surprised you?*
• *Is there anything else you noticed that you'd like to share?*
• *(If applicable) What did we notice that was different/similar from last time?*
Identify a particular moment when you responded to what you noticed.
• *Think together about a time when you and your partner made a decision about what to do next with students? How did it go? What might we try differently next time?*
• *How do you think the instructional decisions you made together related to your focus:*
○ Pressing on existing participation patterns
○ Honoring and making sense of students' diverse ideas
○ Creating opportunities for learning collectively
Summarize learning and plan for future co-learning
• *What new ideas, understandings or questions do you have related to your focus today?*
• *What stands out to you as most important about today?*
• *What questions about teaching do you have now?*
• *How did it feel to try this protocol with your colleague?*
• *When would it be helpful to try this again? What might we want to focus on or try out?*

FIGURE 18.2.   The Co-Debriefing Protocol (Part 4) of the Collaborative Learning Structure

Despite these design decisions, we know that the power dynamics which suggest mentors should be "experts" rather than learners, and that teacher candidates should be sponges rather than resources, are deeply entrenched in the practice and culture of field experiences. We wonder if teacher candidates and mentor teachers might make their way through this protocol without actually participating in more egalitarian ways, and worry the shifts here might remain at the level of language (despite our efforts). We are also aware that the relationship between teacher candidates and mentor teachers is *not* entirely egalitarian. Mentor teachers *do* have more teaching experience and different kinds of expertise than teacher candidates. As we continue to use this CLS with teachers, we better understand how to design features which support each teacher in being a learner and learning from the other, while also acknowledging and leveraging the particular practical knowledge of mentor teachers.

An additional tension arises as we work to design for principles four and five. Design principle four calls us to design a tool that supports teachers in "interrogating problematic aspects of teachers' understanding and enactment of equity-oriented instruction…" and principle five states that tools to support co-learning should "support teachers in being humble and vulnerable as they make their own instruction open to collective inquiry." In the second section, when discussing how a decision they made together went and relates to their co-learning goal, we try to support teachers in understanding their instructional practice as *shared* asking, "*Think together about a time when you and your partner made a decision about what to do next with students? How did it go? What might we try differently next time?" and "How do you think the instructional decisions you made together related to your focus?"* While only one teacher teaches at a time, we hope to diffuse the stress of "performing" instruction for one another by suggesting that the instructional decisions made belong to *both* teachers. We hypothesize that routines which support teachers in *sharing* practice and sharing the implications of instructional decisions will reduce the likelihood of defensiveness and evaluation and therefore support a more vulnerable collaborative interrogation into inequitable practices. Again, we see this as in direct tension with the norms of the teaching profession which suggest that teaching is individual and private work (Lortie, 2002), and that the purpose of being in the classroom together is to either evaluate or observe in an effort to model after. We are deeply curious about what kinds of features might help this CLS challenge these norms. In particular, we look forward to learning more about how the tool can support preservice teachers in sharing authentic observations and ideas in response to a mentor teacher's teaching.

CONCLUSION AND DISCUSSION

This chapter offers provisional routines and a tool aimed at supporting teacher candidates and mentor teachers to learn together in becoming equity-oriented mathematics educators. We have offered a protocol for establishing a shared vision for equity-oriented instruction *and* norms for social routines and interactions among teachers; focusing on classroom practice, including roles and responsibilities of the co-learners. As a part of our work, we have identified several tensions and cautious points that remain unresolved. First, it is important to note that we are currently studying the use of our tool in three teacher education programs, so this chapter offers an illustration of a tool developed from the design principles we generated, not one that has completed the process of rigorous field testing. Second, we wonder about the balance between offering a very structured tool with detailed narrative, including some aspects that are scripted, versus a more open-ended tool and consider what each version affords and what is constrained by the structure of the tool. For example, where do teachers articulate what they notice versus where do we (teacher educators) tell them what and how to notice? Should we name things for the teachers to look for or see what comes up when noticing on their own? We anticipate that many of the teacher candidates and mentor teachers may have a fragile understanding of equity-oriented mathematics teaching practices and explicitness is an important feature for supporting their noticing and learning about equitable teaching practices, but we continue to wonder about the appropriate balance of levels of explicitness offered in the CLS. Additionally, we wonder about the process of using the CLS tools. Our current model is for co-learners to work on one goal for a period of time (e.g., 4–6 weeks) allowing them to become deeply engaged in the practices for a particular CLS. It remains an open question of whether they should work on one goal and dig deeper over time or consider a series of different foci and visit and revisit all goals over time.

Next Steps and Implications

Our next steps include iteratively redesigning the CLSs we have created to date as well as continuing to develop our theory of how co-learning occurs between teacher candidates and mentor teachers. With respect to iteratively redesigning the CLSs we have created to date, we will finish collecting data on the use of three CLSs this year (CLS 1 was used in all three teacher education programs during the 2022–23 school year, CLSs 2 and 3 were used in one teacher education program during the 2022–23 school year). We will spend summer 2023 analyzing our data and revising each CLS based on our findings. During summer 2023, we will also recruit and work collaboratively with six additional teacher education programs to use the CLSs during the 2023–24 school year. During the 2023–24 school year, we will continue to study how the CLSs are used in the three pilot programs from the 2022–23 school year, as well as how they are used in the six additional new teacher education programs. In summer 2024 we will analyze the data collected during the 2023–24 school year and revise the CLSs based on our findings, producing a final version of each CLS, including any new CLSs that we create in the coming two years based on the needs of the teacher education programs participating in this study. We additionally anticipate that this design cycle can serve as a model for other design study projects in the future.

As part of this ongoing design cycle, we are also continuing to develop and refine our theory of co-learning. Through studying the teachers' use of our CLSs, we are adding to and refining our initial ideas about the learning process that the teacher candidates and mentor teachers individually and collectively engage in and experience while using each CLSs as well as identify the critical features of the teachers' collaborative work that support their co-learning. Our study of how dyads use our CLSs includes an investigation of different productive entry points to engage teachers in continually becoming equity-oriented mathematics educators as well as considering how each co-learner engages and takes up their own role (i.e., Do teacher candidates feel comfortable collaborating when debriefing and/or do mentor teachers engage in looking for assets and drawing on the strengths of their assigned teacher candidate?). We are also studying what shifts each teacher makes and which CLS activities prompt more substantial shifts than others and what this implies for understanding a variety of trajectories for supporting teachers in becoming equity-oriented mathematics educators. The findings of our study will serve as the basis for and contribute to our theory of co-learning.

Finally, we hope that our work contributes new framing and tools for reimagining a key component of teacher education, including initial teacher preparation and professional development for practicing teachers who serve as mentor teachers. In particular, we offer a reimagining of the field experience with a dual focus: (1) explicit focus on developing and supporting equity-oriented educators, and (2) using a collaborative learning structure that supports meaningful collaborative routines between teacher candidates and mentor teachers. This (re)imagining posits how the field experience can be a humanizing, assets-focused, and empowering experience for teacher candidates, mentor teachers, and students. The work is designed to lead with equity because of its importance across the work of this project and we expect that it will be useful for co-learners to leverage knowledge of issues of equity into their learning about justification and generalization. Further, we anticipate that teacher candidates and mentor teachers might have more learning needs around issues of equity, race/racism, and justice than they do for justification and generalization. In providing this framing and sharing one of our tools, we hope to provide other mathematics teacher educators with strategies for engaging and supporting teachers across the continuum of professional learning in developing equity-oriented instruction.

ACKNOWLEDGEMENTS

This material is based upon work supported by the National Science Foundation under Grant No. DRL-2010634. Any opinions, findings, and conclusions or recommendations expressed in this material are those of the author(s) and do necessarily reflect the views of the National Science Foundation.

REFERENCES

Aguirre, J. (2009). Privileging mathematics and equity in teacher education: Framework, counter-resistance strategies and reflections from a Latina mathematics educator. In B. Greer, S. Mukhopadhyay, S. Nelson-Barber, & A. Powell (Eds.), *Culturally responsive mathematics education* (pp. 295–319). Routledge.

Aguirre, J. M., Mayfield-Ingram, K., & Martin, D. B. (2013). *The impact of identity in K–8 mathematics teaching: Rethinking equity-based practices.* National Council of Teachers of Mathematics.

Aguirre, J. M., Turner, E. E., Bartell, T. G., Kalinec-Craig, C., Foote, M. Q., Roth McDuffie, A., & Drake, C. (2013). Making connections in practice: How prospective elementary teachers connect to children's mathematical thinking and community funds of knowledge in mathematics instruction. *Journal of Teacher Education, 64*(2), 178–192. https://doi.org/10.1177/0022487112466900

Association of Mathematics Teacher Educators. (2017). *Standards for preparing teachers of mathematics.* https://amte.net/sptm

Baker, C., & Knapp, M. (2019). The decision-making protocol for mathematics coaching: Addressing the complexity of coaching with intentionality and reflection. *Mathematics Teacher Educator, 7*(2), 27–43. https://doi.org/10.5951/mathteaceduc.7.2.0027

Ball, D. L., & Cohen, D. K. (1999). Developing practice, developing practitioners: Toward a practice-based theory of professional education. In G. Sykes & L. Darling-Hammond (Eds.), *Teaching as the learning profession: Handbook of policy and practice* (pp. 3–32). Jossey Bass.

Ball, D. L., Thames, M. H., & Phelps, G. (2008). Content knowledge for teaching: What makes it special. *Journal of Teacher Education, 59*(5), 389–407. https://doi.org/10.1177/0022487108324554

Bartell, T. G. (2013). Learning to teach mathematics for social justice: Negotiating social justice and mathematical goals. *Journal for Research in Mathematics Education, 44*(1), 129–163. https://doi.org/10.5951/jresematheduc.44.1.0129

Bartell, T., Wager, A., Edwards, A., Battey, D., Foote, M., & Spencer, J. (2017). Toward a framework for research linking equitable teaching with the standards for mathematical practice. *Journal for Research in Mathematics Education, 48*(1), 7–21. https://doi.org/10.5951/jresemathclc.48.1.0007

Berry, R. Q., III, Ellis, M., & Hughes, S. (2014). Examining a history of failed reforms and recent stories of success: Mathematics education and Black learners of mathematics in the United States. *Race Ethnicity and Education, 17*(4), 540–568. https://doi.org/10.1080/13613324.2013.818534

Bieda, K. N., & Staples, M. (2020). Justification as an equity practice. *Mathematics Teacher: Learning and Teaching PK–12, 113*(2), 102–108. https://doi.org/10.5951/MTLT.2019.0148

Blanton, M. L., & Kaput, J. J. (2005). Characterizing a classroom practice that promotes algebraic reasoning. *Journal for Research in Mathematics Education, 36*(5), 412–446. https://www.jstor.org/stable/30034944

Borko, H., & Mayfield, V. (1995). The roles of the cooperating teacher and university supervisor in learning to teach. *Teaching & Teacher Education, 11*(5), 501–518. https://doi.org/10.1016/0742-051X(95)00008-8

Boston, M., Dillon, F., Smith, M., & Miller, S. (2018). *Taking action: Implementing effective mathematics teaching practices in grades 9–12.* National Council of Teachers of Mathematics.

Brooks, V., & Sikes, P. (1997). *The good mentor guide: Initial teacher education in secondary schools.* Open University Press.

Bullough Jr., R. V., & Draper, R. J. (2004). Making sense of a failed triad: Mentors, university supervisors, and positioning theory. *Journal of Teacher Education, 55*(5), 407–420. https://doi.org/10.1177/0022487104269804

Cameron, D. H. (2005). Teachers working in collaborative structures: A case study of a secondary school in the USA. *Educational Management Administration & Leadership, 33*(3), 311–330. https://doi.org/10.1177/1741143205054012

Celedon-Pattichis, S., Borden, L. L., Pape, S. J., Clements, D. H., Peters, S. A., Males, J. R., Chapman, O., & Leonard, J. (2018). Asset-based approaches to equitable mathematics education research and practice. *Journal for Research in Mathematics Education, 49*(4), 373–389. https://doi.org/10.5951/jresemathclc.49.4.0373

Civil, M. (1994, April 4–8). *Connecting the home and school: Funds of knowledge for mathematics teaching and learning. Draft* [Paper presentation]. Annual Meeting of the American Educational Research Association, New Orleans, LA. ERIC database. (ED37098)

Civil, M. (2007). Building on community knowledge: An avenue to equity in mathematics education. In N. S. Nassir & P. Cobb (Eds.), *Improving access to mathematics: Diversity and equity in the classroom* (pp. 105–117). Teachers College Press.

Cobb, P., Jackson, K., & Dunlap, C. (2017). Conducting design studies to investigate and support mathematics students' and teachers' learning. In J. Cai (Ed.), *Compendium for research in mathematics education* (pp. 208–236). National Council of Teachers of Mathematics.

Cochran-Smith, M., Feiman-Nemser, S., McIntyre, J., & Demers, K. (Eds.). (2008). *Handbook of research on teacher education: Enduring questions in changing contexts.* Routledge.

Cochran-Smith, M., & Lytle, S. L. (2009). *Inquiry as stance: Practitioner research for the next generation.* Teachers College Press.

Collins, A., Brown, J. S., & Holum, A. (1991). Cognitive apprenticeship: Making thinking visible. *American Educator, 15*(3), 6–11, 38–46.

Darling-Hammond, L. (1999). *Teacher quality and student achievement: A review of state policy evidence.* Center for Teaching and Policy, University of Washington.

Darling-Hammond, L. (2006). Constructing 21st-Century teacher education. *Journal of Teacher Education, 57*(3), 300–314. https://doi.org/10.1177/0022487105285962

Dewey, J. (1965/1904). The relation of theory to practice in education. In M. L. Borrowman (Ed.), *Teacher education in America: A documentary history* (pp. 140–171). Teachers College Press.

Drake, C., Aguirre, J. M., Bartell, T. G., Foote, M. Q., Roth McDuffie, A., & Turner, E. E. (2015). TeachMath learning modules for K–8 mathematics methods courses. *Teachers Empowered to Advance Change in Mathematics Project.* www.teachmath.info

Featherstone, H., Crespo, S., Jilk, L. M., Oslund, J. A., Parks, A. N., & Wood, M. B. (2011). *Smarter together! Collaboration and equity in the elementary math classroom.* National Council of Teachers of Mathematics.

Feiman-Nemser, S. (2012). *Teachers as learners.* Harvard University Press.

Feiman-Nemser, S., & Beasley, K. (2007). Discovering and sharing knowledge: Inventing a new role for cooperating teachers. In D. M. Carroll (Ed.), *Transforming teacher education: Reflections from the field* (pp. 139–160). Harvard Education Press.

Fernandez, M. L., & Erbilgin, E. (2009). Examining the supervision of mathematics student teachers through analysis of conference communications. *Educational Studies in Mathematics, 72*(1), 93–110. https://www.jstor.org/stable/40284610

Furlong, J., & Maynard, T. (1995). *Mentoring student teachers: The growth of professional knowledge.* Routledge.

Gay, G. (2000). *Culturally responsive teaching: Theory, research, and practice.* Teachers College Press.

Gibbons, L. K., Kazemi, E., & Lewis, R. M. (2017). Developing collective capacity to improve mathematics instruction: Coaching as a lever for school-wide improvement. *Journal of Mathematical Behavior, 46*(201500018), 231–250. https://doi.org/10.1016/j.jmathb.2016.12.002

Gutiérrez, R. (2009). Framing equity: Helping students "play the game" and "change the game." *Teaching for Excellence and Equity in Mathematics, 1*(1), 4–8.

Gutiérrez, R. (2013). Why (urban) mathematics teachers need political knowledge. *Journal of Urban Mathematics Education, 6*(2), 7–19. https://doi.org/10.21423/jume-v6i2a223

Gutiérrez, R. (2018). Rehumanizing mathematics. In I. M. Goffney, R. Gutiérrez, & M. Boston (Eds.), *Rehumanizing mathematics for Black, Indigenous, and Latinx students* (Annual Perspectives in Mathematics Education; Vol. 2018). National Council of Teachers of Mathematics.

Guyton, E., & McIntyre, D. J. (1990). Student teaching and school experiences. In W. R. Houston (Ed.), *Handbook of research on teacher education* (pp. 514–534). Macmillan.

Hammerness, K., Darling-Hammond, L., Bransford, J., Berliner, D., Cochran-Smith, M., McDonald, M., & Zeichner, K. (2005). How teachers learn and develop. In L. Darling Hammond & J. Bransford (Eds.), *Preparing teachers for a changing world: What teachers should learn and be able to do* (pp. 358–388). Jossey-Bass.

Henfield, M. S., & Washington, A. R. (2012). "I want to do the right thing but what is it?": White teachers' experiences with African American students. *Journal of Negro Education, 81*(2), 148–161. https://doi.org/10.7709/jnegroeducation.81.2.0148

İnan-Kaya, G., & Rubie-Davies, C. M. (2021). Teacher classroom interactions and behaviours: Indications of bias. *Learning and Instruction, 78*, 1–13. https://doi.org/10.1016/j.learninstruc.2021.101516

Kazemi, E., Franke, M., & Lampert, M. (2009). Developing pedagogies in teacher education to support novice teachers' ability to enact ambitious instruction. In R. Hunter, B. Bicknell, & T. Burgess (Eds.), *Crossing divides: Proceedings of the 32nd annual conference of the Mathematics Education Research Group of Australasia* (Vol. 1, pp. 12–30).

Kirshner, B., & Polman, J. L. (2013). Adaptation by design: A context-sensitive, dialogic approach to interventions. *Teachers College Record, 115*(4), (pp. 215–236). https://doi.org/10.1177/016146811311501

Ladson-Billings, G. (1995). Toward a theory of culturally relevant pedagogy. *American Educational Research Journal, 32*(3), 465–491. https://doi.org/10.3102/00028312032003465

Ladson-Billings, G. (1997). It doesn't add up: African American students' mathematics achievement. *Journal for Research in Mathematics Education, 28*(6), 697–708. https://doi.org/10.2307/749638

Ladson-Billings, G. (2014). Culturally relevant pedagogy 2.0: A. K. A. the remix. *Harvard Educational Review, 84*(1), 74–84. https://doi.org/10.17763/haer.84.1.p2rj131485484751

Lampert, M. (1985). How do teachers manage to teach? Perspectives on problems in practice. *Harvard Educational Review, 55*(2), 178–195. https://doi.org/10.17763/haer.55.2.56142234616x4352

Lampert, M. (2010). Learning teaching in, from, and for practice: What do we mean? *Journal of Teacher Education, 61*(1–2), 21–34. https://doi.org/10.1177/0022487109347321

Leonard, J., Brooks, W., Barnes-Johnson, J., & Berry III, R. Q. (2010). The nuances and complexities of teaching mathematics for cultural relevance and social justice. *Journal of Teacher Education, 61*(3), 261–270. http://dx.doi.org/10.1177/0022487109359927

Leonard, J., & Martin, D. B. (Eds.). (2013). *The brilliance of Black children in mathematics*. IAP.

Little, J. W. (2002). Locating learning in teachers' professional community: Opening up problems of analysis in records of everyday work. *Teaching and Teacher Education, 18*(8), 917–946. https://doi.org/10.1016/S0742-051X(02)00052-5

Little, J. W., & Horn, I. S. (2007). "Normalizing" problems of practice: Converting routine conversation into a resource for learning in professional communities. In L. Stoll & K. S. Louis (Eds.), *Professional learning communities: Divergence, depth and dilemmas* (pp. 79–92). Open University Press.

Lortie, D. (2002). *Schoolteacher: A sociological study* (2nd ed.). University of Chicago Press.

Males, L. M., Otten, S., & Herbel-Eisenmann, B. A. (2010). Challenges of critical colleagueship: Examining and reflecting on mathematics teacher study group interactions. *Journal of Mathematics Teacher Education, 13*(6), 459–471. https://doi.org/10.1007/s10857-010-9156-6

Martin, D. B. (2019). Equity, inclusion, and antiblackness in mathematics education. *Race Ethnicity and Education, 22*(4), 459–478. https://doi.org/10.1080/13613324.2019.1592833

Mintos, A., Hoffman, A. J., Kersey, E., Newton, J., & Smith, D. (2019). Learning about issues of equity in secondary mathematics teacher education programs. *Journal of Mathematics Teacher Education, 22*(5), 433–458. https://doi.org/10.1007/s10857-018-9398-2

Moll, L. C., Amanti, C., Neff, D., & Gonzalez, N. (1992). Funds of knowledge for teaching: Using a qualitative approach to connect homes and classrooms. *Theory Into Practice, 31*(2), 132–141. https://doi.org/10.1080/00405849209543534

National Center for Education Statistics. (2001). Federal programs for education and related activities. In *Digest of education statistics, 2000* (pp. 393–426). https://nces.ed.gov/pubs2001/2001034.pdf

National Center for Educational Statistics. (2021, May). *Characteristics of public school teachers. Condition of Education.* U.S. Department of Education, Institute of Education Sciences. https://nces.ed.gov/programs/coe/indicator/clr

National Council of Supervisors of Mathematics (NCSM) & TODOS: Mathematics for ALL. (2016). *Mathematics education through the lens of social justice: Acknowledgement, actions, and accountability.*

National Council of Teachers of Mathematics. (1989). *National curriculum and evaluation standards.* Author.

National Council of Teachers of Mathematics. (2000). *Principles and standards for school mathematics.* Author.

National Council of Teachers of Mathematics. (2014). *Principles to action.* Author.

National Council of Teachers of Mathematics. (2018). *Catalyzing change in high school mathematics: Initiating critical conversations.* Author.

National Governors Association Center for Best Practices & Council of Chief State School Officers. (2010). *Common core state standards for mathematics.* Author.

National Governors Association Center for Best Practices & Council of Chief State School Officers. (2014). *Common core state standards for mathematics.* Author.

Paris, D. (2012). Culturally sustaining pedagogy: A needed change in stance, terminology, and practice. *Educational Researcher, 41*(3), 93–97. https://doi.org/10.3102/0013189X12441244

Paris, D. (2016). *On educating culturally sustaining teachers.* Teaching Works, University of Michigan.

Sinicrope, R., Eppler, M., Preston, R. V., & Ironsmith, M. (2015). Preservice teachers of high school mathematics: Success, failure, and persistence in the face of mathematical challenges. *School Science and Mathematics, 115*(2), 56–65. https://doi.org/10.1111/ssm.12104

Sleeter, C. (2008). Preparing White teachers for diverse students. In M. Cochran-Smith, S. Feiman-Nemser, D. J. McIntyre, & K. E. Demers (Eds.), *Handbook of research on teacher education: Enduring questions in changing contexts* (3rd ed., pp. 559–582). Routledge.

Smith, M., Steele, M., & Raith, M. L. (2018). *Taking action: Implementing effective mathematics teaching practices in grades 6–8*. National Council of Teachers of Mathematics.

Stein, M. K., Engle, R. A., Smith, M. S., & Hughes, E. K. (2008). Orchestrating productive mathematical discussions: Five practices for helping teachers move beyond show and tell. *Mathematical Thinking and Learning, 10*(4), 313–340. https://doi.org/10.1080/10986060802229675

Stephans, A., Ellis, A., Blanton, M., & Brizuela, B. (2018). Algebraic thinking in the elementary and middle grades. In J. Cai (Ed.), *Compendium for research in mathematics instruction* (pp. 386–420). National Council of Teachers of Mathematics.

Stoltz, A., Goffney, I., Ivy, K., Buli, T., & Shockley, E. T. (2021). *Teacher candidates' implementation of equitable mathematics teaching practices: An examination of divergent paths*. North American Chapter of the International Group for the Psychology of Mathematics Education Conference, Mazatlan, Sinaloa, Mexico.

Stylianides, G. J. (2008). An analytic framework of reasoning and proving. *For the Learning of Mathematics, 28*(1), 9–16. https://www.jstor.org/stable/40248592

Suh, J. M., Birkhead, S., Frank, T., Baker, C., Galanti, T., & Seshaiyer, P. (2021). Developing an asset-based view of students' mathematical competencies through learning trajectory-based lesson study. *Mathematics Teacher Educator, 9*(3), 229–245. https://doi.org/10.5951/MTE.2020.0033

Thompson, J., Hagenah, S., Lohwasser, K., & Laxton, K. (2015). Problems without ceilings: How mentors and novices frame and work on problems-of-practice. *Journal of Teacher Education, 66*(4), 363–381. https://doi.org/10.1177/0022487115592462

Turner, E. E., & Drake, C. (2015). A review of research on prospective teachers' learning about children's mathematical thinking and cultural funds of knowledge. *Journal of Teacher Education, 67*(1), 32–46. https://doi.org/10.1177/0022487115597476

Ukpokodu, O. N. (2011). How do I teach mathematics in a culturally responsive way?: Identifying empowering teaching practices. *Multicultural Education, 18*(3), 47–56.

Valencia, S., Martin, S., Place, N., & Grossman, P. (2009). Complex interactions in student teaching: Lost opportunities for learning. *Journal of Teacher Education, 60*(3), 304–322. https://doi.org/10.1177/0022487109336543

van Es, E. A., Hand, V., & Mercado, J. (2017). Making visible the relationship between teachers' noticing for equity and equitable teaching practice. In E. O. Schack, M. H. Fisher & J. A. Wilhelm (Eds.), *Teacher noticing: Bridging and broadening perspectives, contexts, and frameworks*. Springer International Publishing.

van Es, E. A., & Sherin, M. G. (2008). Mathematics teachers' "learning to notice" in the context of a video club. *Teaching and Teacher Education, 24*(2), 244–276. https://doi.org/10.1016/j.tate.2006.11.005

Wager, A. A. (2012). Incorporating out-of-school mathematics: From cultural context to embedded practice. *Journal of Mathematics Teacher Education, 15*(1), 9–23. https://doi.org/10.1007/s10857-011-9199-3

Wager, A. A., & Foote, M. Q. (2013). Locating praxis for equity in mathematics: Lessons from and for professional development. *Journal of Teacher Education, 64*(1), 22–34. https://doi.org/10.1177/0022487112442549

Warren, C. A. (2018). Empathy, teacher dispositions, and preparation for culturally responsive pedagogy. *Journal of Teacher Education, 69*(2), 169–183. https://doi.org/10.1177/0022487117712487

Wenger, E. (1999). *Communities of practice: Learning, meaning and identity.* Cambridge University Press.

Yosso, T. J. (2005). Whose culture has capital? A critical race theory discussion of community cultural wealth. *Race Ethnicity and Education, 8*(1), 69–91. https://doi.org/10.1080/1361332052000341006

Zeichner, K. (1996). Designing educative practicum experiences for prospective teachers. In K. Zeichner, S. Melnick, & M. L. Gomez (Eds.), *Currents of reform in preservice teacher education* (pp. 215–234). Teachers College Press.

Zeichner, K. (2016). Advancing social justice and democracy in teacher education: Teacher preparation 1.0, 2.0, and 3.0. *Kappa Delta Pi Record, 52*(4), 150–155. https://doi.org/10.1080/00228958.2016.1223986

Zeichner, K., Payne, K. A., & Brayko, K. (2015). Democratizing Teacher Education. *Journal of Teacher Education, 66*(2), 122–135. https://doi.org/10.1177/0022487114560908

CHAPTER 19

CULTIVATING RURAL MATHEMATICS TEACHERS THROUGH PLACE-ATTENTIVE PREPARATION

Jennifer Luebeck
Montana State University

Jayne Downey
Montana State University

In this chapter we highlight a unique model of integrated mathematics teacher preparation and induction for a marginalized and often overlooked population: students and teachers in rural settings. The design of the Montana Rural Teacher Project leverages shared resources and remote learning to overcome barriers to recruitment and retention in rural schools and employs strategies that can potentially be replicated in other educational settings. We discuss the essential elements of responsiveness, communication, collaboration, and partnership and how they contribute to supporting mathematics teacher preparation and induction in complex and challenging contexts. Throughout the chapter, we raise questions about what is non-negotiable in mathematics teacher preparation and induction and explore the boundaries (both real and imagined) that separate the idealistic from the realistic.

The AMTE Handbook of Mathematics Teacher Education: Reflection on Past, Present and Future—Paving the Way for the Future of Mathematics Teacher Education, Volume 5
pages 391–406.

INTRODUCTION

The theme of "Past, Present, and Future" aptly characterizes the current status of mathematics teacher preparation at Montana State University (MSU). Over the past decade mathematics education faculty have updated undergraduate programs in close alignment to recommendations from AMTE's *Standards for Preparing Teachers of Mathematics* (SPTM) (2017) and *The Mathematical Education of Teachers II* (Conference Board of the Mathematical Sciences, 2012). These updates represent creating and revising both courses and signature activities to provide opportunities for:

- Strong content knowledge development (Indicator P.2.1) through a 9-credit elementary sequence and a 4-course secondary content core.
- Engagement in mathematical practices and processes (Indicator P.2.2) through student-centered and technology-supported coursework for teachers.
- Practice-based experiences (Indicator P.3.4) through lesson study, task-based interviews, and analysis of discourse and student work. (AMTE, 2017)

Our efforts have resulted in a well-established and exemplary undergraduate program for secondary mathematics teaching, accompanied by deeply-ingrained perspectives regarding program fidelity. But at present, a pressing demand for mathematics teachers—especially in schools that have difficulty attracting and hiring teachers—has caused us to consider whether our traditional preparation pathway can adequately accommodate the critical need for teachers in our state. We are forced to acknowledge that in the future, providing mathematics teachers for all of Montana's students may call for modified routes to licensure that are better poised to meet urgent needs. At the same time we wonder, will introducing nontraditional teacher preparation pathways potentially undermine our existing programs?

Our concerns are undoubtedly echoed by mathematics teacher educators across the country as they struggle to address severe teacher shortages in a variety of social and economic contexts. For example, since 2012 the Mathematics Teacher Education Partnership, a nationwide network of mathematics teacher preparation programs funded by the Association of Public Land-Grant Universities (APLU), has explored responsive solutions to concerns regarding coursework, clinical experiences, recruitment, and retention. However, these universal issues are less often viewed through a rural lens. We maintain that the rural setting adds a unique layer of geographical, cultural, and historical complexity to educational challenges, especially in terms of access to teacher preparation programs. Access is certainly a concern in mathematics, where both pedagogical priorities and small enrollments preclude offering online sections of the coursework required for licensure.

In light of our experience, as we explore the promise and potential of a uniquely derived route to rural mathematics teaching we will also attempt to illuminate the tensions and tradeoffs that accompany its development. These are represented in authentic episodes from our program and presented as vignettes throughout the narrative. The first three vignettes illustrate the diverse circumstances that surround nontraditional entry into mathematics teaching, while the remainder highlight pedagogical and logistical challenges. At the heart of each vignette is a fundamental dilemma that pits idealism against realism: *How do we balance the desire for a thoughtfully-crafted program that meets rigorous standards with the imperative to address the desperate need for teachers in rural mathematics classrooms, even if those teachers are less than ideally equipped?*

RURAL REALITIES

Montana encompasses over 145,000 square miles and serves roughly the same number of K–12 students. Our very low population density (less than 7 people per square mile) has earned the designation of "frontier" state (National Center for Frontier Communities, 2014) and has resulted in an education system of over 800 schools organized into roughly 400 public school districts. Ninety-six percent of these districts are classified as "small rural"; no state has a higher percentage of rural schools or small rural districts (Showalter et al., 2017). More than half of Montana schools have fewer than 100 students; 65 are considered "one room schools" and only 52 serve more than 500 students (Montana Office of Public Instruction, 2021). While these schools are the strength and social hub for their rural communities, keeping them staffed—especially at the secondary level–has become a challenge of crisis proportions. According to data from the 2017–18 school year, district administrators reported they had difficulty filling, or were unable to fill, 45% of all vacant positions. Rural/remote districts reported almost twice the percentage of difficult-to-fill and unfilled positions as non-rural districts. Mathematics positions were reported as one of the most difficult position types to fill along with music, science, special education, and career-technical education (Yoon et al., 2019).

A recent report found that 56 Montana school districts were without a fully qualified middle and/or high school mathematics teacher (Montana Office of Public Instruction, 2019). Furthermore, even as shortages continue to increase, fewer students are entering preparation programs. Completion numbers across all majors in Montana teacher preparation programs decreased roughly 15% between 2011 and 2017. Those students who do graduate with secondary mathematics credentials have multiple opportunities for employment across the education and business sectors, both within and outside the state. Few choose to submit applications in small and remote districts, further intensifying rural shortages. And research suggests that those who do begin their careers in a rural school may not stay long, especially if they do not have a personal connection to the community and lifestyle (Ulferts, 2016).

CULTIVATING PROSPECTIVE RURAL MATHEMATICS TEACHERS

These scenarios are not unique to Montana; concerns about teacher preparation, recruitment, and retention are paramount across the country. Overcoming recruitment and retention challenges is a nationwide problem shared by institutions of higher education (IHEs), school districts both urban and rural, and AMTE as well as other organizations. With colleagues in the Mathematics Teacher Education Partnership, we are "working to address the significant national shortage of well-prepared secondary mathematics teachers through a coordinated research, development, and implementation effort" (APLU, n.d.). Our faculty also share AMTE's long-term goal to "Provide resources and strategies for recruitment, retention, and diversification of the mathematics teacher pipeline" (AMTE, n.d.).

Our efforts to strengthen Montana's rural mathematics teaching force center around leveraging a potentially untapped resource within rural communities: their residents. What if we could identify and recruit rural citizens with substantial content knowledge and an interest in education, and then offer them a supportive and streamlined pathway to teaching in local schools? Eligible candidates might be teachers licensed for other subject areas and grade levels and seeking an added endorsement, or individuals holding a bachelor's degree in a viable content area but lacking education coursework and experience. Each of these scenarios call for targeted preparation to supplement prior knowledge and experience, as outlined in the first two vignettes.

Tensions and Tradeoffs: Content Knowledge

The lone 7–12 mathematics teacher at Ridgeview High School recently left for a new position in another state. After a fruitless recruitment effort, the superintendent reluctantly placed the students in an online mathematics tutorial program monitored by a 7th grade "teacher of record" and an aide. Gary, a longtime resident of Ridgeview (population <300) with a B.S. in geography expressed willingness to pursue licensure and step in as mathematics teacher. Years ago he completed three semesters of calculus, discrete mathematics, and a probability/statistics course, supplemented by three decades of military service as a pilot, mechanic, and accountant. Gary needed several more mathematics courses to meet formal program requirements, an educational investment he was not willing to make. The position remained empty.

In the scenario above, what additional coursework is non-negotiable? What should be considered "enough" mathematics? The *Mathematical Education of Teachers II* or MET II (CBMS, 2012) prescribes a minimum of a three-course calculus sequence, introductory linear algebra, and statistics with 18 additional semester hours of "advanced mathematics." The expectations stated in the SPTM (AMTE, 2017) are somewhat more open to interpretation:

> Well-prepared beginning teachers of mathematics have solid and flexible knowledge of core mathematical concepts and procedures they will teach, along with knowledge both beyond what they will teach and foundational to those core concepts and procedures.

In truth, there is no satisfying answer to the question "How much is enough?" Mathematics teacher educators must strike a delicate (and often gut-wrenching) balance between upholding the fidelity and integrity of program design and ensuring that every mathematics classroom has an engaged and caring teacher. In Montana's rural schools, waiting for an ideally-prepared teacher might mean having no teacher at all. Is it preferable for students to learn from a tutorial program, or a minimally prepared teacher with 30 years of life experience?

Tensions and Tradeoffs: Will vs. Skill

Birch Creek School (enrollment 56) was without a secondary mathematics teacher until Emma, a community member with an art education degree, volunteered to take on the role and work toward licensure. In her own words: "I decided on adding my math endorsement to my current Art teacher license because there was a need in my school. The math teacher had a baby and did not want to continue working. The district tried to hire a new one with no applicants …. I plan on teaching in [Birch Creek] for years to come" (personal correspondence, 8/27/21). There are others like Emma, citizens of our state who earnestly want to help–they have the will but lack the skill. Some of them don't understand why becoming a teacher is such a process. They are backed by some state policymakers, eager to expedite getting teachers into classrooms, who view teacher educators as "ivory tower" arbiters of quality.

Emma had years of teaching experience and deep connections to her community and students, but had to construct a foundation of mathematical knowledge for teaching from ground level while simultaneously teaching high school mathematics. In this scenario, MSU responded with extensive advising to approve provisional licensure and design a pathway to completion. Emma reached her goal within three years while teaching grades 7–12 mathematics full-time, thanks to a combination of foundation courses transferred from other institutions, pedagogy courses offered virtually during the pandemic, and upper-division substitutes from an online graduate program. She successfully checked all the boxes, but was there potentially a better option?

CANDIDATES, COURSEWORK, AND CAPACITY

Emma's story exemplifies the approach of "growing your own" rural mathematics teachers, an appealing solution that comes with its own set of challenges. Not surprisingly, the first challenge lies in identifying and recruiting teacher candidates. The second has to do with providing access to the coursework required to qualify for mathematics licensure. In Montana, applicants seeking a teaching license or added endorsement must secure a university recommendation verifying completion of an accredited teacher preparation program that includes student teaching and meets other qualifications. In other words, candidates must meet a given institution's program requirements (most often the equivalent of completing a teaching major or minor) through coursework or equivalent experience. While this approach has many strengths, the vast geographical distances in Montana make enrollment and physical attendance in a public or private institution's teacher preparation program a significant barrier for rural candidates. Many live hours from an IHE and can neither relocate nor abandon work and home commitments to attend face-to-face classes.

The third challenge, related to the first two, is concerned with designing accessible pathways to licensure that are not limited by a single institution's capacity to offer courses in multiple formats. As one example, MSU's traditional secondary mathematics education program requires completion of a number of specialized mathematics courses for teachers that develop competency in advanced topics for high school mathematics, mathematical modeling, and algebraic or geometric thinking. The program also requires two semester-long methods courses targeting grades 5–8 and 9–12. At this point in time, virtually none of these specialized courses are readily accessible to remote learners. A candidate could instead apply to the University of Montana, but there they would find different requirements such as Number Theory, Abstract Algebra, and History of Mathematics–also offered only on campus. As a result, substitutions and equivalencies are heavily employed to document mathematical and pedagogical competency and waive required courses. This was the case for Bonnie's pathway to licensure.

Tensions and Tradeoffs: Redefining Coursework

Bonnie, a civil engineer seeking a mathematics teaching license, presented a content background typical for that field: three semesters of calculus, statistics, a few engineering courses with mathematical roots, and a geometry course from a private online provider. Our program requires Modern Geometry, Mathematical Modeling, and Higher Mathematics for Secondary Teaching, all of which cultivate specialized knowledge of mathematics for teaching. Bonnie's tutorial-style geometry course and engineering experience were approved as substitutes for Modern Geometry and Mathematical Modeling, but we required her to complete Higher Mathematics for Secondary Teachers, a course that explores number theory, abstract algebra, and proof in a student- and problem-centered classroom environment. We hoped this experience would fill some gaps in Bonnie's preparation by providing exposure to effective mathematics teaching practices and student-centered learning.

Did Bonnie's preparation adequately develop her knowledge for teaching mathematics? Our program's undergraduate Modern Geometry course is problem-based, student-centered, and technology-infused. Preservice teachers in this course experience first-hand the benefits of collaboration, exploration, justification, and communication. They engage in problem solving and practice discourse as they critique the thinking of their peers. Bonnie's online geometry course most likely did not possess these features. Does it qualify as a substitute?

A PATHWAY TO RURAL PREPARATION:
MASTER OF ARTS IN TEACHING

MSU's undergraduate teaching program, while healthy, does not have the capacity to replicate all of its secondary mathematics requirements in a format accessible to remote teacher candidates. Faculty members have explored and debated teacher preparation alternatives that might help address rural staffing shortages. In 2019, a promising opportunity presented itself: collaborating on the launch of a Master of Arts in Teaching (MAT) teacher preparation program leading to licensure in grades K–8 and in the secondary disciplines of mathematics, science, social studies, and English/language arts. The MAT curriculum purposefully leverages shared resources and remote learning to address limitations presented by distance, time, and isolation. (Note: both co-authors designed and have taught courses in the program.)

The MAT is designed to be *place-attentive*, that is, focused on valuing the "physical place, the diversity of people in and connected to the place, and an understanding of how the place itself affords an agentive tool for educators to use as a teaching tool, framing place with power" (White & Downey, 2021, p. 13). Designed with Montana's rural schools and communities in mind, the MAT program encourages students to recognize rural places as "a source of wealth and strength and as delicate environments that require innate stewardship" (Mokoena & Hlalele, 2021, p. 139) as they attend to, and draw upon, the strengths and assets present in rural schools and communities. Place-attentiveness in the MAT focuses on "getting to know a place and letting it speak" (White & Downey, 2021, p.13); it highlights the importance of local knowledges and diverse perspectives and "the affordances for meaning making when students are given the opportunity to connect curricula with places they care about" (Azano & Callahan, 2021, p. 96).

Secondary-level applicants to the MAT must possess a baccalaureate degree in a related subject area, pass a Praxis content examination, and undergo an assessment of their content background and competencies. Admitted candidates pursue three semesters of full-time graduate study over a 12-month time period. Coursework is delivered virtually with the exception of a weeklong campus-based summer residency. Student teaching and practicum experiences are carried out and supervised in the candidate's local community. Each semester features a clinical field experience, including full-time classroom student teaching in the final semester along with capstone courses. As with traditional programs, successful completion of all coursework and in-school experiences are required for graduation and recommendation for licensure.

MAKING SPACE FOR MATHEMATICS

As a mathematics educator who maintains high standards for preparing preservice teachers, the first author was initially critical of the proposed MAT degree. How could a generalist graduate program, delivered in a compressed virtual format, possibly stand up against a thoughtfully conceived and meticulously balanced four-year degree? The answer to that question seemed obvious, but so was the realization that this customized program had been approved, was already in the design phase, and intended to enroll future mathematics teachers with or without contributions from resident experts. With the realization that contributing was preferable to criticizing, the first author joined the design team.

By necessity, much of the MAT curriculum is focused on pedagogy rather than disciplinary knowledge. A few courses provide limited opportunities for infusing mathematics into investigations and projects (e.g., Content Literacy, Introduction to Curriculum Design, Technology/Instructional Design), but the most significant opportunity by far to develop a candidate's mathematical knowledge for teaching was a 6-credit mathematics methods course offered during the second semester. The first author was commissioned to design this intensive 16-week course for online delivery, and discovered that transforming a traditional methods course into the virtual environment is again an exercise in balancing what is ideal against what is expedient. New tensions arose during the design process.

Tensions and Tradeoffs: Difficult Choices

Funneling the essence of secondary mathematics teacher preparation into one 6-credit course required painful decisions about what to keep and what to let go. Amid a plethora of recommendations from AMTE, CBMS, state agencies, and university researchers, it was necessary to choose what guidance to follow and what to disregard. Our traditional campus program disperses mathematical, pedagogical, and technological knowledge for teaching across four or more content courses and two methods courses. MAT candidates would have none of that broad exposure over time. Given these circumstances, what knowledge and experience should be considered essential? How can students effectively engage with these essentials in a virtual environment?

Eventually, four themes were identified across which to distribute the course material. The Foundations unit includes implementing standards for mathematical content and practice, interpreting learning theories and trajectories, and setting purposeful goals and outcomes. The Teaching and Learning unit focuses on developing skills in planning for instruction and spotlights three inclusive and student-centered strategies: implementing cognitively demanding tasks (Smith & Stein, 2018), orchestrating student discourse, and explorations using dynamic software (e.g., GeoGebra and Desmos). The Assessment unit emphasizes types and uses of formative assessment, applications for analyzing student work, and equitable assessment practices. Each thematic unit incorporates and examines related diversity and equity issues, with two additional weeks distinctly dedicated to these concerns. Finally, during three weeks simply titled "Big Ideas" students investigate concepts, connections, and essential understandings within and across high school mathematics using NCTM's "Developing Essential Understanding" series (Bieda et al., 2012; Lloyd et al., 2010; Peck et al., 2013; Sinclair, 2012).

The four themes and their subtopics emerged from careful consideration and an often uncomfortable "letting go" of well-entrenched convictions regarding the preparation of mathematics teachers. But rather than dwelling on what was reluctantly released, it may be more profitable to describe what was retained. AMTE's SPTM (2017) and other resources informed the following selections.

1. A broad-scope mathematics methods text provides continuity and serves as a resource for fundamental theories about teaching and learning mathematics. In addition, *Principles to Actions: Ensuring Mathematical Success for All* (NCTM, 2014) introduces students to high-leverage teaching practices and foregrounds current concerns within the mathematics education community.

2. Working through the *Five Practices for Orchestrating Productive Mathematics Discussions* (Smith & Stein, 2018) allows students to recognize the value and effectiveness of instructional routines as they explore a set of widely accepted and research-informed practices.

3. Students become familiar with GeoGebra and Desmos and have multiple opportunities to use these tools for mathematical problem solving and when designing tasks and lessons. They also investigate web-based teaching resources (e.g., Three-Act Tasks) and online curriculum materials.

4. The MAT curriculum includes a generalist assessment course; the mathematics methods course further asks candidates to analyze student thinking by looking at student work, develop formative assessments to inform instruction, and explore how assessment practices may produce inequities and reduce access for some learners.

The online mathematics methods course was produced as a series of one-week modules layered over a set of less frequent but more extensive "Tasks of Teaching" projects. This design allowed for efficient modification of content and format; it also turned out to be fortuitous as the structure of the MAT program soon encountered realities that required streamlining of instruction.

Tensions and Tradeoffs: Managing Costs

Initially, teacher educators from each of the four secondary disciplines were contracted to teach the individual methods courses, but this arrangement quickly proved unsustainable given the small enrollments in each discipline. After the first year, a single instructor collectively guided the MAT candidates through a combined secondary methods course, alternating two weeks of general instruction with five weeks of discipline-specific instruction drawn from the original curriculum. Practicing secondary teachers were appointed to monitor and mentor the candidates in the activities of their respective disciplines. Does this modification further diminish the depth and quality of mathematics teacher preparation embedded in the MAT? Or does it add a valuable new dimension by enabling one-on-one interaction with a practicing teacher?

CREATING ROBUST CLINICAL EXPERIENCES

The AMTE SPTM (2017) enumerate five characteristics of effective teacher preparation programs. For better or worse, Standard P2 (*Opportunities to Learn Mathematics*) must be accomplished outside of the MAT program and validated prior to admission. Standard P3 (*Opportunities to Learn to Teach Mathematics*) is enacted in the methods course through efforts to integrate "mathematics, practices for teaching mathematics, knowledge of students as learners, and the social contexts of mathematics teaching and learning" (p. 33). With MAT participants scattered across the state, Standard P4 (*Opportunities to Learn in Clinical Settings*) presents the greatest challenge. Our approach to meeting this standard involves crafting localized and individualized field experiences for each secondary mathematics teacher candidate. Clinical field experiences are integrated into concurrent coursework throughout the program.

The initial summer clinical experience is a weeklong Inclusive Community Camp where teacher candidates interact with middle grades students from a variety of backgrounds, guided by professional youth mentors and program faculty. The candidates apply research-informed teaching and management strategies as they acquire competencies and skills for engaging vulnerable youth in classroom community building. The fall semester field experience is a 100-hour in-school practicum closely coupled with the 16-week methods course. Each candidate is assigned to a rural school near their locale, where they also student teach the following semester. A graduated series of four teaching episodes (a lesson introduction, a mathematical task, and two complete lessons) are required for the practicum and also embedded as assignments in the methods curriculum. Mathematics candidates must document each episode on video, complete reflection and analysis tasks, and debrief the experience with their methods instructor, a clinical mentor, and MAT peers in mathematics as well as the other disciplines. While this process presents logistical challenges, it also produces unexpected benefits.

Tensions and Tradeoffs: Supervision Across the Miles

Relying on virtual supervision of student teachers and feedback based on video-recorded instruction may not be ideal, but there are few alternatives for a program in which clinical experiences may quite literally be "in the field": MAT candidates in the eastern plains of Montana may be located as much as an 8-hour drive from the university. Despite the obstacles presented by distance, a number of benefits can be derived from this scenario. MAT candidates in a rural practicum often spend far more hours in the host school than is typical; they develop awareness of and empathy for the unique circumstances of rural students. They may find it easier to engage with students and observe a wider variety of learning abilities than in a larger school with compartmentalized programs. Digitally recorded teaching episodes enable at least four levels of review and reflection: guided self-critique by the candidate, expert review by a mathematics teacher educator and a second clinical supervisor, immediate in-person feedback from the cooperating teacher, and reactions from fellow candidates in mathematics and other subject areas.

The third and final semester incorporates a full 14-week student teaching experience at the same school site as the 100-hour practicum. During this time, students also enroll in two 3-credit courses: one course exploring equity in linguistic and cultural diversity and a capstone course titled Reflective Inquiry in Rural Education. The latter course engages students in professional inquiry and analysis of theory, research, and practices that support and sustain effective rural teachers from recruitment to retirement. Students have the opportunity to synthesize and deepen their understanding of the complex nature of rural education and develop the knowledge, skills, and dispositions needed to be ready to live, work, and thrive as a teacher in rural classrooms, rural schools, and rural communities.

The workload during this semester represents yet another tradeoff as MAT candidates double down to combine the pressure of student teaching with concurrent coursework that allows the program to be compressed into 12 months. One mathematics candidate at a small but growing high school found the experience both energizing and connected to prior learning:

> My placement at Lone Pine is working out really well, and I love being in the classroom every day. I'm currently teaching 7th grade and 10th grade, which has been a great age range to experience. I'm definitely seeing in practice and trying to incorporate the methods we explored together. I even used the dilations exploration we did last week with my 10th graders!

SUPPORT THROUGH INDUCTION: MONTANA RURAL TEACHER PROJECT

Despite the substantial weight of the MAT methods courses (1/5 of total credits) and the program's ability to integrate relevant clinical experiences, opportunities to encounter and interact with mathematical content and pedagogy are still very limited. Reduced attention to mathematics during teacher preparation can be remediated with increased attention to graduates as they move into classrooms. Ongoing support is an essential factor in the success and retention of newly prepared mathematics teachers, who must continue to deepen their understanding of content and simultaneously develop productive beliefs about their efficacy to teach that content (Ross & Bruce, 2007). Rural teachers and school districts often lack access to such supports, despite facing the same issues of quality and accountability as their counterparts in urban and suburban school districts (Klein, 2015). In the interest of addressing this issue, we sought to extend MAT "preparation" by following graduates into their first years of teaching.

The prospect of ongoing support became reality when we were awarded a United States Department of Education Teacher Quality Partnership grant to support up to 70 rural teachers through degree completion, then continue to uphold them during the first years of teaching. The Montana Rural Teacher Project (MRTP) offers a living wage stipend to teacher candidates throughout their MAT coursework and provides an induction program during the first two years of teaching. In exchange for this assistance, graduates must commit to teaching in eligible rural or high-need schools for three years.

MRTP induction support has three components: one-on-one content mentoring from a veteran mathematics teacher; membership in a virtual community of practice; and participation in a menu of professional learning options. Among these complementary components, the opportunity for individual mentoring is most prized by program graduates navigating their first teaching positions. Assigning mentors revealed a final tension, one of several limiting factors that characterize teaching in remote and rural schools.

Tensions and Tradeoffs: Navigating Depth and Breadth

Many rural mathematics teachers in Montana are solely responsible for grades 9–12, 7–12 or even 5–12. The nearest colleague with a similar assignment may be 100 miles away in the next county. Identifying an experienced, capable, and nearby mathematics mentor for each MRTP graduate quickly became problematic. The issue was resolved simply by removing the "nearby" condition. Mentors are recruited from across the state and assignments are made by prioritizing content knowledge, experience, and circumstances over proximity. We are learning that partnering mentees with mentors from other (possibly distant) districts helps them focus their collaboration on issues related to instruction; in addition, the separation from local politics enhances trust.

MRTP graduates and their mentors are invited to collaborate via email, phone calls and texts, FaceTime and Zoom, and other media of their choice as well as optional site visits. A centralized online platform supports private interactions between mentoring partners as well as public group discussions, access to resources, and submission of reports and queries. This so-called "MRTP Home Base" is a Moodle site housed within the Teacher Learning Hub sponsored by the Montana Office of Public Instruction. Home Base elements are designed to actively engage teachers in collaborative interaction related to content, self, and students as supported by social-cognitive learning theory (Shulman & Shulman, 2004; Timperley et al., 2007).

- "Partners in Practice" activities are collaborative explorations of effective pedagogical practices in mathematics. Mentoring partners complete two activities each semester from among options to implement formative assessment, investigate and interpret standards, examine student work, reflect using self-inventories, and more.
- Through "Community Discussions," mentors and mentees support meaning-making through the diagnostic, prognostic, and motivational frames constructed by participants as they discuss problems and negotiate shared understandings (Bannister, 2018).
- Participants access the "Professional Learning Menu" to select from an array of professional development events and courses sponsored by partner organizations.

The Role of Partnership in Rural Teacher Preparation and Induction

Through the design and implementation process, it has become clear that the success and effectiveness of both the MAT and the MRTP are made possible only through the combined efforts of mathematics educators, general educators, and clinical experience coordinators. Collaborative partnerships within the higher education enterprise have been critical to developing and delivering a high-quality program. However, a far more extensive and inclusive network of K–20 partnerships is needed to ensure the intended long-term outcomes of both projects–the employment and retention of well-prepared mathematics teachers equipped to teach, thrive, and put down roots in rural schools and communities. Myriad complexities are involved in developing strategic teacher preparation and induction pathways, particularly those tailored to address specific staffing shortages, rural or otherwise. It is impossible for any one entity (i.e., higher education, local school districts, or state agencies) to achieve this goal in isolation. Thus, we invested significant time in developing productive collaborative partnerships with sensitivity to the challenges and difficulties that can arise when people work together over long periods of time (Farrah, 2019).

Rather than taking our partnerships for granted, we have deliberately and intentionally cultivated them through mutual commitments to listen, understand, and share knowledge, resources, and power. This type of productive leadership represents "the influence, direction, and creativity derived from deep rural identity and knowledge and enacted to inform and interpret policy and practice in alignment with the vision, commitment and values of rural stakeholders" (White & Downey, 2021, p. 16). Our most successful partnerships in this effort have been *collaborative* (i.e., working cooperatively in order to achieve a shared goal); *reciprocal*, producing mutually beneficial outcomes and benefits; and *trilateral*, defined as three entities committed to working together with a shared purpose, toward a common goal, with mutual trust and respect (Downey & Luebeck, 2023). While not perfect, these partnerships are able to work in harmony because each entity has been "willing to recognize and welcome the unique expertise of the others," enabling them to "contribute from their areas of strength while simultaneously honoring differences of perspective and working to reach mutually agreeable solutions" (Downey & Luebeck, 2023). Within the context of these collaborative partnerships we have made time for tensions to be shared, unpacked, and understood and space to honor each partner's strengths and resources as we navigate the tradeoffs surrounding expectations for content and coursework and move toward reaching our common goals.

CONCLUSION

We must look for new ways to deliver programming, including methods courses, that embraces standards for effective mathematics teacher education programs while opening access for non-typical teacher candidates who are willing to help educate students in rural communities and other difficult-to-staff settings. Designing more readily accessible mathematics teacher preparation programs and offering methods courses in alternative formats aligns with AMTE's goal to build toward equitable practices in teacher education. A virtual model of teacher preparation casts a wider net for attracting candidates, especially in places where distance and other factors limit access to traditional education. But to do this successfully, as a profession we must reconcile the value of maintaining rigorous standards against the imperative of serving schools that struggle to recruit and retain traditionally prepared teachers.

Ultimately, even unique initiatives like the MAT and MRTP fall short of filling the need for mathematics teachers in rural places. Many Montana schools cannot afford to delay hiring for a year while they wait for these programs to produce a qualified mathematics educator who might consider their open position. Instead, a non-educator will be hired on a provisional basis with the hope that they earn licensure before penalties are invoked. Unfortunately, this scenario is playing out nationwide wherever isolation, economics, performance, social issues, and lack of interest affect teacher recruitment and retention. We must engage in conversations and collaborations about ways to simultaneously sustain traditional mathematics teacher preparation, incorporate less orthodox alternatives, and address insufficient solutions. While we are doing so, let us remain mindful that somewhere in a rural school, and so many other schools, there are students waiting for their teacher.

REFERENCES

Association of Mathematics Teacher Educators. (n.d.). *Long-term goal update 2022 conference.* https://amte.net/sites/amte.net/files/Long%20Term%20Goal%20Update%20 2022%20Conference.pdf

Association of Mathematics Teacher Educators. (2017). *Standards for preparing teachers of mathematics.* amte.net/standards.

Association of Public Land Grant Universities. (n.d.). *Mathematics teacher education partnership.* https://www.aplu.org/our-work/2-fostering-research-innovation/mathematics-teacher-education-partnership/index.html

Azano, A., & Callahan, C. (2021). Innovations in providing quality gifted programming in rural schools using place-conscious practices. In S. White & J. Downey (Eds.), *Rural education across the world: Models of innovative practice and impact* (pp. 91–106). Springer Singapore.

Bannister, N. A. (2018). Theorizing collaborative mathematics teacher learning in communities of practice. *Journal for Research in Mathematics Education, 49*(2), 125–139.

Bieda, K., Knuth, E., Ellis, A., & Zbiek, R. M. (2012). *Developing essential understanding of proof and proving for teaching mathematics in grades 9–12.* National Council of Teachers of Mathematics.

Conference Board of the Mathematical Sciences. (2012). *The mathematical education of teachers II*. American Mathematical Society and Mathematical Association of America.

Downey, J., & Luebeck, J. (2023). The power of three: Promising practices for sustaining trilateral partnerships in rural contexts. In S. Hartman & B. Klein (Eds.), *The middle of somewhere: Rural education partnerships that promote innovation and change* (pp. 53–69). Harvard University Press.

Farrah, Y. (2019). Collaborative partnership: Opening doors between schools and universities. *Gifted Child Today, 42*(2), 74–80.

Klein, R. (2015). Connecting place and community to mathematics instruction in rural schools. In U. Gellert, J. Giménez Rodríguez, C. Hahn, & S. Kafoussi (Eds.), *Educational paths to mathematics: A C.I.E.A.E.M. sourcebook* (pp. 33–65). Springer International Publishing.

Lloyd, G., Beckmann Kaziz, S., & Zbiek, R. M. (2010). *Developing essential understanding of functions for teaching mathematics in grades 9–12*. National Council of Teachers of Mathematics.

Mokoena, M., & Hlalele, D. (2021). Thriving school enrichment programs at rural South African schools. In S. White & J. Downey (Eds.). *Rural education across the world: Models of innovative practice and impact* (pp. 129–146). Springer Nature Singapore.

Montana Office of Public Instruction. (2019). *Critical quality educator shortages*. https://leg.mt.gov/content/Committees/Interim/2019-2020/Education/Meetings/Jan-2020/Critical_Educator_Shortage_Report.pdf

Montana Office of Public Instruction. (2021). *Facts about Montana education 2021*. https://opi.mt.gov/Portals/182/Superintendent-Docs-Images/Facts%20About%20Montana%20Education.pdf?ver=2021-06-10-162844-327.

National Center for Frontier Communities. (n.d.). *Definition of frontier*. http://frontierus.org/what-is-frontier/.

Peck, R., Gould, R., Miller, S., & Zbiek, R. M. (2013). *Developing essential understanding of statistics for teaching mathematics in grades 9–12*. National Council of Teachers of Mathematics.

Ross, J. A., & Bruce, C. (2007). Professional development effects on teacher efficacy: Results of randomized field trial. *The Journal of Educational Research, 101*(1), 50–66.

Showalter, D., Klein, R., Johnson, J., & Hartman, S. L. (2017). Why rural matters 2015–16: Understanding the changing landscape. Rural School and Community Trust. https://mrea-mt.org/wp-content/uploads/2017/08/RSCT-Why-Rural-Matters-MT-Specific.pdf

Shulman, L. S., & Shulman, J. H. (2004). How and what teachers learn: A shifting perspective. *Journal of Curriculum Studies, 36*(2), 257–271.

Sinclair, N., Pimm, D., Skelin, M., & Zbiek, R. M. (2012). *Developing essential understanding of geometry for teaching mathematics in grades 9–12*. National Council of Teachers of Mathematics.

Smith, M. S,. & Stein, M. K. (2018). *Five practices for orchestrating productive mathematics discussions* (2nd ed.). Reston, VA: National Council of Teachers of Mathematics.

Timperley, H. S., Wilson, A., Barrar, H., & Fung, I. (2007). *Teacher professional learning and development best evidence synthesis*. Wellington: Ministry of Education.

Ulferts, J. (2016). A brief summary of teacher recruitment and retention in the smallest Illinois rural schools. *The Rural Educator 37*(1), 14–24.

White, S., & Downey, J. (2021). International trends and patterns in innovation in rural education. In S. White & J. Downey (Eds.), *Rural education across the world: Models of innovative practice and impact.* Springer Nature Singapore.

Yoon, S. Y., Mihaley, K., & Moore, A. (2019). *A snapshot of educator mobility in Montana: Understanding issues of educator shortages and turnover.* Regional Educational Laboratory Northwest.

CHAPTER 20

PROSPECTIVE ELEMENTARY TEACHER EDUCATION PROGRAM DESIGN

What We Still Do Not Know About Balancing Content and Pedagogy

Daniel L. Clark

Western Kentucky University

Tuyin An

Georgia Southern University

Prospective elementary teacher (PSET) education programs in the United States vary greatly in the courses employed to prepare their students in terms of the balancing of mathematics content and pedagogy. This chapter explores potential tradeoffs that arise for mathematics teacher educators (MTEs) in the preparation of PSETs' mathematical knowledge for teaching due to the choices PSET education programs make. More specifically, the chapter looks at literature through the lenses of MTEs' course preparation and PSETs' learning trajectories and discusses different models of content and pedagogy integration across/within PSETs' mathematics education

The AMTE Handbook of Mathematics Teacher Education: Reflection on Past, Present and Future—Paving the Way for the Future of Mathematics Teacher Education, Volume 5
pages 407–422.

coursework. Literature-based discussions on future research directions regarding PSET education program and course design are also provided.

INTRODUCTION

Teachers who teach mathematics at the K–5 level are often referred to as *elementary teachers*, although most states in the United States (U. S.) issue either K–6 or K–8 levels of teacher licensure for elementary teachers (Conference Board of the Mathematical Sciences [CBMS], 2012). A similar situation also exists in middle-grades teacher preparation programs that "the majority of middle grades teachers are likely prepared in a program designed as preparation to teach all academic subjects in grades K–8 or in a program to teach mathematics in grades 7–12 or 6–12" (CBMS, 2012, p. 39). The mixed-level licensure system created room for prospective elementary teacher (PSET) education programs to design highly varied mathematics education curricula for PSETs, which could lead to deeper issues such as an imbalanced coverage or inefficient sequencing/integration of the mathematical content and pedagogy, insufficient PSET-particular mathematics content courses, and few or no elementary mathematics content instructors with elementary teaching experience (Greenberg & Walsh, 2008; Masingila et al., 2012). A large scale national survey of PSET preparation programs found, "the majority (56.7%) of schools having mathematics content courses specifically for prospective elementary teachers offer two of these mathematics content courses, while 17.1% offer three, 16.1% offer one, and 9.9% offer from four to 12 of these courses" (Masingila et al., 2012, p. 352). In addition to the variety in teacher preparation programs, mathematics teacher educators (MTEs) hold varying beliefs as to the goals of PSET education programs (An et al., 2021). These differing goals lead to a variety of often contrasting approaches in the enactment of instruction designed to meet these aims.

THEORETICAL FRAMEWORK

As we grapple with the ideas of how to integrate mathematics content and pedagogy in the preparation of PSETs, we can approach the question with a combination of frameworks already in use: 1) the idea that PSETs are essentially relearning mathematics (Castro Superfine et al., 2020), and 2) learning trajectories (Confrey et al., 2014).

Considering that PSETs are relearning mathematics, Castro Superfine et al. (2020) elaborated on Ball et al.'s (2008) model of mathematical knowledge for teaching to create a model of mathematical knowledge for teaching teachers. As the reader can see in Figure 20.1, Ball et al.'s original model is on the top left and constitutes one area of which MTEs need to be aware. Then Castro Superfine et al. have analogous categories to all of Ball et al.'s original categories that MTEs need to be aware of and plan for in their teaching of PSETs.

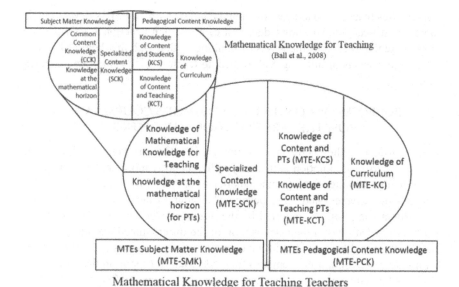

FIGURE 20.1.   Castro Superfine et al.'s (2020) Model of Mathematical Knowledge for Teaching Teachers

We argue in this chapter that MTEs need to not only think of this model individually when planning their courses for PSETs, but that MTEs should consider this model as they design and refine their teacher preparation programs and make hiring choices. Even if an individual faculty member is not an expert in all of these areas, programs should attempt to have PSETs encounter faculty with expertise in each area during their programs.

Turning to learning trajectories, while they have mostly been researched with respect to teaching mathematics to elementary students, we believe a similar notion can apply to PSETs. In general, learning trajectories consist of three pieces: "a) a mathematical goal; b) a developmental path along which children develop to reach that goal; and c) a set of instructional activities, or tasks, matched to each of the levels of thinking in that path that help children develop higher levels of thinking" (Clements & Sarama, 2009, p. 1). As most of the PSETs in our courses are either close to or completely finished with their cognitive development, our use of the word *development* is not quite the same as that used for elementary student learning trajectories. Instead, we are referring to PSETs' development as mathematics teachers. Furthermore, although learning trajectories are sometimes used to understand how an individual student learns about a particular mathematical concept, they have also been used to understand how students make sense of material in sets of standards across grade levels (Weber et al., 2015). Therefore, a PSET learning trajectory could consist of a) the goal to develop effective mathematics teachers, b) a developmental path along which PSETs develop their content and pedagogical knowledge, and c) a set of instructional activities, or tasks, matched to each of the levels of thinking in that path that help PSETs develop higher levels of thinking with respect to teaching elementary mathematics effectively.

PURPOSES OF CONTENT AND METHODS COURSES AND HOW THEY ARE TRADITIONALLY OFFERED

When considering the mathematical training of PSETs, a reasonable (though not universally true) assumption is that they will encounter more content in courses taught by mathematics faculty (mathematicians or MTEs who teach mathematics and statistics courses) and more pedagogy in courses taught by education faculty. This assumption is supported by the *Standards for Preparing Teachers of Mathematics* (SPTM) in their description of the differentiated learning opportunities that should be provided in mathematics content courses (opportunities to learn mathematics) and mathematics methods courses (opportunities to learn to teach mathematics), respectively (Association of Mathematics Teacher Educators [AMTE], 2017). While we will discuss what happens inside courses later in this chapter, we want to consider the distribution of content and methods courses first.

Traditionally, PSET preparation programs offer separate mathematics content and methods courses taught in separate departments (Ball, 1991; Brown & Borko, 1992; Grouws & Schultz, 1996). The methods courses are usually housed in an education department with a focus on PSETs' learning of how to teach, and therefore, are expected to have a significant pedagogy component. The content courses, however, are often housed in a mathematics department and focus more on the mathematics that PSETs need to master as teachers. According to a survey of elementary teacher preparation programs by Masingila et al. (2012), "the vast majority of the responding schools (78.4%) offer mathematics content courses specifically for prospective elementary teachers, and 88.3% of the schools offering these courses do so through a mathematics department" (p. 351–352). The separation of the methods and the content courses makes the integration of content and pedagogy a wide spectrum, from little content component in some methods courses to little pedagogy component in some content courses.

Comparing these existing program structures to existing standards, the CBMS has recommended that "Prospective teachers need mathematics courses that develop a solid understanding of the mathematics they will teach" and "Coursework that allows time to engage in reasoning, explaining, and making sense of the mathematics that [they] will teach" (2012, p. 17). The CBMS goes on to recommend that PSETs "should be required to complete at least 12 semester-hours on fundamental ideas of elementary mathematics, their early childhood precursors, and middle school successors" (p. 18). While focusing mostly on the mathematical content PSETs should learn, the CBMS does also note that the recommended 12 credit hours of mathematics should include training in methods of instruction.

The AMTE (2017) endorses both the 12-credit hour requirement for PSETs and the sharing of those hours between content and methods courses. Masingila and Olanoff (2021) note that 8% of PSET preparation programs that responded to their survey require 12 or more credit hours of mathematics content courses; while that gives the impression that 92% of PSET preparation programs are not meeting the CBMS and AMTE standards, that is likely not the case as methods courses were not included.

There are benefits and constraints to separating content and methods courses, though. The methods courses are usually taught by teacher educators who have elementary teaching background, so PSETs are provided with rich opportunities to build knowledge of learning theories and hone skills of teaching. However, although methods courses should "address deep and meaningful mathematics content knowledge" (SPTM Indicator P.3.1) (AMTE, 2017, p. 33), the depth and breadth of the mathematics topics addressed in some of the methods courses are not sufficient due to the instructors' limited mathematical background. In particular, the SPTM require PSETs to be able to make connections to the mathematical ideas that come before and after the elementary level (AMTE, 2017). For example, "Well-prepared beginning teachers of mathematics at the upper elementary level understand foundational mathematics concepts that they will teach, and they connect those concepts to mathematical practices as well as to the mathematics of Pre-K–2 and the middle level curriculum" (AMTE, 2017, p. 75).

The content courses are usually housed in mathematics departments taught by mathematicians or mathematics educators who have relatively deep understanding of the content that allows them to create opportunities for PSETs to experience mathematics learning from both a student's and a teacher's perspective. However, one commonly seen issue is that many faculty members in mathematics departments have little to no prior elementary teaching experience, often coming from secondary or undergraduate teaching backgrounds if they had any prior teaching experience themselves (Greenberg & Walsh, 2008; Masingila et al., 2012). Thus, they may not have much expertise in the school curriculum and pedagogy appropriate for teaching mathematics at the elementary level.

There is no doubt that PSETs should have deep understanding of both the content and pedagogy. As is well stated in the *Mathematical Education of Teachers II* (MET II) (CBMS, 2012):

> A major advance in teacher education is the realization that teachers should study the mathematics they teach in depth, and from the perspective of a teacher. There is widespread agreement among mathematics education researchers and mathematicians that it is not enough for teachers to rely on their past experiences as learners of mathematics. It is also not enough for teachers just to study mathematics that is more advanced than the mathematics they will teach. (p.23)

Crucially, they go on to recommend that teacher preparation programs should design courses that blend content and pedagogy when possible. In the next section we examine the various explorations that MTEs and researchers have done to find the right mix of content and pedagogy in the past few decades.

MODELS OF CONTENT AND PEDAGOGY INTEGRATION

Edwards (1997) points out that successful integration of content and pedagogy is not a destination but a "cultural journey" in which faculty's view on the culture of the program change is "transforming and energizing" (p. 52); the key is to create quality learning experiences, model a balanced perspective of content and pedagogy, and provide opportunities for PSETs to reflect on both the content and pedagogy components of the program. MTEs agree that the best PSET preparation practice should be based on interdisciplinary partnerships among mathematics content, pedagogy, and field experiences (Ball, 2000; Ball & Bass, 2003). Depending on the specific institutional settings and resources, integration of content and pedagogy in PSETs' coursework can be accomplished through various arrangements of the instructor, the enrollment timeframe, and the course focus. In this section, we introduce a few research-based models of content and pedagogy integration.

One Instructor for Content, Methods, and Field Experience

Having one instructor for teaching all three types of courses (content, methods, and field experience) can help maximize the coherence in PSETs' experience of mathematics learning and teaching. Hart (2002) studied an Urban Alternative Preparation Program housed in the department of early childhood education, in which PSETs entered the program as a cohort (n = 14) and took 12 credit hours of coursework over three consecutive semesters. The content and methods courses were paired up and taught by the same instructor to provide PSETs with seamless learning experiences. The same instructor also supervised PSETs' student teaching in the following semester. The results of the study suggest positive changes in PSETs' beliefs and practice with respect to reform-oriented mathematics teaching after completing Phase I (undergraduate education) of the program. The participants also reached a 100% pass rate on the state certifying board required Praxis tests, which demonstrates their qualified content and pedagogical knowledge and skills as beginning teachers. Although the study was limited by a relatively small sample size and a lack of iterative tests on the data collection instruments, it still informed the PSET education community about a possible solution for content and pedagogy integration. One potential drawback of this solution, though, is the number of qualified instructors. Faculty members with both a master's degree in mathematics and several years of elementary teaching experience are relatively rare.

Concurrent Enrollment for Content, Methods, and Field Experience

Sometimes integration of content and pedagogy can be done by allowing PSETs to enroll in all three courses concurrently to gain coherent experience in learning, teaching, and reflecting. By comparing several different types of course enrollment models with a total of 96 PSETs, Strawhecker (2005) discovered that PSETs' concurrent enrollment in a mathematics methods course, a mathematics content course, and a field experience resulted in the greatest gains in their knowledge of content for teaching mathematics and positive changes in beliefs about teaching elementary mathematics. The study especially highlights the positive impact of the concurrent enrollment in field experience and a methods course (and/or a content course), which indicates the importance of offering PSETs opportunities to learn the pedagogical/content knowledge and apply the knowledge simultaneously in their teaching practice. Similar gains in knowledge were not found for PSETs who had field experiences in semesters prior to, but not during, their methods courses.

Forming Faculty Working Partnerships

In most cases, the content course and methods course instructors (and field experience supervisors) work in separate departments and need to follow their own departments' requirements on the design and teaching of their courses. It might not be convenient to directly combine these courses into one course due to cross-departmental administrative policies and procedures. However, the content and pedagogy integration can be done through an indirect route: forming a faculty working partnership. The instructors can still teach separate courses, but they collaborate on the design or teaching of the courses. Edwards (1997) shared such an example of integrating content and pedagogy that was considered as "innovative" at the time and had worked well over the previous decade. By forming a close working partnership, the instructors of the content and the methods courses were able to offer both courses simultaneously, with the content course taking 5/7 of the combined course credit hours and the methods course taking 2/7 of the credit hours. The instructors co-designed their courses to keep the same sequence of topics in both courses and purposely planned the connections between the two courses for teaching some of the more difficult mathematical concepts, such as learning to count in different bases and developing volume formulas for polyhedra. Since the instructors knew the teaching agenda of each other's courses, they were able to notice PSETs' misconceptions brought from each other's courses and use such spontaneous "teachable moments" to help PSETs develop correct understanding toward mathematical/pedagogical concepts. Edwards cites one example of this where the content instructor overheard PSETs quietly wondering why the methods instructor relied so heavily on base ten blocks. Hearing this, and knowing what was occurring in the methods course, the content instructor was able to elaborate on the importance of base ten blocks to elementary mathematics teaching during that day's lesson. PSET student evaluations at the end of the courses verified the key benefit of this integration model: PSETs were given opportunities to retrace missed mathematical concepts from the perspective of an elementary student in the methods course at the same time the concepts are being studied and enhanced in the content course (Edwards, 1997).

An even closer partnership between the content course instructor and the methods course instructor is reported by Ford and Strawhecker's (2011) study, in which the two instructors (a mathematician and a mathematics educator) co-designed and co-taught the two courses for a cohort of 12 PSETs. The classes were carefully planned so that the related topics were addressed across the courses but from a content-focused and a pedagogy-focused angle, respectively. This proved beneficial as the arrangement allowed, for example, one PSET who was struggling with geometry concepts in her content course to discover useful methods to help her understand them in her methods course. Some classes were led by one of the instructors and assisted by the other instructor; some were completely co-taught. PSETs' feedback indicates a general support for this model of content and pedagogy integration.

By studying collaborations between mathematicians and mathematics educators in a large-scale national Inspiring Mathematics and Science in Teacher Education (IMSITE) project conducted in Australia, Goos and Bennison (2018) discovered that interdisciplinary practices led to the integration of content and pedagogy in two ways: co-developed and co-taught courses and building communities of preservice mathematics teachers (cohort experiences). The study also identified conditions that could support or hinder sustained interdisciplinary collaboration between mathematicians and mathematics educators either within or across institutions, which could guide the design of faculty working partnerships in teacher preparation programs. For example, the supporting conditions within institutions include matching personal qualities (open-mindedness, trust, mutual respect, and shared beliefs and values) and having shared problems. The hindering conditions include physical separation, workload formulas or financial models, cultural differences, and the ambiguous nature of boundaries.

Integrating Content and Pedagogy in One Course

Research has repeatedly shown positive effects on PSET learning by infusing mathematics content in mathematics methods courses and vice versa. Benbow (1993) showed positive effects from integrating the entire content and methods course sequence. After a two-semester sequence of integrated content and methods courses, he found that PSETs were less likely to view mathematics as dependent on memorization, less likely to view mathematics in "right-wrong, one answer-one method terms" (p. 2), more likely to understand the importance of word problems, and better able to develop higher levels of teaching self-efficacy. Although working with secondary preservice teachers, Steele and Hillen (2012) also showed positive results from what they termed a content-focused methods course. They posit three design principles for such a course that amount to selecting a narrow mathematical content focus and weaving that content throughout the methods course. By the end of the methods course, with respect to the selected content topic, the preservice teachers were better able to provide examples and nonexamples, use multiple representations, and select high-cognitive demand tasks. However, more modest modifications to programs can also bear worthwhile fruit.

One such example was shown by Burton et al. (2008). They studied two sections of a methods course that met once a week for three hours. Each class meeting for one of the sections, they focused on an aspect of mathematics content for 20 minutes. Compared to the students in the other section, these PSETs' levels of mathematical content knowledge for teaching showed more improvement. Again, this is a promising result for a rather small intervention.

Integration can also occur in the opposite direction as mathematics content instructors have worked to include pedagogy in the courses that they teach for their PSETs. Appova and Taylor (2019) worked with 10 MTEs to ascertain their reasons for and ways of including the teaching of pedagogical content knowledge in their content courses for PSETs. Topics they often included were the varying affordances of different types of manipulatives, study of elementary student conceptions and misconceptions, and helping PSETs make connections between the content they would be teaching and more advanced mathematics curriculum that their future students would go on to encounter. Having this infusion of methods and pedagogy in content courses can potentially be critical to PSET learning as most PSETs take their content courses before their methods courses (Marzocchi & DiNapoli, 2020).

Integrating content and pedagogy in the context of other levels (non-elementary) of mathematics curricula could also be beneficial for PSETs' learning. By integrating reform-oriented middle school curriculum and pedagogical activities, Lloyd and Frykholm (2000) designed a geometry content course for PSETs. Their study indicates that such integration enhanced PSETs' understanding about connections between mathematics teaching and learning.

Integrating Content and Pedagogy with Other Areas

In addition to integrating the mathematics content and methods courses, integrating courses across the content areas can also have positive impacts on PSETs' learning and teaching. The other school subject most commonly associated with mathematics is science, including technology and engineering (i.e., STEM). As such, there have long been calls to integrate these two disciplines when teaching them to children (Furner & Kumar, 2007); however, asking teachers to teach by integrating different subject matters can be frustrating for all involved if the teachers do not have training on how to do that and have never experienced being taught in that fashion themselves (Hillman et al., 2000).

Rinke et al. (2016) analyzed the outcomes of a novel STEM teacher preparation model in which the two traditional mathematics and science methods courses are combined into one STEM block. The results show STEM block preservice teachers experienced significantly greater gains in teaching efficacy and STEM literacies compared to the traditional-route preservice teachers. Furthermore, when preservice teachers have this kind of integrated methods experience with mathematics and science in their training programs, they ultimately view integrating these subjects in their own teaching as more feasible and efficient (Berlin & White, 2012).

While STEM subjects may seem most relevant for integration with mathematics, combining the learning of mathematics with other subjects can also be beneficial. For example, integrating language arts and mathematics can help students strengthen their ability to communicate mathematically both verbally and in writing (Schram & Rosaen, 1996). Again, if this is something that the field believes would be beneficial for teachers to engage in with elementary students, then PSETs need to have similar experiences in their preparation programs. One example of this is when Hillman et al. (2000) integrated mathematics, reading, music, and special education in their methods courses. Where possible, they engaged in interdisciplinary, team-taught class sessions; however, they also reserved time for individual instructors to teach specific aspects of content and pedagogy within their subject areas.

DISCUSSION

Role of PSET Mathematics Preparation Standards

We presented a variety of content-pedagogy integration models in the previous section, each of which has its advantages and/or disadvantages based on each individual PSET preparation program's unique setting and available resources. We understand that only through rigorous scientific research is it possible to start discussions about the comparison of the effectiveness of these integration models. If we are going to pursue this goal, it is also important to know what characteristics an effective program for preparing beginning mathematics teachers should possess. The SPTM (AMTE, 2017) is such a guiding document that provides a set of standards and related indicators for effective programs, in terms of the Partnership (Standard P.1), Opportunities to Learn Mathematics (Standard P.2), Opportunities to Learn to Teach Mathematics (Standard P.3), Opportunities to Learn in Clinical Settings (Standard P.4), and Recruitment and Retention of Teacher Candidates (Standard P.5). We looked across the standards (mainly P.1, P.2, P.3) and the content-pedagogy integration models presented earlier, trying to envision some of the effective integration practices.

First, as discussed in earlier sections in this chapter, mathematics faculty and education faculty have different expertise and strengths in providing different learning opportunities for PSETs in content courses (Standard P.2 focused) and methods courses (Standard P.3 focused), respectively. However, there are deeply intertwined content and pedagogy areas suggested in both types of courses, such as "Attend to Mathematics Content Relevant to Teaching" (Indicator P.2.1) in a content course and "Address Deep and Meaningful Mathematics Content Knowledge" (Indicator P.3.1) in a methods course (AMTE, 2017, p. 26), which naturally calls for a partnership between the mathematics faculty and education faculty in the design and implementation of the content and methods courses. This partnership is particularly advocated in Standard P.1, "An effective teacher preparation program benefits from an interdisciplinary collaborative partnership that is a shared endeavor focused on the preparation of mathematics teachers who are well prepared to improve Pre-K–12 student learning in mathematics from multiple perspectives" (p. 27). We also agree with the standards that MTEs should provide leadership in the partnership due to its complex nature and actively engage mathematics faculty and research scholars to enhance the program design. Institutional support also plays an important role in the partnership, including supporting faculty's professional development, acknowledging and rewarding the collaborative effort, providing necessary teaching materials, etc. (Indicator P.1.2. Provide Institutional Support).

Depending on each individual PSET preparation program's setting and resources, there could be various models of the content and methods course arrangement, such as the Concurrent Enrollment for Content, Methods, and Field Experience and Integrating Content and Pedagogy in One Course models discussed previously. Regardless of these variations, we believe the essence of the course arrangement is to provide a coherent learning trajectory with multidimensional mathematics learning experiences to help PSETs get fully prepared for their mathematics teaching career. We fully agree with the suggestion in Indicator P.2.3. (Provide Sustained, Quality Experiences), "engagement with rich content opportunities must be coupled with ongoing integrated learning experiences in mathematics, such as analyzing mathematics content within the context of designing a mathematics lesson for a methods class or clinical experience" (AMTE, 2017, p. 32). This suggestion implies the sequence of the content and pedagogy in course arrangement that PSETs should have studied a particular mathematics concept before being able to apply it to pedagogical contexts, either within one integrated course or across content and methods courses (and field experience). We would also like to point out that such well-sequenced integrated learning opportunities can only be achieved through partnerships.

Meanwhile, while the focus of this chapter has been on the integration of content and pedagogy within PSET preparation programs at colleges and universities, why it is important, and how it can be improved, considerations about integrating content and pedagogy are also important when working with practicing teachers. One example of this comes from Hill and Ball (2004). A mathematician and mathematics educator co-designed and co-taught a professional development program that integrated content and pedagogy with positive results for the practicing teachers.

As the field continues to research how to best practice integration of content and pedagogy for PSETs in their preparation programs, more research along these lines may be needed to help practicing teachers who were prepared in programs where this was not a priority. The SPTM explicitly state the following:

> Studying mathematics content is necessary but not sufficient. High-quality early childhood teacher preparation programs *weave together* [emphasis added] the learning of mathematics content, the study of specific mathematics pedagogies and effective mathematics instruction, and, at the core, developmental knowledge of children's mathematical thinking and reasoning. (AMTE, 2017, p.68)

It is this weaving together that we have focused on in this chapter. The extant research has shown that there are multiple ways to conceive of and accomplish this integration. We believe that frameworks like Castro Superfine et al.'s (2020) model of mathematical knowledge for teaching teachers can be used to help the field research and understand how to best integrate content and pedagogy in PSET preparation programs and how to better equip MTEs with sufficient knowledge in carrying out this integration work.

FUTURE DIRECTIONS

In this chapter, we have laid out a variety of methods for content and pedagogy integration. The research that the field chooses to pursue with respect to these various methods should be governed at least in part by the social constraints and emerging opportunities of the current time. For example, financial considerations play an increasingly larger role in decision making every year for both institutions and PSETs (Mitchell et al., 2019). This is true for universities and their teacher preparation programs that regularly have to attempt to accomplish more with fewer monetary resources each year despite increasing tuition, while PSETs must choose preparation programs that will provide them the best training possible for a cost that they can reasonably afford.

With those financial considerations in mind, resources for research may potentially be best directed at the one instructor for content, methods, and field experience model. Teacher preparation programs, particularly smaller programs, may potentially gravitate toward this model because of the greater flexibility it offers with respect to assigning faculty to cover courses and generating student credit hour production. Having one or two multifaceted experts teach all those courses allows other faculty to be deployed elsewhere more efficiently, which would potentially ease a program's financial burdens. As universities are hiring fewer tenure track professors (Stein, 2022), getting the most out of each hire will be crucial.

Research resources could be spent in a variety of ways. First, graduate programs and placement experiences could be co-developed to give more candidates the requisite experience needed to be as competent teaching in all three of those settings as possible. This would involve investing in people to get graduate level mathematics course experience while also allowing them to gain elementary level teaching experience. Mathematicians and mathematics educators who teach in doctoral programs will have to collaborate to make such training experiences possible. In time, candidates who train in such programs will gain experience as faculty and, ultimately, some will become mathematics teacher educator leaders.

This all-in-one model most commonly occurs at institutions with smaller PSET preparation programs. As more people are trained as described above, they will likely find job opportunities at small and medium size institutions. As financial pressures cause institutions to seek to do more with less and students to seek education bargains, studying how smaller preparation programs with all-in-one instructors operate will be crucial. Therefore, it becomes even more important to research the program structures and decisions made by faculty at smaller institutions who are fulfilling all three roles. The mathematics PSET preparation research literature is dominated by the study of PSET preparation programs at large universities with highly ranked teacher preparation and/or mathematics education programs. As more smaller and medium size institutions potentially shift to an all-in-one model with more versatile faculty, they need to be studied more than they have been at present.

All-in-one instructors at smaller institutions likely have fewer financial and time resources to conduct such research on their own. Therefore, in order to accomplish this research, larger institutions will likely have to partner with smaller institutions. This will require going against the ever more present idea that universities and teacher preparation programs are and should be in competition with each other. This could take the form of one-on-one pairings of larger programs with smaller programs to study structures and practices, or having larger institutions coordinate studies of small groups of smaller institutions. Such studies could facilitate interaction and sharing of resources between the smaller institutions that would outlive the studies themselves.

Online PSET preparation is also worthy of study with respect to the blending of content and pedagogy. Financial considerations of both the universities and the students make online education attractive. Furthermore, while online teacher preparation was already becoming increasingly common (Hurlbut, 2018), the pandemic accelerated the amount of online teacher preparation in the market, a good deal of which will not disappear as the pandemic subsides. Therefore, studying how decisions are made regarding the blending of content and pedagogy in online teacher preparation programs that are completely online or that have online components is of increasing importance.

Finally, we propose the creation of a clearinghouse that can track the structures put in place to accomplish PSET mathematics education at institutions across the country. Previous studies (Masingila & Olanoff, 2021; Masingila et al., 2012) have given us significant and useful snapshots of who is teaching mathematics education courses and program structures at a given time. An ongoing clearinghouse would expand upon this work by being a living resource that interacts with the mathematics education community. Teacher preparation programs would be able to easily find other programs like themselves in order to share resources and build new program structures. Teacher preparation programs wanting to change an aspect of their program could easily find other programs who have already made the change or who might be considering a similar change. Having this catalog of resources available in one place would facilitate communication across programs of all types as well as lessen the financial burden of programs looking to improve themselves.

REFERENCES

An, T., Clark, D., Lee, H. Y., Miller, E. K., & Weiland, T. (2021). A discussion of programmatic differences within mathematics content courses for prospective elementary teachers. *The Mathematics Educator, 30*(1), 52–70.

Appova, A., & Taylor, C. E. (2019). Expect mathematics teacher educators' purposes and practices for providing prospective teachers with opportunities to develop pedagogical content knowledge in content courses. *Journal of Mathematics Teacher Education, 22*, 179–204. https://doi.org/10.1007/s10857-017-9385-z

Association of Mathematics Teacher Educators. (2017). *Standards for preparing teachers of mathematics.* https://amte.net/sites/default/files/SPTM.pdf

Ball, D. L. (1991). Teaching mathematics for understanding: What do teachers need to know matter? In M. Kennedy (Ed.), *Teaching academic subjects to diverse learners* (pp. 63–83). Teachers College Press.

Ball, D. L. (2000). Bridging practices: Intertwining content and pedagogy in teaching and learning to teach. *Journal of Teacher Education, 51*(3), 241–247. https://doi.org/10.1177/0022487100051003013

Ball, D. L., & Bass, H. (2003). Toward a practice-based theory of mathematical knowledge for teaching. In B. Davis & E. Simmt (Eds.), *Proceedings of the 2002 Annual Meeting of the Canadian Mathematics Education Study Group* (pp. 3–14). CMESG/GCEDM.

Ball, D. L., Thames, M. H., & Phelps, G. (2008). Content knowledge for teaching: What makes it special? *Journal of Teacher Education, 59*(5), 389–407.

Benbow, R. M. (1993, April). *Tracing mathematical beliefs of preservice teachers through integrated content-methods courses.* Annual Meeting of the American Educational Research Association. https://eric.ed.gov/?id=ED388638

Berlin. D. F., & White, A. L. (2012). A longitudinal look at attitudes and perceptions related to the integration of mathematics, science, and technology education. *School Science and Mathematics, 112*(1), 20–30.

Brown, A., & Borko, H. (1992). Becoming a mathematics teacher. In D. Grouws (Ed.), *Handbook of research of mathematics teaching and learning* (pp. 209–239). New York: MacMillan.

Burton, M., Daane, C. J., & Giesen, J. (2008). Infusing mathematics content into a methods course: Impacting content knowledge for teaching. *Issues in the Undergraduate Mathematics Preparation of School Teachers (IUMPST): The Journal, 1.* https://www.math.ttu.edu/k12/htdocs/journal/1.contentknowledge/burton01/article.pdf

Castro Superfine, A., Prasad, P. V., Welder, R. M., Olanoff, D., & Eubanks-Turner, C. (2020). Exploring mathematical knowledge for teaching teachers: Supporting prospective teachers' relearning of mathematics. *The Mathematics Enthusiast, 17*(2&3), 367–402.

Clements, D. H., & Sarama, J. (2009). Learning trajectories in early mathematics—Sequences of acquisition and teaching. *Encyclopedia on Early Childhood Development* (pp. 1–7). Canadian Language and Literacy Research Network.

Conference Board of the Mathematical Sciences. (2012). *The mathematical education of teaches II.* American Mathematical Society. https://www.cbmsweb.org/the-mathematical-education-of-teachers/

Confrey, J., Maloney, A. P., & Corley, D. (2014). Learning trajectories: A framework for connecting standards with curriculum. *ZDM Mathematics Education, 46*(5), 719–733.

Edwards, N. T. (1997). Integrating content and pedagogy: A cultural journey. *Action in Teacher Education, 19*(2), 44–54. https://doi.org/10.1080/01626620.1997.10462865

Ford, P., & Strawhecker, J. (2011). Co-teaching math content and math pedagogy for elementary pre-service teachers: A pilot study. *Issues in the Undergraduate Mathematics Preparation of School Teachers (IUMPST): The Journal, 2.* https://eric.ed.gov/?id=EJ962626

Furner, J. M., & Kumar, D. D. (2007). The mathematics and science integration argument: A stand for teacher education. *Eurasia Journal of Mathematics, Science and Technology Education, 3*(3), 185–189. https://doi.org/10.12973/ejmste/75397

Goos, M., & Bennison, A. (2018). Boundary crossing and brokering between disciplines in pre-service mathematics teacher education. *Mathematics Education Research Journal, 30,* 255–275.

Greenberg, J., & Walsh, K. (2008). *No common denominator: The preparation of elementary teachers in mathematics by America's education schools.* National Council on Teacher Quality. https://eric.ed.gov/?id=ED506643

Grouws, D. A., & Schultz, K. A. (1996). Mathematics teacher education. In J. Sikula, T. J. Buttery, & E. Guyton (Eds.), *Handbook of research on teacher education* (pp. 442–458). Simon & Schuster Macmillan.

Hart, L. C. (2002). Preservice teachers' beliefs and practice after participating in an integrated content/methods course. *School Science and Mathematics, 102*(1), 4–14.

Hill, H. C., & Ball, D. L. (2004). Learning mathematics for teaching: Results from California's mathematics professional development institutes. *Journal for Research in Mathematics Education, 35*(5), 330–351.

Hillman, S. L., Bottomley, D. M., Raisner, J. C., & Malin, B. (2000). Learning to practice what we teach: Integrating elementary education methods courses. *Action in Teacher Education, 22*(sup2), 91–100. https://doi.org/10.1080/01626620.2000.10463043

Hurlbut, A. R. (2018). Online vs. traditional learning in teacher education: A comparison of student progress. *American Journal of Distance Education, 32*(4), 248–266. https://doi.org/10.1080/08923647.2018.1509265

Lloyd, G. M., & Frykholm, J. A. (2000). How innovative middle school mathematics can change prospective elementary teachers' conceptions. *Education, 120*(3), 486, 575–580.

Marzocchi, A. S., & DiNapoli, J. (2020). Infusing pedagogy in mathematics content courses for future elementary teachers. *Issues in the Undergraduate Mathematics Preparation of School Teachers (IUMPST): The Journal, 2*, 1–10.

Masingila, J. O., & Olanoff, D. (2021). Who teaches mathematics content courses for prospective elementary teachers in the USA? Results of a second national survey. *Journal of Mathematics Teacher Education, 25*(4). https://doi.org/10.1007/s10857-021-09496-2

Masingila, J. O., Olanoff, D. E., & Kwaka, D. K. (2012). Who teaches mathematics content courses for prospective elementary teachers in the United States? Results of a national survey. *Journal of Mathematics Teacher Education, 15*(5), 347–358.

Mitchell, M., Leachman, M., & Saenz, M. (2019). *State higher education funding cuts have pushed costs to students, worsened inequality.* Center on Budget and Policy Priorities. https://www.cbpp.org/sites/default/files/atoms/files/10-24-19sfp.pdf

Rinke, C. R., Gladstone-Brown, W., Kinlaw, C. R., & Cappiello, J. (2016). Characterizing STEM teacher education: Affordances and constraints of explicit STEM preparation for elementary teachers. *School Science and Mathematics, 116*(6), 300–309. https://doi.org/10.1111/ssm.12185

Schram, P. W., & Rosaen, C. L. (1996). Integrating the language arts and mathematics in teacher education. *Action in Teacher Education, 18*(1), 23–38.

Steele, M. D., & Hillen, A. F. (2012). The content-focused methods course: A model for integrating pedagogy and mathematics content. *Mathematics Teacher Educator, 1*(1), 53–70. https://doi.org/10.5951/mathteaceduc.1.1.0053

Stein, M. (2022, April 25). The end of faculty tenure. *Inside Higher Ed.* https://www.insidehighered.com/views/2022/04/25/declining-tenure-density-alarming-opinion

Strawhecker, J. (2005). Preparing elementary teachers to teach mathematics: How field experiences impact pedagogical content knowledge. *Issues in the Undergraduate Mathematics Preparation of School Teachers (IUMPST): The Journal, 4.* https://eric.ed.gov/?id=EJ835513

Weber, E., Walkington, C., & McGalliard, W. (2015). Expanding notions of "learning trajectories" in mathematics education. *Mathematical Thinking and Learning, 17*(4), 253–272.

CHAPTER 21

MAKING AS A WINDOW INTO THE PROCESS OF BECOMING A TEACHER

The Case of Moira

Steven Greenstein
Montclair State University

Doris Jeannotte
Université du Québec à Montréal

Erin Pomponio
Montclair State University

The process of becoming a mathematics teacher is one that entails the development of an integrated base of teacher knowledge. A considerable body of research has been undertaken to identify the knowledge that is essential to effective mathematics teaching and to determine how it could be developed. In this chapter, we aim to make a contribution to this body of research by accounting for the essential role of the *knower's* situated and agentive interactions with the social, material, and conceptual artifacts that mediate such *knowledge* development. We do so through a

The AMTE Handbook of Mathematics Teacher Education: Reflection on Past, Present and Future—Paving the Way for the Future of Mathematics Teacher Education, Volume 5
pages 423–445.

revelatory case study of one prospective teacher named "Moira" as she participates in a constructionist Making experience within a specialized mathematics content course for future elementary teachers. Using both cultural-historical activity theory and figured worlds perspectives, we took a novel approach to the analysis of a range of interactions that mediated Moira's design activity. This analysis of three "moments of becoming" on Moira's trajectory demonstrates the methodological value of this analytic approach by revealing the blended nature of knowledge and identity that constitutes learning in and through collective social practice. In addition, these findings establish the theoretical value of framing Making as mediated learning and offer empirical evidence of the formative power of the Making experience not only for the development of Moira's mathematical, pedagogical, and design knowledge, but also for her identity as a mathematics teacher. We conclude the chapter by considering the implications of these findings for teacher preparation coursework and proposing future directions for research.

INTRODUCTION

The process of becoming a mathematics teacher entails the development of an integrated base of teacher knowledge. A considerable body of research has been undertaken to identify the knowledge that is essential to effective mathematics teaching (e.g., Ball et al., 2008; Shulman, 1986) and to determine how it could be developed (e.g., Scheiner et al., 2019). That said, one risk in prioritizing knowledge over the knower is that doing so suggests an acquisition model of learning that may not account for the essential role of the knower's situated and agentive interactions with the artifacts that mediate their knowledge development. Insight into the complexity of formative interactions can illuminate the processes of teacher identity development and have implications for teacher preparation coursework, specifically in relation to teacher learning through the design of instructional resources where very little research has been done to date (Pepin, 2018). Thus, in this chapter, we aim to contribute to the research on teacher learning by elucidating moments of a teacher becoming as they are mediated by a variety of social, material, and conceptual resources within teacher preparation. We do so through a revelatory case study of a prospective teacher named "Moira" as she participates in a Making experience within a specialized mathematics content course for future elementary teachers.

MAKING IN/AS MATHEMATICS TEACHER PREPARATION

We connect with a body of research that conceives of *Making* in education as the creative practice of designing, building, and innovating with analog and digital tools and materials. Schad and Jones's (2019) review of research on the Maker movement in K–12 education finds that students' learning through Making dominates that literature, with foci that include the improvement of STEM learning outcomes; increasing student motivation, interest, and agentive learning opportunities; and increasing equity by broadening notions of what counts as Making.

Their review also foregrounds how *teachers* learn to design and run makerspaces and integrate maker-centered learning strategies (Clapp et al., 2016) into their curriculum. However, there is almost no research on supporting teacher learning through Making.

The case study we share here is part of a larger study testing the hypothesis that a pedagogically genuine, open-ended, and iterative design experience centered on the Making (Halverson & Sheridan, 2014) of an original physical mathematics manipulative would be formative for the development of the kinds of conceptual and pedagogical thinking that would enable prospective mathematics teachers (PMTs) to support and promote students' mathematical sense-making, understanding, and reasoning. Accordingly, we view this Making experience from a constructionist perspective (Papert, 1991), which posits that meaningful learning happens through the designing and sharing of digital or physical artifacts "that learners care about and have some degree of agency over" (Schad & Jones, 2019, p. 2). In this regard, we see promise in the design experience for supporting future educators to enact the knowledge, skills, and dispositions that AMTE (2017) suggests in its standards are essential to its image of a "well-prepared beginning teacher of mathematics" (p. 6). Relative to our study, this support comes from Making experiences that focus on "essential big ideas across content and processes that foster a coherent understanding of mathematics for teaching" (P.2: Opportunities to Learn Mathematics) in a curricular context where mathematics, pedagogy, and knowledge of students as learners are integrated (P.3: Opportunities to Teach Mathematics).

Framing *teachers as designers* (Kalantzis & Cope, 2010; Pepin et al., 2017) naturally evokes notions of teacher identity, which was central to our analysis of the trajectory of Moira's becoming a teacher. That analysis entailed a view of identity as a collection of "narratives about individuals that are reifying, endorsable, and significant" (Sfard & Prusak, 2005, p. 16), and a framing of learning to teach mathematics through Making as the interplay among discourses of identity, pedagogy, mathematics (see Heyd-Metzuyanim & Sfard, 2012; Sfard, 2008), and design (i.e., narratives about design decision-making; see Greenstein et al., 2020). This framing allowed us to document moments in the process of becoming a teacher through an analysis of changes in discourse resulting from dialectic interactions between a PMT's iteratively evolving design, their knowledge of mathematics, pedagogy, and students that mediate its evolution, and their theoretical link to teacher identity as it is shaped through Making. Now, we take a step further and use a novel theoretical approach to answer the following research question: *How does a Making experience in mathematics teacher preparation mediate the social and conceptual dimensions of the process of becoming a teacher?*

THEORETICAL FRAMEWORK

This research question focuses our investigation on both social and conceptual dimensions of the process of becoming a teacher. Thus, we bring both situated and sociocultural theories of learning to bear on our attention to practice and to the social and conceptual artifacts that mediate it. In particular, this work is grounded in Engeström's (1987) cultural-historical activity theory and Holland et al.'s (1998) concept of figured worlds.

Cultural-Historical Activity Theory (CHAT) assumes that cognition can only be understood when it is studied within the context (Lave & Wenger, 1991) of individual-environment interactions and the resources that mediate them. Leontiev's (1978, 1981) contribution is the distinction he makes between individual actions and a *culturally* and *historically* informed form of social (inter)action called *activity*. Engeström's (1987) depiction of an *activity system* (Figure 21.1) is a useful conceptual tool for situating individual actions and collective activity simultaneously within the contexts in which they arise.

The six nodes of the triangle designate the six core components of an activity system. The double arrows express the dynamic nature of the six nodes and are the basic unit of analysis. Three of these nodes appear in Vygotsky's (1978) original view of mediated activity: The *subject* (e.g., actor, learner) uses *instruments* (i.e., social and material tools and signs) to achieve an *objective* (i.e., the orienting goal of the activity). The *community* consists of the people who participate along with the subject in the activity. Community-subject relations are mediated by *rules* that regulate the subject's relations with other participants and by *divisions of labor* that are horizontal between members of the community with equal power and status, and vertical between members with differences in power and status (Engeström, 1999).

As this work seeks to understand both the social and conceptual dimensions of the process of becoming a teacher, we supplement this perspective on learning by drawing on Holland et al.'s (1998) concept of *figured worlds*, which captures not only the mediating role of artifacts in learning, but also their role in identity formation (see Vågan, 2011). A figured world is defined as "a socially and culturally constructed realm of interpretation in which particular characters and actors

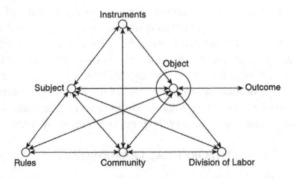

FIGURE 21.1.   Engeström's Conceptualization of an Activity System (1987, p. 78)

are recognized, significance is assigned to certain acts, and particular outcomes are valued over others" (Holland et al., 1998, p. 52). Drawing upon the same culturally and historically informed forms of environmentally-contingent social interaction upon which CHAT theory is framed, figured worlds are perpetuated by participants' interactions as they *figure* who they are in relation to each other.

From a situated perspective, knowing is participating, and learning is marked by progress toward more central participation in a community of practice (Lave & Wenger, 1991). As such, these processes of learning are also processes of identity formation (Vågan, 2011), or "becoming" (p. 44). As figured worlds are defined by the ways in which people participate in them, "the situatedness of identity" (Holland et al., 1998, p. 40) is central to the concept. Furthermore, the figured worlds concept contributes to a situated perspective by emphasizing that *artifacts mediate the formation of identity* (see Vågan, 2011). This insight from a *sociocultural perspective* on mediated activity captures how artifacts provide people with "tools of agency and identity" (p. 45) that enable them to change their identities and re-conceptualize who they are (p. 49). As Urrieta (2007) suggests, "Figured worlds are intimately tied to identity work" (p. 207).

METHODOLOGY

The study shared in this chapter took place in a specialized mathematics content course for PMTs at a mid-sized public university in the Northeastern United States in which over 90% of the students in the class identified as female. Situated in an instructional context in which the teacher educator (TE) practiced an inquiry-oriented pedagogy, the Making experience was defined by the following task: "The purpose of this project is for you to 3D design and print an original physical tool (or 'manipulative') that can be used in teaching a mathematical idea, along with corresponding tasks to be completed by a learner using the tool." These designs aimed to reflect a) a PMT's knowledge of what it means to do mathematics and how we learn with physical tools, b) their knowledge of elementary-level mathematics, and c) their perspective on pedagogy and curriculum in mathematics education. In addition to the design of the tool, the data corpus included: 1) a "Math Autobiography" that calls on students to reflect on their experiences as a student of mathematics and consider how those experiences may have informed their relationship to it; 2) an "Idea Assignment" that describes a PMT's initial thoughts about a manipulative they want to create; 3) a "Project Rationale," which is an account of how their design reflects an understanding of what it means to know and learn mathematics; and 4) a "Final Paper/Reflection" that presents findings from a "Getting to Know You" interview and two problem-solving interviews conducted by the PMT with an elementary-age focus student.

Thirteen PMTs comprising ten groups participated in the study. PMTs worked on their designs during four in-class design sessions that were held in a large design lab (Figure 21.2).

FIGURE 21.2.   Images of the Design Lab

THE REVELATORY CASE STUDY APPROACH

We took a *revelatory* case study approach (Yin, 2009) to better understand the processes at play in design activities that are hypothesized to underpin moments on a trajectory of becoming a teacher. The contextual conditions that may be relevant to these processes provided an added rationale for the case study approach (Yin, 2009). We selected "Moira" for this case study because, more than any other participant, her experience was made "visible" to us through vivid expressions of that experience. In a phrase, she designed aloud. The words, gestures, and other embodied actions of Moira's design conversation (Schön, 1992) gave unprecedented access to elaborate discourses of identity, mathematics, design, and pedagogy (Greenstein et al., 2020). Thus, insights about the phenomenon of becoming a teacher that can be illuminated through thick descriptions (Miles & Huberman, 1994) of Moira's case justified the use of a purposefully sampled (Patton, 2002) single-case study on the grounds of its revelatory nature.

Data Analysis

With our investigation focused on the social and conceptual dimensions of the process of becoming a teacher, we took a constant comparative approach (Strauss & Corbin, 1998) to analyzing the data. After transcribing the video data of sessions in which Moira appeared, we compiled a chronology of experiences that would allow for a time-series analysis (Yin, 2009) of the data. The video transcripts and Moira's written coursework assignments were embedded into the chronology. Next, working individually, the three authors analyzed the data from both CHAT and figured worlds perspectives. The CHAT perspective offered a lens with which to situate Moira's activity in relation to the design context, while the figured worlds perspective captured the mediating role of social, material, and conceptual artifacts on learning as identity formation.

In an iterative fashion, following each review of the chronology, the three researchers took turns re-viewing the video in order to enrich the transcript and refine and elaborate prior interpretations. These "review and reinterpret" cycles continued until there was consensus among the researchers that the data had been saturated (Corbin & Strauss, 2008) and no new insights were being revealed.

Results

At the end of our analysis, we chose three "moments of becoming" from the pool of moments that had been coded. Moments 1 and 2 lead up to Moment 3, which occurs during a follow-up interview with Moira that sought to understand her rationale for a change in her design idea. These moments explain elements of that third moment more deeply. Moment 1 helps us understand Moira's passion for creativity, giving insight into her mathematical and creative knowledge at critical points in her design process. Moment 2, which details the change in Moira's initial design idea, is linked to her problem-solving activity, which appears in Moment 3. Our results present these three moments in chronological order.

MOMENT 1: "I SHOULD NOT BE ALLOWED IN THIS CLASS. I'M HAVING TOO MUCH FUN."

This first vignette focuses on Moira's participation in the first of four design sessions. As the vignette opens, the students are learning to use the core functions of Tinkercad (Autodesk, Inc., 2021; e.g., group, merge, and align) in order to design a cup with their name on it. As the TE demonstrates how to align a cylinder to the center of a cube, Moira has already figured out that the purpose in doing so is to 'subtract' the cylindrical space from the cube to form the negative space of the cup's interior. It becomes evident that Moira has only imagined this process and never carried it out, because when the TE demonstrates how to *merge* the cube and the cylindrical hole to form the negative space of a cup, Moira gasps at the effect. "Oh, I love it," she exclaims. "So pretty." When the TE asks the class, "How do you think we can take that torus and make it into a handle?" another student replies, "Cut it," and within 30 seconds Moira has done so and also rotated the half-torus to fit into place on the side of the cup. Just two minutes later, Moira completes the task by attaching the handle and announces, "I did it!"

Moira continues designing at her own direction by "adding a little me" to one side of the cup and putting "a heart on the front." Again, these actions entail processes that have not yet been demonstrated to students; Moira figures them out entirely on her own. Then, she shares, "I should not be allowed in this class. I'm having too much fun." When the tutorial moves on to demonstrating how to add a figure to the face of the cup, Jenna, one of her classmates, positions Moira as the expert and asks her for assistance. This marks the first of three instances during the tutorial when a different student asks Moira for help.

When the directed tutorial ends, Moira exclaims, "That was so cool! Thank you." There is excitement in the expression on her face. Her mouth is agape and

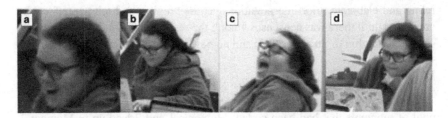

FIGURES 21.3a–21.3d.   Alternating Moments of Thrill and Absorption

she's wide-eyed as she leans in toward her screen (see Figures 21.3a–21.3d): "Oh my god," she exclaims. "I'm so excited."

Learning About Moira Through her Mathematics Autobiography

We learn about Moira's relationship with mathematics through her Math Autobiography, and the connections we make to it help us come to better understand the dynamics of her design activity. In her autobiography, she shares her experiences as a child at home with an affinity for mathematics and as a mathematics student from elementary to middle school and on to college. Features of the figured worlds she's inhabited are invoked through narratives of her history of engagement with mathematics that convey a sense of that relationship as, in a word, tumultuous. In the analysis that follows, we rely on images of these worlds to make inferences about how Moira's participation in them may explain the formation of her identity through agentive activity in the figured world of the design context.

"From a young age," Moira was "fascinated by math." Both at home and at school, the "puzzles and games" and the "math [that] was a way to solve [them]" mediated her relationship with mathematics. Her favorite television show was *Cyberchase*, "a show about three kids solving math problems to save the universe. It fascinated me that math could be so important that it literally saved everyone… That was where my love of math started," she shares, "and I tried to keep that up in school."

At the same time that Moira was fascinated with mathematics at home, she was "performing strongly" in math in elementary school. Her fondest memories were of the "Math-a-thon, which was a 100-problem packet" given to her by a local hospital as part of a fundraiser for cancer research. "The task was an absolute joy for me," she writes, "and I would always sprint through [it]." Moira recalls always being "invited to the reward lunch for top students because of [her] completion of the packet" and for "raising a lot of money along the way." Math wasn't just a "game" she got to play at home, "Math was a game [she] got to play in school, and the Math-a-thon gave [her] a way to play it even longer." Her teachers "were very supportive of [her] love of math," ensuring that she had problems to work on whenever she wanted.

While Moira's "love of math" gave her opportunities to "play," earn "reward[s]," and be positioned as a "top student" in elementary school, sadly her relationship with mathematics began to sour in middle school. That's where she "started doubting [herself] more and more," she explains, "to the point where I was performing extremely poorly... I had really liked math when I was younger, but I was under the impression that I was not a 'math person' by this point, so I just gave up."

Moira's relationship with math was further ruptured when she "started taking theater classes in school. Kids like me were told that we had 'creative minds' and that math and science will probably come difficult to us." This message, that one cannot be both creative and mathematical, is one we find particularly salient in light of a comment she made during the follow-up interview we present in Moment 3: "Making a graph is like seeing the formula that you made become art." Surely, we argue, the mathematics community would regard Moira's vision as a counterexample to the message she'd received. We also find the message salient in our interpretations of her design activity in the first design session when students were given the time and space to "play around." It is in our descriptions below of phenomena that occurred within that space that the constructs of "space of authoring" and "space of play" (Holland et al., 1998) prove useful as we explain what the figured world seemed to have been like for Moira.

Back to the Design Tutorial

Returning to the first design session, the tutorial has ended and students are self-directed in their tinkering. Out loud, Moira asks herself, "Do I wanna make myself? Yes, I do." She waves her arms as she contemplates the making of her body. Then she pauses and concedes, "I don't think I have the skill yet. I have to start small." Not long afterward, she resumes her plans to make a figurine of herself (see Figure 21.4). With a fluency uncharacteristic of a first-time user, it has taken Moira under a minute to import and scale a picture of her face in Tinkercad. Whatever's driving her to forge ahead is unrelenting, even amidst risks in the design process that a novice designer might find themselves unable to navigate.

The assemblage of design-related practices offered a particular *space of authoring* (Holland et al., 1998) for Moira's identity. Iterative enactments of agency involving conceptual (e.g., knowledge from a history of creative experiences), material (e.g., the use of Tinkercad to bring forth her design), and social artifacts (e.g., community relations and interactions) mediated her identity development as she "reconceptualize[d] what and who" (Vågan, 2011, p. 49) she was from one moment to the next. Indeed, from the very first moment, in choosing to make a 3D print of herself, she quite literally brought *her* self into the experience. Through her ongoing participation in a figured world of play, she did so *figur*-atively, as well.

FIGURE 21.4 Moira's Figurine of Herself (right) and a Demonstration of Her Design Intentions Using Cubes (center).

Residing within a space of authoring, a *space of play* (Holland et al., 1998) provides a window into identity production, with "play" conceived as the "medium of mastery, indeed of creation, of ourselves as human actors" (p. 236). As evidenced in her mathematical autobiography, play has always been an essential feature of Moira's identity. And as we observe through her design activity, play is who she is. Almost entirely on her own, she does the trial-and-error tinkering to develop fluency with Tinkercad, she adorns her cup by "adding a little me" to one side, and she imagines she "could make a cat" or a "cartoon of [her] face." She's "thrilled," "excited," and "having too much fun."

What is driving Moira's improvisational play? It appears to have been fueled by creative tendencies and sensibilities that were embraced in elementary mathematics but later suppressed through the messages she received later on. In striking contrast, these creative capacities were not only invited by the TE into the design session, they were "demand[ed]" in response to the "possibilities of a design situation" (Schön, 1992, p. 4), which Moira helped create through her own volition. This is evident very early in the design tutorial when Moira takes it upon herself to move beyond the boundaries of a rather structured design tutorial. The community-subject interactions that mediated her classmates' positioning of her as an expert may have motivated her, as well, for they speak to the figured world that emerged for Moira as she was once again positioned as the "top student" she was in elementary school.

MOMENT 2. "UGH, YOU'RE RIGHT. WE CAN'T FLIP THE FIRST ONE. I'M GONNA WORK ON THIS."

Moment 2 occurs across multiple classes and begins at about the halfway point in the course. At this point, the PMTs have explored various topics in mathematics teaching and learning, such as the cognitive demand of tasks (Smith & Stein, 1998) and CGI (Carpenter et al., 1996) problem types for addition and subtraction. They have also completed the first of two problem-solving interviews with an elementary-aged child. What we will be observing in this moment are the mediating roles of three social and conceptual artifacts in Moira's design activity that will ultimately lead to her final design. First are the expectations set forth by the TE: The tool should be novel, it should be responsive to the needs of an actual learner, and it should be useful for solving problems in a variety of ways. Second, the *culture* of the classroom–the social and conceptual configuration of values reified in actions established and encouraged by the TE–contributes to moments of tension, excitement, confidence, and doubt for Moira as she navigates the expectations. And third, these values are taken up, enforced, and perpetuated by the Maker classroom community in their interactions with Moira. By focusing on the situated nature of Moira's design activity, we demonstrate how these artifacts mediate her developing design as well as her changing identity as a knower, learner, designer, and teacher of mathematics.

Prior to this moment, Moira had just interviewed "Grace," who, she said, struggled with math: "[Grace] likes... adding and subtracting, but dividing fractions, multiplying big numbers, and most concepts of algebra are foreign to her." Mathematical ideas "foreign" to Grace become the basis for Moira's design intentions, which she describes as "want[ing] to do something with fractions." For her initial design action in this session, Moira sketches a circle in her notebook, divides it pizza-like into six equal parts, and in clockwise order, she places one Katie Cube on each of the 6 partitions. Next, she demonstrates her design intentions through an equal sharing problem: "Suppose we have 2 apples and we're dividing them up between 6 friends." Then she distributes a second layer of 6 cubes onto the first (see Figure 21.4).

In a moment of reflection, Moira's engagement in a process of learner-centered design is apparent in what she shares next with her classmate, Aria, about another child, Finn, with whom she is currently working on mathematics: "I want to help Finn, but [the TE] was like, 'They could just do that [CGI problem] with fraction circles, so you're just making something that's already made.'" Thus, the project's 'novelty requirement' is brought to bear on Moira's design thinking. As a result, she decides to abandon the idea. She then begins to imagine the design of a tool for fraction division, a concept inspired by struggles she identified in her work with Grace. "What if it was like dividing fractions," Moira asks Aria, "like a dividing fraction machine?" To demonstrate what she has in mind, she again accesses the Katie Cubes. Rather than using them to represent a quantity, she explains that they will be connected in pairs and that digits will be drawn on each one to match the numerators and denominators of fractions in a fraction division expression (see Figure 21.5). On the right of Figure 21.5 is Moira's revision to her design, which she made once she realized that the flipping of a fraction would turn the digits upside down. Representing the numerators and denominators as

FIGURE 21.5.   A Reconstruction of Moira's Paper Prototype for a "Fraction Machine"

dots instead of digits leverages an affordance of domino-like representations of numbers, making them agnostic with respect to orientation.

Moira's intention in designing such a tool is to make the "keep-change-flip" algorithm tangible to learners. Analogous to the power she attributes to visual representations of ideas, Moira believes that a tool that can make the algorithm *concrete* will make what was once an abstract model of fraction division meaningful to students. Ironically, Moira's design for a machine to help students learn fraction division betrays her aversion to the disconnected, mechanical approach she experienced in her own learning of mathematics. With this in mind, we seize on the irony to suggest that it offers evidence of a moment of becoming for Moira. Through her participation in the figured world of the design experience and in the class more broadly, she's been given a lot to consider relative to the expectations for her design. These considerations–or forces (Holland et al., 1998)–take on increasing salience for her, and as they do so, they contribute to the meaningfulness of the pedagogy that motivates her.

By now, Moira has told Aria multiple times that thinking out loud is vital to her creative process. In return, Aria has been a critical friend to her. Through feedback that is typically offered in rhetorical form, she endorses Moira's ideas while also seeking to gain some understanding of the rationales for her design decisions. Relative to power and status, Moira's interactions with Aria contrast with her interactions with Caragh, a research assistant who soon engages with Moira and informs her thinking.

Caragh's interactions with students tend to be of a pedagogical nature, often provoking students to further reflect on their designs in relation to the project's expectations. Our impression from the manner in which students respond with deference to her questions is that they attribute some authority to her and regard her as a more knowledgeable other (Vygotsky, 1978). We see one example in her interactions with Moira, who interprets Caragh's words and gestures as assessments of the quality of her ideas. Here, Caragh probes Moira's thinking about her keep-change-flip "machine":

Caragh: What kind of conceptual understanding could they gain from that?

Moira: I think conceptually, the thing that they can gain is that... they will understand that multiplication and division work together like addition and subtraction... This shows that you *can* do multiplication and division with fractions, and you can visually see that it's the same thing. That's my theory. <pauses as she takes note of Caragh's reaction> Your face makes me think that I'm bad..."

Moira's response to Caragh's question tells us that Moira finds it valuable for students to be able to visualize key mathematical relationships, which means her tool should have affordances for visualization. Yet, she also experiences dissonance as Caragh's question is a tacit endorsement of the rule that developing conceptual understanding should be a goal of teaching and thus a feature of her manipulative. In its current form, her design is lacking in that affordance.

Shortly thereafter, Moira tells Caragh that she wants to design a tool that builds from a child's current understandings to "figure higher-level concepts out." In particular, she wants to design a tool that can be used to leverage a student's current understanding of multiplication and division of whole numbers to teach them a procedure for multiplying and dividing with fractions. In doing so, she acknowledges the value of that prior understanding. However, the procedure she aims to teach is disconnected from multiplication, division, and fraction concepts. Eventually, the endorsed vision of developing conceptual understanding will be seen as generative for Moira when it induces tensions that eventually provoke an inflection point in her pedagogical thinking and catalyze a change in her design. An upcoming interaction with the TE will give rise to such a tension as Moira is reminded of what it means to use her design to teach for understanding.

Next, Caragh returns the attention to Moira's design and asks her *why* the divisor gets flipped in the "keep-change-flip" algorithm. With apparent trepidation, Moira responds, "I feel like they do it because–because–I gotta think about that. *I* know that [when] dividing fractions, you flip the second one. *But why do you do that?*" "Right," says Caragh, "and I'm wondering if... get[ting] at that concept... changes the way you think about your manipulative..." Moira replies with a bit of laughter, "I don't know if it will... I'm not sure yet."

Moira's response entails a discursive shift from design thinking to mathematical thinking that served as a pivotal moment in her design trajectory as she suddenly realizes that a procedure that she's used for years is actually devoid of meaning. Intent on understanding the algorithm, Moira writes, "Why do we flip the second and multiply?" in her notebook and then looks through her class notes to see if there's something in there that can help her understand it.

Continuing her inquiry over the next fifteen minutes, she has yet to make sense of the algorithm when the TE arrives to check in on her. As she presents her idea, she seems to be anticipating its inevitable rejection, as she's mindful of its lacking capacity to support learning with understanding:

Moira: What we do is, we flip [the divisor] upside down.

TE: <light-heartedly> No. I'm out.

Moira: Why not?!

TE: You're just teaching them to flip and multiply.

Moira: But I'm not, because I'm showing that division and multiplication are inverse properties!

TE: Even if you say the reason we do it is because they're inverses of each other, it doesn't explain why we flip and multiply. Why not flip the first one?

Moira: I'm gonna work on this.

At some point between this moment and the next in-class design session, Moira abandoned her fraction division design idea. Mediated by her transforming understanding of mathematics and aesthetically-driven design choices, Moira ultimately creates a manipulative for comparing fractions that consists of a collection of rings of equal diameter with each ring containing a different number of equally spaced notches that divide the rings into equal partitions of a whole (see Figure 21.6). Unlike the discrete pieces that form fraction circles, she chose to use notched rings in order to differentiate her design from an existing one, as is the "novel" expectation of the project.

As we analyzed the evolution of Moira's design within Moment 2, we came to realize the saliency of several artifacts that mediated her design decisions. A lived

FIGURE 21.6. Moira's Fraction Rings

history of experiences in the figured worlds of other mathematics classrooms informed her interactions with fraction concepts and the concept images (Tall & Vinner, 1981) that were co-determined in them. In addition, pedagogical commitments and her knowledge of content and students (Ball et al., 2008) interacted with design activity that shaped the flow of social interactions distributed across the design environment. In particular, in order of increasing power and status, Aria played the critical friend for Moira by thinking through her design ideas with her, Caragh played the more knowledgeable other by reminding her of a community-endorsed pedagogical vision, and the TE played the role of final arbiter on design ideas as he recentered Moira's inquiry at the intersection of teaching with understanding and the provision of tools to support it. Taken together, we are reminded of Roth's (2012) assertion that we are only able to understand the actions of a subject on the object of their activity when we consider all the relations that mediate every aspect of the activity. Thus, this situated network of individual and collective activities that mediate the distinctive evolution of Moira's manipulative offers concrete markers of the formation of her identity as she emerges as a more central participant in an educational design community.

MOMENT 3: "I JUST FIGURED OUT SO MUCH MATH!"

This third and final vignette takes place in a follow-up interview with Moira and three researchers on the project, the teacher educator (TE) who taught Moira's class, a second researcher (SR), and a third researcher (TR). As we mentioned earlier, Moira abandoned her idea for a "flip-and-multiply" tool and chose to create one for fraction comparison instead (Figure 21.6). She made this decision outside of class, but it wasn't until we were in the midst of data analysis that we realized we were unsure of the rationale for the change. In response, we invited her in for a task-based follow-up interview and asked that she bring her manipulative with her.

The interview begins with us asking Moira what was important in the design of her Fraction Rings. In response, she pointed to the pedagogy promoted in the class: "[Teaching is] supposed to be facilitating learning and having kids create their own ways to learn. So, I wanted to make something that... wasn't just something that I was forcing them to, like, 'Alright, you have to use it this way.'" She also explained the role that content played in her design by adding that she decided to focus on "fractions, because they are the first concept for kids that they just cannot visualize" like they can when they use their fingers to add. Specifically, she "wanted to create something that they could compare fractions with," which is a struggle experienced by the subject of her problem-solving interview. Enacting the agency that the figured world of the interview setting affords her, Moira demonstrates how seeing enables this comparison of two fractions. She confidently places the thirds ring on top of the fourths ring in order to compare one-third and one-fourth. She moves her finger along the space between two notches on the thirds ring and then along the space between two notches on the fourths ring and concludes that one-third is greater than one-fourth, because the 'thirds space' is longer than the 'fourths space.' She concludes by sitting back and surveying the landscape of what she and her tool have just accomplished. These interactions between Moira's emotions and her mathematical thinking remind us that "affect and cognition constitute different manifestations of the [same] material situation" (Roth, 2012, p. 7). Immediately thereafter, the SR positions Moira as competent when she smiles, nods her head in appreciation, and remarks, "That's *really* cool." "Because for kids," Moira adds, "being able to see everything lined up next to each other is something that, for me personally, that was helpful."

From her recognition of the difficulty children have in visualizing fractions, to her intention to design her rings so that children would be "able to see" equivalent fractions, to her use of perceived length as a metric for fraction magnitude, and finally to her own admission of what's "helpful" about "being able to see," what this opening segment reveals to us is Moira's model of *knowing as seeing*, with *doing as enabling seeing*. That is, for Moira, the primary source of mathematical relationships is in visual images, and in this instance, images enable the comparison of fractions. Thus, she designed her tool with affordances for likely manipulations (i.e., doings) that produce images (i.e., seeings) from which comparisons are made and conclusions are drawn (i.e., knowings). This is an epistemological stance that appears to be informing Moira's developing pedagogy as evidenced in the design of her tool.

Next, we noted that Moira's initial design was a tool for fraction division and asked her why she abandoned it. Again, she spoke about how her pedagogical vision informed her design: "Well, it's because we were talking and you [the TE] said, 'You are just giving them a way to solve the problem.' And I realized, you're right... I was just ... presenting the formula in a visual way. So that is why I switched to something else." By pointing to the TE's influence as a catalyst for her design change, two artifacts that mediate her reconsideration are realized. First, Moira's acceptance of the TE's statement, "You are just giving them a way to solve the problem," which is unaccompanied by any explicit suggestion about how to take it, implicates Moira's willingness to take on the responsibility of an open-ended task. Second, a pedagogy endorsed in the class that is only implicitly embedded in the TE's statement implicates community rules that inform the reconsideration. That is, students understood the expectation that their designs should yield tools that could be used to support the endorsed pedagogy.

Adding to her rationale for a change in design, Moira echoes the importance of designing a tool that allows for multiple uses, yet the "flip and multiply" tool only permitted a single use: "This is useful for, like, one time and then you're like, 'Oh, I figured it out.' ... So, then I came up with this [fraction comparison tool]" instead. These reflections reveal how Moira's design change speaks to her pedagogy and the teacher she's becoming. As a teacher, it is important that students have the opportunity to develop their own ways of thinking about fractions with a tool that can be used in a variety of ways. By switching to a design for comparing fractions, Moira could honor the teacher she wants to be.

Next, we transitioned to a problem-solving interview initiated by the SR's question, "Have you thought about using your tool for understanding fraction division?" Moira began by placing the halves ring on top of the fourths ring as she poses the problem, "One half divided by 2." She references one half of the halves ring in relation to the two fourths of the fourths ring that lie below it: "If I split this half in 2, it's two fourths." She then reconsiders her answer: "Or it's one fourth [...]" Moira's change of answer had us unclear about whether she was using her tool to find an answer or whether she was using it to illustrate an answer she had already found mentally, so we sought clarification. The TE placed the eighths ring on top of the halves ring and asked Moira, "So what question does this answer?" "This answers one half divided by 4. There are four equal parts in one half and that's <pointing to one partition of the eighths ring> one eighth." When Moira observed from our body language that we were both pleased and surprised by what she had just demonstrated, she asked, "Did I mess that up?" "No," the TE responded, "I did not expect that." The SR gave a similar response. Both had expected Moira to use the measurement conception of division to interpret the situation as ½ divided by ⅛, yet counter to this expectation, Moira saw that the four ⅛'s on the eighths ring occupied the same space as ½ on the halves ring and interpreted the situation partitively as ½ divided by those four eighths, or ½ divided by 4. Moira, perhaps emboldened by how simultaneously unsure and impressed we seemed to be, offered to demonstrate her solution a second time. She restated the problem, "One half divided by 4," and the TE asked her what it meant. Seemingly now insecure, she paused, smiled, responded, "It means an answer that I'm going to find," and laughed. We laughed, too. She continued, "I need to find something where there are four equal pieces that equal one half." Then she demonstrates that the eighths ring satisfies this criteria: "If I'm dividing one half by four, <counting the four segments in the half of the eighths ring> 1, 2, 3, 4, I've got one piece left, which is one eighth." The TE asks again, "So one half divided by 4 means?" Moira responds, "One half split into 4 pieces."

The researchers' request for Moira to use her tool in a new way opened a window into her mathematical and pedagogical thinking via her instrumented activity. She had not yet developed utilization schemes (Verillon & Rabardel, 1995) for engaging with fraction division using her tool, although her activity demonstrates that she did have the requisite mathematical knowledge to mediate their development. Nonetheless, what's worth noting here is not so much the change in her solution but that the tool mediated the change in mind. Actually, given the apparently tentative quality of her knowledge of fraction/division, it may be more appropriate to propose that the tool is mediating a mind that is changing. As the manipulative transforms from a tool for comparison to a tool for division, Moira is also transformed. The tensions she experiences as she uses the tool in new ways give rise to the change and eventually to an awareness of her increasing authority over the mathematics.

As we've mentioned, Moira's demonstration of the use of her tool in relation to the partitive conception of division surprised us. We had intended to explore the measurement conception with her, which was where we went next. Coincidentally, since Moira was not at all struck by the realization that her tool *could* be used for division–her original design intention–we surmise that it was the measurement conception she had in mind for that design. The outcomes of her subsequent activity confirm our suspicion.

The SR poses the question, "Can you do 1 divided by ⅓ with your tool?" and Moira responds assuredly, "Sure! <then tentatively> I guess. Maybe. I'll try." She pauses and adds, "Let me think. OK." She stacks the "whole" ring on top of the "thirds" ring and pauses. She seems to realize that she cannot take the same partitive (splitting) approach she took earlier. She continues, "So thirds are smaller than 1's. So, I guess I would have to–I don't know. This is, like, complicated to describe, but there are three pieces in 1. <pause> So I guess it would be 3." The TE and SR realize Moira's hesitancy and confirm to each other that "This is a different model of division." Moira agrees, and then continues, "I know I *can* do it and I'm *seeing* it, but I don't know how to describe it." Moira shifts her gaze from the tool that's been enabling her seeing to the researchers and admits, "This is something for me to work on." From this we infer that her desire to know not just *what* 1 divided by ⅓ is but also "how" it is, is evidence that she holds the model of sense making promoted in the course. Critically, Moira doesn't tell us that she doesn't know how to solve the problem; she tells us that she doesn't know how to solve it *yet*. Recall her use of similar language in another instance in Moment 1 as she was considering creating a design of her own body: "I don't think I have the skill yet." Our framing of Moira as a teacher becoming seems to be consistent with how she's self-identifying. What she realizes she doesn't know is what she is intent on learning. And "learning is a process of becoming" (Vågan, 2011, p. 44).

Next, the TE asks Moira what 1 divided by 1/3 means. After referring to the "flipping" algorithm and thus realizing she's back where she just was, she responds, "I don't know, I don't understand dividing." Although Moira was handling pieces of a tool that could conceivably mediate her thinking toward a solution, her knowledge–and therefore her attention–was just too heavily anchored in this algorithm. Our next steps aimed to disrupt this tendency.

In an effort to help Moira connect what we learned of her knowledge of division and fractions, we then asked her to find $6 \div 3$ and explain her solution: "6 divided by 3? 2, because there are two 3's in one 6." Next, we posed "another one: $12 \div 3$," and Moira responded, "There are four 3's in…Oh! So! So! If I am dividing 1 by 1/3, there are 3 thirds in 1, so it's 3! Yes! You *can* do division with these! I just figured out so much math!" "How does this make you feel about your tool?" asks the SR. "Honestly, so impressed with myself," replied Moira. "I did not think that it had this capability. I thought it was only for comparing fractions."

In this moment of becoming, as Moira realizes that she's used her tool to *learn* fraction division, she's also realized her initial objective, which was to design a tool to *teach* fraction division. And yet again, affect and cognition arise as manifestations of the same accomplishment (Roth, 2012). In an act that is a manifestation of community relations (Engeström, 1987) that have the researchers participating alongside Moira in her quest to make sense of fraction division, everyone at the table celebrated along with her. "Wow!" Moira exclaimed, "Fractions make so much sense now. That blew my mind." To convince both Moira and ourselves that she really had constructed a durable scheme for fraction division, we concluded this segment of the interview with the problem, $1/2 \div 1/4$. Without hesitation, she points to two partitions on the fourths ring that are collectively set against one partition on the halves ring and concludes, "There are 2 fourths in 1/2, so 2! Wow. I've learned so much. This is a great day!"

Discussion

> The self is not something ready-made, but something in continuous formation through choice of action…
>
> *—John Dewey (1916, p. 352)*

In a chapter that aims to distinguish self-study from action research, Feldman et al. (2004) consider how existentialism might be used as a theoretical basis for understanding what it means to be a teacher. This consideration coincides with the efforts of other scholars who seek to understand why education research findings have had so little effect on teacher practice. As we looked across our own findings, we were tempted by what the existential approach could tell us about Moira at a larger grain size than the range of mediated interactions she experienced. We share with the reader our appreciation for what this lens can offer through a summative discussion of our findings.

Drawing on the works of John Dewey (1916), Maxine Greene (1973, 1978), and Jean-Paul Sartre (1956), Feldman and his colleagues (2004) suggest that the key to understanding an existential approach to teacher change is the premise that teaching is a "way of being" (p. 972). It is the way teachers are "immersed in educational situations that extend web-like through time and space, and across human relations" (Feldman, 2002, p. 5). A key tenet of existentialism is the idea that "the self emerges from experience… [It] is constructed through *the choices we make* in our experiences" (Feldman et al., 2004, p. 972, emphasis added). By bringing Feldman et al.'s (2004) existential approach to bear on our retrospective reflections across the three moments, we find that its emphasis on changing our practice by changing our selves can further illuminate what was learned about Moira from a figured worlds perspective. To illustrate a compatibility between the two perspectives that would allow for further illumination, we note that whereas a figured worlds perspective posits that artifacts mediate identity formation (Holland et al., 1998), the existential approach takes seriously that teaching is a way of being (and thus becoming), that immersion in formative experiences shapes who we are, and that through the active choices we make in our experiences, we determine who we choose to be. This is precisely what spaces of authoring (Holland et al., 1998)–or self-identifying (Sfard & Prusak, 2005)–are all about. Within them, agency is the "medium of… creation… of ourselves as human actors" (Holland et al., 1998, p. 236). Thus, the existential approach to teacher change enables us to further realize that Moira's immersion in a constructionist Making experience was formative for her. It also provides a way of using this discussion to communicate the significance of our findings.

For Feldman et al. (2004), existentialism is an attractive theoretical basis for the self-study of teacher change because it concerns itself with themes such as passion, emotion, freedom, and human nature. For us, it's the resonance of themes like these with what a figured worlds perspective (and a CHAT analysis) made visible to us as we aimed to respond to the question, *"How does a Making experience in mathematics teacher preparation mediate the process of becoming a teacher?"* These themes of the self are evident in Moment 1, which established how the design setting figured a world for Moira's identity as a creative Maker that enabled her to finally interweave her creativity into her mathematics. That passion seemed to be the driving force for her agentive play as she found her way around Tinkercad in response to the possibilities generated by the design situation.

They also appear in Moment 2, which offered insights into how Moira's immersion in the Making experience was formative for her. Moment 2 occurred in the context of a crisis Moira experienced as she realized that through her own history of learning, she'd never made sense of concepts that were integral to the design of her manipulative. The pedagogical commitments that were fundamental to her image of the teacher she wants to be demanded that she do so, so that she could support a child's meaningful learning of those very same concepts. But she struggled and eventually moved on to a new design idea, leaving that tension unresolved.

Finally, in Moment 3, Moira uses her tool to think through fraction ideas and eventually comes to recognize its potential not only for her own learning, but also for teaching mathematics in a way that aligns with the teacher she's becoming. As she comes to this realization, the nature of the tool is transformed for her, as are her pedagogy and her identity as a mathematics teacher.

CONCLUSION

Guided by the hypothesis that a Making experience would be formative for the development of an inquiry-oriented pedagogy, Moira and her classmates were immersed in a communal design environment of collective social Making. The task that was posed to them was a constructionist one, and it invited the interplay between the iterative design of a shareable manipulative; the application of mathematical, pedagogical, and design knowledge in the manipulative's development; and the mediating role of their application in identity formation through teacher learning. Using CHAT theory and figured worlds perspectives, we took a novel theoretical approach to the analysis of a range of interactions that occur in activity and that mediate identity formation through teacher learning. While Moira's mathematical, pedagogical, and design knowledge were developed as she reasoned through that experience, this story isn't only about what she learned. It's also about how the choices she made through her participation in that experience leveraged that developing knowledge to move her along a trajectory of a teacher becoming. Remarkably, these choices depict an image of Moira's way of being a teacher that stretches "web-like through time and space" (Feldman, 2002, p. 5), all the way back to her experiences in elementary school.

By attending to Moira's identity through the agentive actions she made while Making, the formative power of her experience was revealed to us as we observed the blended nature of knowledge and identity that constitutes learning through collective social practice. These findings contribute to the bodies of research on both teacher learning and identify formation in teacher preparation. Moreover, by further demonstrating the value of accounting for identity formation in research on mathematics teacher education, new opportunities to move the field forward are generated for research into the potential value of constructionist, STEAM-integrated curricular experiences in teacher preparation. In order to better understand how such experiences can support both teacher becoming and K–12 student learning of mathematics, future research could more closely explore the design of these experiences in teacher preparation, the teacher educator's role in their design and implementation, and the subsequent in-service instruction of teachers who participated in them.

REFERENCES

Association of Mathematics Teacher Educators. (2017). *Standards for preparing teachers of mathematics*. amte.net/standards.

Autodesk, Inc. (2021). *Tinkercad* [online]. https://www.tinkercad.com/

Ball, D. L., Thames, M. H., & Phelps, G. (2008). Content knowledge for teaching: What makes it special? *Journal of Teacher Education, 59*(5), 389–407.

Carpenter, T., Fennema, E., & Franke, M. L. (1996). Cognitively guided instruction: A knowledge base for reform in primary mathematics instruction. *The Elementary School Journal, 97*(1), 3–20.

Clapp, E. P., Ross, J., Ryan, J. O., & Tishman, S. (2016). *Maker-centered learning: Empowering young people to shape their worlds*. Wiley.

Corbin, J., & Strauss, A. (2008). *Basics of qualitative research: Techniques and procedures for developing grounded theory*, (3rd ed.). Sage Publications, Inc. https://doi.org/10.4135/9781452230153

Dewey, J. (1916). *Democracy and education*. Macmillan.

Engeström, Y. (1987). *Learning by expanding: An activity-theoretical approach*. Orienta-Konsultit.

Engeström, Y. (1999). Activity theory and individual and social transformation. In Y. Engeström, R. Miettinen, & R. L. Punamaki (Eds.), *Perspectives on activity theory* (pp. 19–38). Cambridge University Press.

Feldman, A. (2002, April). *What do we know and how do we know it? Validity and value in self-study of teacher education practices* [Paper presentation]. Annual Meeting of the American Educational Research Association, New Orleans, LA.

Feldman, A., Paugh, P., & Mills, G. (2004). Self-study through action research. In J. J. Loughran, M. L. Hamilton, V. K. LaBoskey, & T. Russell (Eds.), *International handbook of self-study of teaching and teacher education practices* (pp. 943–977). Springer Netherlands. https://doi.org/10.1007/978-1-4020-6545-3_24

Greene, M. (1973). *Teacher as stranger: Educational philosophy for the modern age.* Wadsworth Publishing Company.

Greene, M. (1978). Teaching: The question of personal reality. *Teachers College Record. 80*(1), 23–35.

Greenstein, S., Jeannotte, D., Fernández, E., Davidson, J., Pomponio, E., & Akuom, D. (2020). Exploring the interwoven discourses associated with learning to teach mathematics in a making context. In A. I. Sacristán, J. C. Cortés-Zavala, & P. M. Ruiz-Arias (Eds.), *Mathematics education across cultures: Proceedings of the 42nd Meeting of the North American Chapter of the International Group for the Psychology of Mathematics Education* (pp. 810–816). Cinvestav/AMIUTEM/PME-NA.

Halverson, E. R., & Sheridan, K. M. (2014). The maker movement in education. *Harvard Educational Review, 84*(4), 495–504, 563, 565.

Heyd-Metzuyanim, E., & Sfard, A. (2012). Identity struggles in the mathematics classroom: On learning mathematics as an interplay of mathematizing and identifying. *International Journal of Educational Research, 51–52*(3), 128–145.

Holland, D., Lachicotte, W., Jr., Skinner, D., & Cain, C. (1998). *Identity and agency in cultural worlds.* Harvard University Press.

Kalantzis, M., & Cope, B. (2010). The teacher as designer: Pedagogy in the new media age. *E-learning and Digital Media, 7*(3), 200–222.

Lave, J., & Wenger, E. (1991). *Situated learning: Legitimate peripheral participation.* Cambridge University Press.

Leontiev, A. N. (1978). *Activity, consciousness, and personality.* Prentice-Hall.

Leontiev, A. N. (1981). *Problems of the development of the mind.* Progress Publishers.

Miles, M. B., & Huberman, A. M. (1994). *Qualitative data analysis.* Sage Publications.

Papert, S. (1991). Situating constructionism. In S. Papert & I. Harel (Eds.), *Constructionism.* MIT Press.

Patton, M. Q. (2002). *Qualitative research and evaluation methods* (3rd ed.). Sage Publications.

Pepin, B. E. U. (2018). Enhancing teacher learning with curriculum resources. In L. Fan, L. Trouche, C., Qi, S. Rezat, & J. Visnovska (Eds.), *Research on mathematics textbooks and teachers' resources: Advances and issues* (pp. 359–374, ICME 13 Monographs). Springer.

Pepin, B., Gueudet, G., & Trouche, L. (2017). Refining teacher design capacity: Mathematics teachers' interactions with digital curriculum resources. *ZDM Mathematics Education, 49*(5), 799–812.

Roth, W. M. (2012). Cultural-historical activity theory: Vygotsky's forgotten and suppressed legacy and its implication for mathematics education. *Mathematics Education Research Journal, 24*, 87–104. https://doi.org/10.1007/s13394-011-0032-1

Sartre, J. P. (1956). *Being and nothingness* (H. Barnes, Trans.). Philosophical Library.

Schad, M., & Jones, W. M. (2019). The Maker movement and education: A systematic review of the literature. *Journal of Research on Technology in Education, 52*(1), 65–78. https://doi.org/10.1080/15391523.2019.1688739

Scheiner, T., Montes, M. A., Godino, J. D., Carrillo, J., & Pino-Fan, L. R. (2019). What makes mathematics teacher knowledge specialized? Offering alternative views. *International Journal of Science and Mathematics Education, 17*(1), 153–172. https://doi.org/10.1007/s10763-017-9859-6

Schön, D. A. (1992, 1992/03/01/). Designing as reflective conversation with the materials of a design situation. *Knowledge-Based Systems, 5*(1), 3–14.

Sfard, A. (2008). *Thinking as communicating: Human development, the growth of discourses, and mathematizing.* Cambridge University Press.

Sfard, A., & Prusak, A. (2005). Telling identities: In search of an analytic tool for investigating learning as a culturally shaped activity. *Educational Researcher, 34*(4), 14–22. http://www.jstor.org/stable/3699942

Shulman, L. S. (1986, February). Those who understand: Knowledge growth in teaching. *Educational Researcher, 15*(2), 4–14.

Smith, M. S., & Stein, M. K. (1998). Selecting and creating mathematical tasks: From research to practice. *Mathematics Teaching in the Middle School, 3*(5), 344–350.

Strauss, A., & Corbin, C. (1998). *Basics of qualitative research: Techniques and procedures for developing grounded theory* (2nd ed.). Sage.

Tall, D., & Vinner, S. (1981). Concept image and concept definition in mathematics with particular reference to limits and continuity. *Educational Studies in Mathematics, 12*(2), 151–169.

Urrieta, L. (2007). Figured worlds and education: An introduction to the special issue. *The Urban Review, 39*(2), 107–116. https://doi.org/10.1007/s11256-007-0051-0

Vågan, A. (2011). Towards a sociocultural perspective on identity formation in education. *Mind, Culture, and Activity, 18*(1), 43–57.

Verillon, P., & Rabardel, P. (1995). Cognition and artifacts: A contribution to the study of thought in relation to instrumented activity. *European Journal of Psychology of Education, 10*(1), 77–101. http://www.jstor.org/stable/23420087

Vygotsky, L. S. (1978). *Mind in society: The development of higher psychological processes.* Harvard University Press.

Yin, R. K. (2009). *Case study research: Design and methods* (4th ed.). Sage.

CHAPTER 22

LEVERAGING LONGITUDINAL AND ANNUAL ANALYSES TO IMPROVE PROGRAM DESIGN

Considering Teacher Education Standards and Professional Recommendations

Jeremy Zelkowski, Martha Makowski
The University of Alabama

Tye G Campbell
Utah State University

Jim Gleason
The University of Alabama

Casedy A. Thomas
University of Virginia

Allison Mudd
Tuscaloosa Central High School, The University of Alabama

Anna Keefe Lewis, Chalandra Gooden, and Felicia A. Smith
The University of Alabama

The AMTE Handbook of Mathematics Teacher Education: Reflection on Past, Present and Future—Paving the Way for the Future of Mathematics Teacher Education, Volume 5
pages 447–475.

Secondary mathematics teacher preparation programs have changed markedly over the last 30 years. In the last decade, program development has been guided by four interrelated standards and recommendation documents: The Conference Board of the Mathematical Sciences' *Mathematics Education of Teachers II*, the Association of Mathematics Teacher Educators' *Standards for Preparing Teachers of Mathematics*, the National Council of Teachers of Mathematics' *Specialized Professional Association Standards*, and the Mathematics Teacher Education Partnership's *Guiding Principles*. However, struggles to make lasting transformations persist. In this chapter, we present a cross-sectional analysis of these four guiding frameworks, and then present our program as a case study of a structure that adheres closely to those recommended practices. In particular, we present a two-year sequenced cohort design, course descriptions, and key assessments that align to many of the recommendations of best practice. We summarize data collection with many validated instruments that have provided both empirical findings and leverage for program changes to meet the spirit of these four national documents. The chapter capsulates more than a decade of program transformation work leading to a well-aligned programmatic structure sequenced to implement the AMTE Standards' vision, among other standards and professional recommendations of secondary mathematics teacher preparation.

INTRODUCTION

Since the release of the National Commission on Excellence in Education in 1983, there have been many efforts to transform teacher education, teacher practice, and teacher preparation at state and national levels (Cochran-Smith et al., 2013; Darling-Hammond, 2017). Such efforts have included requiring additional clinical field components for teacher candidates (TCs), administering portfolio-based assessments, using nationally standardized assessments, writing state and national teaching preparation standards, and creating national professional committees or organizations. To provide guidance in the development of jurisdictional policies and practices, national professional organizations such as the Association of Mathematics Teacher Educators (AMTE) and the National Council of Teachers of Mathematics (NCTM) have created foundational frameworks (AMTE, 2017; NCTM, 2012, 2020) for mathematics teacher education standards. The shifting reforms of teacher preparation have resulted in similar shifts in assessment of TCs for certification, with professional licensure exams becoming the norm. As a result, many teacher preparation programs (TPPs) have pivoted to make preparation for the licensure exams a central component of the curriculum, potentially at the expense of first-year teaching readiness (Darling-Hammond, 2010; Darling-Hammond & Hyler, 2013). Currently, no exact set of courses, experiences, and agreed-upon set of professional licensing examinations in secondary mathematics teacher education exists. Given the politicization of state standards, it is likely that the push for standardized measures will continue to influence teacher education, regardless of professional recommendations or the inclinations of those who do certification work in secondary mathematics (SEMA) TPPs[1]. Despite these pressures, it remains critical that TCs, just as other pre-professionals, should have the opportunity to be well-prepared at the end of a TPP with a strong sense of self-efficacy, readiness, and knowledge. Arguably, designing programs that include internal program measures that make professional examinations a formality for TCs, as opposed to a potential roadblock, is critical for creating pathways to certification that support the goals of both policymakers (e.g., accountability) and the teacher educators (e.g., TC knowledge and readiness development) (Swars et al., 2019; Zelkowski & Gleason, 2018; Zelkowski et al., 2018). In this chapter, we review relevant professional literature on recommendations and best practices for TPPs and then describe our self-study research efforts to meet these demands.

GUIDING FRAMEWORKS FOR SEMA TEACHER PREPARATION

In her work outlining the vision for teacher education, Darling-Hammond (2006) argues that TPPs meeting 21st-century needs should exhibit (a) tight coherence and integration among courses, (b) strong connections between university coursework and clinical field work in schools with extensive and intensely supervised clinical teaching integrating coursework pedagogy that links theory to practice, and (c) proactive relationships with schools and mentor teachers. Meeting this vision requires attention to the mathematical content, methods coursework, overall program structure, and program accreditation.

Within these four domains, four separate professional documents particular to TPPs exist. The Conference Board of the Mathematical Sciences' *Mathematics Education of Teachers II* (2012) provides guidance on the mathematical content preparation of TCs, focusing particularly on how traditional advanced topics should be tailored to make connections to common high school topics. The AMTE's *Standards for Preparing Teachers of Mathematics* (SPTM) (2017) provides guidance for coursework and clinical field experiences, focusing on the need for coursework devoted to both the content and methods needs particular to secondary mathematics. The Mathematics Teacher Education Partnership's (MTEP) *Guiding Principles* (2014) provides professional guidance for TPPs institutionally, again focusing on the needs for TCs in mathematics, methods, field experiences, and program structure. Lastly, to address accreditation, the National Council of Teachers of Mathematics (NCTM) Specialized Professional Association (SPA) *Standards for Teacher Preparation Programs* (2012, 2020) are utilized by the Council for the Accreditation of Educator Preparation (CAEP) for special

[1] Secondary Mathematics (SEMA) Teacher Preparation Programs (TPPs) as we consider for this chapter, is defined as those that certify for all high school grades (9–12) and inclusively, at least middle grades 7–8 but could include grade 6 or grade 5 depending on the jurisdiction certification requirements. Our program in this chapter certifies grades 6–12. TPPs here forward refers to Secondary Mathematics (SEMA) Teacher Preparation Programs (TPPs) as defined.

program accreditation and recognition in the discipline of mathematics teacher preparation.

We acknowledge it may seem odd to include accreditation concerns when considering the education of TCs. However, recent work has noted a lack of "systemic exploration of programs and their intended outcomes" and that which has occurred has been superficial (Tatto, 2018, p. 410). Tatto argues that program alignment with accreditation demands and content area standards are much more likely to generate highly knowledgeable and well-prepared beginning mathematics teachers. Tatto (1998) further indicated TPPs are much more likely to change and/or mold TCs' initial beliefs about mathematics teaching practices in programs where program faculty have a solid shared vision of outcomes using extensively designed, aligned, and rigorous assessments, clinical experiences, and highly focused and connected coursework.

AMTE's SPTM (2017) state, "A [SEMA] program that does not make clear and strong commitments to ensuring that its candidates are well-prepared as described in this volume is not justifiable. Mediocrity cannot be an option" (p. 166). In this chapter, we draw on the recommendations from Tatto (1998) (see Figure 22.1), Darling-Hammond (2006), and the four previously described guiding framework documents to challenge the conventional program structure too often present in TPPs. Further, we provide an overview of one example of a clearly defined and coherent program that aligns with these standards. Figure 22.1 briefly contrasts conventional programs with more coherent programs.

Traditional Conventional Programs	Coherent Effective Programs
Large programs lacking defined learning community opportunities during preparation	Smaller sized program with defined learning community throughout preparation
Structure is not coherent across all sections	Structure is well-sequenced and coherent
Differing views across program faculty in many courses and experiences	Faculty in strong alignment philosophically across major coursework experiences
Cultural norms dictate beliefs and practices that do not align to research findings	Cultural norms are challenged and rebuilt to align to best practices from research
Beliefs about teaching are based on experiences as K–12 student	Beliefs about teaching tend to align with program faculty over time
Teaching practices as a beginning teacher are based on prior beliefs	Teaching practices as a beginning teacher better align to program faculty goals
Program experiences have little influence on beginning teacher practices	Program experiences have significant influence on beginning teacher practices

Note. Adapted from Tatto (1998) and applying Darling-Hammond's (2006) 21st century teacher preparation conceptual framework with considerations of the AMTE (2017) Standards.

FIGURE 22.1. Teacher Education Program Characteristics of Two Different Program Perspectives

RELEVANT DOCUMENT ANALYSIS
AND REVIEW OF LITERATURE

At the national level, the four framework documents are the primary guiding documents that TPPs may utilize when working to transform programs. These documents can also be used to guide data collection to inform and aid in institutional change. In Figure 22.2 we present a cross-sectional document analysis of these four publications, highlighting the areas which seem, nationally, to present the biggest challenges to TPPs in meeting the vision for such programs: mathematics coursework, followed by the mathematics teaching methods courses, and lastly, the clinical experiences and other programmatic matters.

Recent work has noted a lack of "systemic exploration of programs and their intended outcomes" at more than a superficial level (Tatto, 2018, p. 410). Tatto's (1998, 2018) works leave no doubt the importance of internal program structure, evaluation, and TC desired outcomes. However, program evaluation scholarship is time and resource consuming. The MTEP has been clear in their *Guiding Principles* for secondary mathematics TPPs by stating, "The infrastructure of the institution of higher education supports a focus on secondary mathematics teacher preparation that emphasizes shared responsibility and accountability across the institution" (MTEP, 2014, p. 2). The MTEP further states, "Policies and practices at institutions of higher education provide encouragement, support, and rewards for faculty members who provide leadership in mathematics teacher preparation" (MTEP, 2014, p.3). We consider these statements as foundational to the investment in time and resources as necessary to enact such guiding principles of TPPs for the 21st century preparation models to be consistent, rigorous, meaningful, and productive in the quality of future mathematics teachers. We use the next section of this chapter to elaborate on the best-practices in relationship to the AMTE SPTM (2017), programmatic design, and outcomes related to these areas.

Mathematics Content Coursework

CBMS's *MET II* (2012) arguably sets the standard for the mathematics TCs should study and complete, with the AMTE SPTM (2017) related to mathematical content drawing heavily from the document. The MTEP (2014) and NCTM SPA (2012, 2020) give deference to the CBMS and AMTE specifics for mathematics content coursework. In particular, CBMS outlines two pathways: (a) a long sequence (40+ semester-hours) mathematics major, and (b) a short sequence composed of a compressed set of content courses specific to the preparation of high school mathematics teachers. For TCs seeking middle grade certification, CBMS MET II identifies five additional essential topics of study different from high school preparation. Thus, for programs that certify both middle and high school grades, the AMTE SPTM and CBMS MET II together suggest 50+ semester credit hours in mathematics for majors, which in combination with field

	Mathematics Content	Mathematics Methods	Clinical Experiences & Other
CBMS MET II (2012)	A major in mathematics with at least: • Three-course calculus sequence • Intro to statistics • Intro linear algebra • **9 semester-hours focused on high school math from advanced standpoint** – 9 additional semester-hours – Introduction to Proof – Abstract Algebra – Additional course	High School Preparation • **Math methods courses (plural)** Middle School Preparation • **Two math methods courses** Preparation programs focused on both high school and middle school would need at least three given the plurality of both high school and middle school recommendations.	• **Engage in Standards for Mathematical Practice within coursework as learners** • **Use technology and tools strategically during preparation** • Experiences with reasoning and proof • Modeling rich real-world problems • Historical development of math
NCTM SPA for CAEP (2012, 2020)	Standards 1&2: **TCs have knowledge of major** math concepts, algorithms, procedures, applications in varied contexts, and connections within and among essential topics in the mathematical domains of Number, Algebra, Functions, Geometry, Trigonometry, Measurement, Statistics, Probability, Calculus, & Discrete. TCs solve problems, represent mathematical ideas, reason, communicate, prove, use mathematical models, attend to precision, identify elements of structure, and generalize across mathematical domains & essential topics, **using technology appropriately.**	Standards 3&4: TCs apply **knowledge of curriculum & students with the standards for mathematics and their relationship to student** learning within and across mathematical domains. **They incorporate research-based mathematical experiences and include multiple instructional strategies and mathematics-specific technological tools.** TCs exhibit knowledge of adolescent learning, development, and behavior. They use this knowledge to plan and create sequential learning opportunities grounded in mathematics education research for teaching meaningful mathematics to a wide range of students.	Standard 5: **TCs provide evidence demonstrating that as a result of their instruction,** secondary students' conceptual understanding, procedural fluency, strategic competence, adaptive reasoning, and application of major mathematics concepts in varied contexts, assessing learning, analyzing data, and modifying instruction. Standard 6: TCs **Engage in continuous and collaborative learning that draws upon research & promote equitable learning, identity, and community.** Standard 7: TCs **engage in a planned sequence** of field experiences and clinical practice under the supervision of experienced and highly qualified math teacher.

MTEP CPs (2014)	In reference to the **CBMS MET II**: Mathematical habit of mind, knowledge of the discipline, specialized knowledge of mathematics for teaching, and the nature of mathematics.	TCs can design instruction, use instructional methods and strategies, assess and reflect, use technology well, attend to diverse student populations, and **engage in embedded early and sequential intense** clinical experiences with expert supervision.	**Commitments by institutions** of higher education with shared responsibility across the administration, faculty, and partner schools, as well as supporting **partnerships with disciplinary [mathematics] faculty** and [K–12] schools….Institutional focus exists and faculty are supported and rewarded for leadership in programs…
AMTE SPTM (2017)	**High School**: Math major equivalency, including statistics, with **at least three content courses** relevant to teaching high school mathematics incorporating sufficient attention to a data-driven, simulation-based modeling approach to stats. **Middle School**: Alignment to the **CBMS MET II** (2012) and Statistical Education of Teachers (2015)	High School Preparation • **Three mathematics-specific methods courses** Middle School Preparation • **Coursework** focused specifically on teaching middle level mathematics….TCs need "middle-level-focused methods courses…"	**Clinical experiences in both high school and middle schools** in which TCs have support to develop teaching practices that support learning of conceptual knowledge and engagement by students in mathematical practices.

FIGURE 22.2. Cross-Sectional Analysis of Recommendation and Standards Documents

experiences and methods courses, is unrealistic. Fortunately, commonalities in the recommended content for middle and high school certification exist.

After review across middle grades and high school, we conclude that at the very minimum the following content should be covered in a mathematics major designed for both middle and high school certification in mathematics:

- Single-variable Calculus (with multi-variable calculus advised for high school TCs)
- Algebraic Structures (with a focus on rings and fields)
- Number Systems and Operations (integers, real, and complex)
- Functions (both real-valued and abstract settings)
- Linear Algebra (particularly for high school TCs)
- Geometry and Measurement (using synthetic and analytic techniques with transformations)
- Data Analysis (including statistics and probability concepts).

Figure 22.2 presents a more detailed summary of these consolidated content recommendations. Note that this list of courses covers an entire major. To meet the specific needs of TCs, both CBMS and AMTE include a strong recommendation for nine semester-hours of advanced mathematics content connecting major-level topics to middle and high school mathematics. Often, traditional mathematics major courses do not meet these needs, and many institutions do not offer specialized mathematics content courses to their TC majors (CBMS, 2012). Although recommended for high school TCs, we suggest based upon the CBMS and AMTE, that TPPs should consider developing and integrating such advanced courses with middle grades certification in mind as well, particularly programs that concurrently certify middle and high school grades together. Such courses should be primarily designed to develop TCs' Specialized Content Knowledge, while secondarily advancing Horizon Content Knowledge; Knowledge of Content and Curriculum; and Common Content Knowledge (see Ball et al., 2008).

Lastly, the four documents make explicit the recommendation that mathematics courses (e.g., specialized TC content courses, calculus, linear algebra) should integrate experiences with reasoning and proof at various degrees of rigor, content-specific technology, and both mathematical and statistical modeling (AMTE, 2017; CBMS, 2012; MTEP, 2014; NCTM, 2012, 2020). Although beyond the scope of the standards we discuss here, we also note that best practices in teacher education suggest that the required activities in these courses should allow TCs to engage in *doing mathematics* (Stein et al., 2009) both during and outside of class.

Mathematics Teaching Methods Courses

While all the documents reviewed discuss the content of methods courses, the AMTE (2017), NCTM SPA (2012, 2020), and MTEP (2014) go into the most detail. These documents set the minimum number of methods courses at two mathematics-specific classes (e.g., not combined with science or other second-ary disciplines) for middle school and three for high school certification. Within these mathematics methods courses, TCs should begin building their pedagogical content knowledge domains, which Ball et al. (2008) partition into: (a) Knowl-edge of Content and Students, (b) Knowledge of Content and Teaching, and (c) Knowledge of Content and Curriculum. Developing these domains of knowledge requires time and is not accomplished in a single semester or year. In addition, methods coursework is a critical space for TCs to develop in their ability to build mathematical identity in students, engage with the complexity of equity and ac-cess, and develop culturally responsive teaching ideas (Berry & Thomas, 2019; NCTM, 2014; Strutchens et al., 2012). There is little evidence across these four documents that indicates that TPPs can accomplish even basic knowledge acqui-sition and achieve first-year teaching readiness for TCs in less than one year or via two methods courses.

Clinical Field Experiences and Mentor Teachers

The CBMS MET II (2012) includes only small references to clinical expe-riences and defers to the professional organizations with K–12 teacher-focused missions. Our review of the other three documents suggests that clinical experi-ences should be continuous and sequenced in classrooms where the mentor teach-er shares program values and practices with the TPP faculty. Furthermore, mentor teachers need to be partners to have a three-way exchange of teaching, learning, and TC preparation (i.e., Mentor Teacher, Program Faculty, Teacher Candidate) (Zelkowski et al., 2022). Ultimately, a clear partnership and communication line between mentor teacher and TPP program faculty must exist to complete the Triad-of-Learning for TCs, where mentor teachers are strategically selected and paired with TCs by program faculty.

Our analyses of these documents regarding clinical experiences shows that at a very minimum, one full year of clinical experiences should precede full-time stu-dent teaching. That said, we question whether TCs can be well-prepared for full-time student teaching after only a single semester as an effective model for TPPs. The developmental trajectory of TCs is clear. Research supports field experiences that are shorter earlier in TPPs, that build sequentially from observational, to lon-ger, more teaching-oriented experiences with increased expectations of teaching responsibilities (Preston, 2017).

PUTTING THE STANDARDS, PRINCIPLES, AND RECOMMENDATIONS IN PLACE

Together, the four guiding documents articulate a vision for TPPs. While no one design will ever meet the needs of all TPPs, our synthesis points to some specific best practices for program design. In the previous sections, we summarized the minimum number of courses and types of experiences for best practices, program requirements, and assessment within TPPs. We also argue that our document synthesis makes clear that strong, internal program assessment of TC knowledge and readiness is a non-negotiable given the standards expected to be measured for accreditation and certification purposes. The CAEP clearly states in Standard 5 (for all TPPs, regardless of discipline) that expectations of evidence include validated and reliable measures of programmatic data for TC assessment, program analyses, and on-going improvements.

Best Practice Recommendations

As discussed in the previous section, review of the four sets of documents for TPPs present a persistent case for specific coursework, experiences, and assessment. First, TPPs should have three, three-hour semester courses (or equivalent) of advanced perspective mathematics courses primarily designed for TCs'. Second, TCs must have a minimum of three grade-band specific methods courses. For programs that certify grades 6 through 12, we argue there should be either four or five mathematics methods courses, with the lower number occurring in cases where a well-designed course sequence can take place. Third, at a minimum, programs should require that structured clinical experiences, with TC's developmental trajectory advancing from small responsible teaching requirements (e.g., observation, small group instruction) to fully responsible teaching (i.e., full-time student teaching) over several semesters.

Beyond coursework and field experience best practices, the document analyses further suggest teaching dispositions and self-efficacy are knowledge and skill domains critical for TCs' development as future mathematics teachers. Our analyses revealed TPPs must include a focus on knowledge constructs and skills for building students' mathematical identities, understanding students' cultural influences in the learning environment, understanding access and equity issues in school structure/practices, and technology as a tool for learning and doing mathematics.

We believe these domains are non-negotiables for TPPs and we present a brief historical outline of our program transformation and design to where we are at today. To move the field forward and identify recommendations to strengthen program designs and TC-related outcomes, there have been empirical research and evaluation efforts (Zelkowski & Gleason, 2018; Zelkowski et al., 2018; Zelkowski et al., 2021, Zelkowski et al., under review) to generate larger scale quantitative findings regarding professional recommendations for TPPs.

Validated Program Measures

Since 2014, there have been two guiding works about the use of data and validated instruments in teacher education. Worrell et al. (2014, p. 1–2) stated in part,

> The data and methods required to evaluate the effectiveness of teacher education programs ought to be informed by well-established scientific methods that have evolved in the science of psychology, which at its core addresses the measurement

of behavior. Recent work highlights the potential utility of three methods for assessing teacher education program effectiveness: (1) value-added assessments of student achievement, (2) standardized observation protocols, and (3) surveys of teacher performance. ... Validity is the most important characteristic of any assessment and is the foundation for judging technical quality. Alignment of all of the elements of a program improvement effort is essential to determining what data to use, how good the data are, and what should and could be done with the data. Decisions about program effectiveness need to be made using the most trustworthy data and methods currently available.

While validity evidence of instrumentation is critical, it is not an overnight or single semester endeavor. The American Educational Research Association, the American Psychological Association, and the National Council on Measurement in Education's (2014) *Standards for Educational and Psychological Testing* about validity evidence, so eloquently described more recently for mathematics education (Bostic et al., 2019; Krupa et al., 2019; Lavery et al., 2019), is clear that validity evidence takes years to accumulate to build a strong argument for instrument development, use, and interpretation whether the evidence is published research or internally compiled. We list some examples of measures in Appendix A that programs could consider in internally advancing the use of validated measures while including examples within specific semesters below. For space purposes, we elaborate on some instruments that have been most valuable at transforming our program that helped leverage needed changes that were not always received or initially accepted administratively.

THE UNIVERSITY OF ALABAMA CASE:
PROGRAM DESIGN AND SEQUENCING

This chapter reports on the long history of our program's on-going self-study, utilizing the ideas from the literature on self-study of teacher education programs (Dinkelman, 2003; Hamilton & Pinnegar, 2015; Loughran, 2004, 2005). We build from the literature previously discussed in this chapter and elaborate on our program design, modifications, and outcomes.

To begin transforming our two initial certification degree programs (e.g., B.S.E. & M.A.) between 2008 to 2011, we relied solely on CBMS *MET I* (2001; the *MET II* Standards were not yet written) and the CAEP NCTM SPA Standards (prior to 2012 standards) to examine and begin framing needed changes. The documents prompted us to replace our History of Mathematics content course with a new advanced Algebraic Structures course for secondary TCs and realign our Geometry for Teachers course to sequence well from the new course. We piloted the new mathematics course in 2009. On the pedagogical methods and clinical experiences side of the TPP, we used the documents to re-sequence our three existing mathematics methods courses to reflect TCs with advancing clinical field responsibilities. Whether our TCs are pursuing initial certification at the bachelor's or the master's degree level, a four-semester (i.e., fall-spring-fall-spring)

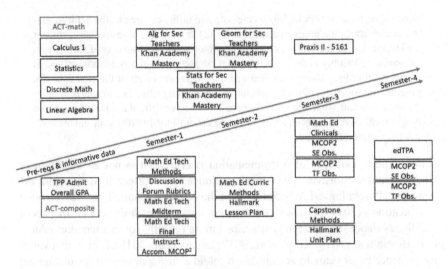

FIGURE 22.3.   Semester Design and Assessments with Validity Evidence

sequenced pathway is required using cross-listed courses depending on a TC's degree sought (see Figure 22.3). We elaborate in detail in the forthcoming sections about the programmatic design of each semesters' courses, field experiences, and key assessments.

As the MTEP *Guiding Principles* emerged in 2014, as well as the CBMS *MET II* and 2012 NCTM *SPA Standards*, we further transformed our program. By 2015, a dispositional mathematics philosophy course was repurposed for mathematics and science TCs to focus on building student identity with access and equity ideas and begin early clinical expectations in a first semester course. By 2016, we began the development of a third advanced mathematics content course for TCs and a greater focus on integrating clinical experiences with coursework by selecting mentor teachers more critically and engaging their mentorship more actively with professional development. By 2017–18, the current structure was fully developed, including the third advanced perspective mathematics course in data analysis, simulation, and probability; five mathematics methods courses; and a three-semester sequence of clinical experiences cumulatively reaching a minimum of 200 field hours (before full-time student teaching in semester four). All field experiences are SEMA program faculty recommended pairings of TCs with MTEP collaborative mathematics teachers and coordinated by the Office of Clinical Experiences. Each field experience is associated with a specific methods course.

We continue to scrutinize the quality of the sequencing and structure of our program on an annual basis via several internal measures and external state-mandated measures (e.g., Praxis II and edTPA). Our continued evaluation using validated measures has allowed us to advocate for program additions, changes, and shifts in sequencing. We discuss this process in more detail in a later section. In the next few sections, we describe our three-semester program sequence leading to the full-time student teaching, including some of the specific measures we use to provide feedback to TCs and ultimately analyze program effectiveness with our cohorted program model.

Semester 1

Instructional Accommodations for Secondary School Learning in Science and Mathematics

The primary goal of the instructional accommodations course is to challenge learners' previously held conceptions about teaching and learning and to help them build productive educational identities as it relates to both the teaching and learning of mathematics. TCs read dispositional articles related to learning theories, unpack what diversity and equity mean in the classroom, learn how to engage in productive reflection, and assimilate teacher and student collaboration. TCs discuss newly developed knowledge in both small and whole groups, where they build on what it means to do and teach mathematics. TCs also learn the essential components of lesson planning, and they implement mini-lesson plans with other TCs acting as students. There are two key assignments in the course. The first key assignment is creating two lesson plans. TCs create detailed lesson plans centered around cognitively demanding tasks, and they learn to anticipate possible student solution responses within their plans. The second key assignment includes five reflective observations completed during their 45-hour clinical field experience in a high school or middle school mathematics classroom. These observations, modified from the *Mathematics Classroom Observation Protocol for Practices* (MCOP[2]) (Gleason et al., 2017) used to evaluate the same TCs in clinical settings, allow TCs to apply their developed understanding of doing mathematics to the practitioner realm in the field. The observations have the TCs evaluate a lesson both from the teacher and student perspectives, concentrating on the cognitive demand of the task, collaborative learning opportunities, multiple representations, technology usage, and student engagement. The TCs write a reflective paper to navigate between what they observed in a real classroom with respect to the required dispositional readings. These two key assignments aim to develop their beliefs about mathematics education and develop a philosophy regarding effective mathematics teaching. In addition to in-course experiences, TCs also complete a classroom management plan that allows them to begin to develop what their future classroom space looks like. They also pick a current educational topic of their choice (e.g., tracking students in mathematics) and develop an informative 15-minute podcast using existing literature and personal experience.

Technologies for Teaching Secondary Mathematics

This course focuses on specific knowledge domains in the TPACK model (Koehler & Mishra, 2009). It has been suggested that it takes teachers upwards of two to five years to have a high self-efficacy with the incorporation of technology to teach mathematics well (Hill & Uribe-Florez, 2020; Niess et al., 2009; Staus et al., 2014). At the same time, technology evolves, changes, and improves, so the course must provide TCs the foundational opportunity to engage technology early to begin building knowledge and self-efficacy. The course focuses on the development of Technology Knowledge (TK), Technology Content Knowledge (TCK), Technology Pedagogical Knowledge (TPK), Pedagogical Content Knowledge (PCK), and TPACK (Zelkowski et al., 2013) by TCs engaging in grades 6–12 lessons while examining the structure of the lessons pedagogically while discussing additional depth of the content.

The course introduces TCs to *Principles to Actions'* (NCTM, 2014) productive beliefs (challenging unproductive) and advances knowledge from the first three chapters of *Teaching Secondary and Middle Grades Mathematics* (Brahier, 2012). The first three chapters are titled: (a) Mathematics as a Process, (b) Principles of Mathematics Education, and (c) Learning Theories and Psychology in Mathematics Education. Within these readings of Brahier, students lead discussion forums in which two validated rubrics are used to evaluate the interpretation of the readings, alignment of talk to the readings, and whether talk is aligned to productive dispositions (Zelkowski & Campbell, 2022). Further, two modules are implemented in the course from the Clinical Experiences Research Action Cluster of the MTEP focused on the Mathematical Practices and Lesson Planning (see Zelkowski et al., 2020).

Within the course, a TCK midterm and final engages TCs in solving mathematical problems with justifications of solutions via a technological tool (e.g., TI-Nspire, Geogebra, spreadsheets). The midterm assessment, with validity evidence from multiple sources (Bostic et al., 2019), has been found to have a strong relationship to other advanced program measures and outcomes, including Praxis II and our MCOP^2 observation protocol (Zelkowski et al., 2018; Zelkowski & Gleason, 2018). It provides good feedback through a seven-level rubric for TCs early in the program, as well as guiding faculty in adjusting instruction as needed.

Algebraic Structures for Secondary Teachers

The first specialized content course for TCs is intended to lay the foundation for the later courses by exposing students to an advanced treatment of foundational ideas that underlie K–12 mathematics, including set theory; concepts of equality and equivalence; number systems and their construction; and the notions of identity, inverse, and function. These initial sections cover content traditionally covered in Number Theory and, to a lesser extent, Real Analysis. After these concepts have been covered, the remainder of the course examines basic algebraic structures, focusing on the interconnections between integers and polynomial functions and connections between the rational numbers and rational functions. This foundational content knowledge is then used to examine real-valued functions that arise in the secondary curriculum as opposed to TCs taking traditional an Abstract/Modern Algebra course and/or an Analysis course.

To ensure TCs have sufficient background knowledge for this sequence, we require TCs to complete the major-level introduction to proof course as a prerequisite, which covers set theory, in order to treat the section on set theory as a review. Notably, during the section on algebraic structures, the course has significant overlap with the content traditionally taught in Abstract Algebra, including basic properties of groups, rings, and fields, with a focus on rings and integral domains. Advanced group theory is integrated into Geometry for Teachers course. Unable to find curriculum resources that fit course goals, we began in 2019 the development of a freely available online textbook focusing on this integrated design (Gleason & Makowski, 2022).

Alongside the specialized course content, TCs fill gaps in their common content knowledge using online mathematics assessments. We have previously used ALEKS and are currently using various units from Khan Academy middle grades, Algebra 1, Algebra 2, and Precalculus courses to help TCs recall content learned during previous years of their education and help prepare them for their standardized certification assessments, which primarily test the Specialized Content Knowledge domains (Ball et al., 2008) of Common Content Knowledge and Horizon Content Knowledge covered in the Khan Academy assignments.

Throughout the entire mathematics content sequence, connections are made to the state and professional curriculum standards TCs will be teaching. Our goal is to help our TCs explicitly connect the deep and abstract mathematical ideas of the courses to the secondary curriculum to better serve their future students. We also integrate opportunities for the TCs to use software such as Desmos and GeoGebra, along with various mathematical manipulatives, to model the mathematics being taught to enhance both their conceptual understanding and their readiness to communicate about the content.

Semester 2

Curriculum in Secondary Mathematics
This course begins by having students engage in critical conversations such as those found in *Catalyzing Change in High School Mathematics* (NCTM, 2018). After having students reflect upon contexts and barriers to equitable mathematics education, TCs dive into the eight effective mathematics teaching practices and identified equitable teaching practices (NCTM, 2014). Prior to midterm, the TCs spend a considerable amount of time with task selection and implementation; TCs examine the importance of cognitive demand, problem solving including reversibility and flexibility, relevance to students' lived experiences, the use of mathematical representations, and multiple entry points to tasks. Building on task selection and implementation, we draw upon scholarship (e.g., Smith & Stein, 2018) to help TCs to understand the power of meaningful discourse for teaching and learning with explanations and justification for creating a discourse commu-

nity in which learners feel willing and comfortable to take risks. Cases of teaching, vignettes, and video are used as representations of pedagogies (Grossman et al., 2009) to assist TCs in gaining mathematics knowledge for teaching (Ball et al., 2008).

Within this course, TCs develop knowledge about what effective and equitable teaching practices are and can demonstrate such knowledge through approximations (Grossman et al., 2009; Thomas, 2021) including lesson planning and teaching. For the first lesson plan assessment, students work in small groups, while later submitting individually composed lessons. While TCs are introduced to writing lesson plans in the prior semester, this course has TCs effectively demonstrate knowledge related to lesson planning. Lesson plans are evaluated with a sophisticated lesson plan rubric with emerging validity evidence (Zelkowski & Gleason, 2018; Zelkowski et al., 2018; Zelkowski et al., 2021, Zelkowski et al., under review). While TCs may not have the opportunity to teach every lesson that they create as part of the course, they are required to complete a 55-hour practicum field experience that is complementary to semester 1 (e.g., middle or high school) and teach a minimum of one full lesson with three to six additional lessons being highly encouraged. TCs complete reflective logs of their experiences throughout the semester to demonstrate awareness and professional growth.

In addition to focusing on the development of lesson planning, the latter part of the course aims to set TCs up for success in their final year in the program by briefly unpacking edTPA components related to program objectives while leveraging TCs to understand and utilize the MCOP² in equitable planning and implementation (see Module-2 in Zelkowski et al., 2020).

Geometry for Secondary Teachers

The geometry course builds on the algebraic structures course, with TCs studying geometry and measurement concepts from axiomatic, synthetic, analytic, and transformational perspectives. The course begins with a review of traditional Euclidean axiomatic geometry, focusing on constructions, triangles, and circles. It then pivots to measurement using standard and non-standard units. Approximately the last two-thirds of the course examines geometry from a transformational perspective drawing on a variety of representations of the plane. To enable TCs to examine relationships between transformations of the plane, the unit starts with an extended treatment of group theory (foci on subgroups, factor groups, group isomorphism theorems). The curriculum of the units draw from sets common in the secondary curriculum. After the group theory unit is complete, the remainder of the course is spent examining each type of transformation of the plane and the relationships between transformations in detail.

Notably, between this course and the algebraic structures course, our TCs cover nearly all the content of a traditional Abstract Algebra course but do so in a way that makes explicit connections to the secondary curriculum. The highly visual nature of the content provides a variety of opportunities for students to use GeoGebra and Desmos to build multiple components of their mathematical knowledge for teaching, particularly their specialized content knowledge.

As with the algebraic structures course, the TCs spend some time working on the related common content knowledge in geometry and trigonometry using computerized assessment units in Khan Academy courses. However, unlike the algebraic structures class, the role of the computerized assessment is lowered, with more focus on in-class activities and traditional assignments where the students grapple with complex questions outside of class.

Data Analysis for Secondary Teachers

The third math course in our advanced mathematics content set is designed to support TCs in teaching for a data centered world and developing statistical literacy. This course focuses on data analysis, using the recommendations in the *Guidelines for Assessment and Instruction in Statistics Education* (GAISE) Reports from the American Statistical Association and the National Council of Teachers of Mathematics (Aliaga et al., 2005; Bargagliotti et al., 2020; Franklin et al., 2007; GAISE, 2016). In particular, the course treats traditional statistics as a set of tools to be used in the discipline of data analysis.

The list of topics covered in this course is fairly traditional, including graphical and numerical methods for representing data, point and interval estimators, and the use of these techniques for hypothesis testing. However, the course does three things that separate it from a traditional statistics course. First, it focuses on data examples of interest to future teachers and secondary students. Second, connections between statistical processes and mathematical processes are routinely made. For example, an Ordinary Least Squares regression is a type of mathematical modeling. Lastly, the course deeply integrates data analysis into the ways the topics are discussed and covered, drawing on the ideas in *Data Analysis Investigative Process* (Bargagliotti et al., 2020). In particular, through a series of projects using real world data, the TCs make decisions about which statistics are appropriate, estimate parameters, and communicate their decisions and conclusions, with the ultimate goals of having them develop conceptual understanding and statistical reasoning. In addition to often being the tool of instruction, write ups of these projects serve as the primary means of assessment in the course.

We integrate technology throughout the course by having the TCs use basic programming in spreadsheets and R to explore data, run simulations, estimate parameters with confidence intervals, and test hypotheses. This use of technology allows the TCs to focus on justifying assumptions and interpreting results, rather than the computational aspects of data analysis.

Semester 3

Advanced Clinical Field Experiences

This course is the final practicum experience prior to the semester 4 full-time student teaching. This practicum experience takes place in the fall semester concurrently with the final mathematics methods course (see below). TCs are required to be in the field 100–120 hours essentially as the course itself. Additionally, TCs are required to teach at least four full lessons throughout the semester but are encouraged to teach at least six to ten additional lessons. Of the four required lessons, the first two are single lessons, such that one is taught in September and the other is taught in October, whereas the final two are consecutive lessons with the same students taught in November. The faculty use the MCOP² to evaluate and provide feedback on three lessons including the one in September, October, and the second lesson in the teaching episode in November. Furthermore, all lesson plans are submitted prior to the observations for feedback and assessment. Practicum goals advance TC knowledge and readiness in planning consecutive lessons prior to the full-time student teaching.

Teaching Secondary School Mathematics

The first part of the course focuses heavily on Boaler's (2016) mathematical mindsets and having TCs develop growth mindsets in how they approach working with students. We discuss how in order for students to develop a love or appreciation for mathematical connections and applications, they must first see themselves as part of the conversation; meaning that they must see themselves as doers of mathematics. TCs reflect upon how they will show up differently with their students, particularly those who have not previously had positive learning experiences in mathematics and who have been historically marginalized.

Building upon conversations of mindset, identity, and access, we engage in what ambitious instruction looks like in practice by drawing upon the *Taking Actions* series (Boston et al., 2017; Smith et al., 2017). The readings and assignments are differentiated based upon whether TCs hope to teach at the middle or high school level. Differentiating allows for meaningful and engaging class discourse across and within the groups. After thoroughly examining standards-based practices and ambitious instruction, we move into conversations of culturally responsive teaching and culturally relevant pedagogy (e.g., Thomas & Berry, 2019) and teaching mathematics for social justice. TCs read *High School Mathematics Lessons: To Explore, Understand, and Respond to Social Injustice*, and adopt or revise lessons to meet the needs of their practicum contexts and their learners (Berry et al., 2020).

Many of the assignments correspond to TC placements. For example, TCs have to video record their teaching within their practicum and self-analyze the instruction to get accustomed to evaluating instruction and learner engagement. In a different assignment, TCs monitor the progress of select students who display various levels of readiness with the content. Over the course of two-weeks, TCs examine the assessments given and plan for how they will provide meaningful feedback and opportunities to learn for the students so that they can help them to reach the intended mathematical goals. The culminating assignment in methods is the unit plan submission. The assignment builds on the TCs' prior lesson planning because they must consider how they will sequence lessons coherently on consecutive days, so students develop conceptual understanding. The unit plan rubric has been shown to have strong relationships to student teaching edTPA overall scores (Zelkowski & Gleason, 2018; Zelkowski et al., 2018, 2021, 2024). The rubric measure provides the final comprehensive feedback and readiness to TCs for the full-time student teaching and concurrent edTPA in the final semester.

Semester 4

Full-time Student Teaching

Semester 4 is a full-time, bell-to-bell, immersion practicum experience for 15 weeks (~475 hours). Like previous practicum experiences totaling just over 200 classroom hours, the faculty strategically pair TCs with a mentor teacher in which strong partnerships exist between mentor teacher and program faculty. Nearly all mentor teachers have engaged in current or previous multi-year professional development projects led by program faculty. TCs complete the entire edTPA portfolio during the first seven weeks. Mentor teachers and supervisors each complete at least four full evaluations of teaching using the MCOP^2 as a validate measure for TCs and program evaluation. TCs take on a much larger role on by teaching a minimum of two sets of five consecutive days of bell-to-bell teaching responsibility. Though, for most TCs, all classes in the second half of full-time student teaching tend to be the responsibility of TCs. Ultimately at the conclusion of the semester and program, TCs have experienced a diverse set of schools, mentor teachers, and most importantly, engaged with students in middle and high schools by growing from observation and small group instruction in the first semester to full-time, multiple consecutive days and weeks of responsible teaching to conclude the semester.

THE UNIVERSITY OF ALABAMA CASE: HISTORICAL DATA TRENDS

Worrell et al. (2014) made 13 specific recommendations for program improvement and accountability leading to the best use of data. We focus the rest of this chapter on how those recommendations guided our program's annual improvements and transformational changes over the last decade. We focus on the *within* program data collection and critical ongoing evaluation.

The CAEP accreditation process occurs every five to seven years at most public institutions. It is rare that annual evaluations/analyses result in substantial and/or structural changes within those five to seven years. Surveys of teacher education faculty tend to reflect only a one- or two-year evaluation and potential changes only in the time leading up to the CAEP required data collection years rather than three or four times during the longer cycles between CAEP reviews. Even more troubling, secondary education disciplines tend to all be housed under similar CAEP data collection, thus providing little information for within-discipline (e.g., mathematics) analyses about pedagogical performances, student impact during student teaching, or comparisons across discipline knowledge in secondary programs.

To move the field forward and identify recommendations to strengthen program designs and outcomes, we have focused empirical research and evaluation efforts (Zelkowski & Gleason, 2018; Zelkowski et al., 2018; Zelkowski et al., 2021, Zelkowski et al., 2022) to generate larger scale quantitative findings regarding professional recommendations. Our program compiles data across all eight of the NCTM SPA assessments that includes course grades in key courses known to be related to program completion and student teaching performance from prior analyses. Beyond the NCTM SPA assessments, the program looks at a more granular level of rubrics, further considering validation research with respect to each. Each cohort has attrition, yet instrumentation provides TCs narrow and specific feedback through courses, in the field, and in preparation for the Praxis II and edTPA for secondary mathematics. The sequence previously presented is detailed more specifically with key measures and internal rubrics (see Figure 22.3).

We do not advocate for or argue for *test prep* in terms of Praxis II or edTPA. Rather, our philosophy is that well-aligned coursework, field experiences, assignments, and assessment, as well as quality mentor teachers, all combined should have TCs ready for state-mandated high-stakes assessments as a chance to easily

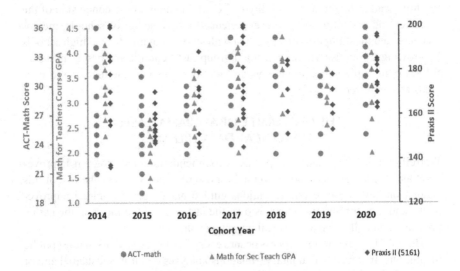

FIGURE 22.4. ACT-math, Math for Teacher Course GPA, and Praxis II Scores Note: Linear model growth rates and standard deviations. ACT-math linear model rate of growth (1.75 points per year or about 40% of a standard deviation). Capstone math GPA linear model rate of growth (0.05 points per year or 7% of a standard deviation). Praxis II linear model rate of growth (0.58 points per year or 4% of a standard deviation)

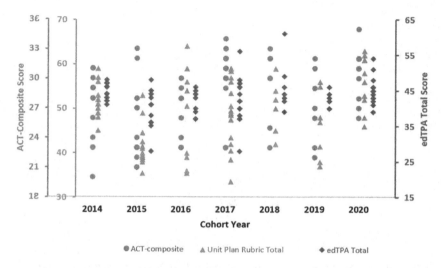

FIGURE 22.5. ACT-composite, Unit Plan Rubric, and edTPA Scores. Note: Linear model growth rates and standard deviations. ACT-composite linear model rate of growth (0.68 points per year or 18% of a standard deviation). Unit Plan linear model rate of growth (1.10 points per year or 13% of a standard deviation). edTPA linear model rate of growth (0.34 points per year or about 7% of a standard deviation)

demonstrate their readiness and knowledge rather than such high stakes assessments acting as a roadblock. We present some historical trends in our data in the Figures 22.4, 22.5, and 22.6 in relation to state-mandated high-stakes testing, prior ability, and observations in the field.

As shown in Figures 22.4 and 22.5, historical upward data trends show not only more selectivity in students enrolling in the program (increasing ACT scores), but rather that high-stakes outcomes of graduates show smaller increases (see notes in figures). We interpret these data in reflecting that our internal measures are much more aligned to high-stakes external measures than the ACT. In Figure 22.6, we consider this as our benchmark critical assessment in terms of readiness for full-time student teaching. That is, TCs do not enter full-time student teaching without successful semester 3 observations with the MCOP[2] at a level predictive of success on edTPA (a minimum 1.5 item mean, grade equivalent to C-, see Zelkowski & Gleason, 2016). Lastly, the annual analyses look for associations between key assessments and outcomes regularly through simple correlational or regression methods, as well as advanced path analyses (Zelkowski et al., 2024). Each association, whether improved or not, results in some small or large systemic curricular change as needed. Any large change requested within the university or state department of education is supported with such data and analyses.

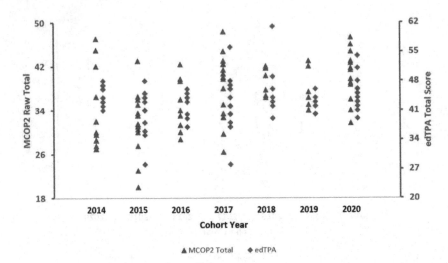

FIGURE 22.6. Mathematics Classroom Observation Protocol for Practices and edTPA Scores. Note: Linear model growth rates and standard deviations. MCOP2 linear model rate of growth (1.42 points per year or 23% of a standard deviation). edTPA linear model rate of growth (0.34 points per year or about 7% of a standard deviation)

LEVERAGING TRANSFORMATIONAL CHANGES THROUGH VALIDATED INSTRUMENT DATA

As this chapter includes the ideas about best practices and recommendations from four guiding frameworks for TPPs, we recognize that not many institutions have such specific coursework in mathematics, multiple semesters of consecutive mathematics methods courses, two years of practicum field experiences, and program faculty from two colleges that share visions of desired outcomes who have built long term relationships with a cadre of mentor teachers. The structure and best practices have not come easily or quickly. From 2008 to 2012, change in our program was minimal and faced challenges with implementation. However, we found that by beginning with a focus on the NCTM SPA accreditation process electively for national recognition, backing up requests with research and program-specific data collection, and starting a formal partnership team with the MTEP in 2012, these works allowed for change more easily and timely. We believe the use of strong internal measures, validation research of said measures, and the analyses of data was critical to this process as well. We encourage other programs and faculty to leverage these key ideas in considering program improvements that may be challenging as was our case from 2008 to 2012. Our recommendations include:

1. Consider efforts to tie research scholarship to program measures, data analyses, and teacher candidate outcomes. This provides faculty oppor-

tunities to double-dip their time efforts to both teaching and research scholarship productivity. The development and validation of instrumentation is labor intensive scholarship and should be recognized by institutional administration as such for the purposes of improving institutional outcomes.

2. Be willing to compromise across campus and with schools to some extent. Making room in a program of study for an advanced perspective mathematics for teachers' course may require the deletion of a general education course, or vice-versa when adding a methods course. Schools and teachers may need stronger voice and input on clinical decisions.

3. Consider external grants from the National Science Foundation and/or state departments of education (e.g., Noyce, IUSE, ITEST, MSP). Grants provide financial resources to both teaching and research institutions for building and improving programs. The funding can be leveraged for additional faculty positions and/or improvements to program curricula. One grant alone being PI'd by co-author Gleason resulted in the creation of the faculty line for co-author Makowski to be hired in the mathematics department as a mathematics educator after repeated years of the requested position being denied.

4. Build a formal partnership team between schools (and districts) in which program faculty, mathematics teachers, and teacher candidates have shared opportunities to engage in the building of common goals and desired outcomes. Our team was initiated in 2012 through the MTEP and subsequent professional development grant projects for the last decade.

DISCUSSION: OVERCOMING
LIMITATIONS AS CLOSING REMARKS

There are clear limitations TPPs need to consider more deeply, i.e., the institutional and administrative commitments for faculty resources and support. First, most TPPs are relatively small, and likely to stay that way for a while given the overall declines in TPP enrollment nationally. Over the course of our transformations, graduating cohorts averaged 13 from 2008 to 2012, sliding to 12 from 2013 to 2020, and dipped to 9 TCs in 2021, though enrollments seem to be increasing once again to the mid-teens and lower 20s for upcoming cohorts. Small programs have little leverage with descriptive data without validated instrumentation. Limitations can be negated or minimized by seeking validated instrumentation or validating existing instruments.

Second, not all institutions have the ability or desire to conduct large and sustained outreach professional development (PD) projects. Our team has had ongoing PD since 2012 with mentor teachers (2012–15; 2015–18; 2019–25). This has given our faculty the ability to build relationships, trust mentor teachers for TC mentorship with shared program values, and ultimately build a field experience teacher cadre for the Triad-of-Learning for TCs. Limitations can be negated by seeking institution administrative support for seeking external funding to improve programs, recruit TCs, and potentially hiring new faculty with needed expertise.

Thirdly, National Science Foundation (NSF) grants are large faculty-hour and institutional commitments with small salary incentives to faculty but large tuition benefits to institutions of higher education. Having completed a Noyce Track 1 (scholarships) from 2012–19, commitments extend well-beyond the grant by mentoring early career teachers two to three years after grant conclusion. Not all TPP institutions wish to make those commitments or have the resources, and thus have an inability to expand enrollments through recruitment and scholarships, nor the ability to work extensively over long periods with mentor teachers such as with Noyce Track 3 (master teacher fellowships) grants. Grant activities at both research and primarily teaching institutions provide TPPs opportunities to overcome such limitations.

Many of these limitations are where we believe TPP administrative teams need to step up the commitment by supporting faculty with the efforts to secure such NSF grants (Noyce Track 1, 2, 3) and state-level awards/contracts, in addition to the ability to collect, analyze, and make decisions with data to transform TPPs into local best-practice structures adhering to the CBMS MET II, AMTE SPTM, MTEP Guiding Principles, and NCTM SPA Standards. Otherwise, institutions should begin to think about what is best for K–12 students; that is, think about as to whether non-coherent programs (Figure 22.1) that cannot provide best practices for all TCs (Figure 22.3) with respect to their mathematical and pedagogical and clinical preparation ought to be preparing next generation mathematics teachers. We have faced administrative closed doors at times and persistence has been challenging in some matters.

Lastly, our instrumentation is always on-going. Some instruments have well-validated histories and are widely available, while others are under review for publication and some in earlier stages of validation work. We recognize these limitations are encouraged by past work that has fueled current and future scholarship. We encourage readers of this chapter to reach out and consider resources that may improve local data collection and evaluation to ultimately transform secondary mathematics teacher preparation programs.

ACKNOWLEDGEMENTS

The authors of this chapter extend their deepest appreciation to area mentor teachers who without their dedication and work with our team and program, teacher candidates could not become who they became as early career mathematics teachers. Your value is immeasurable on the importance of your work with teacher candidates.

This work was supported in part by the National Science Foundation Grant ID#s 1340069, 1726998, 1726362, 1726853, 1849948. Any opinions, findings, and conclusions or recommendations expressed in this chapter are those of the author(s) and do not necessarily reflect the views of the National Science Foundation.

APPENDIX A: TEACHER EDUCATION
MEASURES AND VALIDITY EVIDENCE

Observation Protocols

Mathematics Classroom Observation Protocol for Practices (MCOP²)
>Gleason, J., Livers, S.D., & Zelkowski, J. (2017).https://jgleason.people.ua.edu/mcop2.html
>Mathematics SCAN-M
>Walkowiak, T. A., et al., (2014). Introducing an observational measure of standards-based mathematics teaching practices: Evidence of validity and score reliability. *Educational Studies in Mathematics, 85*(1), 109–128.

Rubrics (NCTM SPA Assessments & more)

>Capstone Unit Plan Rubric
>Lesson Planning Rubric
>Discussion Forum Rubrics aligned to reading and productive teaching beliefs
>TCK Assessment Rubric

Zelkowski and Gleason (2016).
Zelkowski and Gleason (2018).
Zelkowski et al. (2018).
Zelkowski et al. (2020).
Zelkowski et al.. (2021).
Zelkowski & Campbell (2022).
Zelkowski et al. (2024).

Modules from Clinical Experiences Research Action Cluster of the MTEP

>Standards for Mathematical Practice (SMPs); Lesson Planning (LP); Feedback (FB)
>Zelkowski et al. (2020).
>Zelkowski et al. (2022).

Surveys

>TPACK for SEMA TCs
>Zelkowski et al. (2013).
>Note: All references listed by authors are included in the chapter references except where the full citation appears in this appendix.

REFERENCES

Aliaga, M., Cobb, G., Cuff, C., Garfield, J., Gould, R., Lock, R., Moore, T., Rossman, A., Stephenson, B., Utts, J., Velleman, P., & Witmer, J. (2005). *Guidelines for assessment and instruction in statistics education (GAISE): College report.* American Statistical Association. http://www.amstat.org/education/gaise.

Association of Mathematics Teacher Educators (AMTE). (2017). *Standards for preparing teachers of mathematics.* https://amte.net/sites/default/files/SPTM.pdf

Ball, D. L., Thames, M. H., & Phelps, G. (2008). Content knowledge for teaching: What makes it special? *Journal of Teacher Education, 59*(5), 389–407.

Bargagliotti, A., Franklin, C., Arnold, P., Gould, R., Johnson, S., Perez, L., & Spangler, D. (2020). *Pre-K12 guidelines for assessment and instruction in statistics education (GAISE) report II.* American Statistical Association and National Council of Teachers of Mathematics.

Berry, R. Q., III, Conway, B. M., Lawler, B. R., & Staley, J. W. (2020). *High school mathematics lessons: To explore, understand, and respond to social injustice.* Corwin.

Boaler, J. (2016). *Mathematical mindsets: Unleashing students' potential through creative math, inspiring messages and innovative teaching.* Jossey-Bass.

Bostic, J., Krupa, E., & Shih, J. (2019). Introductions: Aims and scope for assessments in mathematics education contexts: Theoretical frameworks and new directions. In J. Bostic, E. Krupa, & J. Shih (Eds.), *Assessment in mathematics education contexts: Theoretical frameworks and new directions* (pp. 1–11). Routledge.

Boston, M., Dillon, F., Smith, M. S., & Miller, S. (2017). *Taking action: Implementing effective mathematics teaching practices in grades 9–12.* National Council of Teachers of Mathematics.

Brahier, D. J. (2012). *Teaching secondary and middle school mathematics* (4th ed.). Pearson.

Cochran-Smith, M., Piazza, P., & Power, C. (2013). The politics of accountability: Assessing teacher education in the United States. *The Educational Forum, 77*(1), 6–27.

Conference Board of the Mathematical Sciences (CBMS). (2001). *The mathematical education of teachers (MET).* American Mathematical Society and Mathematical Association of America.

Conference Board of the Mathematical Sciences (CBMS). (2012). *The mathematical education of teachers II (MET II).* American Mathematical Society and Mathematical Association of America.

Darling-Hammond, L. (2006). Constructing 21st-century teacher education. *Journal of Teacher Education, 57*(3), 300–314.

Darling-Hammond, L. (2010). *Evaluating teacher effectiveness: How teacher performance assessments can measure and improve teaching.* Center for American Progress.

Darling-Hammond, L. (2017). Teacher education around the world: What can we learn from international practice? *European Journal of Teacher Education, 40*(3), 291–309.

Darling-Hammond, L., & Hyler, M. E. (2013). The role of performance assessment in developing teaching as a profession. *Rethinking Schools, 27*(4), 10–15.

Dinkelman, T. (2003). Self-study in teacher education: A means and ends tool for promoting reflective teaching. *Journal of Teacher Education, 54*(1), 6–18.

Franklin, C., Kader, G., Mewborn, D., Moreno, J., Peck, R., Perry, M., & Scheaffer, R. (2007). *Guidelines for assessment and instruction in statistics education (GAISE) report*. American Statistical Association

GAISE College Report ASA Revision Committee. (2016). *Guidelines for assessment and instruction in statistics education college report 2016*. http://www.amstat.org/education/gaise.

Gleason, J., Livers, S. D., & Zelkowski, J. (2017). Mathematics classroom observation protocol for practices (MCOP²): Validity and reliability. *Investigations in Mathematical Learning, 9*(3), 111–129.

Gleason, J., & Makowski, M. (2022). *Mathematical knowledge for secondary teachers*. https://gleasonua.github.io/MKT/

Grossman, P., Hammerness, K., & McDonald, M. (2009). Redefining teaching, re-imagining teacher education, *Teachers and Teaching: Theory and Practice, 15*(2), 273–289.

Hamilton, M. L., & Pinnegar, S. (2015). Considering the role of self-study of teaching and teacher education practices research in transforming urban classrooms. *Studying Teacher Education, 11*(2), 180–190.

Hill, J. E., & Uribe-Florez, L. (2020). Understanding secondary school teachers' TPACK and technology implementation in mathematics classrooms. *International Journal of Technology in Education, 3*(1), 1–13.

Koehler, M., & Mishra, P. (2009). What is technological pedagogical content knowledge (TPACK)? *Contemporary Issues in Technology and Teacher Education, 9*(1), 60–70.

Krupa, E., Bostic, J. & Shih, J. (2019). Validation in mathematics education: An introduction to quantitative measures of mathematical knowledge: Researching instruments and perspectives. In J. Bostic, E. Krupa, & J. Shih (Eds.), *Quantitative measures of mathematical knowledge: Researching instruments and perspectives* (pp. 1–13). Routledge.

Lavery, M., Jong, C., Krupa, E., & Bostic, J. (2019). Developing an instrument with validity in mind. In J. Bostic, E. Krupa, & J. Shih (Eds.), *Assessment in mathematics education contexts: Theoretical frameworks and new directions* (pp. 12–39). Routledge.

Loughran, J. (2004). Understanding self-study of teacher education practices. In J. Loughran & T. Russell (Eds.), *Improving teacher education practices through self-study* (pp. 239–248). Routledge.

Loughran, J. (2005). Researching teaching about teaching: Self-study of teacher education practices. *Studying Teacher Education, 1*(1), 5–16.

Mathematics Teacher Education Partnership (MTEP). (2014). *Guiding principles for secondary mathematics teacher preparation programs*. Association of Public & Land-grant Universities. https://www.aplu.org/projects-and-initiatives/stem-education/SMTI_Library/Updated-Guiding-Principles-for-Secondary-Mathematics-Teacher-Preparation-Programs/File

National Council of Teachers of Mathematics. (2012). *The National Council of Teachers of Mathematics specialized professional association standards for mathematics teacher preparation*. Council for the Accreditation of Educator Preparation.

National Council of Teachers of Mathematics. (2014). *Principles to actions: Ensuring mathematics success for all*. National Council of Teachers of Mathematics.

National Council of Teachers of Mathematics. (2018). *Catalyzing change in high school mathematics: Initiating critical conversations*. National Council of Teachers of Mathematics.

National Council of Teachers of Mathematics. (2020). *The National Council of Teachers of Mathematics specialized professional association standards for mathematics teacher preparation.* Council for the Accreditation of Educator Preparation.

Niess, M. L., Ronau, R. N., Shafer, K. G., Driskell, S. O., Harper, S. R., Johnston, C., Browning, C., Özgün-Koca, S. A., & Kersaint, G. (2009). Mathematics teacher TPACK standards and development model. *Contemporary Issues in Technology and Teacher Education, 9*(1), 4–24. Society for Information Technology & Teacher Education. Retrieved August 26, 2021 from https://www.learntechlib.org/primary/p/29448/.

Preston, C. (2017). University-based teacher preparation and middle grades teacher effectiveness. *Journal of Teacher Education, 68*(1), 102–116.

Smith, M. S., Steele, M. D., & Raith, M. L. (2017). *Taking action: Implementing effective mathematics teaching practices in grades 6–8.* National Council of Teachers of Mathematics.

Smith, M. S., & Stein, M. K. (2018). *Five practices for orchestrating productive mathematics discussions* (2nd edition*).* National Council of Teachers of Mathematics.

Staus, N., Gillow-Wiles, H., & Niess, M. (2014). TPACK development in a three-year online masters program: How do teacher perceptions align with classroom practice? *Journal of Technology and Teacher Education, 22*(3), 333–360.

Stein, M., Smith, M., Henningsen, M., & Silver, E. (2009). *Implementing standards-based mathematics instruction: A casebook for professional development* (2nd ed.). Teachers College Press and the National Council of Teachers of Mathematics.

Strutchens, M., Bay-Williams, J., Civil, M., Chval, K., Malloy, C. E., White, D. Y., D'Ambrosio, B., & Berry, R. Q. (2012). Foregrounding equity in mathematics teacher education. *Journal of Mathematics Teacher Education, 15*(1), 1–7.

Swars, S. A., Smith, S. Z., Smith, M. E., & Myers, K. (2019). A case study of elementary teacher candidates' preparation for high stakes teacher performance assessment. *Journal of Mathematics Teacher Education, 23*(3), 269–291.

Tatto, M. T. (1998). The influence of teacher education on teachers' beliefs about purposes of education, roles, and practice. *Journal of Teacher Education, 49*(1), 66–77.

Tatto, M. T. (2018). The mathematical education of secondary teachers. In M. T. Tatto, M. Rodriguez, W. Smith, M. Reckase, & K. Bankov (Eds.), *Explore the mathematical education of teachers using TEDS-M Data* (pp. 409–450). Springer.

Thomas, C. A. (2021). One university's story on teacher preparation in elementary mathematics: Examining opportunities to learn. *Journal of Mathematics Teacher Education, 24*(6), 641–669.

Thomas, C. A., & Berry, III, R. Q. (2019). Qualitative metasynthesis of culturally relevant pedagogy & culturally responsive teaching: Unpacking mathematics teaching practices. *Journal of Mathematics Education at Teachers College, 10*(1), 21–30.

Worrell, F., Brabeck, M., Dwyer, C., Geisinger, K., Marx, R., Noell, G., & Pianta, R. (2014). *Assessing and evaluating teacher preparation programs.* American Psychological Association.

Zelkowski, J., & Campbell, T. G. (2022). Early mathematics teacher preparation evaluation rubrics for the context of live discussion forms. In A. Lischka, J. Strayer, J. Lovett, R. S. Jones, & E. Dyer (Eds.), *Proceedings of the 44th annual meeting of the North American Chapter of the International Group for the Psychology of Mathematics Education* (pp. 1309–1318). Middle Tennessee State University.

Zelkowski, J., Campbell, T. G., & Gleason, J. (2018). Programmatic effects of capstone math content and math methods courses on teacher licensure exams. In W. M. Smith, B. R. Lawler, J. F. Strayer, & L. Augustyn (Eds.), *Proceedings of the 7th Annual Mathematics Teacher Education—Partnership Conference* (pp. 91–96). Association of Public Land-grant Universities.

Zelkowski, J., Campbell, T. G., & Moldavan, A.M. (2024). The relationships between internal program measures and a high-stakes teacher licensing measure in mathematics teacher preparation: Program design considerations. *Journal of Teacher Education, 75*(1), 58–75. https://doi.org/10.1177/00224871231180214

Zelkowski, J., Campbell, T. G., Moldavan, A. M., & Gleason, J. (2021). Maximizing teacher candidate performances based on internal program measures: Program design considerations. In W. M. Smith & L. Augustyn (Eds.), *Proceedings of the 10th Annual Mathematics Teacher Education—Partnership Conference* (pp. 39–42). Association of Public Land-grant Universities.

Zelkowski, J., & Gleason, J. (2016). Using the MCOP² as a grade bearing assessment of clinical field observations. In B. R. Lawler, R. N. Ronau, & M. J. Mohr-Schroeder (Eds.), *Proceedings of the 5th Annual Mathematics Teacher Education—Partnership Conference* (pp. 129–138). Association of Public Land-grant Universities.

Zelkowski, J., & Gleason, J. (2018). Programmatic effects on high stakes measures in secondary math teacher preparation. In L. Venenciano & A. Redmond-Sanogo (Eds.), *Proceedings of the 45th Annual Meeting of the Research Council on Mathematics Learning* (pp. 169–176). Research Council on Mathematics Learning.

Zelkowski, J., Gleason, J., Cox, D., & Bismarck, S. (2013). Developing and validating a reliable TPACK instrument for secondary mathematics preservice teachers. *Journal of Research on Technology in Education, 46*(2), 173–206.

Zelkowski, J., Yow, J., Ellis, M. E., & Waller, P. (2020). Engaging mentor teachers with teacher candidates during methods courses in clinical settings. In W. G. Martin, B. R. Lawler, A. E. Litchka, & W. M. Smith (Eds.), *The mathematics teacher education partnership: The power of a networked improvement community to transform secondary mathematics teacher preparation* (AMTE Professional Book Series, Vol 4, pp. 211–234). Information Age Publishing.

Zelkowski, J., Yow, J., Waller, P., Edwards, B. P., Anthony, H. G., Campbell, T. G., Keefe, A., & Wilson, C. (2022). Linking the field-based mentor teacher to university coursework: Methods course modules for completing the triad of learning for mathematics teacher candidates. In D. Polly, & E. Garin (Eds.), *Preparing quality teachers: Advances in clinical practice* (pp. 579–608). Information Age Publishing.

EDITOR BIOGRAPHY

Babette M. Benken, Ph.D., is the Richard D. Green Professor of Mathematics Education in the Department of Mathematics and Statistics and the Director of Graduate Studies in the College of Natural Sciences and Mathematics at the California State University, Long Beach. Dr. Benken has investigated multiple areas, including pure mathematics, mathematics and STEM teacher education, models for K–20 teacher development, graduate education, and educational pathways (e.g., remedial education, calculus reform). The overarching focus of this work has been on supporting teacher growth and facilitating access and success for students, often for those who are marginalized and underrepresented in STEM and higher education. She is currently the AMTE Vice President for Publications and has served in this role since 2018.

The AMTE Handbook of Mathematics Teacher Education: Reflection on Past, Present and Future—Paving the Way for the Future of Mathematics Teacher Education, Volume 5
page 477.
Copyright © 2024 by Information Age Publishing
www.infoagepub.com

Printed in the United States
by Baker & Taylor Publisher Services